Lecture Notes in Computer Science   8645

Commenced Publication in 1973
Founding and Former Series Editors:
Gerhard Goos, Juris Hartmanis, and Jan van Leeuwen

T0236419

Lecture Notes in Computer Science   8045

Hendrik Decker   Lenka Lhotská
Sebastian Link   Marcus Spies
Roland R. Wagner (Eds.)

Database and Expert Systems Applications

25th International Conference, DEXA 2014
Munich, Germany, September 1-4, 2014
Proceedings, Part II

 Springer

Volume Editors

Hendrik Decker
Instituto Tecnológico de Informática
Valencia, Spain
E-mail: hendrik@iti.upv.es

Lenka Lhotská
Czech Technical University in Prague, Faculty of Electrical Engineering
Prague, Czech Republic
E-mail: lhotska@fel.cvut.cz

Sebastian Link
The University of Auckland, Department of Computer Science
Auckland, New Zealand
E-mail: s.link@auckland.ac.nz

Marcus Spies
Knowledge Management, LMU University of Munich
Munich, Germany
E-mail: marcus.spies@lrz.uni-muenchen.de

Roland R. Wagner
University of Linz, FAW
Linz, Austria
E-mail: rrwagner@faw.at

ISSN 0302-9743   e-ISSN 1611-3349
ISBN 978-3-319-10084-5   e-ISBN 978-3-319-10085-2
DOI 10.1007/978-3-319-10085-2
Springer Cham Heidelberg New York Dordrecht London

Library of Congress Control Number: 2014945726

LNCS Sublibrary: SL 3 – Information Systems and Application,
incl. Internet/Web and HCI

Typesetting: Camera-ready by author, data conversion by Scientific Publishing Services, Chennai, India

Printed on acid-free paper

Springer is part of Springer Science+Business Media (www.springer.com)

Preface

This book comprises the research articles and abstracts of invited talks presented at DEXA 2014, the 25th International Conference on Database and Expert Systems Applications. The conference was held in Munich, where DEXA had already taken place in 2001. The papers and abstracts show that DEXA faithfully continues to cover an established core of themes in the areas of databases, intelligent systems and applications, but also boosts changing paradigms and emerging trends.

For the 2014 edition, we had called for contributions in a wide range of topics, including

- Acquisition, Modeling and Processing of Data and Knowledge
- Authenticity, Consistency, Integrity, Privacy, Quality, Security
- Big Data
- Constraint Modeling and Processing
- Crowd Sourcing
- Data Clustering and Similarity Search
- Data Exchange, Schema Mapping
- Data Mining and Warehousing
- Data Provenance, Lineage
- Data Structures and Algorithms
- Database and Information System Architecture
- Database Federation, Cooperation, Integration, Networking
- Datalog 2.0, Deductive, Object-Oriented, Service-Oriented Databases
- Decision Support
- Dependability, High Availability, Fault Tolerance, Performance, Reliability
- Digital Libraries
- Distributed, Replicated, Parallel, P2P, Grid, Cloud Databases
- Embedded Databases, Sensor Data, Streaming Data
- Graph Data Management, Graph Sampling
- Incomplete, Inconsistent, Uncertain, Vague Data, Inconsistency Tolerance
- Information Retrieval, Information Systems
- Kids and Data, Education, Learning
- Linked Data, Ontologies, Semantic Web, Web Databases, Web Services
- Metadata Management
- Mobile, Pervasive and Ubiquitous Data
- NoSQL and NewSQL Databases
- Spatial, Temporal, High-Dimensional, Multimedia Databases
- Social Data Analysis
- Statistical and Scientific Databases
- Top-k Queries, Answer Ranking
- User Interfaces

- Workflow Management
- XML, Semi-structured, Unstructured Data

In response to this call, we received 159 submissions from all over the world, of which 37 are included in these proceedings as accepted full papers, and 46 more as short papers. We are grateful to all authors who submitted their work to DEXA. Decisions on acceptance or rejection were based on at least 3 (on average more than 4) reviews for each submission. Most of the reviews were meticulous and provided constructive feedback to the authors. We owe many thanks to all members of the Program Committee and also to the external reviewers, who invested their expertise, interest, and time to return their evaluations and comments.

The program of DEXA 2014 was enriched by two invited keynote speeches, presented by distinguished colleagues:

- Sourav S. Bhowmick: "DB ⋈ HCI"
- Dirk Draheim: "Sustainable Constraint Writing and a Symbolic Viewpoint of Modeling Languages"

In addition to the main conference track, DEXA 2014 also featured 6 workshops that explored a wide spectrum of specialized topics of growing general importance. The organization of the workshops was chaired by Marcus Spies, A. Min Tjoa, and Roland R. Wagner, to whom we say "many thanks" for their smooth and effective work.

Special thanks go to the host of DEXA 2014, the Ludwig-Maximilians University in Munich, Germany, where, under the guidance of the DEXA 2014 general chairpersons Marcus Spies and Roland R. Wagner, an excellent working atmosphere was provided.

Last, but not at all least, we express our gratitude to Gabriela Wagner. Her patience, professional attention to detail, skillful management of the DEXA event as well as her preparation of the proceedings volumes, together with the DEXA 2014 publication chairperson Vladimir Marik, and Springer's editorial assistant, Elke Werner, are greatly appreciated.

September 2014

Hendrik Decker
Lenka Lhotská
Sebastian Link

Organization

General Chair

Marcus Spies   Ludwig-Maximilians Universität München,
                                          Germany
Roland R. Wagner   Johannes Kepler University Linz, Austria

Program Committee Co-chairs

Hendrik Decker   Instituto Tecnológico de Informática, Spain
Lenka Lhotska   Czech Technical University, Czech Republic
Sebastian Link   The University of Auckland, New Zealand

Program Committee

Slim Abdennadher	German University at Cairo, Egypt
Witold Abramowicz	The Poznan University of Economics, Poland
Hamideh Afsarmanesh	University of Amsterdam, The Netherlands
Riccardo Albertoni	Institute of Applied Mathematics and Information Technologies - Italian National Council of Research, Italy
Eva Alfaro	UPV, Spain
Rachid Anane	Coventry University, UK
Annalisa Appice	Università degli Studi di Bari, Italy
Mustafa Atay	Winston-Salem State University, USA
Spiridon Bakiras	City University of New York, USA
Jie Bao	University of Minnesota at Twin Cities, USA
Zhifeng Bao	National University of Singapore, Singapore
Ladjel Bellatreche	ENSMA, France
Nadia Bennani	INSA Lyon, France
Morad Benyoucef	University of Ottawa, Canada
Catherine Berrut	Grenoble University, France
Debmalya Biswas	Iprova, Switzerland
Athman Bouguettaya	RMIT, Australia
Danielle Boulanger	MODEME, University of Lyon, France
Omar Boussaid	University of Lyon, France

Victor Felea	"Al. I. Cuza" University of Iasi, Romania
Stefano Ferilli	University of Bari, Italy
Flavio Ferrarotti	Software Competence Center Hagenberg, Austria
Filomena Ferrucci	University of Salerno, Italy
Vladimir Fomichov	National Research University Higher School of Economics at Moscow, Russian Federation
Flavius Frasincar	Erasmus University Rotterdam, The Netherlands
Bernhard Freudenthaler	Software Competence Center Hagenberg GmbH, Austria
Hiroaki Fukuda	Shibaura Institute of Technology, Japan
Steven Furnell	Plymouth University, UK
Aryya Gangopadhyay	University of Maryland Baltimore County, USA
Yunjun Gao	Zhejiang University, China
Manolis Gergatsoulis	Ionian University, Greece
Bernard Grabot	LGP-ENIT, France
Fabio Grandi	University of Bologna, Italy
Carmine Gravino	University of Salerno, Italy
Sven Groppe	Lübeck University, Germany
William Grosky	University of Michigan, USA
Jerzy Grzymala-Busse	University of Kansas, USA
Francesco Guerra	Università degli Studi Di Modena e Reggio Emilia, Italy
Giovanna Guerrini	University of Genova, Italy
Antonella Guzzo	University of Calabria, Italy
Abdelkader Hameurlain	Paul Sabatier University, France
Ibrahim Hamidah	Universiti Putra Malaysia, Malaysia
Takahiro Hara	Osaka University, Japan
André Hernich	Humboldt-Universität zu Berlin, Germany
Francisco Herrera	University of Granada, Spain
Estevam Rafael Hruschka Jr.	Federal University of Sao Carlos, Brazil
Wynne Hsu	National University of Singapore, Singapore
Yu Hua	Huazhong University of Science and Technology, China
Jimmy Huang	York University, Canada
Xiaoyu Huang	South China University of Technology, China
Michal Huptych	Czech Technical University in Prague, Czech Republic
San-Yih Hwang	National Sun Yat-Sen University, Taiwan
Theo Härder	TU Kaiserslautern, Germany
Ionut Emil Iacob	Georgia Southern University, USA
Sergio Ilarri	University of Zaragoza, Spain
Abdessamad Imine	Inria Grand Nancy, France

Yasunori Ishihara	Osaka University, Japan
Peiquan Jin	University of Science and Technology of China, China
Anne Kao	Boeing, USA
Dimitris Karagiannis	University of Vienna, Austria
Stefan Katzenbeisser	Technische Universität Darmstadt, Germany
Sang-Wook Kim	Hanyang University, South Korea
Hiroyuki Kitagawa	University of Tsukuba, Japan
Carsten Kleiner	University of Applied Sciences & Arts Hannover, Germany
Henning Koehler	Massey University, New Zealand
Solmaz Kolahi	Oracle, USA
Ibrahim Korpeoglu	Bilkent University, Turkey
Harald Kosch	University of Passau, Germany
Michal Krátký	Technical University of Ostrava, Czech Republic
Petr Kremen	Czech Technical University in Prague, Czech Republic
Arun Kumar	IBM Research, India
Ashish Kundu	IBM T.J. Watson Research Center, USA
Josef Küng	University of Linz, Austria
Kwok-Wa Lam	University of Hong Kong, Hong Kong
Nadira Lammari	CNAM, France
Gianfranco Lamperti	University of Brescia, Italy
Anne Laurent	LIRMM, University of Montpellier 2, France
Mong Li Lee	National University of Singapore, Singapore
Alain Léger	FT R&D Orange Labs Rennes, France
Daniel Lemire	LICEF, Université du Québec, Canada
Wenxin Liang	Dalian University of Technology, China
Stephen W. Liddle	Brigham Young University, USA
Lipyeow Lim	University of Hawaii at Manoa, USA
Tok Wang Ling	National University of Singapore, Singapore
Chengfei Liu	Swinburne University of Technology, Australia
Chuan-Ming Liu	National Taipei University of Technology, Taiwan
Hong-Cheu Liu	University of South Australia, Australia
Hua Liu	Xerox Research Labs at Webster, USA
Jorge Lloret Gazo	University of Zaragoza, Spain
Peri Loucopoulos	Harokopio University of Athens, Greece
Jianguo Lu	University of Windsor, Canada
Alessandra Lumini	University of Bologna, Italy
Hui Ma	Victoria University of Wellington, New Zealand
Qiang Ma	Kyoto University, Japan
Stephane Maag	TELECOM SudParis, France

Elio Masciari	ICAR-CNR, Università della Calabria, Italy
Norman May	SAP AG, Germany
Jose-Norberto Mazón	University of Alicante, Spain
Brahim Medjahed	University of Michigan - Dearborn, USA
Alok Mishra	Atilim University at Ankara, Turkey
Harekrishna Mishra	Institute of Rural Management Anand, India
Sanjay Misra	University of Technology at Minna, Nigeria
Jose Mocito	INESC-ID/FCUL, Portugal
Lars Moench	University of Hagen, Germany
Riad Mokadem	IRIT, Paul Sabatier University, France
Yang-Sae Moon	Kangwon National University, South Korea
Reagan Moore	University of North Carolina at Chapel Hill, USA
Franck Morvan	IRIT, Paul Sabatier University, France
Francesc Munoz-Escoi	Polytechnic University of Valencia, Spain
Ismael Navas-Delgado	University of Málaga, Spain
Martin Necasky	Charles University in Prague, Czech Republic
Wilfred Ng	Hong Kong University of Science & Technology, Hong Kong
Javier Nieves Acedo	University of Deusto, Spain
Mourad Oussalah	University of Nantes, France
Gultekin Ozsoyoglu	Case Western Reserve University, USA
George Pallis	University of Cyprus, Cyprus
Christos Papatheodorou	Ionian University & "Athena" Research Centre, Greece
Marcin Paprzycki	Polish Academy of Sciences, Warsaw Management Academy, Poland
Oscar Pastor Lopez	University of Politecnica de Valencia, Spain
Dhaval Patel	Indian Institute of Technology Roorkee, India
Jovan Pehcevski	European University, Macedonia
Jorge Perez	Universidad de Chile, Chile
Olivier Pivert	Ecole Nationale Supérieure des Sciences Appliquées et de Technologie, France
Clara Pizzuti	Institute for High Performance Computing and Networking (ICAR) - National Research Council (CNR), Italy
Jaroslav Pokorny	Charles University in Prague, Czech Republic
Pascal Poncelet	LIRMM, France
Elaheh Pourabbas	National Research Council, Italy
Fausto Rabitti	ISTI, CNR in Pisa, Italy

Claudia Raibulet   Università degli Studi di Milano-Bicocca, Italy
Isidro Ramos   Technical University of Valencia, Spain
Praveen Rao   University of Missouri-Kansas City, USA
Manjeet Rege   University of St. Thomas, USA
Rodolfo F. Resende   Federal University of Minas Gerais, Brazil
Claudia Roncancio   Grenoble University/LIG, France
Edna Ruckhaus   Universidad Simon Bolivar, Venezuela
Massimo Ruffolo   ICAR-CNR, Italy
Igor Ruiz-Agundez   University of Deusto, Spain
Giovanni Maria Sacco   University of Turin, Italy
Simonas Saltenis   Aalborg University, Denmark
Carlo Sansone   Università di Napoli "Federico II", Italy
Igor Santos Grueiro   Deusto University, Spain
Ismael Sanz   Universitat Jaume I, Spain
N.L. Sarda   I.I.T. Bombay, India
Marinette Savonnet   University of Burgundy, France
Raimondo Schettini   Università degli Studi di Milano-Bicocca, Italy
Peter Scheuermann   Northwestern University, USA
Klaus-Dieter Schewe   Software Competence Centre Hagenberg,
                            Austria
Erich Schweighofer   University of Vienna, Austria
Florence Sedes   IRIT, Paul Sabatier University, France
Nazha Selmaoui   University of New Caledonia, New Caledonia
Patrick Siarry   Université Paris 12, LiSSi, France
Gheorghe Cosmin Silaghi   Babes-Bolyai University of Cluj-Napoca,
                            Romania
Hala Skaf-Molli   Nantes University, France
Leonid Sokolinsky   South Ural State University, Russian
                            Federation
Bala Srinivasan   Monash University, Australia
Umberto Straccia   ISTI - CNR, Italy
Darijus Strasunskas   DS Applied Science, Norway
Lena Strömbäck   Swedish Meteorological and Hydrological
                            Institute, Sweden
Raj Sunderraman   Georgia State University, USA
David Taniar   Monash University, Australia
Cui Tao   Mayo Clinic, USA
Maguelonne Teisseire   Irstea - TETIS, France
Sergio Tessaris   Free University of Bozen-Bolzano, Italy
Olivier Teste   IRIT, University of Toulouse, France
Stephanie Teufel   University of Fribourg, Switzerland
Jukka Teuhola   University of Turku, Finland
Taro Tezuka   University of Tsukuba, Japan

Bernhard Thalheim   Christian Albrechts Universität Kiel, Germany
Jean-Marc Thevenin   University of Toulouse 1 Capitole, France
Helmut Thoma   Thoma SW-Engineering, Basel, Switzerland
A. Min Tjoa   Vienna University of Technology, Austria
Vicenc Torra   IIIA-CSIC, Spain
Traian Marius Truta   Northern Kentucky University, USA
Theodoros Tzouramanis   University of the Aegean, Greece
Maria Vargas-Vera   Universidad Adolfo Ibanez, Chile
Krishnamurthy Vidyasankar   Memorial University of Newfoundland, Canada
Marco Vieira   University of Coimbra, Portugal
Jianyong Wang   Tsinghua University, China
Junhu Wang   Griffith University at Brisbane, Australia
Qing Wang   The Australian National University, Australia
Wei Wang   University of New South Wales at Sydney,
                                Australia
Wendy Hui Wang   Stevens Institute of Technology, USA
Gerald Weber   The University of Auckland, New Zealand
Jef Wijsen   Université de Mons, Belgium
Andreas Wombacher   Nspyre, The Netherlands
Lai Xu   Bournemouth University, UK
Ming Hour Yang   Chung Yuan Christian University, Taiwan
Xiaochun Yang   Northeastern University, China
Haruo Yokota   Tokyo Institute of Technology, Japan
Zhiwen Yu   Northwestern Polytechnical University, China
Xiao-Jun Zeng   University of Manchester, UK
Zhigang Zeng   Huazhong University of Science and
                                Technology, China
Xiuzhen (Jenny) Zhang   RMIT University, Australia
Yanchang Zhao   RDataMining.com, Australia
Xiaofang Zhou   University of Queensland, Australia
Qiang Zhu   The University of Michigan, USA
Yan Zhu   Southwest Jiaotong University, China

External Reviewers

Samhaa R. El-Beltagy   Nile University, Egypt
Christine Natschläger   Software Competence Center Hagenberg
                                GmbH, Austria
Badrinath Jayakumar   Georgia State University, USA
Janani Krishnamani   Georgia State University, USA
Kiki Maulana Adhinugraha   Monash University, Australia
Deepti Mishra   Atilim University, Turkey
Loredana Caruccio   University of Salerno, Italy
Valentina Indelli Pisano   University of Salerno, Italy
Claudio Gennaro   ISTI-CNR, Italy

Tymoteusz Hossa	The Poznan University of Economics, Poland
Jakub Dzikowski	The Poznan University of Economics, Poland
Shangpu Jiang	University of Oregon, USA
Nunziato Cassavia	ICAR-CNR, Italy
Pietro Dicosta	Unical, Italy
Bin Wang	Northeastern University, China
Frédéric Flouvat	University of New Caledonia, New Caledonia
Saeed Samet	Memorial University, Newfoundland, Canada
Jiyi Li	Kyoto University, Japan
Alexandru Stan	Romania
Anamaria Ghiran	Romania
Cristian Bologa	Romania
Raquel Trillo	University of Zaragoza, Spain
María del Carmen Rodríguez-Hernández	University of Zaragoza, Spain
Julius Köpke	University of Klagenfurt, Austria
Christian Koncilia	University of Klagenfurt, Austria
Henry Ye	RMIT, Australia
Xumin Liu	RMIT, Australia
Azadeh Ghari-Neiat	RMIT, Australia
Hariton Efstathiades	University of Cyprus, Cyprus
Andreas Papadopoulos	University of Cyprus, Cyprus
Meriem Laifa	Bordj Bouarreridj University, Algeria
Zhifeng Bao	University of Tasmania, Australia
Jianxin Li	Swinburne University of Technology, Australia
Md. Saiful Islam	Swinburne University of Technology, Australia
Tarique Anwar	Swinburne University of Technology, Australia
Mikhail Zymbler	South Ural State University, Russian Federation
Constantin Pan	South Ural State University, Russian Federation
Timofey Rechkalov	South Ural State University, Russian Federation
Ruslan Miniakhmetov	South Ural State University, Russian Federation
Pasqua Fabiana Lanotte	University of Bari, Italy
Francesco Serafino	University of Bari, Italy
Xu Zhuang	Southwest Jiaotong University, China
Paul de Vrieze	Bournemouth University, UK
Lefteris Kalogeros	Ionian University, Greece
Sofia Stamou	Ionian University, Greece
Dingcheng Li	Mayo Clinic, USA
Zhe He	Columbia University, USA
Jun Miyazaki	Tokyo Institute of Technology, Japan
Yosuke Watanabe	Nagoya University, Japan

Jixue Liu	University of South Australia, Australia
Selasi Kwashie	University of South Australia, Australia
Christos Kalyvas	University of the Aegean, Greece
Stéphane Jean	LIAS, France
Brice Chardin	LIAS, France
Selma Khouri	ESI, Algeria
Selma Bouarar	LIAS, France
Rima Bouchakri	ESI, Algeria
Idir Ait Sadoune	Supelec, France

Table of Contents – Part II

Social Computing

Similarity Search

Ranking

Data Mining

Big Data

Approximations

Privacy

Data Exchange

Data Integration

Web Semantics

Repositories

Partitioning

Business Applications

Table of Contents – Part I

Keynote Talks

Data Quality

Social Web

XML Keyword Search

Skyline Queries

Graph Algorithms

Information Retrieval

XML

Security

Semantic Web

Classification and Clustering

Queries

A Unified Semi-supervised Framework for Author Disambiguation in Academic Social Network

Peng Wang[1,3,4], Jianyu Zhao[1], Kai Huang[1], and Baowen Xu[2,3]

[1] School of Computer Science and Engineering, Southeast University, China
[2] State Key Laboratory for Novel Software Technology, Nanjing University, China
[3] State Key Laboratory of Software Engineering, Wuhan University, China
[4] Provincial Key Laboratory for Computer Information Processing Technology,
Soochow University, China
pwang@seu.edu.cn, bwxu@nju.edu.cn

Abstract. This paper addresses the author disambiguation problem in academic social network, namely, resolves the phenomenon of synonym problem "multiple names refer to one person" and polysemy problem "one name refers to multiple persons". A unified semi-supervised framework is proposed to deal with both the synonym and polysemy problems. First, the framework uses semi-supervised approach to solve the cold-start problem in author disambiguation. Second, robust training data generating method based on multi-aspect similarity indicator is used and a way based on support vector machine is employed to model different kinds of feature combinations. Third, a self-taught procedure is proposed to solve ambiguity in coauthor information to boost the performances from other models. The proposed framework is verified on a large-scale real-world dataset, and obtains promising results.

Keywords: author disambiguation, social network, cold-start problem.

1 Introduction

Academic social network, which mainly consists of authors, papers and corresponding relations such as *coauthor* between authors and *publish* between authors and papers, is a kind of popular data used in many social network analyzing and mining works. However, when people use the academic data, for example, building a academic social network from ACM library, they must face the author disambiguation problem: many authors may have same name or one author may refer to multiple similar but not same names. In other words, if authors with same or similar names publish more than one paper, it needs disambiguation to identify them. Actually, the author disambiguation is very frequent in academic publication data. With the quick increasing of academic publications and authors, author disambiguation becomes more important in data clean process for analyzing and mining academic social network.

H. Decker et al. (Eds.): DEXA 2014, Part II, LNCS 8645, pp. 1–16, 2014.

Name disambiguation problem [1, 2], as one of the most important and fundamental problem in many research fields including natural language processing [3, 4], data mining[5], digital library [6–8] and information retrieval [9–11], is about how to resolve the phenomenon of *"multiple names refer to one entity"* called synonym problem and *"one name refers to multiple entities"* known as polysemy problem. Due to different naming standards adopted by various academic databases, conferences and journals, same author may appear in different names. For example,"Anoop Gupta" can be "A Gupta" in different academic libraries. On the other hand, cultural issues like polyphone and common name can cause same name shared by different persons. For instance, "Chen" in Chinese and "Muhammad" in Arabic. Author disambiguation aims to partition authors with same or different names according to whether they correspond to identical person. It is a fundamental issue of the applications involving academic social network. However, since the noisy raw data, different degrees of information loss, inconsistence, heterogeneous, no prior labelling and large scale characteristics of real-world academic social network, author disambiguation is an open problem in past few decades.

This paper proposes a unified semi-supervised framework to solve both synonym and polysemy problems for author disambiguation in academic social network. The main contributions are: (1) Using semi-supervised method to address the cold-start problem. Our framework does not make any assumption on prior knowledge such as number of real authors and true entity of given author name. (2) Introducing a robust method for generating training data based on multi-aspect similarity indicators, and a way to model different kinds of features combination separately using support vector machine. (3) Proposing a simple but effective self-taught algorithm to boost effectiveness of results generated by other models. (4) Verifying the efficiency and effectiveness of our framework on a large-scale and real-world dataset and obtaining promising results.

2   Problem Definition

For convenient and consistent discussion, we use "author" to represent an author name and "person" to represent a real author entity. In an academic social network, let author set be S_a, paper set be S_p, and person set be S_e. An author vertex has a *name* and at least one *organization* attributes. A paper vertex has *title, year, publisher,keywords* attributes. Some attribute values may be missing in practical applications.

Definition 1 (Coauthor Graph). Coauthor graph is an undirected weighted graph $G_{ca} =< V_{ca}, E_{ca} >$. $V_{ca} = \{v|v \in S_a\}$ is the vertex set indicating authors who have coauthor relationship. Edge set $E_{ca} = \{e = (v_i, v_j)|v_i, v_j \in V_{ca} \land v_i, v_j \text{ publish same paper}\}$.

Definition 2 (Publication Graph). Publication graph is a bipartite graph $G_{pb} =< V_{pb}, E_{pb} >$. Vertex set $V_{pb} = S_a \cup S_p$. Edge set $E_{pb} = \{e = (v_i, v_j)|v_i \in S_a \land v_j \in S_p \land v_i \text{ publishes } v_j\}$.

Definition 3 (Academic Social Network). An academic social network G_{sn} consists of G_{ca} and G_{pb}, namely, $G_{sn} = G_{ca} \cup G_{pb}$.

In academic social network, it is easy to determine whether two paper vertices are the same paper but it is very difficult to determine whether two author vertices are the same person. The latter is the author disambiguation problem we discussed in this paper.

Definition 4 (Name-Entity Relation). The name-entity relation is a mapping from S_a to S_e : $g : S_a \rightarrow S_e$.

For an author $a \in S_a$, if the corresponding entity e, then $g(a) = e$ and $e \in S_e$. In author disambiguation problem, any given author corresponds to a unique person.

Definition 5 (Partition of Author Set). Given author set S_a, there exists a partition Ω satisfying following conditions:

$$\Omega = \{C_1, C_2, \ldots, C_n\}$$

$$(C_1 \cup C_2 \cup \ldots \cup C_n = S_a) \wedge (C_1 \cap C_2 \cap \ldots \cap C_n = \emptyset)$$

For example, for author set $\{Wei\ Pan, W\ Pan\}$, two reasonable partitions are $\Omega_1 = \{\{Wei\ Pan, W\ Pan\}\}$ and $\Omega_2 = \{\{Wei\ Pan\}, \{W\ Pan\}\}$.

Definition 6 (Author Disambiguation). Author disambiguation is a function f, which maps S_a to partition Ω, satisfying following conditions:

$$f : S_a \rightarrow \Omega \ s.t. \ \forall C \in \Omega, \forall a_i, a_j \in C \ has \ g(a_i) = g(a_j)$$

$$\Omega = \underset{all\ \Omega}{\mathrm{argmin}} |\Omega|$$

Namely, we hope to have minimum partition in which every set has authors correspond to an identical person.

Definition 7 (Disambiguation Transitivity Property). Given authors $a_i, a_j, a_k \in S_a$, following deduction holds:

$$(g(a_i) = g(a_j)) \wedge (g(a_j) = g(a_k)) \rightarrow (g(a_i) = g(a_k))$$

Definition 8 (Polysemy Problem). For two authors (with same name) correspond to different persons, they should be in different sets in the partition:

$$g(a_i) \neq g(a_j) \wedge a_i = a_j \ satisfies \ a_i \in C_i \wedge a_j \in C_j \wedge C_i \neq C_j$$

Definition 9 (Synonym Problem). For two authors (with different name) correspond to identical person, they should be in same set in partition:

$$g(a_i) = g(a_j) \wedge a_i \neq a_j \ satisfies \ a_i \in C_i \wedge a_j \in C_j \wedge C_i = C_j$$

Author disambiguation in academic social network aims to find function f which best approximate g.

3 Overview of the Framework

This paper proposes a unified semi-supervised framework for solving the author disambiguation problem in academic social network. Fig. 1 depicts our framework. The first stage (1-3) uses multiple rules to pre-process and separate the dataset, and generate features depict various aspects of authors. Topics and coauthor information play important roles in addressing cold start problem. The second stage (4-5) addresses the cold start problem by applying community detection on a graph, which is built from coauthor relationship and topic similarity. A self-taught model is used to improve effectiveness of other models. We also find that attributes like affiliation, year, conference and journal could help supervised disambiguation a lot, despite that they may be noisy or absent. The third stage (6-7) is about improving effectiveness. By referencing disambiguation results from second stage, we generate training data for supervised learning models, and use these models to segregate given author set. Finally in stage four, we blend all the outputs generate by each individual model to get final results.

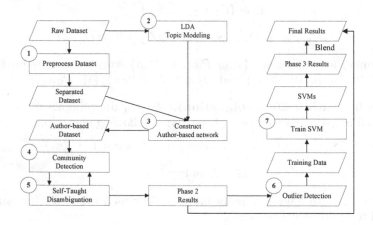

Fig. 1. Overview of the framework

4 Disambiguation Candidate Determining by Heuristics

Usually, there are over millions authors in an academic social network, thus it is a time-consuming process to complete disambiguation in pair-wise manner. This is unacceptable in practice. We first propose some rules to eliminate author pairs which clearly need not to be disambiguated, while also take care of synonym and polysemy problem.

Definition 10 (Author Signature). Signature of an author is defined as concatenation of the initial letter of first name and first 4 letters of last name. If the length of last name is less than 4, it uses the whole last name in signature.

For example, the signature of "Anoop Gupta" is "a gupt", while "Chun Li" will be shorten as "c li".

Definition 11 (Compatible Property). Two authors who correspond to identical person should have identical signature.

Compatible property is used for divide-and-conquer large-scale author sets into small ones. Namely, it partitions all authors into many small disambiguation candidate sets. When we create a set for each possible author signature, we actually create a set for each possible length-4 last name prefix. This does not violate the compatible property, since author signature can assure that all authors will be put in small sets. The length of last name's prefix controls average size of small author sets. Length 4 is determined empirically, since it balances average size of small sets and possibility of authors being separated due to name variations, which leads potential loss of polysemy cases. After pre-processing, models can process each small set individually, thus following processes can be parallel in order to improve efficiency. In implementation, after partitioning using compatible property, average size of author sets is 13.12, which is practically acceptable for quadratic time disambiguation. Meanwhile, each small set would be processed in parallel during subsequent processes. For two different persons with identical author signature, we may still wrongly classify them as identical person. For example, in some cultures like Chinese, it is more probable to have more than one person with identical name in same research area. We propose some heuristics to label name-pairs , that will be useful for disambiguation.

Heuristic 1 (HighFreq). *For two authors whose signatures with frequency higher than threshold T1, this pair is D1-Compatible.*

In practice, we set T1 to 10. For example, signature "a gupta" appear more than 10 times in dataset. This heuristic is used to treat the common names in different cultures. For example, "Kumar", "Miller" and "Jone".

Heuristic 2 (Phonetic). *For either name that can be expressed in Chinese, Cantonese or Korean phonetic parts, this pair is D2-compatible.*

Characters in above cultures that read phonetically identical can be very different. Common family name is more likely to appear in these cultures, making them more likely to prone to disambiguation problem. Heuristic 2 is used to treat this problem individually. To judge whether a term is a phonetic part, we create a table with all available consonants and vowels. If a term can be expressed by one of combinations of consonant and vowel, it is a phonetic part. For example, "chen" can be expressed in concatenation of "ch" and "en", "yu" can be split into "y" and "u". We also extend these rules to separate cases with two characters. For instance, "zhihua" can be separated into "zhi hua" then determined to be D2-Compatible.

Heuristic 3 (Signature). *For either name that is identical to its signature, this pair is D3-Compatible.*

This cases is very likely to happen in publication that prefer shortened author names. For example, "c li" is identical to its signature. D3-Compatible also means the family name is less than 4. Applying this heuristic can give special treatment to those cases missed by D2 and D1.

Heuristic 4 (Edit Distance). *For two names that do not satisfy D2- and D3-Compatible, and their firstname-lastname string's edit distance is greater than threshold T2, this pair is D4-Compatible.*

In practice, we use Levenshtein Distance and set T2 to 3. Treatment of this heuristic is given to cases of unicode compatibility, OCR recognition errors and typos. For example, "ö" and "o" are compatible; "m" can be recognized as "nn"; "ll" can be missed as "l". The reason we opt out D2- and D3-Compatible is that family names in those type are less likely to vary. For instance, "shen" and "sheng" can be completely different in Chinese. Making models less conservative for D4-Compatible make polysemy problem more likely to be resolved in academic social network.

Heuristic 5 (Mid Name). *For two names with different middle names or middle initials, they are not compatible.*

It is reasonable to treat "A Y Chen" and "A N Chen" to be different person, since their middle name initials are different. However, names in western culture tend to have more than one middle names and will be eliminated differently in publications. Practically, we also use some other rules like judging substring of middle initial string to resolve this problem.

Heuristic 6 (Mid Name Missing). *For two names that satisfy compatible property, if one has middle name but the other has not, this pair is D5-Compatible. Signatures can be extend to other signatures by adding middle name, is also classified into D5-Compatible.*

The reason for this heuristic is to treat cases that publication completely eliminate middle name. For example, "A Chen" would be "A N Chen" or "A Y Chen". By treating this type more conservatively, the overall false positive answers can be greatly eliminated.

Heuristic 7 (Active Year). *For two names with identical signature, if similarity of their active years (defined in later) is greater than T3, this pair is D6-Compatible.*

Practically, T3 is set to 0.5. Intuition of this heuristic is that authors that active in different age are less likely to be identical.

Heuristic 8 (Common Rule). *For two authors with identical signature, and they are not D1- to D6-Compatible, this pair is D7-Compatible.*

It is possible for a pair to satisfy more than one heuristic, we should define priority for these compatibility types.

Definition 12 (Compatibility Priority). *For compatibility types described above, their priority is defined as follows:*

$$D3 = D4 = D5 = D6 > D2 > D1 > D7$$

According to these heuristics, we apply different parameter values to different compatibility types in order to improve overall effectiveness.

5 Unsupervised Disambiguation Model

5.1 Modelling Domain Features

Domain information is very useful in determining whether two authors are identical. Although we cannot reason that two authors in similar domain correspond to same person, but it is unlikely for same person to publish in very distinct research areas. Information like title, journal/conference name and keyword depict domain characteristics of a given author. We concatenate all these text features to get a virtual documents. By generating virtual document for each individual author, we have a virtual corpus to model domain features. We utilize two classic approaches in text modeling: TF/IDF-weighted vectors and topic vectors inference by Latent Dirichlet Allocation [12]. Similarity generated by calculating cosine value of vectors.

For each virtual document, stop words and term with length less than 3 are removed. All letters are transformed into lowercases. For term which mix number and letters, we use "[MixAlpha]" to represent such term.

We use dampened TF/IDF to weight significance of terms appear in corpus. Given term t, document d and corpus D, the TF/IDF weight of term t is:

$$tfidf(t, d, D) = tf(t, d) \times idf(t, D)$$

$$tf(t, d) = 0.4 + \frac{0.6 \times f(t, d)}{max\{f(t, d) : w \in d\}}$$

$$idf(t, d) = \log \frac{|D|}{|\{d \in D : t \in d\}|}$$

$f(t, d)$ represents frequency of term t in document d.

Similarity based on TF/IDF-weighted vectors is greatly affected by shared terms. If two virtual documents shared no term, the similarity will be zero. However, this does not mean two authors are in distinct areas. Polysemy could cause this problem. One defeat of TF/IDF is that it doesn't take advantages of correlations of terms. Topic model like LDA is a way to use correlations of terms. We employ classic LDA model to inference topic distribution of author's virtual documents. A collapsed Gibbs Sampler is used to inference parameters of LDA [13].

Since all documents for testing are included in corpus, we can get topic vectors of all virtual documents. This is a way to avoid test time topic inference, thus improve the efficiency of overall disambiguation process.

There is another benefit of LDA. Numbers of published papers may vary a lot for different authors. This means some authors have few papers. In other words, terms used by authors with few papers will appear to be "orthogonal" to those authors with lots of papers, even they are in same research area. LDA assumes shared topic parameters of virtual documents, this allows short virtual documents to borrow statistic power of long documents.

5.2   Community Detection

Another information that depicts author's domain characteristic is coauthor pattern. It is common for a person to cooperate consistently within a set of coauthors. Given a set of compatible authors and their coauthors, we can build a graph by adding coauthor and compatible relation as edge. We assume that two compatible authors will very likely to be identical if they are in the same community.

Definition 13 (Compatible Graph). Compatible graph is an undirected weighted graph $G_{cp} =< V_{cp}, E_{cp} >$. $V_{cp} = \{v|v \in S_n\}$ is the vertex set, indicating authors that are compatible. Edge set $E_{cp} = \{e = (v_i, v_j)|v_i, v_j \in V_{cp} \land v_i, v_j\ satisfy\ D1\ or\ D4 - Type\}$.

Definition 14 (Coauthor-Compatible Graph). Coauthor-Compatible Graph, CCG for brevity, is an undirected weighted graph $G_{cc} =< V_{cc}, E_{cc} >$, in which:

$$V_{cc} = V_{cp} \cup \{v|v \in V_{ca} \land \exists v_i, v_j \in V_{cp}\ s.t.\ (v_i, v), (v_j, v) \in E_{ca}\}$$

$$E_{cc} = \{e = (v_i, v_j)|v_i, v_j \in V_{cc}\}$$

Definition 15 (CCG Edge Weight). Given two authors v_i and v_j, topic vectors inferenced by LDA are t_i and t_j, the weight of edge between v_i and v_j is $\cos(t_i, t_j)$.

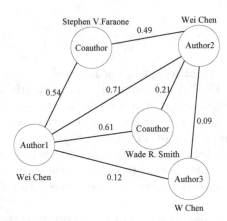

Fig. 2. An example of coauthor-compatible graph

We can learn from the graph that if two authors are identical, they are likely to share coauthors and be connected edges with high topic similarity. As CCG example shown in Fig. 2, since Author1 and Author2 share two coauthors, and they have high topic similarity, they are identical, while Author3 is another person. On the other hand, although "W Chen" is compatible with "Wei Chen", but they shared no coauthor and are connected with low weight edges. Therefore, it is reasonable to guess "W Chen" is not "Wei Chen". Author 1, 2 and two coauthors actually form a "community". It is very probable for two compatible authors in same community to be identical. We use the fast unfolding community detection algorithm [14] to generate "circles" of given CCG.

However, there is one trap of directly using CCG as input of community detection. The core of community detection is to optimize modularity, which actually affected by total weight in given graph. Suppose we have three authors compatible with one another, but share low and close similarity. No matter the similarity is 1 or 0.05, community detection will tend to think three authors form a community, though former one really makes sense while latter one does not. To address this problem, we need to eliminate edges with low weight. As we discussed before, distribution of topic similarity for different compatible types varies. Thus we set different thresholds for different compatible types.

5.3 Self-taught Algorithm

Since the author disambiguation problem is cold-start, coauthors also need to be disambiguated. In CCG, shared coauthors contribute greatly in forming communities. If we do not separate coauthors, it is very possible that community detection algorithm would miss potential shared coauthors. Fig. 3 gives one such case. If Author1 and Author2 share low similarity and we have no idea whether "Stephen Faraone" and "S Faraone" are identical, thus we will likely to give a false negative answer. Our framework propose a simple but effective algorithm called self-taught as treatment of the issue.

The detail of self-taught procedure is showed in Algorithm 1. The key idea of self-taught algorithm is to merge known duplicated authors as one (line 4), and apply community detection on merged CCG to see if more duplicated authors can be discovered (line 8-11). This procedure run iteratively until no more duplicates can be found.

6 Supervised Disambiguation Model

6.1 Multi-aspect Similarity Features

Besides topic information, we also build multi-aspect similarity features using title, conference/journal name, keywords, publication year and author affiliation.

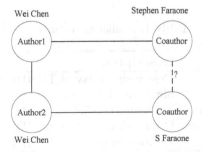

Fig. 3. An example for impact of coauthor disambiguation

Algorithm 1. Self-Taught algorithm

Input: CCG; First phase disambiguation results
Output: Improved disambiguation results

1 **begin**
2   **while** *True* **do**
3   Construct Union-Find set S based on initial results ;
4   Merge identical authors ;
5   Construct G_{ca} ;
6   Construct G_{cp} ;
7   Construct CCG ;
8   Community detection, build S' based on new results ;
9   **if** $S = S'$ **then**
10 | break ;
11   **end**
12   **end**
13 **end**

Unlike virtual documents used for topic modeling, we extract multi-aspect features from virtual publications. Different from virtual documents, which only contains bag of words, virtual publications have attributes like other publication, including title, keyword, and affiliation, etc. We concatenate data in same attribute from all papers of a given author to generate corresponding attributes of the virtual publications. For instance, given an author, title of his virtual publication is generated by concatenate all titles of his papers. If two authors are identical, their virtual publications may share similar attributes. We can get a similarity value for each attribute of given two publications. Table 1 shows four kinds of features we used and corresponding similarity functions. The cosine similarity means the similarity of two TF/IDF weighted vectors. Tanimoto Coefficient is a classic for measuring similarity between sets. Given two set A and B, the Tanimoto Coefficient $s_{tanimoto}$ is:

$$s_{tanimoto} = \frac{|A \cap B|}{|A^2| + |B^2| - |A \cap B|}$$

Table 1. Similarity features

ID	Feature Name	Description
0	Title Similarity	Cosine similarity of TF/IDF weighted vectors
1	Year Similarity	Similarity for virtual publications
2	Affiliation Similarity	Tanimoto coefficient of affiliations
3	Keyword Similarity	Cosine similarity of TF/IDF weighted vectors
4	Venue Similarity	Cosine similarity of TF/IDF weighted vectors

The reason we use Tanimoto Coefficient for affiliation is that semantic of term in affiliation is more concise. On the other hand, term in affiliation sparsely appears, which is not intuitive for TF/IDF to model.

Year attribute of virtual publications is the average of all publication years of papers. We define a segmented function for comparing virtual publication year.

Definition 16 (Year Similarity). *Given active year y_i and y_j, similarity s_{year} is defined as follow:*

$$s_{year} = \begin{cases} 0 & 10 < |y_i - y_j| \text{ or either one is missing} \\ 0.25 & 5 < |y_i - y_j| \leq 10 \\ 0.50 & 2 < |y_i - y_j| \leq 5 \\ 0.75 & |y_i - y_j| = 2 \\ 1 & otherwise \end{cases}$$

Notice that the five features in Table 1 can be combined to multi-aspect similarity feature vectors. Each feature vector corresponds to two virtual publications. However, we know that not all papers have complete information. We give each attribute an identity, and encode feature vector using combination of identities of shared attributes from two virtual publications. For example, if publication A has title, year, venue, keywords, publication B has title, year, venue and affiliation, the feature vector of theirs are encode as "124". Although there are total 31 kinds of encodes theoretically, we only find 20 feature encodes for the practical academic social network.

6.2 Training SVM on Robust Data

Training data for supervised disambiguation is generated from unsupervised disambiguation results using heuristics. Two authors that marked as identical are retrieved from results. Multi-aspect feature vectors made from virtual publications of the two authors is marked as positive cases. If signatures of two authors are different, they unlikely to be identical. Feature vectors for these two authors is marked as negative case.

The training data generated using this heuristic is sure to be noisy. There is no guarantee that results given by unsupervised disambiguation is correct. On the other hand, not any pair of authors' virtual publications are guaranteed to look alike (or this supervised disambiguation effort is nonsense). Those scenarios make positive cases seem like negative ones. For negative cases, there is also coincidence that two different authors' virtual publications are very similar, which makes negative cases look like positive ones.

By treating these noisy cases as outliers, we can use local outlier factor [15] detection method to refine training data. During calculating local outlier factor of each training case, those with high factor value can be removed.

After getting robust training data, we use LibSVM [16] to train models for each feature combinations and tune parameters separately. In disambiguation process, system first generates multi-aspect similarity features for given authors; Then it retrieves a model according to code of shared attributes and use it to tell whether two authors are identical or not.

Blending results from different models boost the effectiveness. We use transient property to merge results from individual model. Results in our approach is made of pairs, which indicates two identical authors. Treating pairs as a merge queries, we use set union algorithm [17] to quickly perform this process.

7   Experimental Evaluation

7.1   Dataset and Criteria

We implement the proposed framework in Python. PostgreSQL is used to host the dataset. All experiments are obtained on a single machine with Intel i5 quad-core 3.10 GHz CPU and 4GB memory.

We use the Microsoft Academic Search (MAS) dataset [18], which is a snapshot of real academic social network data use in Microsoft Academic Search. Table 2 shows detail statistics in dataset. Additionally, we find that only 21% publications contain complete attributes, and only 21.8% authors verified/denied related publication records. Comparing to other datasets used in many previous related works, MAS dataset has several challenging issues need to be resolved: (1) *Mixing synonym and polysemy problems*: Both two problems universally exist; (2) *Noisy data with missing and inconsistent attributes*: Dataset are integrated from heterogeneous data sources. There is no guarantee that all records have complete information; (3) *Absence of labelled data*: Neither the number of real author entity nor their relevant information is provided in MAS dataset. This kind of problem is known as cold-start problem. Notice that train-deleted and train-confirmed table only tell parts of information about whether an author published a paper, which are not directly associate with authors' real identities.

Table 2. Statistics of the MAS dataset

Table	#Record	Description
Author	247,203	names and organization
Conference	4,545	short name, full name and url
Journal	15,151	short name, full name and url
Paper	2,257,249	title, year, venue, keyword
PaperAuthor	12,775,821	author-paper relation.
Train-Deleted	112,462	author denied papers
Train-Confirmed	123,447	author verified papers

Mean F-score is used to evaluate quality of results. P is precision, R is recall. TP, FP and FN refer to true positive, false positive and false negative results. Then $P = \frac{|TP|}{|TP \cup FP|}, R = \frac{|TP|}{|FP \cup FN|}, F = 2\frac{P \cdot R}{P+R}$.

7.2   Experimental Results

We use title similarity and author compatibility to produce baseline result. By title similarity, we mean that if two authors have publications which edit distance of two titles is less than average word length in titles, those two compatible

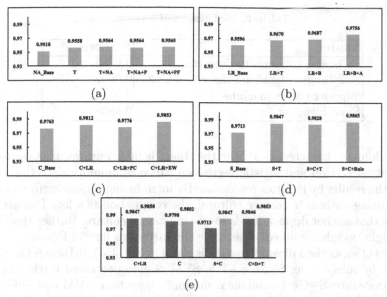

Fig. 4. Performance of different models : (a) Rule-based disambiguation result; (b) Logistic regression disambiguation result; (c) Community detection disambiguation result; (d) SVM disambiguation model result; (e) Self-taught disambiguation result.

authors is identical. To further refine, if family name frequency of compatible author is higher than 10, we will not merge them by title similarity rule. NA_Base is based on author name compatibility with full family name and author affiliation; T uses title similarity; T+NA combines NA_Base and T; T+NA+P uses name compatibility we defined; T+NA+PF further combines family name frequency rule. As Fig. 4 (a) shows, title similarity rule gives better results than simply using author affiliation. Using prefix rather than full family name involves polysemy case produced by typo and Unicode compatibility, which gives better results than solely title similarity, but the improvement is not significant.

In Fig. 4 (b), we try logistic regression on training data generated by rule-based results. LR_Base is the results produced by simply using logistic regression; LR+T merges results by LR_Base and title similarity rule; LR+B is the results produced by applying different parameters on different multi-aspect feature codes; LR+B+A further blends in author affiliation information. The performance is better than baseline results, but it reaches a bottleneck at 0.9756.

Fig. 4 (c) depicts results based on community detection. C_Base is the results by community detection itself; C+LR combines community detection and logistic regression; C+LR+PC further introduces phonetic rules. Previous strategy is to treat weight for edges between authors and coauthor differently but fixed according to compatibility type, C+LR+EW otherwise weights all edges as topic similarity. CCG captures important features of identical author by using shared coauthor and topic similarity. Simply using CCG and community detection yields 0.9763 F-score, which beats best results by logistic regression. This indicates coauthor relationship and domain feature is effective in describe

Table 3. Best results of models

Model	F-Score
CCG and community detection	0.9763
Self-Taught algorithm (based on CCG)	0.9802
Support vector machine	0.9713
Final blending	0.9868

author's domain characteristics. Note that there is little improvement combining community detection and logistic regression, since community detection covers most of the results by logistics regression. By introducing phonetic rules to separated treating authors in eastern cultural, the results boosts a lot. This strongly indicates that author duplication is severe in eastern culture. Rather than using fixed weight, weighting all edges as topic similarity has 0.9853 F-score.

Results of supervised disambiguation is given in Fig. 4(d). S_Base is the results produces by solely using SVM; S+T is SVM results processed further by self-taught procedure; S+C+T combine community detection, SVM and self-taught procedure; We try some rules to separate authors merged in blending phase, given as S+C+T+Rule. The rule we use for separation is simple, if a set of authors considered to be identical has incompatible author names, they will be rolled back to state before blending. This clearly indicates transient assumption do generate error, but still gives effective results than not blending models.

Fig. 4 (e) gives results of applying self-taught procedure: C is the community detection results; S+C is the results combining SVM and community detection; C+D+T is results of combining community detection and LDA weight. Red histogram is without self-taught and blue one is the results with self-taught. It is clear that self-taught procedure improve results of other models differently, which is align to the assumption that coauthors also need disambiguation.

Table 3 gives best results for each model. Note that SVM itself does not give a very good results, but it boost final blending due to its effectiveness of capturing aspects that have not been considered by other models.

Table 4 gives running time for key procedures of our framework. Offline procedures only need to be run one time and can be updated online in a new disambiguation. Online procedures are those required to be run in each disambiguation. The most time-consuming procedure is LDA topic inference. Since topic distribution is mainly inferred from word correlation in initial corpus, a very drastic variation on LDA model is not frequent when new information is added to the system. Topic inference can be done offline when an author's information is updated. Therefore, this procedure does not affect disambiguation efficiency. In real-world application, a disambiguation iteration is fired only when there are enough data increment or predefined duration has passed. Our approach can gives out final disambiguation results in less than 1.5 hour, so it is also reasonable in real-world application.

Table 4. Performance of main procedures

Procedure	Type	Avg. Time
Create Signature Subset	Offline	2min
Generate Virtual Documents	Offline	32min
Topic Inference	Offline	34hours
Community Detection	Online	24min
Self-Taught Procedure	Online	20min
Generate Training Data	Online	6min
Supervised Disambiguation	Online	30min

8 Conclusion

This paper propose a unified semi-supervised framework for author disambiguation in academic social network. Our approach is capable of dealing with both synonym and polysemy problems simultaneously. By using coauthor information and domain features, our model addresses cold-start problem of large-scale real-world dataset. A self-taught procedure is also proposed as treatment of coauthors' disambiguation issue.

Acknowledgments. This work was supported by the National Key Basic Research and Development Program of China (2014CB340702)the National Natural Science Foundation of China (61170071, 91318301, 61321491), and the foundation of the State Key Laboratory of Software Engineering (SKLSE).

References

1. Chang, C.H., Kayed, M., et al.: A survey of web information extraction systems. IEEE Trans. on Knowledge and Data Engineering 18(10), 1411–1428 (2006)
2. Ferreira, A.A., Gonalves, M.A., Laender, A.H.: A brief survey of automatic methods for author name disambiguation. ACM SIGMOD Record 41(2), 15–26 (2012)
3. Cortes, C., Vapnik, V.: Support-vector networks. Machine Learning 20(3), 273–297 (1995)
4. Yin, X., Han, J., Yu, P.S.: Object distinction: Distinguishing objects with identical names. In: Proceedings of ICDE 2007, Istanbul, Turkey (2007)
5. Kanani, P., McCallum, A.: Efficient strategies for improving partitioning-based author coreference by incorporating web pages as graph nodes. In: Proceedings of AAAI 2007 Workshop on Information Integration on the Web, Canada (2007)
6. Qian, Y., Hu, Y., Cui, J., Zheng, Q., et al.: Combining machine learning and human judgment in author disambiguation. In: Proceedings of the CIKM 2011, Glasgow, UK (2011)
7. Tang, J., Fong, A.C.M., et al.: A unified probabilistic framework for name disambiguation in digital library. IEEE Trans. on Knowledge and Data Engineering 24(6), 975–987 (2012)
8. Gurney, T., Horlings, E., Besselaar, P.V.D.: Author disambiguation using multi-aspect similarity indicators. Scientometrics 91(2), 435–449 (2012)

9. Tan, Y.F., Kan, M.Y., Lee, D.: Search engine driven author disambiguation. In: Proceedings of JCDL 2006, USA (2006)
10. Minkov, E., Cohen, W.W., Ng, A.Y.: Ucontextual search and name disambiguation in email using graphs. In: Proceedings of SIGIR 2006 (2006)
11. Bekkerman, R., McCallum, A.: Disambiguating web appearances of people in a social network. In: Proceedings of WWW 2005 (2005)
12. Blei, D.M., Ng, A.Y., Jordan, M.I.: Latent dirichlet allocation. Journal of Machine Learning Research 3, 993–1022 (2003)
13. Darling, W.M.: A theoretical and practical implementation tutorial on topic modeling and gibbs sampling. In: Proceedings of ACL 2011 (2011)
14. Blondel, V.D., Guillaume, J.L., et al.: Fast unfolding of communities in large networks. Journal of Statistical Mechanics: Theory and Experiment 2008(10), 10008 (2008)
15. Breunig, M.M., Kriegel, H.P., et al.: Lof: identifying density-based local outliers. ACM Sigmod Record 29(2), 93–104 (2000)
16. Chang, C.C., Lin, C.J.: Libsvm: a library for support vector machines. ACM Transactions on Intelligent Systems and Technology 2(3), 27 (2011)
17. Tarjan, R.E., Leeuwen, J.V.: Worst-case analysis of set union algorithms. Journal of the ACM 31(2), 245–281 (1984)
18. Roy, B.S., Cock, D.M., Mandava, V., et al.: The microsoft academic search dataset and kdd cup 2013. In: KDD Cup 2013 Workshop, Chicago, USA (2013)

Designing Incentives for Community-Based Mobile Crowdsourcing Service Architecture

Mizuki Sakamoto[1], Hairihan Tong[1], Yefeng Liu[1],
Tatsuo Nakajima[1], and Sayaka Akioka[2]

[1] Department of Computer Science and Engineering, Waseda University
{mizuki,tong.hairihan,yefeng,tatsuo}@dcl.cs.waseda.ac.jp
[2] Department of Network Design, Meiji University
akioka@meiji.ac.jp

Abstract. Good design strategies for designing social media are important for their success, but current designs are usually ad-hoc, relying on human intuition. In this paper, we present an overview of three community-based mobile crowdsourcing services that we have developed as case studies. In community-based mobile crowdsourcing services, people voluntarily contribute to help other people anytime and anywhere using mobile phones. The task required is usually trivial, so people can perform it with a minimum effort and low cognitive load. This approach is different from traditional ones because service architecture designers need to consider the tradeoff among several types of incentives when designing a basic architecture. We then extract six insights from our experiences to show that motivating people is the most important factor in designing mobile crowdsourcing service architecture. The design strategies of community-based mobile crowdsourcing services explicitly consider the tradeoff among multiple incentives. This is significantly different from the design in traditional crowdsourcing services because their designers usually consider only a few incentives when designing respective social media. The insights are valuable lessons learned while designing and operating the case studies and are essential to successful design strategies for building future more complex crowdsourcing services.

Keywords: Social Media, Mobile Crowdsourcing, Design Strategy, Motivation, Community-Based Approach, Case Studies.

1   Introduction

Social media have become increasingly popular in our daily life. We use *Facebook*[1] and *Twitter*[2] to expand our social interactions with both friends and non-friends every day. This situation changes our daily life. For example, collaborating with unknown people on social media has become common, and we can exploit their knowledge to support our daily activities. A crowdsourcing service is a promising approach to

---

[1] https://www.facebook.com/
[2] https://twitter.com/

H. Decker et al. (Eds.): DEXA 2014, Part II, LNCS 8645, pp. 17–33, 2014.

exploit our social power and to enhance our human ability and possibilities [7]. We can exploit the collective knowledge in the entire world because crowdsourcing services such as *InnoCentive*[3] allow us to divide a task into micro tasks, and ask them to be performed by respective professionals. Additionally, crowdfunding services such as *Kickstarter*[4], which is one of the variations of crowdsourcing services, can be widely used to collect money or solicit funds. These trends of new social media have been changing the way we work, think, and solve problems.

Many studies have already been performed to analyze the existing social Q&A[5] systems and monetary rewards-based crowdsourcing services [2, 9, 11, 12]. In these social media services, motivating people by offering incentives is essential to their successes. For example, *Amazon Mechanical Turk*[6] uses monetary rewards as the economic incentive, and *Yahoo! Answers*[7] and *Foursquare*[8] adopt social facilitation and self-respect as the social incentives.

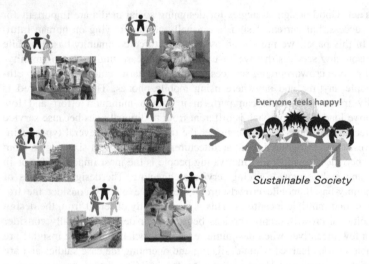

Fig. 1. Community-based Mobile Crowdsourcing Service Architecture

We propose a new concept, i.e., a community-based mobile crowdsourcing service architecture. In our approach, as shown in Fig. 1, people voluntarily contribute to help other people anytime and anywhere using mobile phones. The task required is usually trivial, so people can perform it with a minimum effort and low cognitive load. This approach is different from traditional ones because service architecture designers need to consider the tradeoff among several types of incentives when

---

[3] https://www.innocentive.com/
[4] https://www.kickstarter.com/
[5] Social Q&A is an approach for people to seek and share information in participatory online social sites.
[6] https://www.mturk.com/
[7] http://answers.yahoo.com/
[8] https://www.foursuare.com/

designing a basic architecture. We developed three case studies of community-based mobile crowdsourcing services and operated these services. The first one is *MoboQ* [13], the second one is *MCfund* [19, 28], and the third one is *BianYi*.

We extracted six insights from our experiences with the three case studies. Their design strategies explicitly consider the tradeoff among multiple incentives. This is significantly different from the design in traditional crowdsourcing services because their designers usually consider only a few incentives when designing respective social media. Thus, the tradeoff among the incentives was not explicitly addressed in their design and operation experiences, and these insights were not addressed in the papers describing in respective case studies.

The design of more complex future social media needs to consider how to choose the correct incentive from a choice of incentives to motivate the people participating in the social media. The six insights contain useful lessons learned from our experiences in designing and operating the case studies. We think that the lessons describing the tradeoff among incentives are useful information for future social media designers.

The remainder of the paper is organized as follows. In Section 2, we discuss some related work. In Sections 3 through 5 we introduce three community-based mobile crowdsourcing services as case studies. Section 6 discusses the six insights, describing how to motivate people's activities in social media. Section 7 concludes the paper.

2   Related Work

UbiAsk [12] is a mobile crowdsourcing platform that is built on top of an existing social networking infrastructure. It is designed to assist foreign visitors by involving the local population to answer their image-based questions in a timely fashion. Existing social media platforms are used to rapidly allocate micro-tasks to a wide network of local residents. The basic concept of *UbiAsk* is similar to that of *MoboQ* because it assumes that the local crowd will answer questions from foreign visitors without monetary rewards.

Amazon Mechanical Turk is currently the best-known commercial crowdsourcing service. It uses monetary rewards to encourage people to work on micro tasks [11]. This means that the system only considers economic incentives. However, as shown in [1], monetary rewards are not always the best way to motivate people to perform micro tasks. Instead, contributors appreciate many intangible factors, such as community cooperation, learning new ideas and entertainment.

The non-monetary motivations were represented successfully in examples such as *Yahoo!Answers* and *Answers.com*[9]. Moreover, if money is involved, quality control becomes a major issue because of the anonymous and distributed nature of crowdworkers [9]. Although the quantity of work performed by participants can be increased, the quality cannot, and crowdworkers may tend to cheat the system to increase their overall rate of pay if monetary rewards are adopted.

---

[9] http://www.answers.com/

More recently, digital designers have begun to adopt ideas from game design to seek to incentivize desirable user behaviors. The idea of taking entertaining and engaging elements from computer games and using them to incentivize participation in other contexts has been studied in a variety of fields. In education, the approach is known as serious gaming, and in human computing, it is sometimes called games with a purpose [2]. Most recently, digital marketing and social media practitioners have adopted this approach under the term *gamification*. The idea is to use game mechanics such as on-line games to make a task entertaining, thus engaging people to conscientiously perform tasks. *Foursquare* is a typical pervasive service that uses a gamification approach [17].

Participatory sensing is the process whereby community members use their smart phones that they carry everyday to collect and analyze data in their surrounding worlds [6]. Ubiquitous sensing technologies are easily deployable using their mobile phones and online social networking with a citizenry. Motivating people's participation is an important aspect of successful participatory sensing and some work tried to include gamification-based techniques to encourage the participants.

MINDSPACE, proposed in the UK, is a framework that adopts some concepts from behavioral economics to affect human attitudes and behaviors [8]. However, using public policy to modify behavior has a significant disadvantage; it takes a long time to formulate the policies. Therefore, it is effective to use technologies that have strong effects on our mind to navigate human attitude and behavior. In fact, information technologies can be used to solve health problems by increasing people's self-efficacy.

Maslow claims that human motivation is based on a hierarchy of needs [14]. In *Maslow's hierarchy*, the basic needs include such needs as physiological needs and food. Other needs are safety, attachment, esteem, cognitive needs and aesthetic needs. At the highest level, when all other needs are satisfied, we can start to satisfy self-actualization needs. Because people value products when they satisfy their needs, satisfying needs is closely related to defining values. For example, *Boztepe* proposes four values: utility value, social significance value, emotional value and spiritual value [3].

Persuasion can attempt to influence a person's beliefs, attitudes, intentions, motivations, or behaviors. *Cialdini* defined six influence cues: *Reciprocity, Commitment and consistency, Social proof, Liking, Authority*, and *Scarcity* [4]. The key finding is that people automatically fall back on a decision making based on generalizations in a recent complicated world where people are overloaded with more information than they can process.

It is difficult to find crowdfunding-based social media designed in a research community. *Muller* et al. [15] present an experiment in enterprise crowdfunding. Employees allocate money for employee-initiated proposals in an enterprise Intranet site, including a medium-scaled trial of the system in a large multinational company. The results show that communities in a large company propose ideas, participate, and collaborate and that their activities are encouraged through crowdfunding. The approach details a new collaboration opportunity and shows that crowdfunding is a promising method for increasing activity within communities.

3   Case Study I: Using Crowds for Sensing Context Information

The location-based real-time question-answering service, *MoboQ*, is built on top of a microblogging platform [13]. In *MoboQ*, end users can ask location- and time-sensitive questions, such as whether a restaurant is crowded, a bank has a long waiting line, or if any tickets are left for an upcoming movie at the local cinema, i.e., questions that are difficult to answer with ordinary Q&A services. *MoboQ*, as shown in Fig. 2, analyzes the real-time stream of the microblogging service *Sina Weibo*[10], searches for the *Weibo* users who are most likely at the given location at this moment based on the content of their microblog posts, and pushes the question to those strangers. Note that the answerers in this system are *Sina Weibo* users, not *MoboQ* users, and may not even be aware of the existence of *MoboQ*. This design takes advantage of the popularity and furious growth rate of *Weibo*. The real-time nature of microblogging platforms also makes it possible to expect a faster response time than with traditional Q&A systems. To some extent, *MoboQ* utilizes the *Weibo* users as local human sensors and allows a questioner to extract context information at any given location by asking the local *"human sensors"* what is happening around them.

The main components of *MoboQ* are as follows:

i.   Communication Module: This module consists of the REST(Representational State Transfer)[11] Web Service, with an open API to client applications and the *Sina Weibo* API. It handles the communications between an asker from *MoboQ* and an answerer from *Sina Weibo*.

ii.   Ranking Engine: The ranking engine searches and selects the best candidates on the *Weibo* platform to answer a question.

iii.   Client Applications: Each client application includes a Web site, mobile web, and a native mobile application to present the questions and answer the user in an accessible and interactive form.

The *MoboQ* server, which comprises the communication module and ranking engine, is implemented in Ruby on Rails[12]. The mobile web is implemented using HTML5[13] technology.

Because *MoboQ* is a Q&A system between people who are likely to be complete strangers, it is a challenge to motivate the potential answerers from *Weibo* to answer the strangers' questions. From a design point of view, we concentrate on two aspects simultaneously: 1) how to establish trust among the answerer, the *MoboQ* platform, and the asker and 2) how to provide appropriate incentives to the candidate answerers.

---

[10] http://weibo.com/
[11] http://www.w3.org/TR/2002/WD-webarch-20020830/
[12] http://rubyonrails.org/
[13] http://dev.w3.org/html5/

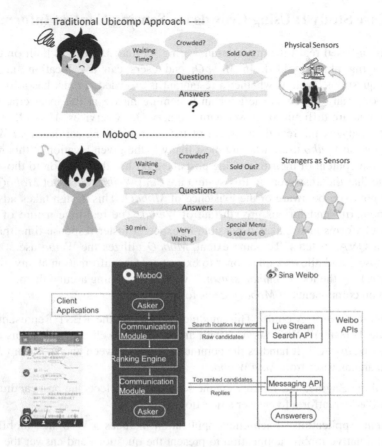

Fig. 2. An Overview of MoboQ

Early studies suggest that lack of trust is identified as one of the greatest barriers inhibiting Internet transactions. To support online trust building, the content of the *MoboQ* query message is uniquely designed, i.e., it includes the reason why the candidate answerer is selected (e.g., "Hi, we found that you just visited #Location#."), the URL of the asker's profile page on *MoboQ* site to show that the asker is a real person, and another URL to the question's page on *MoboQ* to help the answerer learn about the service. All necessary information is open to the candidate answerers, and we hope that this information helps the answerer understand that this is not a spam message but a real question from a real person who is seeking help.

We also utilize findings from social psychology as incentives in our system. Social incentives, such as social facilitation and social loafing, are two commonly cited behaviors that can affect contributions on social media [16]. The social facilitation effect refers to the tendency of people to perform better on simple tasks while someone else is watching, rather than while they are alone or working alongside other people. The social loafing effect is the phenomenon of people making less effort to achieve a goal when they work in a group than when they work alone because they feel that their contributions do not count or are not evaluated or valued as much when they work as a group. This is considered one of the main reasons that groups are less productive than the combined performance of members working alone. Different mechanisms are employed in *MoboQ* to take advantage of positive social facilitation and avoid negative social loafing as follows:

i. A public thank you message is provided to publicly and prominently display individuals' efforts;
ii. The query is a public message, thus individuals should know that others can easily evaluate their work; and
iii. Every question is sent to up to 15 people in a separate message so the unique value of each individual's contribution can be evaluated.

MoboQ was designed and implemented during the autumn of 2011 and released early in 2012 in China. Until the beginning of October 2012, we collected 15,224 questions. With regards to answers, we received 29,491 total answers from 162,954 *Weibo* candidate answerers. This indicates an 18.0% reply rate for individual *Weibo* strangers. Because *MoboQ* sends one question to up to 15 candidate answerers, the overall average response rate for a question is 74.6%, where 28% of the answers arrived within 10 minutes, 51% arrived within 20 minutes, and 72% arrived within an hour. This result demonstrates that the approach is feasible in real-world conditions.

4   Case Study II: Using Crowdfunding to Achieve a Sustainable Society

Crowdfunding is an emerging new way of funding new ideas or projects by borrowing funding from crowds. In this concept, a person proposes a new project, explains the importance and the target amount of money, and shows what people who fund the project will receive when the mission is successfully completed. When the total amount of funds from contributors exceeds the target amount, the project starts. After successful completion of the project, each contributor receives the benefits according to his or her funding level. If the benefit offers high scarcity value, it has a high incentive for the contributors. However, the existing crowdfunding platforms such as *Kickstarter* require participants to contribute real money; thus, only people who have extra money can participate.

We propose a new approach, named *micro-crowdfunding (MCfund)*, to increase people's awareness of the importance of sustaining our society. It will help motivate people in urban areas to participate in achieving a sustainable society [19, 23, 28]. In *MoboQ*, completing micro tasks is motivated through the use of social incentives, and then individuals complete the tasks through their own spirit of reciprocity for strangers. This incentive is not strong enough to complete the more complex micro tasks referenced in *MCfund*. In *MCfund*, completing micro tasks is motivated within a community whose members are known to each other. An economic incentive is also used to motivate the community members to complete the tasks, but the incentive is not in the form of a monetary reward. Instead, *MCfund* increases people's awareness of the meaning behind completing micro tasks, thereby increasing their intrinsic motivation to complete the tasks. Using mobile phones is also a key factor in lowering the hurdle for contributing to the community. Its members increase their activities in the face of smaller incentives because activities can be performed anytime and anywhere by accessing the services through mobile phones.

The main characteristics of the approach are as follows:

i.   The *crowdfunding* concept is adopted to allow people to choose among the small, common resources to which they would like to contribute to maintain sustainability;

ii.   The currency used in the proposed approach is based on the *aging-money* concept, which encourages people to participate in *MCfund* before the money's value is gradually degraded;

iii.   The interaction in *MCfund* is lightweight. People in a community can easily propose new micro tasks, called missions, in *MCfund* and fund them from their smart phones through a simple interaction; and

iv.   The participants can share information and details about a mission and receive appropriate feedback for the activities that they perform.

Fig. 3 shows an overview of *MCfund* activities. In *MCfund*, a member of a community related to a small common resource, called a mission organizer, proposes a new mission when he or she is aware that an activity must be completed to maintain the sustainability of a resource. Typical examples of such common resources are a public sink at a floor of a building or a public shelf used by a university laboratory. The proposal includes the mission's summary, which specifies the necessary activities and the total amount of money required to achieve the mission. The mission proposal is simply done by touching the common resource with the mission organizer's smart phone and sending a photo showing the resource's current status. In the next step, when other members, called mission investors, receive requests to fund the mission, they decide whether they want to fund the mission based on the delivered photo. If some of them would like to fund the mission, they simply click on the requests on their phones to notify the mission organizer that they want to fund the mission. When the total submitted funds exceed the target amount, the mission can be executed by any member who can access the resource in his or her spare time. Such a member is called a mission performer. The mission is usually a very simple task such as cleaning up a public sink or putting a shelf in order. After completion, the mission performer takes a photo of the resource to show the mission's completed status and sends it to the mission organizer. Finally, the mission organizer verifies the quality of the achievement, and a completion notification of the mission containing a photo of the resource is delivered to all members who have funded the mission. The system is implemented using HTML5 technology.

We conducted a small experiment to evaluate how the social and economic incentives designed for *MCfund* are useful using a role-playing evaluation method. We found that money can be used to increase the awareness of the importance of social sustainability through the support of missions by investing funds. This increase occurs because the awareness of the importance of the mission is high if participants

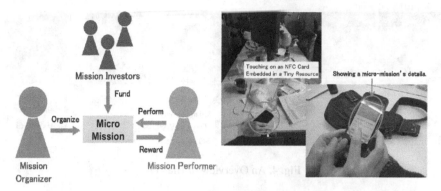

Fig. 3. An Overview of MCfund Activities

frequently invest funds. Additionally, the *aging money* encourages people to invest more before the value of their money decreases. The results show that participating in activities to achieve social sustainability is most important to increase the awareness of sustainability.

5   Case Study III: Crime Mapping via Social Media

A crime map is a tool that visualizes crime information based on the geographical location of crimes. In earliest time, the police use a crime map to recognize the inherent geographical component of crimes by sticking pins into maps displayed on walls, where each pin on the map represents a crime incident. With the progress of GIS and the Internet, a crime map places the 'pins', which are shown as color spots, on a digital map that can be available on a mobile phone. Thus, the police have recently used crime maps for crime analysis, and ordinary citizens have used these maps to obtain an understanding of neighborhood crime or even to receive alert notices when a crime occurs in the place in which he or she is interested.

Nowadays, social media services, especially microblogs, have become an indispensable part of people's daily life, where they can share events around them, such as what they have seen and what events have occurred around them, by posting text messages, geo-information, photos, or even videos on social media.

The BianYi system, as shown in Fig. 4, that we developed is a crime map that is automatically created from crime data published on microblogs. Unlike similar crime maps, such as *SpotCrime*[14] or *WikiCrimes*[15], which are not responsible for the crime data uploaded in maps and make no representation or warranty about the accuracy or content of any data contained, *BianYi* obtains crime data by analyzing microblogs posted during 2010 to 2013 from *Sina Weibo* and *Tencent Weibo*[16], which serve as open crime databases.

---

[14] http://spotcrime.com/
[15] http://www.wikicrimes.org/
[16] http://t.qq.com/

Fig. 4. An Overview of BianYi

   The current version of *BianYi* consists of the following two modules: a crowdmap module and an information flow monitoring module.

Crowdmap module: The module offers a user interface of the system, which visualizes the crime data on a digital map and shows the user how serious each crime is through interactions with the crime data. We have implemented the module using the open-source crowdmapping services of the *Ushahidi*[17] platform, which also supports a mobile phone.

Information flow monitoring module: This module is a core component of data collection from social media. It determines the type of keywords to be used to search APIs, what type of microblogs should be collected, and how the contents collected from microblogs are converted to crime data. This module can be broken down into the following two sub-modules:

Key conversations extraction module: Sensitive crime information is collected through the topic search function of the *Sina Weibo* APIs and the microblog search function of the *Tencent Weibo* API. *BianYi* focuses on the 'steal' type of crime, so keywords such as 'steal', 'thief', and 'pickpocket' are used as the parameters.

Data cleaning module: This module primarily provides the functionality of filtering the contents of the microblogs to input to the crime database format of the *Ushahidi* platform. Specifically, in the first step, the module filters data that are extracted at the key conversations extraction steps from the microblogs and excludes spam, such as repeated records, and unavailable geo-messages. In the second step, the module matches all of the microblogs' attributes to the crime incident database columns, e.g., a crime description and geo-information in the micro-text to a crime description, and a microblog post-created time to an incident-occurred time.

The initial version of *BianYi* did not obtain as much sensitive data as we expected. The limitations of the *Sina Weibo* API, such as the fact that the microblog search API service was closed for a time and that only 200 search result records related to one topic were available, may be part of the reason for this lack of data. We also found that users were not willing to send crime data to microblogs that contained geo-information. Using the crime type 'steal' for search APIs may be another component of the lack of data because people did not know where and when things were stolen,

---

[17] http://www.ushahidi.com/

in a common sense. In fact, it is difficult for people to record non-current location neither GIS(Geographic Information System) nor texts. A new method is required for
users to map the location timely and accurately. One solution is for microblogs to
prepare a specific interface to report crime data from users.

The crime data in *BianYi*, as extracted from microblogs, have been obtained from
verified individual users, organizational users, and media users, such as police offices,
TV programs, and news media, as verified by the Sina Corporation and Tencent Corporation. Because trust is a significant factor for a crowdsourcing service, the functionality of verified account features in Chinese microblogging services will be a key
factor for users to evaluate the crime reports' reliability.

The second version of *BianYi* adds the natural language processing module to the
data cleaning module to extract the location information from the microblog content
and to convert the information into geo-information using *Google Map* APIs[18]. In this
version, we still experience some problems when extracting location information from
the text in the microblogs. The primary reason for these problems is that the location
information described in the microblogs is not as clearly described by the microblog
users as it can be translated for *Google Map* APIs.

6   Six Insights Extracted from Our Experiences

In this section, we present six insights that were extracted from our experiences with
developing and operating our community-based mobile crowdsourcing services. We
found that choosing the correct incentives is not easy. *Nakajima* and *Lehdonvirta* [16]
proposed five incentives and discussed the importance of choosing the correct incentives, but our experience shows that incentives are not used as independent design
variables. The appropriateness of the incentives changes according to users' current
situations, personalities, and cultural backgrounds. The facts shown in this section did
not presented in the past papers of respective case studies. Thus the following insights
are main contributions of this paper.

6.1   Mobility and Curiosity

All of the case studies described in this paper use mobile phones to access crowdsourcing services. Thus, people can access them anytime and anywhere. The most
important factor when using mobile phones is that people can use the services when
they have a little time, e.g., in their spare time. In urban cities, such as Tokyo, people
usually have a lot of spare time while waiting for trains, food, or to start meetings or
events. These days, many young people in Tokyo watch their mobile phones in their
spare time. If a micro task defined in a crowdsourcing service is lightweight and can
be completed in a short time, such as *MoboQ*, the possibility of a user trying to complete such a task is higher. However, people need to have an incentive to perform the
task, even if the task is lightweight.

---

[18] https://developers.google.com/maps/

A useful incentive to perform a task is curiosity. If people have enough curiosity, they are willing to perform the proposed task, but there is a tradeoff between their curiosity and the time required to complete the task. For example, in *MCfund*, if people do not have an interest in sustainability, they do not want to become either a mission organizer or a mission performer. A task that offers less curiosity needs to be completed within a smaller time before they lose their curiosity. When people do not have enough curiosity to complete a task, a service needs to offer other incentives to encourage them.

This factor shows that we should be careful when evaluating a new service because most people usually feel curiosity for the service that they use first. This means that the type of curiosity that encourages people should be carefully considered during the evaluation. However, curiosity is the first incentive that a social media designer uses to encourage people.

6.2   Using Social Incentives and a Gamification-Based Approach

Social incentives can be classified into three types. The first type is self-respect. Typically, self-respect is stimulated via badges and leaderboards to increase a user's self-esteem. The gamification-based approach typically uses this incentive. The second type is to use reciprocity. People help other people. This trait encourages users to complete tasks to support other people. The trust relationship among people is the key to success when using reciprocity. The third type is social facilitation. Visualizing one's differences from others is a useful technique when using social facilitation. When someone's performance is not sufficiently high, using visualization to show how their performance is less than that of others can be effective.

Using self-respect is the most popular technique that is used to encourage people to complete a task in existing social media; thus, we applied it in the early design. However, it did not work well, so the current prototypes do not use that approach. It is important to understand why the approach did not succeed. Using a badge or a leaderboard based on people's self-respect is also based on a feeling of rarity, such as a *Mr. Bill* badge or becoming a mayor in *Foursquare*. The rarity is a type of economic incentive. In most previous studies, monetary rewards are discussed as the economic incentive, but for people, rare objects can be considered the same as a monetary reward. That is, if people do not feel the value of the objects, they may want to exchange them for other objects. If the object can be used similar to money, it can be exchanged among people. The economic incentive may not work every time, as shown by *Antikainen* et al. [1]. The above experience indicates that the game mechanics used to offer virtual rare objects do not work successfully in some cases. This claim is consistent with our experience described in [12].

In the case studies, the social media developed for research purposes usually do not have many users to start, as shown by the case of the user study in the early phase. Therefore, when designing the case studies, we decided to use reciprocity or social facilitation as much as possible. These incentives can work well even if the community size is small, but for self-respect to be effective, a large number of participants in the region are required to make some virtual objects scarce.

6.3   Game-Based Psychological Incentive

Achievement is often used in games, and it is a useful incentive to motivate people. By setting a goal, people try to achieve the goal. When a task is very simple, people can complete it without exerting a great effort. However, if the complexity of a task is increased, as in *MCfund*, more effort is required; therefore, showing an explicit goal becomes important. In games, the goal is divided into several sub-goals, and a player can achieve the sub-goals step-by-step because each subsequent sub-goal is usually achieved with a little increased effort. In the real world, we may not be able to offer such easily achievable sub-goals because the progress required to achieve the goal in the real world is not usually linear. For example, daily, constant exercise does not reduce people's weight linearly; this is one reason why people give up exercising. In *MCfund*, each mission is currently independent, so it is not easy to achieve the final goal of social sustainability. A mission organizer should be conscious of his or her mission as a sub-goal to achieve a sustainable society. One solution is to incorporate a fictional story in the real world. This approach can define the sub-goals in the fictional story, and then they are easier to achieve than real-world sub-goals. Of course, maintaining consistency with the real world is important, to feel a sense of reality, when achieving the sub-goals [22, 26]. Fictional stories may contain ideological messages that make us aware of the important social issues while we are living in the daily life [24, 26], that offer a promising possibility towards better human navigation. Specifically, Japanese popular culture such as games and animations offers tremendous examples including ideological messages to increase the tension in the drama [27]. Many young Japanese create their own stories to explain their daily activities as a type of myth or destiny. They also propose virtual festivals, and many of them join the festival and achieve their goals together. Thus, relying on the creative community to create such a story to coordinate various missions to achieve a common goal is a promising approach.

6.4   Intrinsic Motivation

To motivate people to complete a task, they must be made aware of the importance of the task. Specifically, if the task is related to solving serious social problems, such as environment sustainability or human well-being, intrinsic motivation is essential to encourage people. For example, as shown in *BianYi*, if users believe that offering more criminal information decreases the number of crimes, then their motivation to inform on criminals will be increased. Helping other people by investing money is also useful for increasing their awareness of the importance of the task. People usually believe that money can be used to increase the economic incentive, but we found that money is also useful for increasing the awareness of important issues. Money can be exchanged between individuals; therefore, it can be used as a tool to remind them of the importance of completing a task. However, the gamification-based approach leads only a few people to participate heavily in the target activities [3]. Thus, that approach may not be suitable for increasing the awareness of the activities' importance.

It is difficult to remind users why the task is important without a large amount of information, e.g., the sustainability in *MCfund* or a secure daily life in *BianYi*. Learning the importance of the task to perform takes a long time and requires a large cognitive load. Therefore, the approach is not suitable for mobile-based social media. A story that is used to teach the importance of the task is the key to efficiently increasing intrinsic motivation. If the story is popular and includes a clear ideological message, a character or object appearing in the story becomes a metaphor to remind users of the importance of performing the task through the ideological message embedded in the story, as shown by *Sakamoto* et al. [21, 28].

The gamification-based approach is more effective if people feel that the gamification performs activities, as discussed by *Denny* [5]. It is thus desirable to intrinsically motivate people, even when gamification is adopted.

6.5 Personal and Cultural Differences

The effects of the users' personalities and their cultural differences are important factors in the design of social media. The types of effective incentives differ for different personalities and cultural backgrounds. It is important to vary the incentives used in social media according to the preference of the target audience. As shown by *Kimura* and *Nakajima* [10], using social facilitation or social loafing depends on the user's cultural background. In a collective culture, social facilitation works well, and social loafing becomes a more serious problem when designing social media. The success of *MoboQ* may thus operate differently in China, where these social incentives are more effective.

Using reciprocity depends on the users' personalities, as shown by *Sakamoto* et al. [20]. As shown above, reciprocity is an effective social incentive for designing social media, but the trust relationship among users is an important factor when using this incentive. Reciprocity works well if people believe that others have a feeling of gratitude for their help. In our case studies, especially in *MCfund*, we carefully designed it to maintain the trust relationship between people by showing why a person needs support from others and that his or her claim is not dishonest. Additionally, after completing a task, people receive gratitude for their effort. This reduces the effect of personality, so the social media works well for most people.

6.6 Integrating Multiple Incentives towards the Same Goal

As described in [20], the personality of each person affects his or her perceived value. For example, one person may perceive value in jewelry, but another person may not perceive this value. This means that each person may need a different incentive on achieving the same goal as shown in Fig. 5. A community-based mobile crowdsourcing service can be customized according to his or her personality, so that each community member still needs to use the same service that multiple persons with different personalities can use simultaneously. Because each person perceives a different value for his or her most important incentive, the service must offer multiple values to satisfy all of them. As described in [24], a participatory design helps to incorporate multiple values into one service, and the potential for more people to prefer this service increases.

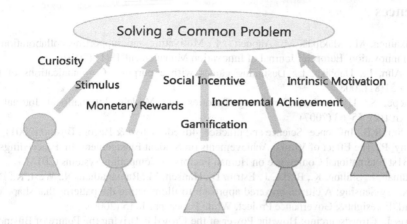

Fig. 5. Using Multiple Incentives to Navigate Human Behavior

The *value-based design framework* proposed in previous studies [18] is a promising direction to integrate multiple incentives. The value set is a good tool to design incentives because we can determine which incentive is effective and which incentive can be replaced with a different incentive, according to humans' personalities and their cultural backgrounds. Incorporating fictionality with transmedia storytelling is also a promising technique to integrate multiple values by embedding the values into respective story fragments in a unified story represented as transmedia storytelling [25]. When a character or an item appeared in a fictional story is pervasively embedded in a service with each fragment of the story, the character or good can be used as a leitmotif of the story to remind the story's ideological messages and they make it possible to navigate human behavior more effective [21].

7   Conclusion and Future Direction

The paper described three case studies of community-based mobile crowdsourcing services. Unlike traditional crowdsourcing services, they adopt various incentives to motivate people to join crowdsourcing activities. We extract six insights from experiences with the services as design strategies to build future crowdsourcing services. The approach enables us to develop crowdsourcing services that take into account personal and cultural differences by incorporating various types of incentives for different types of people.

Social media will become a significant part of our daily life. The insights described in this paper are important for building future successful community-based mobile crowdsourcing services. In the next step, we need to investigate a more systematic framework to design incentives. The tradeoff described in this paper is essential to developing the framework.

References

1. Antikainen, M., Mäkipää, M., Ahonen, M.: Motivating and supporting collaboration in open innovation. European Journal of Innovation Management 13(1) (2010)
2. von Ahn, L., Dabbish, L.: Designing games with a purpose. Communications of the ACM 51(8) (2008)
3. Boztepe, S.: User value: Competing Theories and Models. International Journal of Design 1(2), 55–63 (2007)
4. Cialdini, R.B.: Influence: Science and practice, 4th edn. Allyn & Bacon, Boston (2001)
5. Denny, P.: The Effect of Virtual Achievements on Student Engagement. In: Proceedings of the 31st International Conference on Human Factors in Computing Systems (2013)
6. Goldman, J., Shilton, K., Burke, J., Estrin, D., Hansen, M., Ramanathan, N., West, R.: Participatory Sensing: A citizen-powered approach to illuminating the patterns that shape our world. Foresight & Governance Project, White Paper, pp. 1–15 (2009)
7. Howe, J.: Crowdsourcing: How the Power of the Crowd is Driving the Future of Business. Random House Books (2009)
8. Institute of Government, MINDSPACE: Influencing Behaviour through Public Policy. Cabinet Office (2010)
9. Harper, F.M., Moy, D., Konstan, J.A.: Facts or Friends? Distinguishing Informational and Conversational Questions in Social Q&A Sites. In: Proceedings of the 29th International Conference on Human Factors in Computing Systems (2009)
10. Kimura, H., Nakajima, T.: Designing Persuasive Applications to Motivate Sustainable Behavior in Collectivist Cultures. PsychNology Journal 9(1), 7–28 (2011)
11. Kittur, A., Chi, E.H., Suh, B.: Crowdsourcing User Studies With Mechanical Turk. In: Proceedings of the 26th International Conference on Human Factors in Computing Systems (2008)
12. Liu, Y., Lehdonvirta, V., Alexandrova, T., Nakajima, T.: Drawing on Mobile Crowds via Social Media. ACM/Springer Multimedia Systems Journal 18(1), 53–67 (2012)
13. Liu, Y., Alexandrova, T., Nakajima, T.: Using Stranger as Sensor: Temporal and Geosensitive Question Answering via Social Media. In: Procecddings of International World Wide Web Conference (2013)
14. Maslow, A.H.: Motivation and Personality. Harper and Row, New York (1970)
15. Muller, M., Geyer, W., Soule, T., Daniel, S., Cheng, L.-T.: Crowdfunding inside the enterprise: Employee-initiatives for innovation and collaboration. In: Proceedings of the 31st International Conference on Human Factors in Computing Systems (2013)
16. Nakajima, T., Lehdonvirta, V.: Designing Motivation Using Persuasive Ambient Mirrors. Personal and Ubiquitous Computing 17(1), 107–126 (2013)
17. René, G.: Breaking Reality: Exploring Pervasive Cheating in Foursquare. Transactions of the Digital Games Research Association 1(1) (2013)
18. Sakamoto, M., Nakajima, T., Alexandrova, T.: Digital-Physical Hybrid Design: Harmonizing the Real World and the Virtual World. In: Proceedings of the 7th International Conference on the Design & Semantics of Form & Movement (2012)
19. Sakamoto, M., Nakajima, T.: Micro-Crowdfunding: Achieving a Sustainable Society through Economic and Social Incentives in Micro-Level Crowdfunding. In: Proceedings of International Conference on Mobile and Ubiquitous Multimedia (2013)
20. Sakamoto, M., Alexandrova, T., Nakajima, T.: Analyzing the Effects of Virtualizing and Augmenting Trading Card Game based on the Player's Personality. In: Proceedings of The Sixth International Conference on Advances in Computer-Human Interactions (2013)

21. Sakamoto, M., Alexandrova, T., Nakajima, T.: Augmenting Remote Trading Card Play with Virtual Characters used in Animation and Game Stories - Towards Persuasive and Ambient Transmedia Storytelling. In: Proceedings of the 6th International Conference on Advances in Computer-Human Interactions (2013)
22. Sakamoto, M., Nakajima, T., Akioka, S.: Designing Enhanced Daily Artifacts based on the Analysis of Product Promotions using Fictional Animation Stories. In: Yoshida, T., Kou, G., Skowron, A., Cao, J., Hacid, H., Zhong, N. (eds.) AMT 2013. LNCS, vol. 8210, pp. 266–277. Springer, Heidelberg (2013)
23. Sakamoto, M., Nakajima, T.: Gamifying Social Media to Encourage Social Activities with Digital-Physical Hybrid Role-Playing. In: Meiselwitz, G. (ed.) SCSM 2014. LNCS, vol. 8531, pp. 581–591. Springer, Heidelberg (2014)
24. Sakamoto, M., Nakajima, T., Akioka, S.: A Methodology for Gamifying Smart Cities: Navigating Human Behavior and Attitude. In: Streitz, N., Markopoulos, P. (eds.) DAPI 2014. LNCS, vol. 8530, pp. 593–604. Springer, Heidelberg (2014)
25. Sakamoto, M., Alexandrova, T., Nakajima, T.: Introducing Virtuality for Enhancing Game-related Physical Artifacts. International Journal of Smart Home 8(2), 137–152 (2014)
26. Sakamoto, M., Nakajima, T.: A Better Integration of Fictional Stories into the Real World for Gamifying Daily Life. In: Proceedings of the 1st International Symposium of Simulation and Serious Games (2014)
27. Sakamoto, M., Nakajima, T.: The GamiMedia Model: Gamifying Content Culture. In: Rau, P.L.P. (ed.) CCD 2014. LNCS, vol. 8528, pp. 786–797. Springer, Heidelberg (2014)
28. Sakamoto, M., Nakajima, T.: A Community-based Crowdsourcing Service for Achieving a Sustainable Society through Micro-Level Crowdfunding. In: Proceedings of International Conference on Internet, Politics, Policy 2014: Crowdsourcing for Politics and Policy (2014)

Arguing Prism: An Argumentation Based Approach for Collaborative Classification in Distributed Environments

Zhiyong Hao, Li Yao, Bin Liu, and Yanjuan Wang

National University of Defense Technology, Changsha, Hunan, P.R. China, 410073
{haozhiyongphd,liyao6522,binliu0314,nudtwyj}@gmail.com

Abstract. This paper focuses on collaborative classification, concerning how multiple classifiers learned from distributed data repositories, can come to reach a consensus. Interpretability, in this context, is favored for the reason that they can be used to identify the key features influencing the classification outcome. In order to address this problem, we present Arguing Prism, an argumentation based approach for collaborative classification. The proposed approach integrates the ideas from modular classification rules inductive learning and multi-agent dialogue games. In particular, argumentation is used to provide an interpretable classification paradigm in distributed environments, rather than voting mechanisms. The results of experiments reveal that Arguing Prism performs better than individual classifier agents voting schemes. Moreover, an interpretable classification can be achieved without losing much classification performance, when compared with ensemble classification paradigms. Further experiment results show that Arguing Prism out-performs comparable classification paradigms in presence of inconsistent data, due to the advantages offered by argumentation.

Keywords: Argumentation, Prism algorithm, Inconsistency tolerance.

1 Introduction

Recently, learning from distributed data has become a focus of intensive research, mainly due to the inherent distributed scenarios such as multi-sensor classification. In the task of distributed multi-sensor classification, spatially remote sensors often collect data measurements at different times, and may use different sensing modalities [8]. Hence, each sensor site may have its own local training data, which should not be brought together into a single 'data warehouse' due to high communication cost or proprietary reasons. In these cases, the ability to understand the overall classification model is an important requirement [2], as well as high accuracy, and this trend accelerates in recent years [3]. The main reason is that an interpretable classification model can be used to identify the key features influencing the classification outcome.

The well-known interpretable classifiers are the decision trees in supervised learning. Many decision tree algorithms working in distributed environments have been designed in the last decade, such as [4]. More recently, Andrzejak et al. [1] presented algorithms for classification on distributed data, by which not only the individual

H. Decker et al. (Eds.): DEXA 2014, Part II, LNCS 8645, pp. 34–41, 2014.

learner on each data site is interpretable, but also the overall model is also interpretable using merging decision trees rather than voting schemes. However, the tree structure is deemed as the main drawback of the 'divide and conquer' approach which results in unnecessary and confusing trees, especially for the rules extracting from different decision tree on distributed data sites. An alternative approach to inducing classification rules is the 'separate and conquer', which induces a set of modular rules that do not necessarily fit into a decision tree representation.

This paper presents and evaluates an interpretable classification approach in distributed environments based on argumentation using Prism algorithms, called Arguing Prism. The main aim of Arguing Prism is to perform tasks of interpretable classification to new instances, using distributed classifiers. It is noted that all these rule based classifiers comprise of modular rules, i.e. they do not normally fit into a decision tree representation. The key ideas are (1) that argumentation provides an interpretable conflict resolution mechanism for different classifiers, and (2) that arguments can be generated using inductive learning dynamically.

2   Preliminaries

Classification is a common task in machine learning, which has been solved using various approaches [5]. On one hand, some approaches, such as support vector machine (SVM), can obtain accurate classification models, but they must be regarded as 'black boxes'. On the other hand, classification rules inductive learning methods, such as decision trees, can provide transparent and comprehensible classifiers. Thus, these classifiers can offer understandable and comparable accurate models for end users.

Generally, classification rule induction algorithms can be categorized into two different approaches, the 'divide and conquer' and the 'separate and conquer'. The former approach induces classification rules in the intermediate form of decision trees, whereas the later one induces a set of modular rules directly. Rules from different decision tree classifiers on distributed sites could not be merged directly into a single decision tree using the 'divide and conquer' approach. In such cases, a single decision tree construction for the overall classification model will first need to introduce additional feature value tests that are logically redundant to force the rules into a form suitable for merging into a tree structure. This will inevitably lead to a decision tree that is unnecessary large and confusing. While the basic Prism algorithm generates the rules concluding each of the possible classes in turn, and each rule is generated term by term, which is in the form of 'attribute-value' pairs (called 'feature-value' pair in the current paper). The 'feature-value' term added at each step is chosen to maximize the separation between the classes.

Actually, combining the rules of different classifiers induced from distributed data sites have been investigated in recent years, such as [1]. But in these approaches, little changes of individual data sites may lead to a completed new overall classification model, thus it may be very 'expensive' for the classification tasks in dynamic environments. The following sections provide an overview of how the proposed argumentation based approach can cope with collaborative classification task dynamically. In particular, several classifier agents engage in argumentation dialogues, each induces modular rules from their own training instances, and further generates arguments.

3   Argumentation Based Approach

This section presents Arguing Prism, an argumentation based approach for collaborative classification in distributed environments. The scenarios envisaged are those where there are several options for a new instance to be classified, and each of these options is advocated by one or more classifier agents on distributed data sites. The intuition behind Arguing Prism is to provide an interpretable classification paradigm, whereby multiple classifier agents can reach a consensus decision about the classification tasks with interpretability requirements. In particular, we propose applying Prism family of algorithms to induce 'modular' classification rules from the data on distributed sites, and from which the arguments used by each classifier agent are derived. In the following subsections, the realization of this approach is described in more detail. We first present an overview of our multi-agent argumentation model, called Arena [12], in subsection 3.1.

3.1   The Arena Model for Arguing Prism

Arena [12] is a dialectical analysis model for multiparty argument games to evaluate rules learned from different past instances. The model provides a novel way that can transform the multiparty arguments games into two-party argument games using ideas from the Arena Contest of Chinese Kungfu. As investigating by Yao et al., Arena has a capability in learning and performs well, and thus it provides a feasible way to evaluate the rules from different classifier agents for classification.

The Arena model is used here to allow any number of classifier agents to engage in a dialogue process, the aim of which is to classify a new instance collaboratively. Each classifier agent formulate arguments for one advocated classification or against the classification advocated by other agents, using 'modular' classification rules induction algorithms, rather than associate rule mining algorithms in [12].

As already stated, each participant agent has its own local repository of data in forms of data instances. These agents produce reasons for or against certain classifications by inducing rules from their own datasets using Prism family of algorithms (subsection 3.3). The antecedent of every classification rule represents a set of reasons for believing the consequent. The classification rules induction provides for several different types of moves, which can be employed to perform the argumentation dialogues, further detailed in the following subsection.

3.2   Interpretable Collaborative Classification within Arena

As indicated above, the Arena model used for Arguing Prism allows a number of classifier agents to argue about the classification of a new instance. Each classifier agent argues for a particular classification or against other classifications. Arguments for or against a particular classification are made with reference to an agent's own rule set induced from individual data repository. Each data instance consists of a set of feature-value pairs and a single class-value pair indicating a particular classification. Arguing Prism uses arguments differing form Dung's framework in that the arguments are generated from classifier agent's rule set. Let us define the classification arguments considered by Arguing Prism:

Definition 1 (Classification Argument). A Classification Argument= $<CLR,$ $L\_confidence>$ is a 2-tuples, where CLR is a classification rule induced from the data repository of an individual agent, $L\_confidence$ is the confidence of a rule argument following the Laplace probability estimation procedure. It is noted that the $L\_confidence$ should be higher than a given threshold for a legal classification argument. Further details for the arguments are described in subsection 3.3.

Each participant agent can employ one of the following types of moves (i.e. speech acts) to generate arguments:

(1) Proposing moves. This type of moves allows a new classification argument with its $L\_confidence$ higher than a given threshold to be proposed.
(2) Attacking moves. Moves intended to show that a classification argument proposed by some other agent should not be considered decisive with respect to the current instance to be classified.
(3) Refining moves. Moves that enable a classification rule to be refined to meet an attack.

The arguments exchanged via the moves described above are stored in a central data structure, called *dialectical analysis tree* [12], which is maintained by the referee. Having introduced the legal moves in the Arena model, the realization of these moves is detailed through the argumentation dialogue protocol in Arena. Assuming that we have a new instance to be classified, and a number of classifier agents participating in the Arena model, the argumentation dialogue protocol operates as follows:

Before the start of the dialogue, the referee randomly selects one participate agent as a *master* to begin.

(1) At the first round, the *master* proposes a new classification argument, such that its $L\_confidence$ is higher than a given threshold. If the *master* fails to play an opening move, then the referee selects anther one to commence the dialogue. If all the classifier agents fail to propose an opening move, the dialogue terminates with failure.
(2) In the second round, the other participate agents attempt to defend or attack the proposing argument, using any kinds of move described previously. If all the agents fail to play a move, the dialogue terminates, and the instance is classified according to the class prompted by the *master*. Otherwise, the *dialectical analysis tree* is updated with submitted moves.
(3) The argumentation process continues until the *master* is defeated, then another round of argumentation begins, the protocol moves to (1).
(4) If one subsequent round passed without any new moves being submitted to the *dialectical analysis tree*, or if numerous rounds have passed without reaching an agreement, the referee terminates the dialogue.

More details of the realization of Arguing Prism proposed will no longer be demonstrated here because of the limited space. Once an argumentation dialogue has terminated, the status of the *dialectical analysis tree* will indicate the 'winning' class and its corresponding explanation. This interpretable classification paradigm is the essential advantage provided by the argumentation based approach proposed in the current paper.

3.3   Generating Arguments Using Prism

Having introduced the legal moves in Arguing Prism before, the realization of arguments for these moves using Prism algorithms is described in this subsection. A suite of several algorithms is developed for the moves described in subsection 3.2, respectively.

A classifier agent can generate a classification argument using any inductive learning algorithms capable of learning modular classification rules. For the purpose of simplicity, we use Prism family of algorithms in this paper. In particular, we use PrismGLP, which is Prism algorithm with Global Largest class Priority. Moreover, an information-theoretic pre-pruning, named J-Pruning [10], is incorporated with PrismGLP, to reduce the overfitting on training data. Thus, the algorithm implemented in the current paper for generating arguments is called J-PrismGLP.

Since rules for generating arguments are derived through inductive learning techniques, their validity may not be ensure. Thus only   the classification arguments satisfying certain confidence criteria are accepted for the collaborative classification task. This is the reason that the *L_confidence* is exploited in the definition of Classification Argument. The *L_confidence* is defined following [9]:

Definition 2 (*L_confidence*) The *L_confidence* of a Classification Argument <*CLR*, *L_confidence* > for a classifier agent is:

$$L\_confidence = \frac{|\,instances\ covered\ by\ the\ CLR\,|+1}{|\,instances\ covered\ by\ the\ antecedent\ of\ the\ CLR\,|+2}$$

where the instances are all from the individual local data repository of that agent. *L_confidence* is mainly used to prevent estimations too close to 0 or 1 when very few instances are covered by inductive learning algorithms. In this case, the Classification Argument for proposing moves can be generated.

In addition, in order to construct attacking moves for argumentation, we revised Prism to generate negative rules, called Prism_N. The algorithm has been implemented following the basic framework of Prism. For refining moves, the generating algorithms are just like that of proposing moves, differing on the *L_confidence*, in order to increase the *L_confidence* of a rule.

4   Experiments

This section presents an empirical evaluation of the proposed approach. For the evaluation, six datasets (Car evaluation, Congress vote, Balance scale, Chess, Soybean, Lymphograpy, represented as 'D1' to 'D6' successively) from UCI repository are used. In subsection 4.1, Arguing Prism's classification performance is empirically evaluated in terms of precision and recall on the datasets. Furthermore, we have investigated Arguing Prism's tolerance to inconsistent data in subsection 4.2. The threshold of *L_confidence* for a legal classification argument is set to be 0.65. The results presented throughout this section are obtained using TCV test, for each dataset, we report the average results for each of the different classification paradigms.

4.1   Arguing Prism's Classification Performance

We conducted experiments to compare the operation of Arguing Prism, against two classification paradigms. One is the operation that classifier agents simply perform inductive learning on training instances individually, and then make decisions on test instances via simple voting scheme. The other is the operation that centralizing all the instances and then performing ensemble classification. The comparison with ensemble methods is undertaken for the reason that both approaches 'pool' results to perform a better classification. For ensemble classification, we choose to apply Bagging, combined with Prism algorithm (J-PrismGLP), and their corresponding WEKA [7] implementation is used.

The results of our experiments are presented in Figure 1 and 2. The classification performance is measured using precision and recall. It is shown that Arguing Prism can greatly increase the precision over the operation of individual classifier agents voting, reaching the levels close to those of ensemble classification (see Figure 1), but decreases slightly for recall (see Figure 2). In general, Arguing Prism has provided an efficient mechanism for collaborative classification since its performance is indistinguishable from ensemble classification. But, via Arguing Prism, an interpretable classification can be achieved through argumentation rather the voting schemes.

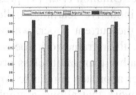   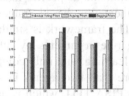

Fig. 1. Precision obtained using different approaches

Fig. 2. Recall obtained using different approaches

4.2   Experimenting with Inconsistent Data

The ability to handle inconsistent data is seen as important requirements for real world applications. Thus, we consider inconsistent datasets containing conflict instances, which are characterized by identical values for all features, yet belong to different classes. In order to evaluate Arguing Prism's tolerance to inconsistent data with respect to the two classification paradigms described in subsection 4.1, we introduce some conflict data into the experiment data sets, with different inconsistent rate. For each level of inconsistent rate in the experimental datasets, the results of Arguing Prism compared with the other two classification paradigms are shown in Figure 3.

In general, it can be observed in Figure 3 that for all the classification paradigms, the accuracy of classifications decreases with an increasing inconsistent rate in the datasets. However, in most cases, Arguing Prism out-performs the other classification paradigms in presence of inconsistent data. This tolerance to inconsistent may due to the effective mechanism of conflict resolution offered by argumentation based approach.

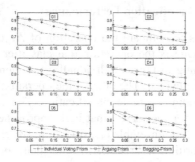

Fig. 3. Real datasets study (the horizontal axe represents the inconsistent rate and the vertical represents the accuracy)

5   Related Work

Concerning argumentation based multi-agent classification, the idea that argumentation may be helpful to machine learning was first discussed with the reason that argumentation can provide a formalization for reasoning with incomplete information [6]. However, they did not develop any specific framework for the integration of argumentation and a specific machine learning technique. Ontañón *et al* [9] presented an argumentation based framework for multi-agent inductive learning, called A-MAIL. The framework is similar to that proposed in the current paper, where constructing rule arguments by rule induction. However, their framework focused on concept learning, and only comprised of two agents, rather than multi-class problem with more than two agents in Arguing Prism. Another approach similar to Arguing Prism is PISA [11], which is an argumentation framework that allows a number of agents to use experiences to argue about the classification of a specific problem. But they used associate rules mining techniques for generating arguments rather than modular classification rules inductive learning algorithms in the current paper.

6   Conclusion

This work has presented Arguing Prism, an argumentation based approach for collaborative classification in distributed environments, using Prism algorithms. Empirical evaluation on several real world datasets shows that Arguing Prism out-performs individual classifier agents voting paradigm, and only slightly below the ensemble classification. This means that an interpretable collaborative classification model can be achieved without losing much classification performance. Arguing Prism has also been investigated in terms of its tolerance to inconsistent data, and it can out-perform comparable classification paradigms due to the advantages offered by argumentation.

Acknowledgments. This research is supported by the National Natural Science Foundation of China (No.71371184) and the Specialized Research Fund for the Doctoral Program of Higher Education of China (No.20124307110023).

References

1. Andrzejak, A., Langner, F., Zabala, S.: Interpretable models from distributed data via merging of decision trees. In: 2013 IEEE Symposium on Proceedings of the Computational Intelligence and Data Mining (CIDM), pp. 1–9 (2013)
2. Cano, A., Zafra, A., Ventura, S.: An interpretable classification rule mining algorithm. Information Sciences 240, 1–20 (2013)
3. Cao, J., Wang, H., Kwong, S., Li, K.: Combining interpretable fuzzy rule-based classifiers via multi-objective hierarchical evolutionary algorithm. In: 2011 IEEE International Conference on Proceedings of the Systems, Man, and Cybernetics (SMC) (2011)
4. Caragea, D., Silvescu, A., Honavar, V.: A framework for learning from distributed data using sufficient statistics and its application to learning decision trees. International Journal of Hybrid Intelligent Systems 1(1), 80–89 (2004)
5. Fisch, D., Kühbeck, B., Sick, B., Ovaska, S.J.: So near and yet so far: New insight into properties of some well-known classifier paradigms. Information Sciences 180(18), 3381–3401 (2010)
6. Gómez, S.A., Chesñevar, C.I.: Integrating defeasible argumentation and machine learning techniques: A preliminary report. In: Proceedings of the V Workshop of Researchers in Comp. Science (2003)
7. Hall, M., Frank, E., Holmes, G., Pfahringer, B., Reutemann, P., Witten, I.H.: The WEKA data mining software: an update. ACM SIGKDD Explorations Newsletter 11(1), 10–18 (2009)
8. Landgrebe, D.A.: Multispectral land sensing: where from, where to? IEEE Transactions on Geoscience and Remote Sensing 43(3), 414–421 (2005)
9. Ontañón, S., Plaza, E.: Multiagent inductive learning: an argumenta-tion-based approach. In: Proceedings of the 27th International Conference on Machine Learning (ICML 2010), pp. 839–846 (2010)
10. Stahl, F., Bramer, M.: Jmax-pruning: A facility for the information theoretic pruning of modular classification rules. Knowledge-Based Systems 29, 12–19 (2012)
11. Wardeh, M., Coenen, F., Bench-Capon, T.: Multi-agent based classification using argumentation from experience. Autonomous Agents and Multi-Agent Systems 25(3), 447–474 (2012)
12. Yao, L., Xu, J., Li, J., Qi, X.: Evaluating the Valuable Rules from Different Experience Using Multiparty Argument Games. In: Proceedings of the Proceedings of the 2012 IEEE/WIC/ACM International Joint Conferences on Web Intelligence and Intelligent Agent Technology, vol. 2, pp. 258–265 (2012)

Rank Aggregation of Candidate Sets
for Efficient Similarity Search

David Novak and Pavel Zezula

Masaryk University, Brno, Czech Republic
{david.novak,zezula}@fi.muni.cz

Abstract. Many current applications need to organize data with respect to mutual similarity between data objects. Generic similarity retrieval in large data collections is a tough task that has been drawing researchers' attention for two decades. A typical general strategy to retrieve the most similar objects to a given example is to access and then refine a *candidate set* of objects; the overall search costs (and search time) then typically correlate with the candidate set size. We propose a generic approach that combines several independent indexes by aggregating their candidate sets in such a way that the resulting candidate set can be one or two orders of magnitude smaller (while keeping the answer quality). This achievement comes at the expense of higher computational costs of the ranking algorithm but experiments on two real-life and one artificial datasets indicate that the overall gain can be significant.

1   Introduction

For many contemporary types of digital data, it is convenient or even essential that the retrieval methods are based on mutual similarity of the data objects. The reasons may be that either similarity corresponds with the human perception of the data or that the exact matching would be too restrictive (various multimedia, biomedical or sensor data, etc.). We adopt a generic approach to this problem, where the data space is modeled by a data domain $\mathcal{D}$ and a black-box *distance* function δ to assess dissimilarity between each pair of objects from $\mathcal{D}$.

The field of metric-based similarity search has been studied for almost two decades [17]. The general objective of metric accesses methods (MAMs) is to preprocess the indexed dataset $\mathcal{X} \subseteq \mathcal{D}$ in such a way that, given a query object $q \in \mathcal{D}$, the MAM can effectively identify objects x from $\mathcal{X}$ with the shortest distances $\delta(q, x)$. Current MAMs designed for large data collections are typically approximate [17,15] and adopt the following high-level approach: dataset $\mathcal{X}$ is split into *partitions*; given a query, partitions with the highest "likeliness" to contain query-relevant data are read from the disk and this data form the *candidate set* of objects x to be *refined* by evaluation of $\delta(q, x)$. The search costs of this schema consist mainly of (1) the I/O costs of reading the candidate partitions from the disk and of (2) CPU costs of refinement; thus, the overall costs typically strongly correlate with the candidate set size.

In this work, we propose a technique that can significantly reduce the candidate set size. In complex data spaces, the data partitions typically span relatively large areas of the space and thus the candidate sets are either large or imprecise.

H. Decker et al. (Eds.): DEXA 2014, Part II, LNCS 8645, pp. 42–58, 2014.

The key idea of our approach is to use several independent partitionings of the data space; given a query, each of these partitionings generates a ranked set of candidate objects and we propose a way to effectively and efficiently aggregate these rankings. The final candidate set determined in this way is retrieved from the disk and refined. The combination of several independent indexes has been used before [11,8,14] suggesting to simply union the multiple candidate sets. Contrary to this approach, we propose not to *widen the candidate set* but to *shrink it* significantly by our aggregation mechanism; to realize this aggregation efficiently, we employ an algorithm previously introduced by Fagin et al. [10].

In order to make our proposal complete, we propose a specific similarity index that is based on multiple independent pivot spaces and that uses our rank aggregation method. The index encodes the data objects using multiple prefixes of *pivot permutations* [16,6,12,8,1] and builds several trie structures. We suppose the index fits in main memory; the large memories of current common HW configurations are often used only for disk caching – our explicit use of the memory seems to be more efficient. Our approach also requires that the final candidate objects are accessed one-by-one (not partially sequentially but randomly); current SSD disks without seeks can be well exploited by this approach.

The experiments conducted on three diverse datasets show that our approach can reduce the candidate set size by *two orders of magnitude*. The response times depend on the time spared by this candidate set reduction (reduced I/O costs and δ-refinement time) versus the overhead of the rank aggregation algorithm. To analyze this tradeoff, we have run experiments on an artificial dataset with adjustable object sizes and tunable time of δ evaluation; the results show that our approach is not worthwhile only for the smallest data objects with the fastest δ function. Most of the presented results are from trials on two real-life datasets (100M CoPhIR [5] and 1M complex visual signatures [4]); for these, our approach was two- to five-times faster than competitors on the same HW platform.

In Section 2, we define fundamental terms and analyze related approaches. Further, we propose our *PPP-Encoding* of the data using multiple pivot spaces (Section 3.1), ranking within individual pivot spaces and, especially, our rank aggregation proposal (Section 3.2); index structure *PPP-Tree* supporting the rank aggregation algorithm is proposed in Section 3.3. Our approach is evaluated and compared with others in Section 4 and the paper is concluded in Section 5.

2   Preliminaries and Related Work

In this work, we focus on indexing and searching based on mutual object distances. We primarily assume that the data is modeled as a metric space:

Definition 1. Metric space *is an ordered pair* $(\mathcal{D}, \delta)$, *where* $\mathcal{D}$ *is a domain of objects and* δ *is a total* distance function $\delta : \mathcal{D} \times \mathcal{D} \longrightarrow \mathbb{R}$ *satisfying postulates of non-negativity, identity, symmetry, and triangle inequality [17].*

Our technique does not explicitly demand triangle inequality. In general, the distance-based techniques manage the dataset $\mathcal{X} \subseteq \mathcal{D}$ and search it by the *nearest neighbors query* K-$NN(q)$, which returns K objects from $\mathcal{X}$ with the smallest distances to given $q \in \mathcal{D}$ (ties broken arbitrarily). We assume that the search answer A may be an approximation of the precise K-NN answer A^P and the result quality is measured by $recall(A) = precision(A) = \frac{|A \cap A^P|}{K} \cdot 100\%$.

During two decades of research, many approximate metric-based techniques have been proposed [17,15]. Further in this section, we focus especially on techniques based on the concept of *pivot permutations* and on approaches that use several independent space partitionings. Having a set of k pivots $P = \{p_1, \ldots, p_k\} \subseteq \mathcal{D}$, Π_x is a *pivot permutation* defined with respect to object $x \in \mathcal{D}$ if $\Pi_x(i)$ is the index of the i-th closest pivot to x; accordingly, sequence $p_{\Pi_x(1)}, \ldots, p_{\Pi_x(k)}$ is ordered with respect to distances between the pivots and x (ties broken by order of increasing pivot index). Formally:

Definition 2. *Having a set of k pivots $P = \{p_1, \ldots, p_k\} \subseteq \mathcal{D}$ and an object $x \in \mathcal{D}$, let Π_x be permutation on indexes $\{1, \ldots, k\}$ such that $\forall i : 1 \leq i < k$:*

$$\delta(x, p_{\Pi_x(i)}) < \delta(x, p_{\Pi_x(i+1)})$$
$$\vee \left(\delta(x, p_{\Pi_x(i)}) = \delta(x, p_{\Pi_x(i+1)}) \wedge \Pi_x(i) < \Pi_x(i+1) \right). \quad (1)$$

Π_x *is denoted as* pivot permutation *(PP) with respect to* x.

Several techniques based on this principle [6,8,7,12] use the PPs to group data objects together (data partitioning); at query time, query-relevant partitions are read from the disk and refined; the relevancy is assessed based on the PPs. Unlike these methods, the MI-File [1] builds inverted file index according to object PPs; these inverted files are used to rank the data according to a query and the candidate set is then refined by accessing the objects one-by-one [1]. In this respect, our approach adopts similar principle and we compare our results with the MI-File (see Section 4.3).

In this work, we propose to use *several* independent pivot spaces (sets of pivots) to define several PPs for each data object and to identify candidate objects. The idea of multiple indexes is known from the Locality-sensitive Hashing (LSH) [11] and it was also applied by a few metric-based approaches [8,14]; some metric indexes actually define families of *metric LSH functions* [13]. All these works benefit from enlarging the candidate set by a simple union of the top results from individual indexes; on the contrary, we propose such rank aggregation that can significantly reduce the size of the candidate set in comparison with a single index while preserving the same answer quality.

3 Data Indexing and Rank Aggregation

In this section, we first formalize a way to encode metric data using several pivot spaces, then we show how to determine candidate set by effective aggregation of candidate rankings from individual pivot spaces, and finally we propose an efficient indexing and searching mechanism for this approach.

3.1 Encoding by Pivot Permutation Prefixes

For a data domain $\mathcal{D}$ with distance function δ, object $x \in \mathcal{D}$ and a set of k pivots $P = \{p_1, \ldots, p_k\}$, the pivot permutation (PP) Π_x is defined as in Definition 2.

In our technique, we do not use the full PP but only its prefix, i.e. the ordered list of a given number of nearest pivots:

Notation: Having PP Π_x with respect to pivots $\{p_1, \ldots, p_k\}$ and object $x \in \mathcal{D}$, we denote $\Pi_x(1..l)$ the pivot permutation prefix (PPP) of length l: $1 \leq l \leq k$:

$$\Pi_x(1..l) = \langle \Pi_x(1), \Pi_x(2), \ldots, \Pi_x(l) \rangle. \tag{2}$$

The pivot permutation prefixes have a geometrical interpretation important for the similarity search – the PPPs actually define *recursive Voronoi partitioning* of the metric space [16]. Let us explain this principle on an example in Euclidean plane with eight pivots $p_1^1, \ldots, p_8^1$ in the left part of Figure 1 (ignore the upper indexes for now); the thick solid lines depict borders between standard *Voronoi cells* – sets of points $x \in \mathcal{D}$ for which pivot p_i^1 is the closest one: $\Pi_x(1) = i$. The dashed lines further partition these cells using other pivots; these sub-areas cover all objects for which $\Pi_x(1) = i$ and $\Pi_x(2) = j$, thus $\Pi_x(1..2) = \langle i, j \rangle$.

The pivot permutation prefixes $\Pi_x(1..l)$ form the base of the proposed PPP-Encoding, which is actually composed of λ PPPs for each object. Thus, let us further assume having λ independent sets of k pivots $P^1, P^2, \ldots, P^\lambda$, $P^j = \{p_1^j, \ldots, p_k^j\}$. For any $x \in \mathcal{D}$, each of these sets generates a pivot permutation Π_x^j, $j \in \{1, \ldots, \lambda\}$ and we can define the PPP-Encoding as follows.

Definition 3. *Having λ sets of k pivots and parameter l : $1 \leq l \leq k$, we define PPP-Code of object $x \in \mathcal{D}$ as a λ-tuple*

$$PPP_l^{1..\lambda}(x) = \langle \Pi_x^1(1..l), \ldots, \Pi_x^\lambda(1..l) \rangle. \tag{3}$$

Individual components of the PPP-Code will be also denoted as $PPP_l^j(x) = \Pi_x^j(1..l)$, $j \in \{1, \ldots, \lambda\}$; to shorten the notation, we set $\Lambda = \{1, \ldots, \lambda\}$. These and other symbols used throughout this paper are summarized in Table 1.

Figure 1 depicts an example where each of the $\lambda = 2$ pivot sets defines an independent Voronoi partitioning of the data space. Every object $x \in \mathcal{X}$ is encoded by $PPP_l^j(x) = \Pi_x^j(1..l)$, $j \in \Lambda$. Object x_5 is depicted in both diagrams and, for instance, within the first (left) partitioning, the closest pivots from x_5 are $p_7^1, p_4^1, p_8^1, p_5^1$, which corresponds to $PPP_4^1(x_5) = \Pi_{x_5}^1(1..4) = \langle 7, 4, 8, 5 \rangle$.

first pivot space partitioning ($j = 1$)　　　　second pivot space partitioning ($j = 2$)

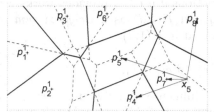 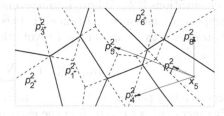

Fig. 1. Principles of encoding data objects as PPP-Codes $PPP_l^{1..\lambda}(x)$ with two pivot sets ($\lambda = 2$) each with eight pivots ($k = 8$) and using pivot permutation prefixes of length four ($l = 4$). Object x_5 is encoded by $PPP_4^{1..2}(x_5) = \langle \langle 7, 4, 8, 5 \rangle, \langle 7, 8, 4, 6 \rangle \rangle$.

Table 1. Notation used throughout this paper

Symbol	Definition
$(\mathcal{D}, \delta)$	the data domain and metric distance $\delta : \mathcal{D} \times \mathcal{D} \to \mathbb{R}$
$\mathcal{X}$	the set of indexed data objects $\mathcal{X} \subseteq \mathcal{D}$; $|\mathcal{X}| = n$
k	number of reference objects (pivots) in one pivot space
$\Lambda = \{1, \ldots, \lambda\}$	Λ is the index set of λ independent pivot spaces
$P^j = \{p_1^j, \ldots, p_k^j\}$	the j-th set of k pivots from $\mathcal{D}$; $j \in \Lambda$
Π_x^j	pivot permutation of $(1 \ldots k)$ ordering P^j by distance from $x \in \mathcal{D}$
$\Pi_x^j(1..l)$	the j-th PP prefix of length l: $\Pi_x^j(1..l) = \langle \Pi_x^j(1), \ldots, \Pi_x^j(l) \rangle$
$PPP_l^{1..\lambda}(x)$	the PPP-Code of $x \in \mathcal{D}$: $PPP_l^{1..\lambda}(x) = \langle \Pi_x^1(1..l), \ldots, \Pi_x^\lambda(1..l) \rangle$
d	measure that ranks pivot permutation prefixes $d(q, \Pi(1..l))$
$\psi_q^j : \mathcal{X} \to \mathbb{N}$	the j-th ranking of objects according to $q \in \mathcal{D}$ generated by d
$\Psi_{\mathbf{p}}(q, x)$	the overall rank of x by the $\mathbf{p}$-percentile of its $\psi_q^j(x)$ ranks, $j \in \Lambda$
R	size of candidate set – number of objects x refined by $\delta(q, x)$

3.2   Ranking of Pivot Permutation Prefixes

Having objects from $\mathcal{X}$ encoded by λ pivot permutation prefixes as described above, we first want to rank individual PPPs $\Pi_x^j(1..l)$, $j \in \Lambda$ with respect to query $q \in \mathcal{D}$. Formally, we want to find a function $d(q, \Pi_x(1..l))$ that would generate a "similar" ranking of objects $x \in \mathcal{X}$ as the original distance $\delta(q, x)$.

In the literature, there are several ways to order pivot permutations or their prefixes to obtain such function d. A natural approach is to also encode object q by its permutation Π_q and to calculate "distance" between Π_q and $\Pi_x(1..l)$. There are several standard ways to measure difference between full permutations that were also used in similarity search: Spearman Footrule, Spearman Rho or Kendall Tau measure [6,1]. The Kendall Tau seems to slightly outperform the others [6] and can be generalized to work on prefixes [9]. Details of these measures are out of scope of this work.

The query object $q \in \mathcal{D}$ in the the ranking function d can be represented more richly than by permutation Π_q, specifically, we can use directly the query-pivot distances $\delta(q, p_1), \ldots, \delta(q, p_k)$. This approach has been successfully applied in several works [7,8,12] and it can outperform the measures based purely on pivot permutations [12]. Taking into account functions d used in previous works, we propose to measure "distance" between q and $\Pi(1..l)$ as a *weighted sum* of distances between q and the l pivots that occur in $\Pi(1..l)$:

$$d(q, \Pi(1..l)) = \sum_{i=1}^l c^{i-1} \delta(q, p_{\Pi(i)}). \tag{4}$$

where c is a parameter $0 < c \le 1$ to control influence of individual query-pivot distances (we set $c = 0.75$ in the experimental part of this work). Detailed comparison of different measures d is not crucial for this work and the rank aggregation approach proposed below is applicable with any function $d(q, \Pi_x(1..l))$.

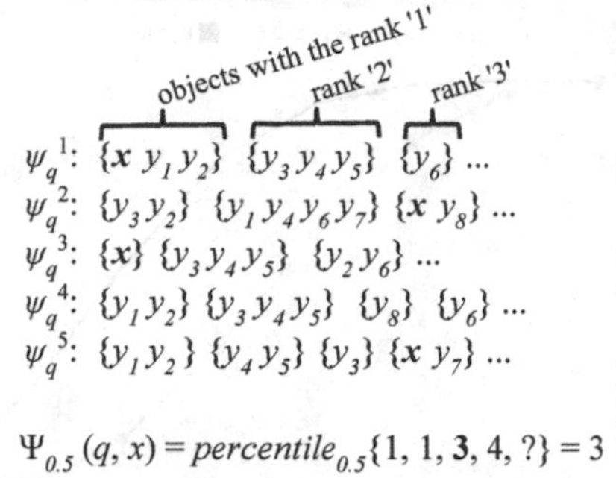

$$\overbrace{\text{objects with the rank '1'}}^{} \quad \overbrace{\text{rank '2'}}^{} \quad \overbrace{\text{rank '3'}}^{}$$

$$\psi_q^1: \quad \{x\ y_1 y_2\} \quad \{y_3 y_4 y_5\} \quad \{y_6\} \ ...$$
$$\psi_q^2: \quad \{y_3 y_2\} \quad \{y_1 y_4 y_6 y_7\} \quad \{x\ y_8\} \ ...$$
$$\psi_q^3: \quad \{x\} \quad \{y_3 y_4 y_5\} \quad \{y_2 y_6\} \ ...$$
$$\psi_q^4: \quad \{y_1 y_2\} \quad \{y_3 y_4 y_5\} \quad \{y_8\} \quad \{y_6\} \ ...$$
$$\psi_q^5: \quad \{y_1 y_2\} \quad \{y_4 y_5\} \quad \{y_3\} \quad \{x\ y_7\} \ ...$$

$$\Psi_{0.5}(q,x) = percentile_{0.5}\{1, 1, 3, 4, ?\} = 3$$

Fig. 2. Rank aggregation by $\Psi_{\mathbf{p}}$ of object $x \in \mathcal{X}$, $\lambda = 5$, $\mathbf{p} = 0.5$

A measure d together with $q \in \mathcal{D}$ naturally induces ranking ψ_q of the indexed set $\mathcal{X}$ according to growing distance from q. Formally, ranking $\psi_q : \mathcal{X} \to \mathbb{N}$ is the smallest numbering of set $\mathcal{X}$ that fulfills this condition for all pairs $x, y \in \mathcal{X}$:

$$d(q, \Pi_x(1..l)) \leq d(q, \Pi_y(1..l)) \Rightarrow \psi_q(x) \leq \psi_q(y). \tag{5}$$

Let us now assume that x is encoded by $PPP_l^{1..\lambda}(x)$ codes composed of λ PPPs $\Pi_x^j(1..l)$, $j \in \Lambda$ and that we have a mechanism able to provide λ sorted lists of objects $x \in \mathcal{X}$ according to rankings ψ_q^j, $j \in \Lambda$. Figure 2 (top part) shows an example of five rankings ψ_q^j, $j \in \{1, \ldots, 5\}$. These rankings are partial – objects with the same PPP $\Pi(1..l)$ have the same rank (objects from the same Voronoi cell). This is the main source of inaccuracy of these rankings because, in complex data spaces, the Voronoi cells typically span relatively large areas and thus the top positions of ψ_q contain both objects close to q and more distant ones. Having several independent partitionings, the query-relevant objects should be at top positions of most of the rankings while the "noise objects" should vary because the Voronoi cells are of different shapes. The objective of our rank aggregation is to filter out these noise objects. Namely, we propose to assign each object $x \in \mathcal{X}$ the **p**-percentile of its ranks, $0 \leq \mathbf{p} \leq 1$:

$$\Psi_{\mathbf{p}}(q, x) = percentile_{\mathbf{p}}(\psi_q^1(x), \psi_q^2(x), \ldots, \psi_q^\lambda(x)). \tag{6}$$

For instance, $\Psi_{0.5}$ assigns median of the ranks; see Figure 2 for an example – positions of object x in individual rankings are: 1, 3, 1, *unknown*, 4 and median of these ranks is $\Psi_{0.5}(q, x) = 3$. This principle was used by Fagin et al. [10] for a different purpose and they propose MEDRANK algorithm for efficient calculation of $\Psi_{\mathbf{p}}$. This algorithm does not require to explicitly find out *all* ranks of a specific object, but only $\lceil \mathbf{p}\lambda \rceil$ first (best) ranks (this is explicit in the Figure 2 example). Details on the MEDRANK algorithm [10] are provided later in Section 3.3.

Now, we would like to show that the $\Psi_{\mathbf{p}}$ aggregation actually improves the ranking in comparison with a single ψ_q ranking by increasing the probability that objects close to q will be assigned top positions (and vice versa). Also, we would like to find theoretically suitable values of **p**.

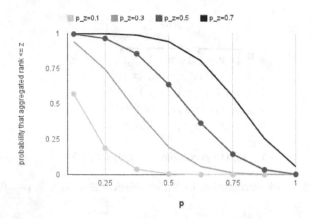

Fig. 3. Development of $Pr[\Psi_{\mathbf{p}}(q,x) \le z]$ for $\lambda = 8$, selected p_z and variable $\mathbf{p}$

Let $x \in \mathcal{X}$ be an object from the dataset and p_z be the probability such that $p_z = Pr[\psi_q(x) \le z]$, where $z \ge 1$ is a position in ψ_q ranking. Having λ independent rankings $\psi_q^j(x)$, $j \in \Lambda$, we want to determine probability $Pr[\Psi_{\mathbf{p}}(q,x) \le z]$ with respect to p_z. Let X be a random variable representing the number of ψ_q^j ranks of x that are smaller than z: $|\{\psi_q^j(x) \le z, j \in \Lambda\}|$. Assuming that the probability distribution of p_z is the same for each of $\psi_q^j(x)$, we get

$$Pr[X = j] = \binom{\lambda}{j} \cdot (p_z)^j \cdot (1 - p_z)^{\lambda-j}.$$

In order to have $\Psi_{\mathbf{p}}(q, x) \le z$, at least $\lceil \mathbf{p}\lambda \rceil$ positions of x must be $\le z$ and thus

$$Pr[\Psi_{\mathbf{p}}(q, x) \le z] = \sum_{j=\lceil \mathbf{p}\lambda \rceil}^{\lambda} Pr[X = j].$$

The probability distribution of $Pr[\Psi_{\mathbf{p}}(q,x) \le z]$ depends on p_z and $\mathbf{p}$; Figure 3 shows development of this distribution for variable $\mathbf{p}$ and selected values of p_z ($\lambda = 8$). We can see that the rank aggregation increases the differences between individual levels of p_z (for non-extreme $\mathbf{p}$ values); e.g. for $\mathbf{p} = 0.5$, probabilities $p_z = 0.1$ and $p_z = 0.3$ are suppressed to lower values whereas $p_z = 0.5$ and $p_z = 0.7$ are transformed to higher probabilities. The probability $p_z = Pr[\psi_q(x) \le z]$ naturally grows with z but, more importantly, we assume that p_z is higher for objects close to q then for distant ones. Because ψ_q^j are generated by distance between q and Voronoi cells (5) and these cells may be large, there may be many distant objects that appear at top positions of individual ψ_q although having low probability p_z. The rank aggregation $\Psi_{\mathbf{p}}(q, x)$ for non-extreme values of $\mathbf{p}$ would push away such objects and increase the probability that top ranks are assigned only to objects close to q.

3.3 Indexing of PPP-Codes and Search Algorithm

So far, we have proposed a way to encode metric objects by PPP-Codes and to rank these codes according to given query object. In this section, we propose an index built over the PPP-encoded data and an efficient search algorithm.

The $PPP_l^{1..\lambda}(x)$ code is composed of λ PPPs of length l. Some indexed objects $x \in \mathcal{X}$ naturally share the same prefixes of $PPP_l^j(x)$ for individual pivot spaces $j \in \Lambda$, therefore we propose to build λ *PPP-Trees* – dynamic trie structures that keep the l'-prefixes of the PPP-Codes only once for all objects sharing the same l'-prefix, $l' \leq l$. Schema of such PPP-Tree index is sketched in Figure 4. A leaf node at level l', $l' \leq l$ keeps the unique ID_x of the indexed objects x together with part of their PPPs $\Pi_x(l'+1..l)$ (in Figure 4, these are pairs $\Pi(4..l), ID$).

Similar structures were used in PP-Index [8] and in M-Index [12]. In order to minimize the memory occupation of the PPP-Tree, we propose a dynamic leveling that splits a leaf node to the next level only if this operation *spares* some memory. Specifically, a leaf node with n' objects at level l', $1 \leq l' < l$ spares $n' \cdot \lceil \log_2 k \rceil$ bits because the memory representation of $\Pi_x(l'+1..l)$ would be shorter by one index if these n' objects are moved to level $l' + 1$. On the other hand, the split creates new leaves with certain memory overhead; thus we propose to split the leaf iff

$$n' \cdot \lceil \log_2 k \rceil > b \cdot \text{NodeOverhead}$$

where b is the potential branching of the leaf, $b \leq n'$ and $b \leq k - l' + 1$. The actual value of b can be either precisely measured for each leaf or estimated based on the statistics of average branching at level l'. Value of NodeOverhead depends on implementation details. Finally, we need λ of these PPP-Tree structures to index the dataset represented by the PPP-Codes; an object x in all λ trees is "connected" by its unique ID_x.

Having these PPP-Tree indexes, we can now propose the PPPRank algorithm that calculates the aggregated ranking $\Psi_\mathbf{p}(q, x)$ of objects $x \in \mathcal{X}$ for given query $q \in \mathcal{D}$. The main procedure (Algorithm 1) follows the idea of the MedRank algorithm [10] and the PPP-Tree structures are used for effective generation of individual λ rankings (subroutine GetNextIDs).

Given a query object $q \in \mathcal{D}$, percentile $0 \leq \mathbf{p} \leq 1$ and number R, PPPRank returns IDs of R indexed objects $x \in \mathcal{X}$ with the lowest value of $\Psi_\mathbf{p}(q, x)$; recall

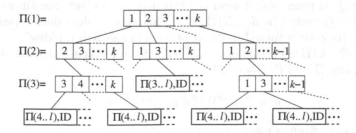

Fig. 4. Schema of a single dynamic PPP-Tree

Algorithm 1. PPPRANK($q, \mathbf{p}, R$)

Input: $q \in \mathcal{D}$; percentile $\mathbf{p}$; candidate set size R
Output: IDs of R objects $x \in \mathcal{X}$ with lowest $\Psi_{\mathbf{p}}(q, x)$

```
    // calculate query-pivot distances necessary for GetNextIDs routines
1   calculate δ(q,pⁱⱼ), ∀i ∈ {1,...,k} and ∀j ∈ Λ
    // initialize λ priority queues Qⱼ of nodes from PPP-Trees
2   Qⱼ ← {root of j-th PPP-Tree}, j ∈ Λ
    // S is a set of ''seen objects'': IDₓ with their frequencies fₓ
3   set S ← ∅     // A is answer list of object IDs
4   list A ← ⟨⟩
5   while |A| < R do
6       foreach j ∈ Λ do
7           foreach IDₓ in GETNEXTIDS(Qⱼ,q) do
8               if IDₓ ∉ S then
9                   add IDₓ to S and set fₓ = 1
10              else
11                  increment fₓ
12      foreach IDₓ in S such that fₓ ≥ ⌈pλ⌉ do
13          move IDₓ from S to A
14  return A
```

---

that this aggregated rank is defined as the $\lceil \mathbf{p}\lambda \rceil$-th best position from $\psi_q^j(x)$ ranks, $j \in \Lambda$ (6). In every iteration (lines 6–11), the algorithm accesses next objects of all rankings ψ_q^j as depicted in Figure 2 (line 7 of Algorithm 1); set S carries the already seen objects x together with the number of their occurrences in the rankings (frequencies f_x). GETNEXTIDS(Q_j, q) always returns next object(s) with the best ψ_q^j rank and thus, when an object x achieves frequency $f_x \geq \lceil \mathbf{p}\lambda \rceil$, it is guaranteed that any object y achieving $f_y \geq \lceil \mathbf{p}\lambda \rceil$ in a subsequent iteration of PPPRANK must have higher rank $\Psi_{\mathbf{p}}(q, y) > \Psi_{\mathbf{p}}(q, x)$ [10].

Idea of the GETNEXTIDS(Q_j, q) algorithm is to traverse the j-th PPP-Tree using a priority queue Q_j in a similar way as generally described in [17] or as used by the M-Index [12]. Every PPP-Tree node corresponds to a PPP $\Pi(1..l')$: $1 \leq l' \leq l$ (the root corresponds to empty PPP $\langle\rangle$) and individual objects x, stored in leaf cells are represented by their PPPs $\Pi_x(1..l)$. Queue Q_j is always ordered by $d(q, \Pi(1..l'))$ (where d generates ranking ψ_q^j (5)). In every iteration, the head of Q_j is processed; if head is a leaf node, its objects identifiers ID_x are inserted into Q_j ranked by $d(q, \Pi_x^j(1..l))$. When object identifiers appear at the head of Q_j, they are returned as "next objects in the j-th ranking".

This GETNEXTIDS routine can work only if the following property holds for function d, any Π and l', $1 \leq l' < l$:

$$d(q, \Pi(1..l')) \leq d(q, \Pi(1..l'+1)). \tag{7}$$

This property is fulfilled by the function d (4) and also by some generalizations of standard distances between permutations like Kendall Tau [9].

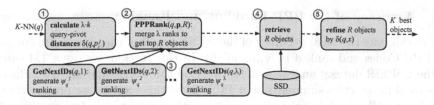

Fig. 5. Search pipeline using the PPP-Encoding and PPPRANK algorithm

Search Process Review Schema in Figure 5 reviews the whole search process. Given a K-NN(q) query, the first step is calculating distances between q and all pivots: $\delta(q, p_i^j)$, $i \in \{1, \ldots, k\}$, $j \in \Lambda$ (step 1, see line 1 in Algorithm 1). This enables initialization of the GETNEXTIDS(Q_j, q) procedures (steps 3), which generate the continual rankings ψ_q^j that are consumed by the main PPPRANK($q, \mathbf{p}, R$) algorithm (step 2). The candidate set of R objects x is retrieved from the disk (step 4) and refined by calculating $\delta(q, x)$ (step 5). The whole process can be parallelized in the following way: The λ steps 3 run fully in parallel and step 2 continuously reads their results; in this way, the full ranking $\Psi_{\mathbf{p}}(q, x)$ is generated item-by-item and is immediately consumed by steps 4 and then 5.

4   Efficiency Evaluation

We evaluate efficiency of our approach in three stages: Influence of various parameters on behavior of the rank aggregation (Section 4.1), overall efficiency of the proposed algorithm (Section 4.2), and comparison with other approaches (Section 4.3). We use three datasets – two of them are real-life, and the third one is artificially created to have fully controlled test conditions:

CoPhIR. 100 million objects each consisting of five MPEG-7 global visual descriptors extracted from an image [5,2]. The distance function δ is a weighted sum of partial descriptor distances [2]; each object has about 590 B on disk and the computation of δ takes around 0.01 ms;

SQFD. 1 million visual feature *signatures* each consisting of, on average, 60 cluster centroids in a 7-dimensional space; each cluster has a weight and such signatures are compared by *Signature Quadratic Form Distance* (SQFD) [4] which is a cheaper alternative to Earth Movers Distance. Each object occupies around 1.8 kB on disk and the SQFD distance takes around 0.5 ms;

ADJUSTABLE. 10 million float vectors uniformly generated from $[0, 1]^{32}$ compared by Euclidean distance; the disk size of each object can be artificially set from 512 B to 4096 B and time of δ computation can be tuned between 0.001 ms and 1 ms.

In all experiments, the accuracy of the search is measured as K-NN recall within the top R candidate objects $x \in \mathcal{X}$ identified by $\Psi_{\mathbf{p}}(q, x)$. All values are averaged over 1,000 randomly selected queries outside the dataset and all pivot sets P^j were selected independently at random from the dataset.

4.1   Accuracy of the PPP-Encoding and Ranking

Let us evaluate the basic accuracy of the K-NN search if the objects are encoded by PPP-Codes and ranked by $\Psi_\mathbf{p}(q,x)$ (6). In this section, we use a 1M subset of the CoPhIR dataset and we focus entirely on the trends and mutual influence of several parameters summarized in Table 2. In this section, we present results of 1-NN recall, which has the same trend as other values of K (see Section 4.2).

Graphs in Figure 6 focus on the influence of percentile $\mathbf{p}$ used in $\Psi_\mathbf{p}$. The left graph shows average 1-NN recall within the top $R = 100$ objects for variable $\mathbf{p}$, selected values of k and fixed $l = 8$, $\lambda = 4$. We can see that, as expected, the higher k the better and, more importantly, the peak of the results is at $\mathbf{p} = 0.75$ (just for clarification, for $\lambda = 4$, $\Psi_{0.75}(q,x)$ is equal to the third $\psi_q^j(x)$ rank of x out of four). These measurements are in compliance with the theoretical expectations discussed in Section 3.2.

The right graph in Figure 6 shows the *probe depth* [10] – average number of objects that had to be accessed in each ranking $\psi_q^j(x)$, $j \in \Lambda$ in order to discover 100 objects in at least $\lceil \mathbf{p}\lambda \rceil$ rankings (and thus determine their $\Psi_\mathbf{p}(q,x)$). Naturally, the probe depth grows with $\mathbf{p}$, especially for $\mathbf{p} \geq 0.75$. We can also see that finer space partitioning (higher k) results in lower probe depth because the Voronoi cells are smaller and thus objects close to q appear in $\lceil \mathbf{p}\lambda \rceil$ rankings sooner. The general lessons learned are the following: the more pivots the better (for both recall and efficiency), ideal percentile seems to be around 0.5–0.75, which is in compliance with results of Fagin et al. [10].

In general, we can assume that the accuracy will grow with increasing values of k, l, and λ. The left graph in Figure 7 shows influence of the number of pivot

Table 2. Parameters for experiments on 1M CoPhIR

param.	description	default
λ	number of pivot spaces	4
k	pivot number in each space	128
l	length of PPP	8
$\mathbf{p}$	percentile used in $\Psi_\mathbf{p}$	0.75
R	candidate set size	100 (0.01 %)

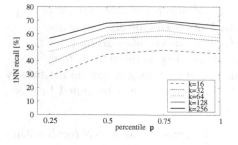 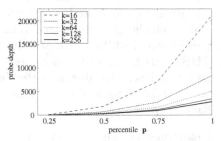

Fig. 6. CoPhIR 1M: 1-NN recall within the top $R = 100$ objects (left) and average probe depth of each ψ_q^j, $j \in \Lambda$ (right) for various number of pivots k and percentile $\mathbf{p}$

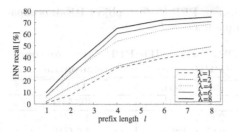 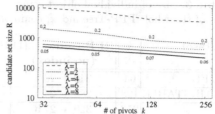

Fig. 7. CoPhIR 1M: 1-NN recall within the top $R = 100$ objects influenced by prefix length l (left); candidate set size R necessary to achieve 80% of 1-NN recall

spaces λ and of the prefix length l; we can see a dramatic shift between $\lambda = 2$ and $\lambda = 4$. Also, the accuracy grows steeply up to $l = 4$ and then the progress slows down. The right graph in Figure 7 adopts an inverse point of view – how does the aggregation approach reduce the candidate set size R necessary to achieve given recall (80 %, in this case); please, notice the logarithmic scales. The small numbers in the graph show the reduction factor with respect to $\lambda = 1$; we can see that R is reduced down to about 5 % using $\lambda = 8$.

4.2 Overall Efficiency of Our Approach

The previous sections shows that the ranking aggregation can significantly reduce the size of candidate set size R with respect with a single pivot space ranking. Let us now gauge the overall efficiency of our approach by standard measures from the similarity search field [17,14]:

I/O costs number of 4 kB block reads; in our approach, it is practically equal to the candidate set size R (step 4 in Figure 5);

distance computations (DC) number of evaluations of distance δ; equal to $\lambda \cdot k + R$ (steps 1 and 5);

search time the wall-clock time of the search process running parallel as described above.

All experiments were conducted on a machine with 8-core Intel Xeon @ 2.0 GHz with 12 GB of main memory and SATA SSD disk (measured transfer rate up to 400 MB/s with random accesses); for comparison, we also present some results on the following HDD configuration: two 10,000 rpm magnetic disks in RAID 1 array (random reads throughput up to 70 MB/s). All techniques used the full memory for their index and for disk caching; caches were cleared before every batch of 1,000 queries. The implementation is in Java using the MESSIF framework [3].

As a result of the analysis reported in Section 4.1, the indexes use the following parameters: $\lambda = 5$, $l = 8$, $\mathbf{p} = 0.5$ (3 out of 5). Our approach encodes each object by a PPP-Code and a PPP-Tree index is built on these codes. Table 3 shows the sizes of this representation for individual datasets and various numbers of pivots k used throughout this section. The third column shows the bit size of a single PPP-Code representation without any indexing (plus size of the object ID

Table 3. Size of individual object representation: PPP-Code + object ID(s); occupation of the whole PPP-Tree index; common parameters: $l = 8$, $\lambda = 5$

dataset	k	PPP-Code + ID (no index)	PPP-Code + IDs (PPP-Tree)	PPP-Trees memory occupation
SQFD	64	240 + 20 b	161 + 100 b	32.5 MB
ADJUSTABLE	128	280 + 24 b	217 + 120 b	403 MB
CoPhIR	256	320 + 27 b	205 + 135 b	4.0 GB
	380	360 + 27 b	245 + 135 b	4.5 GB
	512	360 + 27 b	258 + 135 b	4.6 GB

unique within the dataset). The fourth column shows PPP-Code sizes as reduced by PPP-Tree; on the other hand, this index has to store object IDs λ-times (see Section 3.3) and we can see that the memory reduction by PPP-Trees and the increase by multiple storage of IDs are practically equal. The last column shows the overall sizes of the PPP-Tree indexes for each dataset.

From now on, we focus on the search efficiency. In Section 4.1, we have studied influence of various build parameters (actually fineness of the partitioning) to the answer recall, but the main parameter to increase the recall at query time is the candidate set size R. Figure 8 shows development of recall and search time with respect to R on the full 100M CoPhIR dataset ($k = 512$). We can see that our approach can achieve very high recall while accessing thousands of objects out of 100M. The recall grows very steeply in the beginning, achieving about 90 % for 1-NN and 10-NN around $R = 5000$; the time grows practically linearly.

Table 4 (top) presents more measurements on the CoPhIR dataset. We have selected two values of $R = 1000$ and $R = 5000$ (10-NN recall 64 % and 84 %, respectively) and we present the I/O costs, computational costs, and the overall search times on both SSD and HDD disks. All these results should be put in context – comparison with other approaches. At this point, let us mention metric structure M-Index [12], which is based on similar fundamentals as our approach: it computes a PPP for each object, maintains an index structure similar to

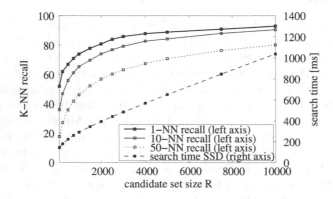

Fig. 8. Recall and search time on 100M CoPhIR as the candidate set grows; $k = 512$

Table 4. Results on CoPhIR ($k = 512$) and SQFD ($k = 64$) with PPP-Codes and M-Index (512 and 64 pivots)

technique		cand. set R	recall 10-NN	recall 50-NN	I/O costs	# of δ comp.	time on SSD [ms]	time on HDD [ms]
CoPhIR	PPP-Codes	1000	65 %	47 %	1000	3560	270	770
		5000	84 %	72 %	5000	7560	690	3000
	M-Index	110000	65 %	54 %	15000	110000	320	1300
		400000	85 %	80 %	59000	400000	1050	5400
SQFD	PPP-Codes	100	70 %	45 %	100	420	160	230
		1000	95 %	89 %	1000	1320	250	550
	M-Index	3200	65 %	53 %	1500	2200	650	820
		13000	94 %	89 %	6100	6500	750	920

our single PPP-Tree (Figure 4), and it accesses the leaves based on a scoring function similar to d (4); our M-Index implementation shares the core with the PPP-Codes and it stores the data in continuous disk chunks for efficient reading. Comparison of M-Index and PPP-Codes shows precisely the gain and the overhead of PPPRANK algorithm, which aggregates λ partitionings.

Looking at Table 4, we can see that M-Index with 512 pivots needs to access and refine $R = 110000$ or $R = 400000$ objects to achieve 65 % or 85 % 10-NN recall, respectively; the I/O costs and number of distance computations correspond with R. According to these measures, the PPP-Codes are one or two orders of magnitude more efficient than M-Index; looking at the search times, the dominance is not that significant because of the PPPRANK algorithm overhead. Please, note that the M-Index search algorithm is also parallel – processing of data from each "bucket" (leaf node) runs in a separate thread [12].

In order to clearly uncover the conditions under which the PPP-Codes overhead is worth the gain of reduced I/O and DC costs, we have introduced the ADJUSTABLE dataset. Table 5 shows the search times of PPP-Codes / M-Index while the object disk size and the DC time are adjusted; the results are measured on 10-NN recall level of 85 %, which is achieved at $R = 1000$ and $R = 400000$ for the PPP-Codes and M-Index, respectively. We can see that for the smallest objects and fastest distance, the M-Index beats PPP-Codes but as values of these two variables grow, the PPP-Codes show their strength. We believe that this table well summarizes the overall strength and costs of our approach.

Table 5. Search times [ms] of PPP-Codes / M-Index on ADJUSTABLE with 10-NN recall 85 %. PPP-Codes: $k = 128$, $R = 1000$, M-Index: 128 pivots, $R = 400000$.

PPP-Code / M-Index [ms]	size of an object [bytes]			
	512	1024	2048	4096
δ time 0.001 ms	**370** / **240**	**370** / 410	**370** / 1270	**370** / 1700
0.01 ms	**380** / 660	**380** / 750	**380** / 1350	**380** / 1850
0.1 ms	**400** / 5400	**400** / 5400	**420** / 5500	**420** / 5700
1 ms	**1100** / 52500	**1100** / 52500	**1100** / 52500	**1100** / 52500

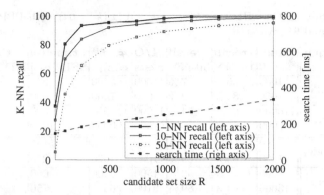

Fig. 9. Recall and search time on 1M SQFD dataset as candidate set grows; $k = 64$

The SQFD dataset is an example of data type belonging to the lower middle part of Table 5 – the signature objects occupy almost 2 kB and the SQFD distance function takes 0.5 ms on average. Graph in Figure 9 presents the PPP-Codes K-NN recall and search times while increasing R (note that size of the dataset is 1M and $k = 64$). We can see that the index achieves excellent results between $R = 500$ and $R = 1000$ with search time under 300 ms. Again, let us compare these results with M-Index with 64 pivots – the lower part of Table 4 shows that the PPP-Code aggregation can decrease the candidate set size R down under 1/10 of the M-Index results (for comparable recall values). For this dataset, we let the M-Index store precomputed object-pivot distances together with the objects and use them at query time for distance computation filtering [12,17]; this significantly decreases its DC costs and search times, nevertheless, the times of PPP-Codes are about 1/5 of the M-Index.

These results can be summarized as follows: The proposed approach is worthwhile for data types with larger objects (over 512 B) or with the time-consuming δ function (over 0.001 ms). For the two real-life datasets, our aggregation schema cut the I/O and δ computation costs down by one or two orders of magnitude. The overall speed-up factor is about 1.5 for CoPhIR and 5 for the SQFD dataset.

4.3   Comparison with Other Approaches

Finally, let us compare our approach with selected relevant techniques for approximate metric-based similarity search. We have selected those works that present results on the full 100M CoPhIR dataset – see Table 6.

M-Index. We have described the M-Index [12] and compared it with our approach in the previous section, because it shares the core idea with PPP-Codes which makes these approaches well comparable. Table 6 shows also variant when four M-Indexes are combined by a standard union of the candidate sets [14]; we can see that this approach can reduce the R necessary to achieve given answer recall but the actual numbers are no close to results of the PPP-Codes.

Table 6. Comparison with other approaches on 100M CoPhIR dataset

technique	overall # of pivots	cand. set R	recall 10-NN	I/O costs	# of δ comp.
PPP-Codes	2560	5000	84 %	5000	7560
M-Index	512	400000	85 %	59000	400512
M-Index (4 indexes)	960	300000	84 %	44000	301000
PP-Index (8 indexes)	800	∼333000	86 %	∼49000	∼334000
	8000	∼52000	82 %	∼7670	∼60000
MI-File	20000	1000	88 %	∼20000	21000

PP-Index. The PP-Index [8] also uses prefixes of pivot permutations to partition the data space; it builds a slightly different tree structure on the PPPs, identifies query-relevant partitions using a different heuristic, and reads these candidate objects in a disk-efficient way. In order to achieve high recall values, the PP-Index also combines several independent indexes by merging their results [8]; Table 6 shows selected results – we can see that the values are slightly better than those of M-Index, especially when a high number of pivots is used (8000).

MI-File. The MI-File [1] creates inverted files according to pivot-permutations; at query time, it determines the candidate set, reads it from the disk one-by-one and refines it. Table 6 shows selected results on CoPhIR; we can see that extremely high number of pivots (20000) resulted in even smaller candidate set then in case of PPP-Codes but the I/O and computational costs are higher.

5 Conclusions and Future Work

Efficient generic similarity search on a very large scale would have applications in many areas dealing with various complex data types. This task is difficult especially because identification of query-relevant data in complex data spaces typically requires accessing and refining relatively large candidate sets. As the search costs typically closely depend on the candidate set size, we have proposed a technique that can identify a more accurate candidate set at the expense of higher algorithm complexity. Our index is based on encoding of each object using multiple pivot spaces. We have proposed a two-phase search algorithm – it first generates several independent candidate rankings according to each pivot space separately and then it aggregates these rankings into one that provably increases the probability that query-relevant objects are accessed sooner.

We have conducted experiments on three datasets and in all cases our aggregation approach reduced the candidate set size by one or two orders of magnitude while preserving the answer quality. Because our search algorithm is relatively demanding, the overall search time gain depends on specific dataset but the speedup measured for all datasets under test was distinct, growing with the size of the indexed objects and complexity of the similarity function.

Our approach differs from others in three aspects. First, it transfers a large part of the search computational burden from the similarity function evaluation towards the search process itself and thus the search times are very stable across different data types. Second, our index explicitly exploits large main memory in comparison with an implicit use for disk caching. And third, our approach reduces the I/O costs and it fully exploits the strength of the SSD disks without seeks or, possibly, of a fast distributed key-value store for random data accesses.

Acknowledgments. This work was supported by Czech Research Foundation project P103/12/G084.

References

1. Amato, G., Gennaro, C., Savino, P.: MI-File: Using inverted files for scalable approximate similarity search. In: Multimedia Tools and Appl., pp. 1–30 (2012)
2. Batko, M., Falchi, F., Lucchese, C., Novak, D., Perego, R., Rabitti, F., Sedmidubsky, J., Zezula, P.: Building a web-scale image similarity search system. Multimedia Tools and Appl. 47(3), 599–629 (2010)
3. Batko, M., Novak, D., Zezula, P.: MESSIF: Metric Similarity Search Implementation Framework. In: Thanos, C., Borri, F., Candela, L. (eds.) Digital Libraries: R&D. LNCS, vol. 4877, pp. 1–10. Springer, Heidelberg (2007)
4. Beecks, C., Lokoč, J., Seidl, T., Skopal, T.: Indexing the signature quadratic form distance for efficient content-based multimedia retrieval. In: Proc. ACM Int. Conference on Multimedia Retrieval, pp. 1–8 (2011)
5. Bolettieri, P., Esuli, A., Falchi, F., Lucchese, C., Perego, R., Piccioli, T., Rabitti, F.: CoPhIR: A Test Collection for Content-Based Image Retrieval. CoRR 0905.4 (2009)
6. Chávez, E., Figueroa, K., Navarro, G.: Effective Proximity Retrieval by Ordering Permutations. IEEE Tran. on Pattern Anal. & Mach. Intel. 30(9), 1647–1658 (2008)
7. Edsberg, O., Hetland, M.L.: Indexing inexact proximity search with distance regression in pivot space. In: Proceedings of SISAP 2010, pp. 51–58. ACM Press, NY (2010)
8. Esuli, A.: Use of permutation prefixes for efficient and scalable approximate similarity search. Information Processing & Management 48(5), 889–902 (2012)
9. Fagin, R., Kumar, R., Sivakumar, D.: Comparing top k lists. In: Proc. of the 14th Annual ACM-SIAM Symposium on Discrete Alg., Phil., USA, pp. 28–36 (2003)
10. Fagin, R., Kumar, R., Sivakumar, D.: Efficient similarity search and classification via rank aggregation. In: Proceedings of ACM SIGMOD 2003, pp. 301–312. ACM Press, New York (2003)
11. Gionis, A., Indyk, P., Motwani, R.: Similarity search in high dimensions via hashing. In: Proceedings of VLDB 1999, pp. 518–529. Morgan Kaufmann (1999)
12. Novak, D., Batko, M., Zezula, P.: Metric Index: An Efficient and Scalable Solution for Precise and Approximate Similarity Search. Information Systems 36(4), 721–733 (2011)
13. Novak, D., Kyselak, M., Zezula, P.: On locality- sensitive indexing in generic metric spaces. In: Proc. of SISAP 2010, pp. 59–66. ACM Press (2010)
14. Novak, D., Zezula, P.: Performance Study of Independent Anchor Spaces for Similarity Searching. The Computer Journal, 1–15 (October 2013)
15. Patella, M., Ciaccia, P.: Approximate similarity search: A multi-faceted problem. Journal of Discrete Algorithms 7(1), 36–48 (2009)
16. Skala, M.: Counting distance permutations. Journal of Discrete Algorithms 7(1), 49–61 (2009)
17. Zezula, P., Amato, G., Dohnal, V., Batko, M.: Similarity Search: The Metric Space Approach. Springer (2006)

Predicting Pair Similarities for Near-Duplicate Detection in High Dimensional Spaces

Marco Fisichella, Andrea Ceroni, Fan Deng, and Wolfgang Nejdl

L3S Research Center, Hannover, Germany
{fisichella,ceroni,deng,nejdl}@L3S.de

Abstract. The problem of near–duplicate detection consists in finding those elements within a data set which are closest to a new input element, according to a given distance function and a given closeness threshold. Solving such problem for high–dimensional data sets is computationally expensive, since the amount of computation required to assess the similarity between any two elements increases with the number of dimensions. As a motivating example, an image or video sharing website would take advantage of detecting near–duplicates whenever new multimedia content is uploaded. Among different approaches, near–duplicate detection in high–dimensional data sets has been effectively addressed by SimPair LSH [11]. Built on top of Locality Sensitive Hashing (LSH), SimPair LSH computes and stores a small set of near-duplicate pairs in advance, and uses them to prune the candidate set generated by LSH for a given new element. In this paper, we develop an algorithm to predict a lower bound of the number of elements pruned by SimPair LSH from the candidate set generated by LSH. Since the computational overhead introduced by SimPair LSH to compute near-duplicate pairs in advance is rewarded by the possibility of using that information to prune the candidate set, predicting the number of pruned points would be crucial. The pruning prediction has been evaluated through experiments over three real–world data sets. We also performed further experiments on SimPair LSH, confirming that it consistently outperforms LSH with respect to memory space and running time.

Keywords: Indexing methods, Query, Near–duplicate detection, Locality Sensitive Hashing, high–dimensional data sets.

1   Introduction

Near–duplicate detection, also known as similarity search, is an old and well–established research topic which finds applications in different areas, from information retrieval to pattern recognition, from multimedia databases to machine learning. The near–duplicate detection problem can be informally stated as follows: given an object and a collection, find those elements in the collection which are most similar to the input one. Objects are usually represented through a set of features, which are exploited to assess similarities among them. Note that a similarity metric has to be defined over the feature space in order to evaluate

H. Decker et al. (Eds.): DEXA 2014, Part II, LNCS 8645, pp. 59–73, 2014.
© Springer International Publishing Switzerland 2014

similarities. Depending on the application domain, objects can be, for instance, documents, images or videos. However, once features have been extracted, they are represented just as points in a multidimensional space, independently from their original nature. In the rest of this paper we will refer to *objects* and *features* respectively as *points* and *dimensions*.

The near–duplicate detection problem, although it has been studied for decades, is still considered a challenging problem. The main reason is the well known *curse of dimensionality* [6]. It has been shown that the computational complexity of the near–duplicate detection exponentially increases with the number of point dimensions [1]. Moreover, any partitioning and clustering based indexing approach degenerates to a linear scan approach when the problem dimensionality is sufficiently high, i.e. more than 10 dimensions. This fact was both theoretically and empirically proved in [23].

In this work, we study near–duplicate detection within the context of *incremental range searches*. Given a point as a query, the problem consists in retrieving all the similar points from the data set before processing the query. We further introduce the term *incremental* to specify that in this scenario the data set is built in an incremental manner: whenever a point has to be inserted in the data set, the similar points previously inserted are retrieved and taken into account before actually performing the insertion. If a similar–enough point is already present in the collection, then the insertion will not be performed. Clearly, such a decision is taken by considering a pre–defined similarity threshold which allows to determine whether two points are similar or not. In addition to the previously mentioned *curse of dimensionality*, which equally affects any instance of the near–duplicate detection problem, a key issue of incremental range searches is fast query response. This is so because the near–duplicate detection is performed online during the insertion process, which has to be completed within the lowest possible amount of time.

Among others, one effective approach to near-duplicate detection in high-dimensional data sets is SimPair LSH [11]. Working on top of LSH, SimPair LSH substantially speeds up the running time of LSH by exploiting a small, constant amount of extra space. The key intuition consists in computing and storing a small set of near–duplicate pairs in advance, using them to prune the candidate set generated by LSH for a given new element. In this paper, we present an algorithm to predict a lower bound of the number of elements pruned by SimPair LSH from the candidate set. This would allow to predict whether the computational overhead introduced by SimPair LSH (see Paragraph 3.4 in [11]) is rewarded by the number of elements pruned. This and other contributions of this paper are summarized as follows:

- We present and evaluate an algorithm to predict a lower bound of the number of elements pruned from the candidate set by SimPair LSH.
- We describe an efficient procedure to maintain and update similar pair information in dynamic scenarios, i.e. when data sets are incrementally built with bursty sequences of queries.

- We perform additional experiments, on 3 real-world data sets, confirming that SimPair LSH consistently outperforms LSH under different aspects.

The remainder of this paper is the following. We first describe related works in Section 2 and we recall the basic behavior of SimPair LSH in section 3. We then present an algorithm to predict the number of points pruned by SimPair LSH from the candidate set in Section 4. In Section 6, we show that SimPair LSH outperforms LSH via extensive experiments. Finally, we give overall comments and conclusions in Section 7.

2 Related Work

As already stated, this paper is a substantial extension of the approach presented in [11], whose description is recalled in Section 3. In the rest of this section we give an overview of the current state of the art related to Locality Sensitive Hashing and near–duplicate detection.

2.1 Locality Sensitive Hashing (LSH)

Locality Sensitive Hashing (LSH) [16,13] was proposed by Indyk and Motwani and finds applications in different areas including multimedia near–duplicate detection (e.g., [7,21]). LSH was first applied in indexing high–dimensional points for Hamming distance [13], and later extended to L_p distance [9] where L_2 is Euclidean distance, the one used in this paper.

LSH exploits certain hash functions to map each multi–dimensional point into a scalar; the employed hash functions have the property that similar points have higher probability to be mapped together than dissimilar points. When LSH is used for indexing a set of points to speed up similarity search, the procedure is as follows: (i) select k hash functions h randomly and uniformly from a LSH hash function family H, and create L hash tables, hereafter called *buckets*; (ii) create an index (a hash table) by hashing all points p in the data set P into different buckets based on their hash values; (iii) when a query point q arrives, use the same set of hash functions to map q into L buckets, one from each hash table; (iv) retrieve all points p from the L buckets, collect them into a candidate set C, and remove duplicate points in C; (v) for each point p in C compute its distance to q and output those points that are similar to q.

In LSH, the probability that two points p_1 and p_2 are hashed into the same bucket is proportional to their distance c, and it can be computed as follows:

$$p(c) = Pr\left[h\left(p_1\right) = h\left(p_2\right)\right] = \int_0^r \left(\frac{1}{c}f\left(\frac{t}{c}\right)\left(1 - \frac{t}{r}\right)\right) dt \qquad (1)$$

where $f(t)$ is the probability density function of the absolute value of the normal distribution. Having $p(c)$, we can further compute the collision probability, namely success probability, under H:

$$P(c) = Pr[H(p_1) = H(p_2)] = 1 - (1 - p(c)^k)^L \qquad (2)$$

2.2 Other LSH-Based Approaches

Since it was proposed, LSH has been extended in different directions. Lv et al. [20] proposed multi-probe LSH, showing experimentally that it significantly reduces space cost while achieving the same search quality and similar time efficiency compared with LSH. The key idea of multi-probe LSH is that the algorithm not only searches for the near–duplicates in the buckets to which the query point q is hashed, but it also searches the buckets where the near–duplicates have slightly less chance to appear. The benefit is that each hash table can be better utilized since more than one bucket of a hash table is checked, which decreases the number of hash tables. However, multi-probe LSH does not provide the important search quality guarantee as LSH does.

Another extension of LSH is LSH forest [5] where multiple hash tables with different parameter settings are constructed such that different queries can be handled with different settings. In the theory community, a near-optimal LSH [3] has been proposed. However, currently it is mostly of theoretical interest because the asymptotic running time improvement is achieved only for a very large number of input points [1].

Inspired by LSH technique is the distributed similarity search and range query processing in high dimensional data [15]. Authors consider mappings from the multi-dimensional LSH bucket space to the linearly ordered set of peers that jointly maintain the indexed data and derive requirements to achieve high quality search results and limit the number of network accesses. Locality preserving properties of proposed mappings is proved.

In [1], authors observe that despite decades of research, current solutions still suffer from the "curse of dimensionality", i.e. either space or query time exponential effort in the dimensionality d is needed to guarantee an accurate result. In fact, for a large enough dimensionality, current solutions provide little improvement over a brute force linear scanning of the entire data set, both in theory and in practice. To overcome this limitation, in one of the latest work Indyk et al. [2] lead to a two-level hashing algorithm. The outer hash table partitions the data sets into buckets of bounded diameter. Then, for each bucket, they build the inner hash table, which uses the center of the minimum enclosing ball of the points in the bucket as a center point. By this approach, authors claim to achieve a better query time and space consumption with respect to their previous work [16]. In a recent research conducted in [8], authors proposed two new algorithms to speed up LSH for the Euclidean distance. These algorithms are able to reduce the hash index construction time.

Finally, it is important to note that SimPair LSH is orthogonal to LSH and to other LSH variants described above and, more significant, it can be applied in those scenarios, taking advantage of the improvements achieved in those works.

2.3 Similarity Search and Duplicate Detection on Streaming Data

The applications we consider have certain data stream characteristics such as continuous queries and fast response requirements, although they are still mainly traditional Web applications. Within the past decade, data streaming processing has been a popular topic where the applications include sensor data processing, real-time financial data analysis, Internet traffic monitoring and so on.

Gao et al. [12] and Lian et al. [19] studied the problem of efficiently finding similar time series on streaming data, and they achieved efficiency by accurately predicting future data. Their methods are for time series data and cannot be used for the type of applications we consider. Koudas et al. [18] studied the problem of finding k nearest neighbors over streaming data, but they were concerned about low-dimensional case. Deng and Rafiei [10] studied the problem of detecting duplicates for streaming data, where no similarity is involved.

Recently, Sudaram et al. [4] presented the Parallel LSH (PLSH) workflow, a system to handle near–duplicate queries on large amounts of text data, over one Billion Tweets, based on an efficient parallel LSH implementation.

3   SimPair LSH

In this section we recall the behavior of SimPair LSH, since it is used as a starting point in the rest of the paper. Hereafter, unless noted otherwise, the term *LSH* denotes the original LSH indexing method.

3.1   Problem Statement (Incremental Range Search)

Near–duplicate. Given a point $q \in \mathbb{R}^d$, any point $p \in \mathbb{R}^d$ such that $d(q, p) < \tau$ is a near–duplicate of q, where $\tau \in \mathbb{R}$ is a similarity threshold. Among the available distance functions d which might be exploited, the Euclidean distance has been chosen in this work, since it has been widely used in different applications. However, SimPair LSH can be easily extended to other distance functions, e.g. L1 and Hamming distance, since LSH can be applied in those cases as well.
Incremental Range Search. Given a point $q \in \mathbb{R}^d$ and a set P of n points $p_i \in \mathbb{R}^d$, find all the near–duplicate of q in P according to a given distance function $d(\cdot, \cdot)$ and a given similarity threshold τ before q is inserted into P.

3.2   The SimPair LSH Algorithm

The main intuition behind SimPair LSH resides in how LSH works. Given a query point q and a set of points p, LSH fills a candidate set C for q with *all* the points p stored in the same buckets in which q has been hashed. The near–duplicates of q are then found by comparing it with all the points in C. Such approach suffers from two facts. First, in order to increase the probability of including in C all the near–duplicates of q, a large number of hash tables has to be created. This can lead to an higher number of points in C, which means an higher number of comparisons. Second, in case of high-dimensional points, the number of operations required to compare two points increases. SimPair LSH speeds up the search by pre-computing and storing in memory a certain number of pair-wise similar points in the data set. They are exploited at query time to prune the candidate set C, which results in a lower number of comparisons.

Formally, SimPair LSH algorithm works as follows. Given a set P of n points $p_i \in \mathbb{R}^d$, SimPair LSH creates L buckets as in LSH. In addition, given a similarity threshold θ, the set SP of similar pairs $(p_1, p_2) : d(p_1, p_2) < \theta$ is built and stored along with the computed distances $d(p_1, p_2)$. Whenever a query point $q \in \mathbb{R}^d$ comes, SimPair LSH retrieves all points in the buckets to which q is hashed. Let this set of points be the candidate set C. Instead of linearly scanning through all the points p in C and computing their distances to q as in LSH, SimPair LSH checks the pre-computed similar pair set SP whenever a distance computation $d(q, p)$ is done. Depending on the value of $d(q, p)$ and on the value of a given similarity threshold τ, SimPair LSH can behave in 2 different ways:

- If $d(q, p) <= \tau$, then SimPair LSH searches in SP for all points $p\prime$: $d(p, p\prime) \leq \tau - d(q, p)$. It checks if $p\prime$ is in the candidate set C or not: if yes, then it marks $p\prime$ as a near–duplicate of q without computing the distance $d(p\prime, q)$.
- If $d(q, p) > \tau$, SimPair LSH searches in SP for all those points $p\prime$: $d(p, p\prime) < d(q, p) - \tau$. It checks if $p\prime$ is in the candidate set C or not. If yes, then it removes $p\prime$ from C without the distance computation.

For a more detailed description of SimPair LSH, please refer to [11].

4   Pruning Prediction

In this Section we propose an algorithm to predict a lower bound of the number of elements pruned by SimPair LSH with respect to LSH. As we saw in Section 3, SimPair LSH requires the maintenance and the access to SP. That is translated in a computational overhead which might be compensated by pruning elements in the candidate set. Finally, predicting the prunes in advance gives us the possibility to know if the overhead is rewarded.

4.1   The Intuition

According to the pruning analysis presented in [11], a lower bound of the number of prunes given a query q can be estimated. The key intuition is as follows: take a few sample points p from C and for each p compute $d(q, p)$. Based on the sample, estimate the distribution of different $d(q, p)$ for all p in C. From $d(q, p)$ we can derive an upper bound of $d(q, p\prime)$ according to the triangle inequality. Thus, we can estimate a lower bound of the probability that $p\prime$ appears in C and, accordingly, a lower bound of the number of prunes.

Again, by using the small fraction of sample points obtained from C, SimPair LSH can check SP and find if there is any *close enough* point $p\prime$ of p such that $d(p\prime, p) < |d(q, p) - \tau|$. Since the distance $d(q, p)$ is known, an upper bound of $d(q, p\prime)$ can be derived according to the triangle inequality. Knowing $d(q, p\prime)$, one can know the probability that $p\prime$ appears in C, which leads to a prune.

4.2   The Pruning Prediction Algorithm

The algorithm for predicting a lower bound of the number of prunes is described in Algorithm 1. First, let us consider the probability that two points p_1 and p_2 are hashed into the same bucket. Such a probability, according to Equation 1, is proportional to their distance $d(p_1, p_2)$, which will be referred to as simply d in the rest of the section. Second, in our algorithm, the full distance range is cut into multiple intervals. Then, given a distance value d, the function $I(d)$ can be used to determine which interval d falls in. Counter $Count[]$ is a histogram for storing the number of points in a particular interval. The interval is determined by the collision probabilities of pair–wise distances. For a fixed parameter r in Equation 1 the probability of collision $P(d)$ decreases monotonically with $d = ||p_1 p_2||_p$, where p is 2 in our case, where we consider *Gaussian* distribution. The optimal value for r depends on the data set and the query point. However, in [9] was suggested that $r = 4$ provides good results and, therefore, we currently use the value $r = 4$ in our implementation.

Under this hash function setting, there is not much difference for distances d within the range $[0, 1]$ in terms of hash collision probability. Thus, we use fewer intervals. In contrast, for distances within the range $[1, 2.5]$, the hash collision probabilities differ significantly. We use more intervals for this distance range.

The policy that we use to split a distance range into intervals is defined as follows. Fix the number of intervals first (e.g., 100); assign one interval for range $[0, x_1]$ and one for range $[x_2, \infty]$, within both of which the hash collision probabilities are similar; assign the rest of intervals to range $[x_1, x_2]$ by cutting the range evenly. Since the function $I(d)$ is only determined by the hash function, the intervals cutting can be done off–line before processing the data set.

4.3   Sampling Accuracy

SimPair LSH only takes a small fraction (e.g., 10%) of points from C. Increasing the number of points in the sample whose distances to q are within an interval and estimating the value in C will generate some errors.

Lemma 1. *A uniform random sample gives an unbiased estimate for the number of points with certain property, and the relative accuracy is inversely proportional to the number of points, and proportional to the sample size and to the true number being estimated. The standard deviation of the ratio between the estimate and the true value is $\sqrt{\frac{n}{R}(\frac{1}{x} - \frac{1}{n})}$, where n is the total number of points to be sampled, R is the sample size and x is the true value to be estimated, i.e. the number of points with the property.*

Proof. Let us assume that R different points are randomly taken into the sample, and each point with the property has a probability $\frac{x}{n}$ to stay in the sample. Let X_i be an indicator random variable indicating if the i-th point being sampled is with the property or not. That is,

$$X_i = \begin{cases} 1, & \text{if the sampled point has the property} \\ 0, & \text{otherwise} \end{cases}$$

Algorithm 1: Prediction of the number of prunes.

Input: A set C with n d-dimensional points $\boldsymbol{p}$; a distance function $d(\cdot, \cdot)$, a distance threshold τ defining near–duplicates; a query point $\boldsymbol{q}$; a distance interval function $I(d)$; the set of similar point pairs SP.

Output: Number of prunes $PruneNumber$

begin

    Construct a sample set S by selecting s points from C uniformly at random;

    for *each point $\boldsymbol{p}$ in S* **do**

        Compute the distance $d(\boldsymbol{q}, \boldsymbol{p})$;

        search for $\boldsymbol{p\prime}$ in SP where $d(\boldsymbol{p\prime}, \boldsymbol{p}) < |d(\boldsymbol{q}, \boldsymbol{p}) - \tau|$;

        for *each $\boldsymbol{p\prime}$ found* **do**

            increment the counter $Count[I(d(\boldsymbol{q}, \boldsymbol{p}) + |d(\boldsymbol{q}, \boldsymbol{p}) - \tau|)]$ by 1;

    Scale up the non-zero elements in the counter by a factor of $|C|/|S|$ and store them back to Count[]; PruneNumber = 0;

    for *each non-zero element in $Count[i]$* **do**

        Let d_i be the maximum distance of interavl i;

        Let $P(d_i)$ be the probability that 2 points with distance d_i are hashed to the same value, according to Equation 1;

        $PruneNumber \mathrel{+}= Count[i] \cdot P(d_i)$;

    Output PruneNumber;

It can be easily shown that $Pr(X_i = 1) = \frac{x}{n}$ and $Pr(X_i = 0) = 1 - \frac{x}{n}$. Given the observed value $Y = \frac{n}{R} \sum_{i=1}^{R} X_i$, since $E[Y] = \frac{n}{R} \sum_{i=1}^{R} E[X_i] = x$ then it is possible to deduce that Y is an unbiased estimate. Thus, we can conclude that:

$$VAR[Y] = \frac{n^2}{R^2} VAR\left[\sum_{i=1}^{R} X_i\right] = \frac{n}{R} x \left(1 - \frac{x}{n}\right)$$

$\square$

5   Similar Pair Maintenance and Updating

Since the gain from SimPair LSH lies on the fact that similar pair information (i.e. SP) is stored in memory, it is important to maintain it properly and efficiently. Moreover, it is crucial to handle sequences of queries by efficiently updating the similarity pair information. In case data are static and not created by the incremental process, one can build the similar pair set SP offline, before queries arrive, by using existing similarity join algorithms (e.g., [17]). Since it is out of the scope of this work, we will not discuss this in detail.

5.1   Maintenance

Data Structure. The set SP can be implemented as a two dimensional linked list. The first dimension is a list of points; the near–duplicates of each point

q in the first-dimension list are stored in another linked list (i.e. the second dimension), ordered by the distances to q. Then, a hash index is built on top of the first-dimension linked list to speed up the look-up operations.

Bounding the Size of SP. Since the total number of all similar pairs for a dataset of n objects can be $O(n^2)$, an underlying issue is how to restrict the size of SP. To achieve this purpose we set θ (the similarity threshold for the similar point pairs stored in SP) to τ (the similarity threshold for the similarity search). However, in case the data set size is large, SP can be too big to fit in memory. Thus we impose a second bound on the size of SP: it must not be greater than a constant fraction of the index size (e.g., 10%). To satisfy the latter constraint we can reduce the value of θ within the range $(0, \tau]$. Clearly, a bigger value of θ will generate a larger set of SP increasing the chances of finding $p\prime$ of p in C, and thus triggering a prune. Another possible solution consists in removing a certain number of similar pairs having the largest distances when the size of SP is above the space bound. We first estimate the number of similar pairs, called k pairs, to be removed based on the space to be released. The k pairs to be removed should be those whose distances are the largest, since they have less chance to generate a prune. Thus, we should find the top–k pairs with the largest distances. To obtain such top–k list, we can create a sorted linked list with k entries, namely L_k; we then take the part with the largest distances of the first second-dimension linked list in SP and fill L_k (assuming L_k is shorter than the first second-dimension linked list; if not, we go to the next second-dimension linked list). Then, we scan the second second-dimension linked list and update L_k. After scanning the second-dimension linked lists, the top–k pairs with the largest distances are found, and we can remove them from SP. Note that real–time updates are not necessary for these operations, and they can be buffered and executed when the data arrival rate is slower.

5.2   Updating

Point Insertions. In a continuous query scenario, each point q issues a query before inserted into the database. That is, as an application requirement, all near–duplicates of each newly arrived point need to be found before the new point updates SP. Hence, to maintain the similar pair list, we only needs to add the near–duplicates of q just found into SP and there is no similarity search involved. To add near–duplicates of q into SP, we first sort the near–duplicates based on their distances to q, and store them in a linked list. We then insert q and the linked list into SP, and we update the hash index of SP in the meanwhile. Besides, we need to search for each near–duplicate of q and insert q into the corresponding linked lists of its near–duplicates. Within the linked list into which q is to be inserted, we use a linear search to find the right place q should be put based on the distance between q and the near–duplicate. Note that we could have built indices for the second-dimension linked lists, but this would have increased the space cost. Also, real–time updates of SP are not necessary unlike the query response; we can buffer the new similar pairs and insert them into SP when the data arrival rate is lower.

Point Deletions. When a point q has to be deleted from the data set, we need to update SP. First, we search for all the near–duplicates of q in SP and remove q from each of the second-dimension linked lists of the near–duplicates. We then need to remove the second dimension linked list of q from SP.

Buffering SP Updates. As a matter of fact, the nature of the data arrival rate is bursty, i.e. there can be a large number of queries at one time and very few queries at another time. Then, as mentioned earlier, the updates to SP (inserting a point and reducing size) can be buffered and processed later in case the processor is overloaded because of bursts of the data arrival rate. This buffering mechanism guarantees the real–time response to the similarity search query. Note that the operation of deleting a point cannot be buffered since this can lead to inconsistent results.

6   Experiments and Evaluations

We evaluated our work on different real–world data sets and under different evaluation criteria. In the following sections we describe the data sets, as well as the criteria and results of our evaluation. Note that we only report experiments that have not been performed in [11].

The experiments were ran on a machine with an Intel T2500 2GHz processor, 2GB memory under OS Fedora 9. The algorithms were all implemented in C. Both the data points and the LSH indices were loaded into the main memory. Each index entry for a point takes 12 bytes memory. The source code of the original LSH was obtained from E2LSH[1] and was used without any modification.

6.1   Data Sets

We experiment the effectiveness of SimPair LSH on the same three real–world image data sets used in [11], which are hereafter summarized.

Flickr Images. We sent 26 random queries to Flickr and we retrieved all the images within the result set. After removing all the results with less than 150 pixels, we obtained approximately $55,000$ images.

Tiny Images. We downloaded a publicly available data set with 1 *million* tiny images [22]. Due to the high memory cost of LSH for large data sets, we randomly sampled $50,000$ images from the entire dataset. This allowed us to vary the number of hash tables within a larger range. The random sampling operation also reduced the chance that similar pairs appear in the data set, since the images retrieved from the result set of a query have higher chance to be similar to each other.

Video Key-Frames. We sent 10 random queries to Youtube and obtained around 200 video clips from each result set. We then extracted key frames of the

---

[1] http://web.mit.edu/andoni/www/LSH/manual.pdf

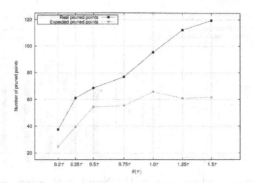

Fig. 1. Predicted prunes vs. $\theta(\tau)$

videos in the following way. Sequentially scan HSV histograms (whose dimensionality is equal to 162) of frames in a video: if the euclidean distance between two consecutive histograms is above 0.1, then keep the second histogram; otherwise skip it. In the end, we obtained 165,000 key–frame images. The choice of setting the distance threshold to 0.1 was driven by an empirical tuning. After an human evaluation, the distance value of 0.1 resulted to be the boundary line between equal and different images: image pairs having distance greater than 0.1 were mostly perceived as different, while those having distance lower than 0.1 where mostly categorized as identical. In other words, starting from a distance equal to 0.1, a human can see that two images are similar but not identical.

For all the image data sets described above, we removed duplicates and converted each image within a data set into a d-dimensional vector ($d \in \{162, 512\}$) by using the standard HSV histogram methods [14]. Each element of the vector represents the percentage of pixels within a given HSV interval.

6.2 Results

In this Section we report and discuss the results of our evaluation. In order to validate our approach, we randomly selected 100 objects from the data set as query objects. We tried other values for the number of queries, from 100 to 1000, without noting any change in performances. Thus, we kept it to 100 in the following experiments. The results presented in the rest of this section represent values averaged over the query objects.

Unless explicitly stated, the results that we report in the rest of this section were achieved on the Flickr data set, and the parameters were set as follows: distance threshold for near–duplicates $\tau = 0.1$, dimensionality of the data set $d = 162$, distance threshold used for filling the similar pair list SP $\theta = 0.1$, number of buckets $L = 136$, and success probability $\lambda = 95\%$. For the other data sets we achieved similar results.

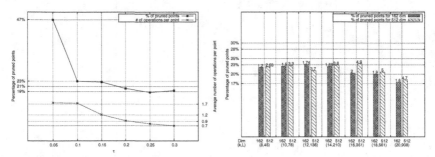

(a) Number of prunes and costs vs. τ　(b) Number of prunes and costs vs. d

Fig. 2. Number of prunes with different values of τ and d

Pruning Prediction. We test our pruning prediction algorithm (Section 4) by predicting the number of prunes with different values of similar pair threshold θ. The parameter settings are as follows: distance threshold for near–duplicates $\tau = 0.1$, the dimensionality of the data set $d = 162$, $k = 20$, $L = 595$, and the success probability $\lambda = 90\%$. The predicted lower bound of pruned points with different θ is shown in Figure 1, where the y-axis shows the average number of points pruned per query, and the x-axis shows the values of θ based on τ.

It is possible to observe that the pruning prediction always underestimates the actual number of pruned points, providing a reliable lower bound. Moreover, the difference between lower bound and real number of pruned points is almost constant (between 20%-30%) for low values of θ, i.e. $\theta < \tau$. This means that the trend of the prediction is similar to the real profile of pruned points: apart from a scale value (i.e. the difference between prediction and real number), the actual number of pruned points can be estimated quite precisely. These facts do not hold for higher values of θ ($\theta > \tau$), where the difference between lower bound prediction and real number of pruned points is more varying. This analysis motivates our choice of setting $\theta = \tau$ in the rest of the experiments.

Number and Cost of Prunes. We computed the number of distance computations saved by SimPair LSH, along with the corresponding time and space costs, by varying different parameters that were not considered in [11].

We varied the value of τ to see how it affects the pruning and costs. The other parameters have the values specified at the beginning of the section. The results are shown in Figure 2a, where the y-axises have the same meaning as in the previous figures and the x-axis represents the values of τ. It is possible to observe from the figure that τ has no significant impact on the pruning effectiveness. The high percentage of pruned points when $\tau = 0.05$ is because the size of the candidate set is small, and the variance of the percentage is higher. As previously mentioned, we fixed $\tau = 0.1$ in the rest of the experiments.

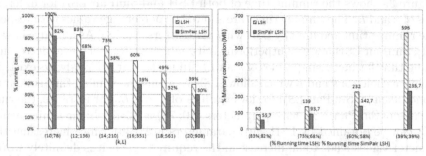

(a) % Running time vs. LSH parame-(b) Memory consumption (MB) vs. %
ters (k, L) running time for LSH and SimPair LSH

Fig. 3. % Running time (based on LSH greatest time) and memory consumption

We experimented two different vector dimensionalities d, 162 and 512, to
investigate the influence of d in the pruning and its costs. The other parameters
are the same as in the previous experiments. The results are shown in Figure 2b,
where the left y-$axis$ shows the percentage of pruned points as in the previous
experiments. The numbers on top of the bars show the average operation cost
per point. From the figure, it is possible to conclude that the operation cost is
larger when the dimensionality is higher. This is partially because the number
of similar pairs in SP is relatively larger when d is large, which has a similar
effect as increasing θ (when d is large, there are more similar pairs although the
similarity threshold does not change).

Time-Space Joint Analysis. The gain in time that SimPair LSH achieved
on different data sets has been already showed in [11]. However, LSH can save
running time as well by increasing the number of hash tables, i.e. increasing the
k and L parameters. In this section we want to show that, to achieve the same
gain in time, the additional space required by LSH for higher values of k and L is
greater than the space required by SimPair LSH. We show this fact by reporting
experiments done on the Flickr image data set, having $n = 55,000$.

The amount of memory required by LSH can be derived from L. Since each
hash table stores the identifiers of the n points in P, each one occupying 12
bytes (as implemented in E2LSH), the LSH space cost is equal to $12nL$. Thus,
in order to compute the extra space needed by LSH to achieve roughly the
same improvement in running time of SimPair LSH, it is sufficient to check the
value of L. Since the size of C dominates the time required by LSH to scan
through the candidate set, the running time being saved can be represented as
the reduction of $|C|$. Recall that bigger values of L correspond to the decrease in
size of C. Regarding SimPair LSH, it requires additional space for keeping the
set of similar point pairs SP, besides the memory for storing the LSH indices.
As already specified, we restricted it to be at most a constant fraction of the
LSH indices (10% in our experiments).

Figure 3a shows the running time of both LSH and SimPair LSH for different values of k and L. The running times are showed as percentage values with respect to the running time required by LSH with parameters $(10; 78)$. As already showed in [11], SimPair LSH always outperforms LSH, which anyway achieves decreasing running time for increasing values of k and L. In particular, there always exists in principle a choice of k and L which allows LSH to achieve a running time similar to the one taken by SimPair LSH for lower values of k and L. For instance, the running time of SimPair LSH with parameters $(10; 78)$ is 82%, while a similar value (83%) can be obtained by LSH with parameters $(12; 136)$. Since increasing the parameter values leads to an increase in the required space, it is definitely worth comparing the memory consumption of LSH and SimPair LSH for achieving almost the same running time. This is shown in Figure 3b. Different running times of both LSH and SimPair LSH are grouped in couples along the x–axis, while the corresponding memory costs to achieve them are plotted along the y–axis. The running time values are taken from Figure 3a and they are coupled so that similar values of different approaches belong to the same couple. As an example, the running time equal to 82% in the $(83\%, 82\%)$ couple is the one achieved by SimPair LSH with parameters $(10;78)$, while 83% is the running time achieved by LSH with parameters $(12;136)$. Figure 3b shows that, for every running time couple, LSH always leads to an higher memory consumption (caused by higher values of k and L). For instance, SimPair LSH requires 142.7 MB to achieve a running time equal to 58%, while LSH needs 232 MB to reach a similar running time (60%). This fact is accentuate in the $(39\%,39\%)$ couple, where the space required by SimPair LSH is the 39.8% of the one taken by LSH.

7   Conclusions

In this paper, we presented and evaluated an algorithm to predict a lower bound of the number of elements pruned by SimPair LSH. In our experiment, the pruning prediction algorithm always provides a reliable lower bound of the actual number of pruned points. Moreover, the difference between lower bound and real number of pruned points is almost constant for a sub set of setups, meaning that the profile of the prediction is similar to the real profile of pruned points apart from a scale value. Thanks to that, our algorithm allows to predict whether the computational overhead introduced by SimPair LSH is rewarded by the number of elements pruned.

Moreover, we described the procedure to maintain and update similar pair information, and we performed further experiments on SimPair LSH, confirming the robustness of SimPair LSH and its superiority with respect to LSH under different setups. In particular, in our evaluations we experimented that LSH can save running time by increasing the number of hash tables, i.e. increasing the k and L parameters. Since increasing the parameter values k and L leads to an increase in the required space, we compared the memory consumption of LSH and SimPair LSH for achieving almost the same running time. Our conclusion is that LSH requires significantly more space than SimPair LSH (e.g., from 40% to 60% more of memory consumption).

Acknowledgement. This work was partly funded by the DURAARK (GA No:600908) project under the FP7 pro- gramme of the European Commission.

References

1. Andoni, A., Indyk, P.: Near-optimal hashing algorithms for approximate nearest neighbor in high dimensions. Commun. ACM 51(1) (2008)
2. Andoni, A., Indyk, P., Nguyen, H.L., Razenshteyn, I.: Beyond locality-sensitive hashing. CoRR, abs/1306.1547 (2013)
3. Andoni, A., Indyk, P., Patrascu, M.: Near-optimal hashing algorithms for approximate nearest neighbor in high dimensions. In: FOCS (2006)
4. Bahmani, B., Goel, A., Shinde, R.: Efficient distributed locality sensitive hashing. In: CIKM (2012)
5. Bawa, M., Condie, T., Ganesan, P.: Lsh forest: self-tuning indexes for similarity search. In: WWW (2005)
6. Bellman, R.E.: Adaptive control processes - A guided tour (1961)
7. Chum, O., Philbin, J., Isard, M., Zisserman, A.: Scalable near identical image and shot detection. In: CIVR (2007)
8. Dasgupta, A., et al.: Fast locality-sensitive hashing. In: KDD (2011)
9. Datar, M., Immorlica, N., Indyk, P., Mirrokni, V.S.: Locality-sensitive hashing scheme based on p-stable distributions. In: SCG (2004)
10. Deng, F.: Approximately detecting duplicates for streaming data using stable bloom filters. In: SIGMOD (2006)
11. Fisichella, M., Deng, F., Nejdl, W.: Efficient incremental near duplicate detection based on locality sensitive hashing. In: Bringas, P.G., Hameurlain, A., Quirchmayr, G. (eds.) DEXA 2010, Part I. LNCS, vol. 6261, pp. 152–166. Springer, Heidelberg (2010)
12. Gao, L., Wang, X.S.: Continuous similarity-based queries on streaming time series. IEEE Trans. on Knowl. and Data Eng. 17(10) (2005)
13. Gionis, A., Indyk, P., Motwani, R.: Similarity search in high dimensions via hashing. In: VLDB (1999)
14. Gonzalez, R.C., Woods, R.E.: Digital Image Processing, 3rd edn. (2006)
15. Haghani, P., Michel, S., Aberer, K.: Distributed similarity search in high dimensions using locality sensitive hashing. In: EDBT (2009)
16. Indyk, P., Motwani, R.: Approximate nearest neighbors: towards removing the curse of dimensionality. In: STOC (1998)
17. Jacox, E.H., Samet, H.: Metric space similarity joins. ACM Trans. Database Syst. 33(2) (2008)
18. Koudas, N., Chin, B., Kian-lee, O., Zhang, T.R.: Approximate nn queries on streams with guaranteed error/performance bounds. In: VLDB (2004)
19. Lian, X., Chen, L.: Efficient similarity search over future stream time series. IEEE Trans. on Knowl. and Data Eng. 20(1) (2008)
20. Lv, Q., Josephson, W., Wang, Z., Charikar, M., Li, K.: Multi-probe lsh: efficient indexing for high-dimensional similarity search. In: VLDB (2007)
21. Teixeira, T., et al.: Scalable locality-sensitive hashing for similarity search in high-dimensional, large-scale multimedia datasets. CoRR, abs/1310.4136 (2013)
22. Torralba, A., Fergus, R., Freeman, W.: Tech. rep. mit-csail-tr-2007-024. Technical report, Massachusetts Institute of Technology (2007)
23. Weber, R., Schek, H.-J., Blott, S.: A quantitative analysis and performance study for similarity-search methods in high-dimensional spaces. In: VLDB (1998)

Fast Phonetic Similarity Search over Large Repositories

Hegler Tissot, Gabriel Peschl, and Marcos Didonet Del Fabro

Federal University of Parana, C3SL Labs, Curitiba, Brazil
{hctissot,gpeschl,marcos.ddf}@inf.ufpr.br

Abstract. Analysis of unstructured data may be inefficient in the presence of spelling errors. Existing approaches use string similarity methods to search for valid words within a text, with a supporting dictionary. However, they are not rich enough to encode phonetic information to assist the search. In this paper, we present a novel approach for efficiently perform phonetic similarity search over large data sources, that uses a data structure called *PhoneticMap* to encode language-specific phonetic information. We validate our approach through an experiment over a data set using a Portuguese variant of a well-known repository, to automatically correct words with spelling errors.

Keywords: Phonetic Similarity, String Similarity, Fast Search.

1 Introduction

A large amount of unstructured data is being produced by different kinds of information systems [8], as free text from the medical records. String similarity algorithms are used to identify concepts when text is loaded with misspellings [4]. Edit Distance (ED) [9] and Jaro-Winkler (JW) distance [14] are two well known functions that can be used to compare the elements of some input data source with an existing dictionary. Princeton WordNet (PWN) is a lexical database that provides an intuitive combination of dictionary and thesaurus to support text analysis [10]. However, PWN should be modified in order to support similarity search.

The existing string similarity algorithms coupled with a supporting dictionary may be very inefficient, in particular when the analyzed text has spelling errors [11], because they do not necessarily handle application aspects related to spelling errors. In these cases, it is necessary to use phonetic similarity metrics. Phonetics are language-dependent [12] and solutions for this sort of problems must be designed for each specific language. In addition, similarity algorithms are often slow when executed over large databases, though fast search methods have been implemented.

In this paper, we present an approach of fast phonetic similarity search (FPSS) over large repositories. First, we define a novel string similarity metric. Second,

H. Decker et al. (Eds.): DEXA 2014, Part II, LNCS 8645, pp. 74–81, 2014.

we present an indexed data structure called *PhoneticMap*, which is used by our novel fast similarity search algorithm. Finally, we integrate the previous contributions with PWN to implement the fast phonetic search. We validate our approach through an experiment in which we try to promote an automatic correction of spelling errors using the Portuguese language.

This article[1] is organized as follows: Section 2 proposes a method to search for phonetic similarities; Section 3 describes the experiments; Section 4 refers to the related work and Section 5 concludes with final remarks and future work.

2 Fast Phonetic Similarity Search

In this section we describe our approach to perform fast phonetic similarity search. We present novel string and phonetic similarity functions that uses *PhoneticMaps* to support finding similar words in the PWN repository.

2.1 String Similarity

We present a novel algorithm to calculate string similarity. The $String_{sim}$ function illustrated in Figure 1 measures the similarity between two input strings w_1 and w_2, resulting a similarity value between 0 (completely different) and 1 (exactly equal).

in:	String w_1, String w_2
out:	Number *similarity*
1:	$g_1 \leftarrow CharsFound(w_1, w_2)$;
2:	$g_2 \leftarrow CharsFound(w_2, w_1)$;
3:	$\Omega \leftarrow 0.975$;
4:	$p_1 \leftarrow \Omega^{PositionPenalty(w_1,w_2)}$;
5:	$p_2 \leftarrow \Omega^{PositionPenalty(w_2,w_1)}$;
6:	$similarity \leftarrow avg(g_1 \times p_1, g_2 \times p_2)$;
7:	$\Upsilon \leftarrow 0.005$;
8:	$S_{MAX} \leftarrow MAX(length(w_1), length(w_2))$;
9:	$s_{min} \leftarrow min(length(w_1), length(w_2))$;
10:	if $(S_{MAX} > s_{min})$ then
11:	$\quad b \leftarrow 1 + (S_{MAX} - s_{min}) \times \Upsilon$;
12:	$\quad f \leftarrow ln(S_{MAX} - s_{min} + 1)$;
13:	$\quad c \leftarrow \frac{S_{MAX} - s_{min}}{2}$;
14:	$\quad similarity \leftarrow similarity \times (\frac{1}{(b^f)^c})$;
15:	end if;
16:	return *similarity*;

Fig. 1. $String_{sim}$: A proposed string similarity function pseudocode

---

[1] Extended version at http://www.inf.ufpr.br/didonet/articles/2014_FPSS.pdf

$String_{sim}$ function calculates the average percentage between w_1 characters found in w_2 and w_2 characters found in w_1 (lines 1–6). $CharsFound(a, b)$ return the number of characters of a found in b, not taking into account the characters' position. For each character found in a different string position, a reduction penalty is calculated based on the constant Ω (lines 3–5). $PositionPenalty(p, q)$ returns the number of characters of p found in q but not in the same string position. Penalty calculated based on Ω guarantees, for example, that strings "ba" and "baba" will NOT result a $similarity = 1$. When the lengths of both strings (S_{MAX} and s_{min}) are different, there is a result adjustment in order to provide another penalty in the similarity level, based on the difference on the length of words and the factor Υ (lines 7–15). Ω (=0.975) and Υ (=0.005) were manually adjusted after testing the proposed function in an application that searches for similar names of people and companies.

2.2 Phonetic Similarity

When considering phonemes, a straightforward string comparison of characters may not be enough. In order to support indexing phonemes for a fast search, we present a structure called *PhoneticMap* and we define the *PhoneticMap Similarity*. Given a word w, the generic function *PhoneticMap(w)* results a PhoneticMap tuple $M = (w, P, D)$, where: w is the word itself, $P = \{p_1, p_2, \ldots, p_n\}$ is a set of n phonetic variations of word w, and $D = \{d_1, d_2, \ldots, d_n\}$ is a set of n definitions, where d_i is the definition of variation p_i. Given two *PhoneticMaps* M_1 and M_2, $PhoneticMapSim(M_1, M_2)$ is a generic function that results a similarity value (ranging from 0=different to 1=equal) between M_1 and M_2.

$PhoneticMap(w)$ and $PhoneticMapSim(M_1, M_2)$ are language-dependent. We develop two variations to support the Portuguese language. The function $PhoneticMap_{PT}(w)$ returns a map of 11 entries that encodes phonetic information. Table 1 describes a *PhoneticMap* generated for a Portuguese word. The function $PhoneticMapSim_{PT}(M_1, M_2)$ calculates the phonetic similarity between PhoneticMaps M_1 and M_2 as the string similarity weighted average between some phonetic variations of M_1 and M_2 (Formula 1), where: a) $S_w = String_{sim}((M_1.w, M_2.w))$, and b) $S_{(i)} = String_{sim}((M_1.p_i, M_2.p_i))$. We manually adjusted weights used in $PhoneticMapSim_{PT}$, in order to give more importance to similarities of consonant phonemes.

$$\frac{1 \times S_w + 2 \times S_{(1)} + 5 \times S_{(2)} + 1 \times S_{(3)} + 3 \times S_{(5)} + 2 \times S_{(7)} + 2 \times S_{(9)}}{1 + 2 + 5 + 1 + 3 + 2 + 2} \quad (1)$$

2.3 Phonetic Search

To perform a fast phonetic similarity search (FPSS), we propose a method for indexing PhoneticMaps (using single column indexes in a relational database) and phonetically searching the words. FPSS must locate phonetically similar

Table 1. $PhoneticMap_{PT}$("arrematação")

Entry i	Definition d_i	Phonetic variation p_i
w	Word	*arrematação*
1	Word with no accents	*arrematacao*
2	Word phonemes	*aRematasao*
3	Vowel phonemes only	*aeaaao*
4	Vowel phonemes (reverse)	*oaaaea*
5	Consonant phonemes	*Rmts*
6	Consonant phonemes (reverse)	*stmR*
7	Articulation manner	*EABC*
8	Articulation manner (reverse)	*CBAE*
9	Articulation point	*FACD*
10	Articulation point (reverse)	*DCAF*

words in the repositories based on the indexed phoneme variations, returning not only similar words but also the similarity level of each one. Given a word w and a minimum desirable similarity level l, $PhoneticSearch(w, l)$ is a generic function that results a set of tuples (r, s), where r is a phonetically similar word, and s is the similarity level resulted between $PhoneticMap(w)$ and $PhoneticMap(r)$, where $s \geq l$. Similarity level ranges from 0 to 1. We develop $PhoneticSearch_{PT}$ function, an extended version of the $PhoneticSearch$ function. Figure 2 shows $PhoneticSearch_{PT}$ pseudocode[2].

$PhoneticSearch_{PT}$ returns a set of similar words in Portuguese for a given input word w, considering the minimum desirable similarity level l. Additional parameters p and s set the number of extended consonant phonemes that can be considered as prefix and suffix when searching for similar words. p and s have default values 0 (zero). When $p > 0$, then $PhoneticSearch_{PT}$ uses the reverse indexed PhoneticMaps entries to locate similar words (entries 4, 6, 8 and 10 described in Table 1). Function $DBPhoneticMapSearch(i, v, e)$ finds records in the $PhoneticMap$ table, searching for $PhoneticMap$ entry i equals to value v (exact match), or entry i like value v with up to e characters added ("like" match), when $e > 0$. $PhoneticSearch_{PT}$ results a exact match when $l = 1$ (lines 2–3). Otherwise, it creates a dataset combining results of different $DBPhoneticMapSearch$ executions (line 5). In lines 6–8, phonetic variations 4, 6, 8, and 10 are used whether it is necessary to perform search over the reverse PhoneticMap entries ($p > 0$). After creating a result set of candidate words, the phonetic similarity between each found word and the search word is calculated (line 10). Words that does not satisfy the minumum similarity level l are removed from the result set (lines 10-11).

---

[2] The approach is presented as an instance of Portuguese language. However, it is tailored to be adapted for different languages, as English and Spanish.

```
in: String w, Number l, Integer p, Integer s
out: Dataset result
 1 :  pm ← PhoneticMap_PT(w);
 2 :  if l = 1 then
 3 :      result ← DBPhoneticMapSearch(0, pm.w);
 4 :  else;
 5 :      result ←
              DBPhoneticMapSearch(1, pm.p_1) ∪
              DBPhoneticMapSearch(2, pm.p_2) ∪
              DBPhoneticMapSearch(3, pm.p_3, s) ∪
              DBPhoneticMapSearch(5, pm.p_5, s) ∪
              DBPhoneticMapSearch(7, pm.p_7, s) ∪
              DBPhoneticMapSearch(9, pm.p_9, s);
 6 :      if p > 0 then
 7 :          result ← result ∪
                  DBPhoneticMapSearch(4, pm.p_4) ∪
                  DBPhoneticMapSearch(6, pm.p_6, p) ∪
                  DBPhoneticMapSearch(8, pm.p_8, p) ∪
                  DBPhoneticMapSearch(10, pm.p_10, p);
 8 :      end if;
 9 :      foreach (fWord in result)
10 :          if PhoneticMapSim_PT(pm, PhoneticMap_PT(fWord))
                  < l then
11 :              result.remove(fWord);
12 :          end if;
13 :  end if;
14 :  return result;
```

Fig. 2. $PhoneticSearch_{PT}$ pseudocode

3 Experiments

In this section we describe the experiments conducted to validate our approach. First, we compare our string similarity algorithms with two well-known ones. Second, we compare the performance of our full search method with a search using the indexed *PhoneticMap*s.

3.1 String Similarity

We performed an experiment to verify the efficiency of $String_{sim}$ in automatic error correction compared with other functions. We extracted a set of 3,933 words containing spelling errors from a sample of medical record texts in Portuguese. Each word was manually annotated with the correct spelling form (*reference* words). We used the $String_{sim}$ function to search for the 10 most similar words for each incorrect word, based on the returned similarity values. We used a

Portuguese version of PWN dictionary containing 798,750 distinct words. The resultsets for each word were ranked from 1 (most similar) to 10 (less similar). We store the rank in which each *reference* word is found in each resultset. The two previous steps were repeated using Edit Distance (ED) and Jaro-Winkler (JW) functions. Lastly, we compared the results of $String_{sim}$ against ED and JW, as shown in Table 2. $String_{sim}$ had more reference words with top-1 ranking, which is the objective of the approach. In 75.5% of cases (2,970 words), both functions find the reference word in the dictionary as a top-1 ranking (the most similar). For the remaining cases, $String_{sim}$ performs better (finds the reference word in a better rank) than ED in 16.9% of cases (666 searches with better ranking) while ED is better than $String_{sim}$ in only 5.8% (230 searches). These results are similar to those found when comparing $String_{sim}$ against Jaro-Winkler function.

Table 2. $String_{sim}$ (SS) x Edit Distance (ED)

SS Rank	ED Rank					Not Found
	1	2	3	4-5	6-10	
1	2970	420	51	30	25	26
2	127	51	37	15	18	13
3	32	12	8	8	3	7
4-5	17	8	7	4	6	3
6-10	14	1	2	4	4	0
Not Found	2	0	0	1	1	6

3.2   Full and Fast Similarity Search

We compared the performance of full and fast similarity search methods. One PhoneticMap for each PWN entry (798,750 words) and 11 single-column indexes were created – one for the *Word* entry and one for each of the 10 phonetic variations in the *PhoneticMap*. The same set of 3,933 were used. A *Full Search* was executed – each input word was compared with each dictionary entry using the $String_{sim}$ (Figure 1), searching for words with a similarity level ≥ 0.8; the spent search time and the number of found words were computed in the result – $PhoneticMapSim_{PT}$ function was not used in the *Full Search* due to its high processing time (60 seconds in average). A *Fast Search* was executed – each input word was submitted twice to $PhoneticSearch_{PT}$, with two different set of parameters: a) similarity level ≥ 0.9, and parameters p and s both equal to 0 (similar words might have the same number of consonant phonemes); and b) similarity level ≥ 0.8, and parameters p and s both equal to 1 (similar words could have one additional consonant phonemes as prefix or suffix); *Full Search* and *Fast Search* results were compared based on the total amount of spent time to execute each search, and the number of words obtained in the result.

3.3   Comparing Results

We observed that *Fast Search* can be 10-30 times faster than *Full Search*. Although a *Full Search* is complete in terms of the resulting words, both search methods did not use the same similarity function, and they do not return the same number of similar words. Even with a different result in the fast method, $PhoneticSearch_{PT}$ is able to find the reference word for each spelling error. Table 3 compares accuracy between $String_{sim}$ (SS) and $PhoneticSearch_{PT}$ (PS). In 80.6% of cases (3.170 words), both functions find the reference word as a top-1 ranking. $String_{sim}$ performs better in 10.2% of cases (402 searches) while PS is better in 8.1% (317 searches).

Table 3. $PhoneticSearch_{PT}$ x $String_{sim}$

PS			SS Rank			*Not*
Rank	1	2	3	4-5	6-10	*Found*
1	3170	189	50	31	14	5
2	143	37	10	7	1	0
3	57	13	1	2	0	1
4-5	46	9	6	4	5	0
6-10	47	6	0	0	2	1
Not Found	59	7	3	1	3	3

4   Related Work

Edit Distance (ED) (or Levenshtein Distance) [9] calculates the minimum number of operations (single-character edits) required to transform string w_1 into w_2. ED can be also normalized to calculate a percentage similarity instead of the number of operations needed to transform one string to another. Jaro-Winkler [14] is another example of string distance function. [5] presents a survey with the existing works on text similarity. [3] compares different string distance metrics for name-matching tasks, including edit-distance like functions, token-based distance functions and hybrid methods. In addition, other examples of string similarity functions can be found in the literature, as in [7,13,1]. Soundex is a phonetic matching scheme initially designed for English that uses codes based on the sound of each letter to translate a string into a canonical form of at most four characters, preserving the first letter [15]. As the result, phonetically similar entries will have the same keys and they can be indexed for efficient search using some hashing method. However, Soundex fails to consider only the initial portion of a string to generate the phonetic representation, which impairs the phonetic comparison when words have more than 4-5 consonants [6]. Fast Similarity Search [2] is an ED-based algorithm designed to find strings similarities in a large database.

5   Conclusions and Future Work

We presented an approach of fast phonetic similarity search coupled with an extended version of the WordNet dictionary. Our main contribution is the definition of an indexed structure, called *PhoneticMap* that stores phonetic information to be used by a novel string similarity search algorithm. The experiments showed that the algorithm has good precision results and that it executes faster than one version not using th *PhoneticMap*. We also presented a string similarity algorithm based on the notion of penalty. We plan to use our solution to address the problem of dealing with spelling errors in an information extraction system. We also plan to explore methods to optimally tune the parameters involved in the proposed hybrid similarity metrics, and adapt it to other languages, as English and Spanish.

Acknowledgments. This work is partially financed by CAPES.

References

1. Allison, L., Dix, T.I.: A Bit-String Longest-Common-Subsequence Algorithm. In: IPL, vol. 26, pp. 305–310 (1986)
2. Bocek, T., Hunt, E., Stiller, B., Hecht, F.: Fast similarity search in large dictionaries. Department of Informatics, University of Zurich (2007)
3. Cohen, W.W., Ravikumar, P., Fienberg, S.E.: A comparison of string distance metrics for name-matching tasks. In: IIWeb, pp. 73–78 (2003)
4. Godbole, S., Bhattacharya, I., Gupta, A., Verma, A.: Building re-usable dictionary repositories for real-world text mining. In: CIKM, pp. 1189–1198. ACM (2010)
5. Gomaa, W.H., Fahmy, A.A.: A Survey of Text Similarity Approaches. In: IJCA, vol. 68, pp. 13–18. Foundation of Computer Science, New York (2013)
6. Hall, P.A.V., Dowling, G.R.: Approximate String Matching. ACM Comput. Surv. 12, 381–402 (1980)
7. Hamming, R.: Error Detecting and Error Correcting Codes. Bell System Technical Journal BSTJ. 26, 147–160 (1950)
8. Jellouli, I., Mohajir, M.E.: An ontology-based approach for web information extraction. In: CIST, p. 5 (2011)
9. Levenshtein, V.I.: Binary codes capable of correcting insertions and reversals. Soviet Physics Doklady 10, 707–710 (1966)
10. Miller, G.A.: WordNet: a lexical database for English. Commun. ACM 38, 39–41 (1995)
11. Stvilia, B.: A model for ontology quality evaluation. First Monday 12 (2007)
12. Mann, V.A.: Distinguishing universal and language-dependent levels of speech perception: Evidence from Japanese listeners' perception of English. Cognition 24, 169–196 (1986)
13. Paterson, M., Dancik, V.: Longest Common Subsequences. In: Privara, I., Ružička, P., Rovan, B. (eds.) MFCS 1994. LNCS, vol. 841, pp. 127–142. Springer, Heidelberg (1994)
14. Winkler, W.E.: String Comparator Metrics and Enhanced Decision Rules in the Fellegi-Sunter Model of Record Linkage. In: Proceedings of the Section on Survey Research, pp. S.354–S.359 (1990)
15. Zobel, J., Dart, P.W.: Phonetic String Matching: Lessons from Information Retrieval. In: SIGIR, pp. 166–172. ACM (1996)

Named Entity Oriented Related News Ranking

Keisuke Kiritoshi and Qiang Ma

Kyoto University, Kyoto, Japan 606-8501
{kiritoshi@db.soc.,qiang@}i.kyoto-u.ac.jp

Abstract. To support the gathering of information from various viewpoints, we focus on descriptions of named entities (persons, organizations, locations) in news articles. We propose methods to rank news articles based on analyzing difference in descriptions of named entities. We extend the stakeholder mining proposed by Ogawa et al. and extract descriptions of named entities in articles. Then, four ranking measures (*relatedness, diversity, difference of polarity, diffeence of detailedness*) are calculated by analyzing the presence or absence of named entities, the coverage of topics and the polarity of descriptions. We carry out user study and experiments to validate our methods.

Keywords: News, Analyzing differences, Biases, Rankings.

1 Introduction

News is never free from bias due to the intentions of editors and sponsors. If users read only one article about an event, they can have a biased impression of that event. Helping users better understand a news event by revealing the differences between news articles is very important.

Many methods and systems to analyze differences between news articles have been proposed[1]-[6]. For instance, Ogawa et al. propose a method which compares news articles by analyzing the descriptions of stakeholders, such as persons, organizations and locations [1]. However, this method has two shortcomings. First, the target users are limited to those who are aware of news bias. Second, they do not provide a scoring mechanism to support searching for and ranking news articles. In this paper, we propose a method to rank news articles by measuring the differences between articles to support user obtaining information from different viewpoints.

Currently, the input of our method is a news article and other related ones. We suppose that a user is browsing the news article by using Google News (*http://news.google.com*) service and the related articles reporting on the same event or topic are provided by Google News.

Revealing which kinds of difference exist between related news articles is important to help users understand the news. We carried out a user survey with 10 undergraduate and graduate students. Based on the survey, we propose four ranking criteria (*relatedness, diversity, difference of polarity and difference of detailedness*) and calculate them by analyzing the descriptions of named entities.

H. Decker et al. (Eds.): DEXA 2014, Part II, LNCS 8645, pp. 82–96, 2014.

Descriptions of named entities (persons, organization, and locations) are extracted from news articles by extending the stakeholder mining method [1]. We introduce a notion of *core entity* to denote the most important entity in a reported event.

Relatedness is the degree to which two news articles report on the same event or topic. *Relatedness* is estimated by comparing the mentioned (core) entities in news articles.

Diversity is a measurement of how many different viewpoints are described in two news articles. If an article includes more named entities, excepting named entities of the original article and core entities, we assume that it reports on some different aspects regarding the same event and assign it a higher *diversity* score.

Diffrence of polarity is a measurement of the difference in description polarities of named entities.

Difference of detailedness is a measurement of the difference in detailed descriptions of entities in news articles. We extract topics regarding named entities by using LDA (Latent Dirichlet Allocation)[9] and calculate *difference of detailedness* by comparing topic coverage and text lengths between news articles.

These four criteria can be used separately or in an integrated manner to rank news articles. In our experiments, we compared and evaluated the ranking methods based on these four measures.

The major contributions of this paper can be summarized as follows:

- We have carried out a user survey to reveal the important criteria for helping users understand news articles (see also Section 3).
- Based on the user survey, we propose four criteria to measure differences between related news articles (Section 4.3). We also propose a named entity mining method (Section 4.2) that extends stakeholder mining proporsed by Ogawa et al.[1] to enable the computation of these four criteria.
- We have performed experiments to compare and validate the ranking methods based on the four criteria (Section 5).

2   Related Work

Ogawa et al. [1] study the analysis of differences between news articles by focusing on named entities and they propose a stakeholder mining method. They find a stakeholder who is mentioned in two news articles. They then present a graph constructed based on description polarity and interest in the stakeholder. This graph helps users analyze the differences between news articles. This method uses text documents, but Xu et al. propose a stakeholder mining method that utilizes multimedia news[2].

NewsCube[3] presents various aspects of a news event and presents these using an aspect viewer to facilitate understanding of the news. The Comparative Web Browser (CWB)[4] searches for news articles that include descriptions similar to those in the article being read by the user. This enables the user to read news articles while comparing articles.

Table 1. News Events for User Survey

Event1	Obama administration's remark about NSA surveillance
Event2	Wrestling selected as a candidate to be an Olympic event
Event3	Osaka mayor Hashimotofs remark about comfort women
Event4	Mayor Hashimoto's news conference on his reflection concerning the above remark
Event5	Demotion to the minor leagues of major league player, Munenori Kawasaki

TVBanc[5] compares news articles based on a notion of topic structure. TVBanc gathers related news from various media and extracts pairs of topics and viewpoints to reveal the *diversity* and bias of news articles. Ishida et al.[6] propose a system which reveals differences between news providers to enhance users' news understanding. They analyze the subject-verb-object (SVO) construction in the descriptions of entities and extract characteristic descriptions of each news provider.

These conventional systems visualize and highlight the differences between news articles to help users better understand the articles. In contrast, we propose methods to measure the differences between news articles by using descriptions of named entities.

3   User Survey

3.1   Summary

We carried out a user survey to find out which kinds of difference among news articles are important for supporting users' news understanding. The subjects were 10 students from Kyoto University. We selected 5 controversial news events for this survey. For each topic, we used one article in Japanese as the original one, meaning it was the one first read by a user, and then five Japanese and five English-language articles as related ones. We asked the 10 subjects to rank these related articles according to their usefulness and their difference from the original article. We also asked the subjects to describe their criteria used in this ranking. The events for our user survey are shown in Table 1.

3.2   Analysis

Table 2 shows the subjects' top three criteria for each of their rankings. We summarized these differences into four categories: relatedness, viewpoint, polarity, and detailedness. We also found that these four factors are evaluated by analyzing the descriptions of named entities.

(1) *Relatedness* is a criteria for estimating whether the news articles are reporting on same news event and the same entities. There are two kinds of *relatedness*.

a) *Relatedness* at the event level: We can evaluate this kind of *relatedness* by comparing the named entities mentioned in the event with those in a news

Table 2. Major differences for each news event

Event1	Critical, much detailed information, high relatedness, positive opinion
Event2	Positive opinion, much detailed information, polarity of description, viewpoints
Event3	Viewpoints
Event4	Positive opinion
Event5	High relatedness, viewpoints

article. For example, if the article is strongly related to Event 1 in Table 1 and the article has named entities such as "Obama", "America" and "NSA", these entities may be strongly related to this event.

b) *Relatedness* at the level of two articles: We can evaluate this kind of *relatedness* by comparing the named entities mentioned in two article. For example, in the news event of example a), article a contains named entities such as "Obama", "NSA", "Bush", "Republican Party" and article b has named entities such as "Obama", "NSA", "America", "Bush", "Republican Party", "FILA". In this example, articles a and b report about criticism towards Bush about NSA surveillance. Article a is therefore strongly related to article b.

(2) We found that similarity of viewpoint and detailedness were highly ranked from users' questionnaires. In some cases, information from more viewpoints means more detailed information.

Similar to *relatedness*, there are two kinds of difference in viewpoints.

a) Differences of viewpoints at the event level: The article reporting the Event 3 from Table 1 from various viewpoints included many different named entities corresponding to named entities often mentioned in the other articles, such as "Hashimoto", "South Korea" and "Osaka". For instance, in articles reporting the reactions of other countries, the names of those countries appeared. In addition, in articles discussing the political effects on mayor Hashimoto, persons in the same political party and other parties usually appeared. Therefore, we can compare the entities mentioned in one article and those in the other articles to find different viewpoints.

b) Differences of viewpoint at the level of two articles: For example, when we compare articles regarding Event 2 in Table 1, article a contains named entities such as "Yoshida Saori", "IOC", "FILA" and "Saint Petersburg". On the other hand, article b contains named entities such as "IOC", "FILA", "MLB", "IBAF" and "Hideki Matsui". Article b reports this event from a different viewpoint more focused on baseball. Therefore, we can analyze differences between articles from the differences in named entities between articles.

(3) Regarding the polarities of descriptions of named entities, for example, a description could be "There are negative effects on Japan". The word "negative effects" qualify the named entity "Japan". In this case, the polarity of descriptions of the named entity "Japan" tilts towards a minus. Therefore, we can judge the polarities of named entities from syntax trees and the positive or negative degrees of words in descriptions of named entities.

(4) We analyze the detailedness, such as the remarks of persons from the lengths of descriptions of named entities and the number of topics about named entities. For example, reporting on a person's remarks, article a might quote one sentence and article b might quote all sentences. In this case, we consider articles b to be more detailed than article a. Because named entities often appear before and after quotes, we consider quotes about named entities along with the description of each named entity. Article b in this example has a longer description and more topics than article a.

In short, we represent differences between news articles by using relatedness, viewpoints, polarities and detailedness. In the following sections, we propose a named entity mining method and describe how these four criteria are calculated to rank news articles.

4   Named Entity Mining

We extend the stakeholder mining method [1] to extract descriptions on named entities for ranking news. The named entity mining and ranking methods consist of four steps as follows.

1. Extracting named entities and descriptions of named entities
2. Extracting core entities
3. Calculating ranking measures (*relatedness, diversity, difference of polarity, difference of detailedness*)
4. Ranking by combining measures

4.1   Extracting Named Entities and Descriptions of Named Entities

We use a language tool StanfordCoreNLP[7] to analyze articles. We extract words with a NamedEntityTag of PERSON, ORGANIZATION and LOCATION as named entities.

We extract descriptions of named entities based on a relationship structure constructed by StanfordCoreNLP. StanfordCoreNLP provides grammatical relations between words as follows:

$$type(governor, dependent)$$

type is a relationship between two words, *governor* and *dependent*. We obtain a tree structure by considering *governor* as the parent and *dependent* as the child. We use conversion operations proposed by Ogawa et al. [1] to generate a tree structure suitable for computing the description polarity of named entities. Table 3 shows these conversion operations. Figure 1 shows the tree structure of the following sentence:

Kane said depressed America and Japan talk with China which is rapidly growing.

We consider descriptions on named entities as sets of sub-trees, the root of each being a verb and its descendants containing the target named entities.

Table 3. Conversion operations proposed by Ogawa et al.

type	Operation
conj	Delete this relationship and change a parent of *govenor* to a parent-dependent. If both *govenor* and *dependent* are verbs, change every child of *govenor* except for *dependent* to children of *dependent*.
appos	Carry out the same operation as *conj*.
rcmod	Replace *govenor* and *dependent*.
cop	Carry out the same operation as *rcmod*.

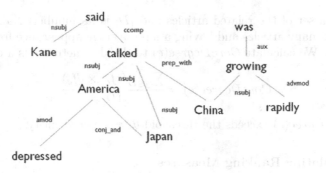

Fig. 1. An example of a tree structure

Suppose e is a named entity in article a. S_e is a set of sentences containing e. v is a verb which is an ancestor of e. For $s_1 \in S_e$, $V_{s_1}(e)$ a sub-tree whose root is v. In other words, $V_{s_1}(e)$ is a description of e in s_1. A set of descriptions of all sentences included by S_e is defined as follows:

$$V(e) = \{V_{s_i} | i = 1, 2, \cdots, n\} \tag{1}$$

where $V(e)$ means the descriptions of named entity e. For example, we can let a sentence regarding Figure 1 be s_1. When the named entity "Japan" is extracted, v is "talked" and the set of descendants V_{s_1} consists of descendants of "talked". Therefore, the description of "Japan" in s_1, $V_{s_1}(Japan)$ is expressed as follows:

$$V_{s_1}(Japan) = \{talked, America, Japan, China, depressed\} \tag{2}$$

4.2 Extracting Core Entities

Core entities in an event are named entities which have a high frequency of appearance.

We explain here the processes to extract core entities. Let e be a named entity in article j, and the appearance frequency of e in j, tf_{e_j} is calculated as follows:

$$tf_{e_j} = \frac{|e|}{n_j} \tag{3}$$

where $|e|$ is the frequency of e and n_j is the number of total terms in article j. Also, in all the articles related to an event, let d be the number of articles including named entity e, so the frequency of description df_e about e is calculated as follows:

$$df_e = \frac{d}{|D|} \tag{4}$$

where D is a set of the related articles and $|D|$ is its number. Named entities appearing in many articles and having a high average appearance frequency are core entities. We calculate $CoreDegree(e)$ to decide whether e is a core entity:

$$CoreDegree(e) = \frac{\sum_{j \in D}(tf_{e_j} \times df_e)}{|D|} \tag{5}$$

when $CoreDegree(e)$ exceeds the threshold θ, e is a core entity.

4.3   Calculating Ranking Measures

In Section 4.3, we explain how to calculate the four ranking criteria. Hereafter, suppose the original article is o, the set of related articles is A and a related article is $a \in A$. E_{Core} is a set of core entities and E_a are the named entities of article a.

Relatedness. *Relatedness* is a measure to estimate whether news articles describe the same event. In this paper, we calculate *relatedness* by comparing sets of named entities in articles. As mentioned, *relatedness* is calculated from the following two aspects.

(1) *Relatedness* to events: We define the *relatedness* of events $rel_{eve}(E_{Core}, a)$ based on how many core entities are mentioned in article a. Let $|E|$ be the number of set E, and $rel_{eve}(e, a)$ of article a is calculated as follows:

$$rel_{eve}(E_{Core}, a) = |E_{Core} \cap E_a| \tag{6}$$

(2) *Relatedness* to the original article: We define *relatedness* to the original article $rel_{uni}(a, o)$ based on the number of named entities mentioned in both the original article o and the related one a. $rel_{uni}(a, o)$ is calculated as follows:

$$rel_{uni}(a, o) = |E_o \cap E_a| \tag{7}$$

The integrated *relatedness* $rel(e, a, o)$ is calculated from $rel_{eve}(e, a)$, $rel_{uni}(a, o)$ with weight α as follows:

$$rel(E_{Core}, a, o) = \alpha \times rel_{eve}(E_{Core}, a) + (1 - \alpha) \times rel_{uni}(a, o) \tag{8}$$

Diversity. *Diversity* is a measure to estimate how many different viewpoints there are of events and in the original article. We calculate this measure by comparing named entities from two aspects, as follows:

(1) *Diversity* in the original article: We define the *diversity* of article a with respect to the original article o, $div_{dif}(a, o)$, based on differences in the named entities mentioned in the articles. $div_{dif}(a, o)$ is defined as the number of named entities included only in article a.

$$div_{dif}(a, o) = w \, |E_{ao} - E_{Core}| + w_{Core} \, |E_{ao} \cap E_{Core}| \qquad (9)$$

$$E_{ao} = E_a - E_o \qquad (10)$$

where w and w_{Core} are weight parameters. Because core entities are more important, generally w_{Core} is bigger than w fundamentally.

(2) *Diversity* regarding the event: We define *diversity* of the target article a regarding the event e, $div_{eve}(E_{Core}, a)$, based on named entities mentioned in a and the core entities. $div_{eve}(E_{Core}, a)$ is defined as follows:

$$div_{eve}(E_{Core}, a) = |E_a - E_{Core}| \qquad (11)$$

The integrated *diversity* $div(E_{Core}, a, o)$ is calculated from $div_{dif}(a, o)$, $div_{eve}(E_{Core}, a)$ with weight β as follows:

$$div(E_{Core}, a, o) = \beta \times div_{dif}(a, o) + (1 - \beta)div_{eve}(E_{Core}, a) \qquad (12)$$

Difference of Polarity. We calculate the *difference of polarity* based on the polarities of named entities. The polarity of a named entity is estimated from descriptions of named entities and is expressed as a positive or negative score.

We extract descriptions of named entities through our named entity mining method and use an emotional word dictionary to assign a polarity score to each description. In the work described here, we used SentiWordNet[8] as the emotional word dictionary.

The sum score of emotional words in descriptions of a named entity is the final description polarity of the named entity. After that, we compare the polarity of each entity in articles o and a. Let $sup_a(e), sup_o(e)$ respectively be the polarity of named entity $e \in E_a \cup E_o$, (if $\{e\} \cap E_o = \emptyset$, let $sup_o(e) = 0$; if $\{e\} \cap E_a = \emptyset$, let $sup_a(e) = 0$) *difference of polarity* $sup(a, o)$ is defined as follows:

$$sup(a, o) = \sum_{e \in \{E_a \cup E_o\}} w_f \times |(sup_a(e) - sup_o(e))| \qquad (13)$$

where,

$$w_f = \begin{cases} w_{Core} & (e \in E_{Core}) \\ w & (\text{others}) \end{cases} \qquad \begin{matrix} (14a) \\ (14b) \end{matrix}$$

where weight w_{Core} is greater than weight w because core entities have a strong effect on a user's positive or negative impression in articles.

Difference of Detailedness. *difference of detailedness* is a measure to estimate how much detailed information about named entities the related article has compared to the original one. In this paper, the *difference of detailedness* is calculated by comparing topic coverage and text length of the topics in news articles. We use LDA (Latent Dirichlet Allocation)[9] to extract a topic from the description of named entities using the Stanford topic modeling toolbox[11]. The difference of detailedness for named entity e between articles a, o is calculated through the following steps.

1. We extract all descriptions $S(e)$ of named entity e from articles $A \cup o$ by applying our entity mining method. We then apply LDA to obtain the topics $T(e)$ for entity e.
2. $S_o(e), S_a(e)$ are defined as sets of descriptions of named entity e in articles a and o, respectively. Based on the results of LDA, we give topic probabilities $p_o(s_o, t)$, $p_a(s_a, t)$ to sentences $s_o \in S_o(e)$ and $s_a \in S_a(e)$, respectively. ($t \in T(e)$).
3. If the topic probability $P_o(s_o, t)$ is more than threshold γ, t is the topic for descriptions of named entity e in article o. We apply this operation to all sentences in $S_o(e)$ and get a topic set $T_o(e)$ and a topic set $T_a(e)$.
4. Let at_i, ot_i be the numbers of words in topic $t_i \in (T_a(e) \cup T_o(e))$ in article a and o, respectively (if $\{t_i\} \cap T_o(e) = \emptyset$, let $ot_i = 0$ and if $\{t_i\} \cap T_a(e) = \emptyset$, let $at_i = 0$).

   To comparing topic coverage and text lengths, the *difference of detailedness* for article a regarding article o on named entity e $f_{ao}(e)$ is calculated as follows:

$$f_{ao}(e) = \sum_{t_i \in \{T_a(e) \cup T_o(e)\}} (-1)^\delta \log \left(\frac{|at_i - ot_i|}{at_i + ot_i + 1} + 1 \right) \qquad (15)$$

where,

$$\delta = \begin{cases} 0 & (at_i \geq ot_i) & (16a) \\ 1 & (at_i < ot_i) & (16b) \end{cases}$$

Finally, we define the *difference of detailedness* between article a, o $det(a, o)$ as the total *difference of detailedness* for all the named entities. Therefore, $det(a, o)$ is calculated as follows:

$$det(a, o) = \sum_{e \in \{E_a \cup E_o\}} w_f \times f_{ao}(e) \qquad (17)$$

where,

$$w_f = \begin{cases} w_{Core} & (e \in E_{Core}) & (18a) \\ w & (others) & (18b) \end{cases}$$

4.4 Ranking

We usually rank articles by using ranking measures separately. However, articles which are less related to the event but have high scores for other measures can be highly ranked. Therefore, we also propose a ranking method using these four criteria in an integrated manner. We will discuss this issue in our experiment.

5 Experiments

We carried out experiments to evaluate ranking methods based on the four criteria by using nDCG. First, we evaluated four measures separately. Next, we evaluated the methods with the criteria in a combined manner.

In the experiments, we used the five events used for the user survey and for each news event we selected seven English-language articles, one as the original article and the others as related articles. nDCG (Normalized Discounted Cumulative Gain)[10] was the evaluation measure in our experiments. The nDCG score regarding a ranking of the top p is defined as follows:

$$nDCG_p = \frac{DCG_p}{IDCG_p} \tag{19}$$

where DCG_p is a weighted score. Let score of the i-th be rel_i, so DCG_p is expressed as follows:

$$DCG_p = rel_1 + \sum_{i=2}^{p} \frac{rel_i}{\log_2 i} \tag{20}$$

where $IDCG_p$ is a value which applies the ideal ranking arranged in descending order of scores to expression (19).

5.1 Experiment on the Extraction of Core Entities

We evaluated the method of extracting core entities. In this experiment, we compared core entities extracted by the proposed method with those selected manually by a user. We varied the threshold θ (See also Formula (5)) and calculated recall, precision and the F-measure. For an event, recall R, precision P and F-measure F are calculated as follows:

$$R = \frac{A}{C} \tag{21}$$

$$P = \frac{A}{N} \tag{22}$$

$$F = \frac{2A}{N+C} \tag{23}$$

where A is the number of core entities in the obtained named entities, N is the number of obtained named entities and C is the number of core entities in the event. Table 5 shows scores calculated depending on threshold θ.

The highest average F-measure was 0.808, and in this case θ was 0.0020. In the remaining experiments, we used $\theta = 0.0020$ to extracted core entities.

Table 4. Core entity's average recall, precision and F-measure

threshold θ	average of recall	average of precision	average of F-measure
0.0010	1.00	0.406	0.571
0.0015	1.00	0.456	0.623
0.0020	0.950	0.714	0.808
0.0025	0.883	0.698	0.770
0.0030	0.883	0.698	0.770
0.0035	0.883	0.718	0.784
0.0040	0.833	0.753	0.787
0.0045	0.767	0.783	0.763
0.0050	0.600	0.833	0.633
0.0055	0.533	0.800	0.580
0.0060	0.483	0.833	0.574
0.0065	0.383	0.833	0.508

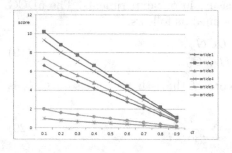

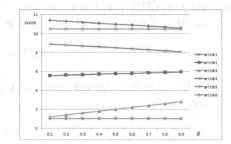

Fig. 2. Relevance between the threshold and the *relatedness* score

Fig. 3. Relevance between the threshold and the *diversity* score

5.2 Experiments on Ranking Based on Four Criteria

We evaluated the ranking based on each criterion. We calculated nDCG for the top 3 results. We varied each parameter (α (Formula 8), β (Formula 12), w_f (Formula 13) and γ (Section 4.3)) to compare the nDCG values for each parameter. For the parameter of *difference of polarity* w_f, we set $w_{Core} = 1 - w$ and varied w (Formula 14). For the parameter of *difference of detailedness* w_f (Formula 17), the only named entities for which we could obtain enough descriptions to acquire topic sets were core entities. Therefore, we defined $w_f = w_c = 1$ (Formula 18) in this experiment.

We calculated the average nDCG for each event. Tables 6-9 show nDCG for the ranking of *relatedness, diversity, difference of polarity* and *difference of detailedness*, respectively. Figures 2 and 3 show the relationships between the threshold and the scores for *relatedness* in event 3 and *diversity* in event 5, respectively. In each figure, the vertical axis shows the score for each criteria and the horizontal axis shows the parameter value.

From the experiment results, we found the following.

Table 5. nDCG in rankings of *relatedness*

event1	event2	event3	event4	event5	average
0.775	0.855	0.587	1.00	1.00	0.843

Table 6. nDCG in rankings of *diversity*

event1	event2	event3	event4	event5	average
0.811	0.660	0.710	0.899	0.740	0.764

As shown in Figure 2, *relatedness* did not greatly change and the order of the rankings did not change for $0.1 \leq \alpha \leq 0.9$. Therefore, in this experiment the effect of weight α was minor. The average nDCG for the top three rankings was 0.843, which is a high score.

As shown in Figure 3, *diversity* also did not change very much and order of the rankings again did not change for $0.1 \leq \beta \leq 0.9$. Therefore, in this experiment the effect of weight β was also small. The average nDCG for the top three rankings was 0.764, which again is a high score. There are some reasons for there being no changes in the rankings for either *relatedness* or *diversity*.

For the ranking based on *relatedness*, one of the main reasons is that some articles used in our experiments included all the core entities and the other articles did not include any core entities. The articles including all the core entities had the same scores for *relatedness* to events. Therefore, the *relatedness* to events did not affect the order of the rankings in these cases. For the ranking based on *diversity*, a similar cause is a primary reason.

In the experiment for the *difference of polarity*, the highest evaluation value was 0.335 and $w = 0.8$ in that case. The greater w became, meaning as the effect of the polarity of named entities other than core entities increased, the greater the evaluation value became. This was due to the following.

Table 7. nDCG in rankings of *difference of polarity*

parameter w	event1	event2	event3	event4	event5	average
0.1	0.274	0.484	0.174	0.262	0.362	0.311
0.2	0.274	0.484	0.174	0.262	0.362	0.311
0.3	0.350	0.484	0.174	0.262	0.362	0.326
0.4	0.350	0.484	0.174	0.262	0.362	0.326
0.5	0.350	0.484	0.174	0.214	0.362	0.317
0.6	0.350	0.484	0.087	0.214	0.362	0.299
0.7	0.350	0.484	0.087	0.214	0.362	0.299
0.8	0.439	0.484	0.087	0.214	0.449	0.335
0.9	0.439	0.484	0.087	0.214	0.449	0.335

Table 8. Evaluation for the *difference of detailedness*

thresholdγ	event1	event2	event3	event4	event5	average
0.1	0.955	0.955	0.738	0.758	0.803	0.842
0.2	0.955	0.955	0.738	0.758	0.955	0.872
0.3	0.955	0.955	0.903	0.758	0.955	0.905
0.4	0.803	0.955	0.738	0.758	0.955	0.842
0.5	0.955	0.834	0.738	0.758	0.955	0.848
0.6	0.955	0.834	0.903	0.911	0.955	0.912
0.7	0.803	0.924	0.903	0.955	0.682	0.853
0.8	0.955	0.924	1.000	0.955	0.682	0.903
0.9	0.955	0.924	0.738	0.879	0.911	0.881

Relatedness **to the event**

Users do not require articles where the event is different from that in the original article. Therefore, these articles were ranked lowly in the subjects' manual rankings. However, in this experiment on the *difference of polarity*, articles 6 and 7 were highly rank even though they were not related to the event covered by the original article. Therefore, the accuracy became low.

Relationship of Entities

We define the *difference of polarity* in terms of the polarity of named entities, but we should also consider the relationship between named entities. For example, when an emotional word is a verb, both subjects and objects are given a correspondingly positive or negative score. However, in some cases only the subject or the object should be so scored. For example, in the sentence "Kane blamed Mary", only the subject "Mary" should be given a negative score based on "blamed".

In the experiment on the *difference of detailedness*, the highest evaluation value was 0.912 and $\gamma = 0.8$ in that case. This value is acceptably high.

5.3   Experiment on Combined Ranking Methods

Articles not related to the event covered by the original article should be ranked lowly because users do not pay attention to these articles. Therefore, we considered rankings when the criteria were used in an integrated manner. In this experiment, we multiplied *relatedness* by each of the three other measures to obtain three integrated ranking scores, respectively. Tables 10-12 show nDCG for the rankings of the top three results.

The highest average nDCG obtained for a combination of *diversity* and *relatedness* was 0.925, a 16.1% improvement compared to ranking based only on *diversity*. The highest average nDCG for a combination of *difference of polarity* and *relatedness* was 0.746, a 41.1% improvement compared to *difference of polarity* alone. The highest average nDCG for a combination of *difference of detailedness* and *relatedness* was 0.663, a worse compared to *difference of detailedness* alone. From these results, we can say that *relatedness* and *difference*

Table 9. Evaluation for *diversity* multiplied by *relatedness*

weightβ	event1	event2	event3	event4	event5	average
0.1	0.982	0.901	0.567	0.655	0.720	0.765
0.2	0.982	0.901	0.567	0.655	0.720	0.765
0.3	1.00	1.00	0.758	0.955	0.911	0.925
0.4	1.00	1.00	0.758	0.955	0.911	0.925
0.5	1.00	1.00	0.758	0.955	0.911	0.925
0.6	1.00	1.00	0.758	0.955	0.911	0.925
0.7	1.00	1.00	0.758	0.955	0.911	0.925
0.8	1.00	1.00	0.758	0.955	0.911	0.925
0.9	1.00	1.00	0.758	0.955	0.911	0.925

Table 10. Evaluation for *difference of polarity* multiplied by *relatedness*

weight$w$	event1	event2	event3	event4	event5	average
0.1	0.669	0.637	0.536	0.593	0.913	0.669
0.2	0.669	0.682	0.775	0.593	1.000	0.744
0.3	0.669	0.682	0.674	0.593	1.000	0.723
0.4	0.669	0.682	0.674	0.593	1.000	0.723
0.5	0.592	0.682	0.674	0.593	1.000	0.708
0.6	0.592	0.682	0.674	0.738	1.000	0.737
0.7	0.592	0.682	0.674	0.738	1.000	0.737
0.8	0.637	0.682	0.674	0.738	1.000	0.746
0.9	0.637	0.682	0.674	0.738	1.000	0.746

Table 11. Evaluation for *difference of detailedness* multiplied by *relatedness*

weight$w$	event1	event2	event3	event4	event5	average
0.1	0.726	0.153	0.738	0.605	0.484	0.541
0.2	0.726	0.153	0.738	0.605	0.484	0.541
0.3	0.726	0.153	0.855	0.605	0.305	0.529
0.4	0.726	0.153	0.903	0.605	0.726	0.623
0.5	0.955	0.153	0.903	0.758	0.242	0.602
0.6	0.955	0.305	0.903	0.911	0.242	0.663
0.7	0.803	0.305	0.855	0.911	0.153	0.605
0.8	0.803	0.305	1.000	0.726	0.153	0.597
0.9	0.803	0.305	0.738	0.726	0.242	0.563

of detailedness can be used alone for news ranking, but *diversity* and *difference of polarity* should be used in combination with *relatedness*.

6 Conclusion

Based on a user survey, criteria were determined for estimating differences between related news articles by analyzing the descriptions of named entities. Experimental results show that *relatedness* and detailedness are two criteria that can be used alone for ranking related news articles, but *diversity* and *difference of polarity* should be combined with *relatedness*.

Our future work can be summarized as follows: Currently, we used only six articles about each event in the experiments. A large-scale experiment is planned. The best combination of ranking criteria is another issue for future research. We also plan to develop a prototype system to support users' understanding of the news.

Acknowledgement. This work is partly supported by KAKENHI(No.25700033) and SCAT Reseach Funding.

References

1. Ogawa, T., Ma, Q., Yoshikawa, M.: News Bias Analysis Based on Stakeholder Mining. IEICE Transactions 94–D, 578–586 (2011)
2. Xu, L., Ma, Q., Yoshikawa, M.: A Cross-Media Method of Stakeholder Extraction for News Contents Analysis. In: Chen, L., Tang, C., Yang, J., Gao, Y. (eds.) WAIM 2010. LNCS, vol. 6184, pp. 232–237. Springer, Heidelberg (2010)
3. Park, S., Kang, S., Song, S.C.J.: Newscube: delivering multiple aspects of news to mitigate media bias. In: Proceedings of the 27th International Conference on Humanfactors in Computing Systems, pp. 443–452 (2009)
4. Nadamoto, A., Tanaka, K.: A Comparative Web Browser (CWB) for Browsing and Comparing Web Pages. In: Proceedings of the 12th International Conference on World Wide Web, pp. 727–735. ACM (2003)
5. Ma, Q., Yoshikawa, M.: Topic and viewpoint extraction for diversity and bias analysis of news contents. In: Li, Q., Feng, L., Pei, J., Wang, S.X., Zhou, X., Zhu, Q.-M. (eds.) APWeb/WAIM 2009. LNCS, vol. 5446, pp. 150–161. Springer, Heidelberg (2009)
6. Ishida, S., Ma, Q., Yoshikawa, M.: Analysis of news agencies' descriptive features of people and organizations. In: Bhowmick, S.S., Küng, J., Wagner, R. (eds.) DEXA 2009. LNCS, vol. 5690, pp. 745–752. Springer, Heidelberg (2009)
7. De Marneffe, M.-C., Manning, C.D.: StanfordCoreNLP (2008), http://nlp.stanford.edu/software/dependenciesmanual.pdf
8. Esuli, A., Sebastiani, F.: Sentiwordnet: A publicly available lexical resource for opinion mining. In: Proceeding of the 5th Conference on Language Resources and Evaluation, pp. 417–422 (2006)
9. Blei, D.M., Ng, A.Y., Jordan, M.I.: Latent Dirichlet Allocation. The Journal of Machine Learning Research 3, 993–1022 (2003)
10. Jarvelin, K., Kekalainen, J.: Cumulated Gain-based Evaluation of IR Techniques. ACM Trans. Inf. Syst. 20, 422–446 (2002)
11. Ramage, D.: Stanford topic modeling toolbox 0.4 (2013), http://nlp.stanford.edu/software/tmt

Semantic Path Ranking Scheme
for Relational Keyword Queries

Zhong Zeng[1], Zhifeng Bao[2], Gillian Dobbie[2],
Mong Li Lee[1], and Tok Wang Ling[1]

[1] National University of Singapore
[2] HITLab Australia & University of Tasmania
[3] University of Auckland
{zengzh,leeml,lingtw}@comp.nus.edu.sg, zhifeng.bao@utas.edu.au,
g.dobbie@auckland.ac.nz

Abstract. Existing works on keyword search over relational databases typically do not consider users' search intention for a query and return many answers which often overwhelm users. We observe that a database is in fact a repository of real world objects that interact with each other via relationships. In this work, we identify four types of semantic paths between objects and design an algorithm called *pathRank* to compute and rank the results of keyword queries. The answers are grouped by the types of semantic paths which reflect different query interpretations, and are annotated to facilitate user understanding.

1 Introduction

Keyword search over relational databases enables users to pose queries without learning the query languages or database schemas, and has become a popular approach to access database information [1,5,3,8,9,2,6,7,4,11]. Existing works use a data graph where each node denotes a tuple and each undirected edge denotes a foreign key-key reference [6,7,11]. An answer to a keyword query is a minimal connected subgraph of tuples which contains nodes that match keywords in the query. Since the keywords in a query may match nodes which are connected by many paths in the data graph, many answers are returned, with possibly complex subgraphs whose meanings are not easy to understand.

One approach to address the above problem is to rank the query answers. The methods range from simple heuristic rules such as ranking the answers based on their sizes [5], to using the TF-IDF model [3,8,9], and the Random Walk model [6,7,11]. Another approach is to organize query answers into clusters so that users can explore the relevant clusters first [10]. However, none of these approaches consider the semantics of the answers.

We observe that a relational database is, in fact, a repository of real world objects that interact with each other via relationships. When a user searches for some target object, s/he is interested in objects that are related in some way to the target object. In this work, we utilize the compact Object-Relationship-Mixed (ORM) data graph [12] to model tuples in a relational database, and

H. Decker et al. (Eds.): DEXA 2014, Part II, LNCS 8645, pp. 97–105, 2014.

identify four types of semantic paths where a pair of nodes in the graph can be connected. These semantic paths form different interpretations of the query answers. Based on these paths, we develop an algorithm to compute and rank the answers of keyword queries. We group the query answers by the types of semantic paths to reflect different query interpretations, and annotate each answer to facilitate user understanding. Experimental results demonstrate that our semantic path-based approach is able to rank answers that are close to users' information needs higher compared to existing ranking methods.

2   Motivating Example

Fig. 2 shows the ER diagram of the student registration database in Fig. 1. Based on the ER model, we see that the database comprises of **Student**, **Course**, **Lecturer** and **Department** objects that interact with each other via the relationships **Enrol**, **Teach**, **PreReq** and **AffiliateTo**. Suppose a user issues the query Q = {Java DB}. The keywords match two tuples in the Course relation, i.e., $<cs421, DB, l1>$ and $<cs203, Java, l1>$, corresponding to course objects with identifier $cs421$ and $cs203$. Based on the ER diagram, these two objects are related via the relationships PreReq, Enrol and Teach as follows:

a. Pre-requisite of a course (Pre-Req), e.g., $cs203$ is a pre-requisite of $cs421$.
b. Students who are enrolled in both courses (Enrol), e.g., student $s1$ (**Mary Smith**) is enrolled in both courses $<cs421, DB, l1>$ and $<cs203, Java, l1>$.
c. Lecturers who teach both **Java** and **DB** (Teach), e.g., lecturer $l1$ (**Steven Lee**) teaches both courses $<cs421, DB, l1>$ and $<cs203, Java, l1>$.

Each of these relationships suggest objects that interest the user. We can annotate the answers by the relationships as shown in Fig. 3. We will explain the these annotations in the next Section.

Student

SID	Name
s1	Mary Smith
s2	John Depp

Enrol

TupleID	SID	Code	Grade
e1	s1	cs421	B
e2	s1	cs203	A
e3	s2	cs203	B

Lecturer

StaffID	Name
l1	Steven Lee

Department

DeptID	Name
d1	CS
d2	IS

Course

Code	Title	StaffID
cs421	DB	l1
cs203	Java	l1

PreReq

TupleID	Code	PreqCode
p1	cs421	cs203

AffiliateTo

TupleID	StaffID	DeptID
a1	l1	d1
a2	l1	d2

Fig. 1. Example student registration database

3   Preliminaries

The work in [12] utilizes database schema constraints to classify relations into object relation, relationship relation, component relation and mixed relation. An object (relationship) relation captures single-valued attributes of an object (relationship). Multivalued attributes of an object (relationship) are captured in

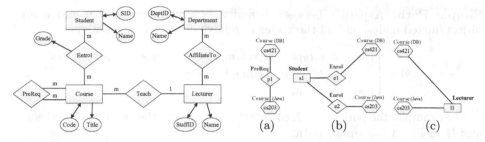

Fig. 2. ER diagram of Fig. 1

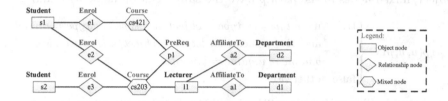

Fig. 3. Annotated answers with node types for Q = {Java DB}

Fig. 4. ORM data graph of the database in Fig. 1

the component relations. A mixed relation contains information of both objects and relationships, which occurs when we have a many-to-one relationship.

The *Object-Relationship-Mixed (ORM) data graph* [12] is an undirected graph $G(V, E)$. Each node $v \in V$ has an *id*, a *type* $\in \{object, relationship, mixed\}$, and a relation name *name*. A node also has a set *tupleIDs* containing the ids of tuples from an object (or relationship, or mixed) relation together with tuples from its associated component relations. Fig. 4 shows the ORM data graph of Fig. 1 comprising object nodes (rectangles), relationship nodes (diamonds) and mixed nodes (hexagons). Each node has an id and a relation name, e.g., the mixed node with id $cs421$ occurs in relation Course.

Given a keyword query $Q = \{k_1, k_2, \cdots, k_n\}$, where $k_i, i \in [1, n]$ denotes a keyword, we say k *matches* a node v in the ORM data graph if k occurs in some tuple in v. We call v a matched node for k.

We now analyze the various ways a pair of object/mixed nodes can be connected in the graph. Let u and v be two nodes in the ORM data graph G, and P be the set of paths between u and v. Each path $p \in P$ is a sequence of connected nodes $<u, \cdots, v>$. The length of a path p is given by the number of nodes in p, denoted by $|p|$. We can form a new path $p = <v_a, \cdots, v_b>$ by joining two paths $p_1 = <v_a, \cdots, v_c>$ and $p_2 = <v_c, \cdots, v_b>$ over a common node v_c; we say that p can be decomposed into sub-paths p_1 and p_2. We call the paths between two object/mixed nodes *semantic paths* since they capture the semantics of objects and relationships. These paths can be classified into one of the following types:

Simple Path. A path p between u and v is a simple path if u and v are object/mixed nodes, and all the nodes in p have distinct relation names.

$$sp(u,v,p) = \begin{cases} \text{true} & \text{if } u.type = object/mixed \text{ and } v.type = object/mixed \\ & \text{and } \forall b \in p, \ b.name \text{ is distinct} \\ \text{false} & \text{otherwise} \end{cases}$$

For example, the path $p=<s2,e3,cs203,l1>$ between the two object nodes $s2$ and $l1$ in Fig. 4 is a simple path.

Recursive Path. A path p between u and v is a recursive path if u and v are both object nodes or mixed nodes and have the same relation name, and all the object/mixed nodes in the path p have the same relation name as u and v.

$$rp(u,v,p) = \begin{cases} \text{true} & \text{if } (u.type = v.type = object \text{ or } u.type = v.type = mixed) \\ & \text{and } \forall b \in p \text{ such that } b.type = object/mixed, \text{we have} \\ & b.name = u.name = v.name \\ \text{false} & \text{otherwise} \end{cases}$$

For example, the path $p=<cs421,p1,cs203>$ between mixed nodes $cs421$ and $cs203$ in Fig. 4 is a recursive path.

Palindrome Path. A path p between u and v is a palindrome path if both u and v have the same relation name, and we can find some object/mixed node $c \in p$ such that the nodes in the paths from c to u, and c to v have the same sequence of relation names.

$$pp(u,v,p) = \begin{cases} \text{true} & \text{if } u.name = v.name \text{ and } \exists \text{ object/mixed node } c \in p \\ & \text{s.t. } p \text{ can be decomposed into 2 sub-paths} \\ & p_1 =< u,b_1,\cdots,b_j,c >, p_2 =< c,b'_j,\cdots,b'_1,v > \text{ where} \\ & \text{both } p_1,p_2 \text{ are simple paths, and } b_i.name = b'_i.name \\ & \forall b_i \in p_1, b'_i \in p_2, 1 \le i \le j \\ \text{false} & \text{otherwise} \end{cases}$$

For example, the path $p=<cs203,e2,s1,e1,cs421>$ between the mixed nodes $cs421$ and $cs203$ in Fig. 4 is a palindrome path as it can be decomposed into two simple sub-paths $p_1=<cs203,e2,s1>$ and $p_2=<s1,e1,cs421>$.

Complex Path. Any path that does not satisfy the conditions for the above three semantic path types is a complex path. A complex path is essentially a combination of simple paths and recursive paths, and has a path length $|p| \ge 3$.

The path $p=<s2,e3,cs203,p1,cs421,l1>$ in Fig. 4 is a complex path as it is a combination of two simple paths and one recursive path.

4   Proposed Ranking Scheme

In this section, we describe our method called *pathRank* to compute and rank keyword query answers. We first generate Steiner trees that contain all the query keywords. Then we augment the relationship matched nodes in the Steiner trees with their associated object and mixed nodes. Finally we rank the Steiner trees based on type of semantic paths they contain. We give the highest score to simple and palindrome paths because they are more intuitive and informative. Complex paths have the lowest scores since they require more user effort to understand.

Let $Obj(k)$ and $Rel(k)$ be the sets of object and relationship nodes that match keyword k in the ORM data graph. Note that if k matches the object part of a mixed node u, then we add u to $Obj(k)$. Otherwise, if k matches the relationship part of u, we add u to $Rel(k)$.

Given two nodes u and v that match keyword k_i and k_j respectively in a Steiner tree T, we have the following cases:

a. Both $u \in Obj(k_i)$ and $v \in Obj(k_j)$. We determine the type of the path between u and v as described in Section 3.
b. Either $u \in Rel(k_i)$ or $v \in Rel(k_j)$. Without loss of generality, suppose $u \in Rel(k_i)$ and $v \in Obj(k_j)$. Let S_u be the set of object/mixed nodes that are directly connected to u, and p' be the path between v and some node $s \in S_u$ that has the highest score. Then the type of the semantic path between u and v is given by the type of path p'.
c. Both $u \in Rel(k_i)$ and $v \in Rel(k_j)$. Let S_u and S_v be the sets of object/mixed nodes that are directly connected to u and v respectively. Let p' be the path between $s \in S_u$ and $t \in S_v$ that has the highest score. Then the type of the semantic path between u and v is given by the type of path p'.

Let V be the set of matched nodes in a Steiner tree T and $C_2^{|V|}$ be the number of node pairs in V. The score of T w.r.t keyword query Q is defined as follows:

$$score(T, Q) = \begin{cases} \frac{\sum_{u,v \in V, u.id < v.id} pathscore(u,v,p)}{num(u,v,p) * C_2^{|V|}} & |V| > 1 \\ 1 & |V| = 1 \end{cases}$$

where $num(u, v, p)$ is the number of object/mixed nodes in the path p between nodes u and v in V, and $pathscore(u, v, p)$ is the score of the path p. Note that our proposed ranking scheme considers the semantic paths between matched nodes as well as the number of participating objects in the Steiner tree.

Algorithm 1 shows the details of *pathRank*. We first classify the matched nodes for each keyword and generate the Steiner trees (Lines 3-5). For each tree T, we check every matched node v for keyword k. If $v \in Rel(k)$, i.e., the keyword matches a relationship node or the relationship part of a mixed node, then we add the object/mixed nodes that are directly connected to v in the ORM data graph and the associated edges into T (Lines 6-11). Next, we determine the score of the path between matched nodes u and v in T (Lines 12-31).

5   Performance Study

We implement the algorithms in Java, and carry out experiments on an Intel Core i7 3.4 GHz with 8GB RAM. We use a subset of the real world ACM Digital

---

Algorithm 1: *pathRank*

Input: keyword query $Q = \{k_1, ...k_n\}$, $maxSize$, ORM data graph G

Output: answer set *Answer*

1   $Answer \leftarrow \emptyset$;
2   **for** $i = 1$ *to* n **do**
3   Let $Obj(k_i)$ be the set of object/mixed nodes in G that match k_i;
4   Let $Rel(k_i)$ be the set of relationship/mixed nodes in G that match k_i;
5   $Answer = generateSteinerTree(Q, G, maxSize)$;
6   **foreach** *Steiner Tree* $T \in Answer$ **do**
7   Let V be the set of matched nodes in T;
8   **foreach** $v \in V$ **do**
9   **if** $v \in Rel(k)$ **then**
10   add object/mixed nodes that are directly connected to v in G into T;
11   add the associated edges in G into T;
12   **foreach** $u, v \in V$ **do**
13         $S_u \leftarrow \emptyset$; $S_v \leftarrow \emptyset$;
14   **if** $u \in Obj(k_i)$ **then**
15   add u into S_u;
16   **else if** $u \in Rel(k_i)$ **then**
17   add object/mixed nodes that are directly connected to u in T into S_u;
18   **if** $v \in Obj(k_j)$ **then**
19   add v into S_v;
20   **else if** $v \in Rel(k_j)$ **then**
21   add object/mixed nodes that are directly connected to v in T into S_v;
22         $score = 0$;
23   **foreach** $s \in S_u, t \in S_v$ **do**
24            $z = pathscore(s, t, p)$;
25   **if** $score < z$ **then**
26               $score = z$;
27         $pathscore(u, v, p) = score$;
28   Let num be the number of object/mixed nodes between u and v;
29         $T.score \mathrel{+}= pathscore(u, v, p) * num$;
30      $T.score = T.score/(|V| * (|V| - 1)/2)$;
31   **return** $Sort(Answer)$;

Library publication dataset from 1995 to 2006. There are 65,982 publications and 106,590 citations. Fig. 6 shows the ER diagram for this dataset.

We compare our semantic path ranking method (`Path`) with the following ranking schemes used in state-of-the-art relational keyword search such as Discover [5], BANKS [6] and SPARK [9]:

a. Number of nodes in answer (`Size`) [5].
b. Node prestige and proximity (`Prestige`) [6].
c. TF-IDF similarity between query and answer (`Tf-idf`) [3,9].

Fig. 7 shows the keyword queries used in our experiments. We show these queries together with the ER diagram of the database to 10 users and obtain their possible search intentions. For each search intention, we generate the SQL statements to retrieve the results from the database to form the ground truth.

Fig. 5 shows the average precision of four ranking methods for the queries in Table 7 when we vary k. We observe that `Path` is able to achieve a higher average precision compared to `Size`, `Prestige` and `Tf-idf` for most of the queries. In fact, the superiority of `Path` increases significantly as k decreases. All the ranking schemes are able to retrieve the relevant answers when k is equal to 50. However, `Size`, `Prestige` and `Tf-idf` start to miss relevant answers as k decreases. Note that for query $Q9$ and $Q10$, all the ranking schemes achieve an average precision

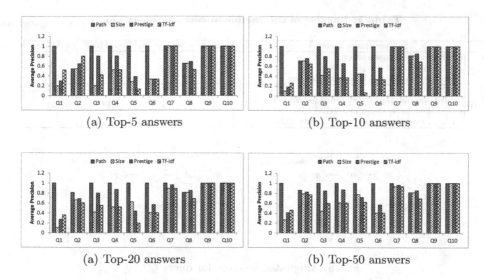

(a) Top-5 answers   (b) Top-10 answers

(a) Top-20 answers   (b) Top-50 answers

Fig. 5. Average precision of ranking schemes for top-k answers

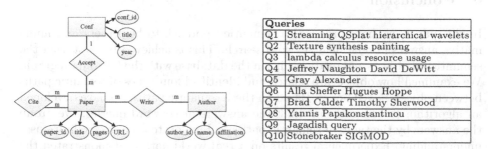

Queries	
Q1	Streaming QSplat hierarchical wavelets
Q2	Texture synthesis painting
Q3	lambda calculus resource usage
Q4	Jeffrey Naughton David DeWitt
Q5	Gray Alexander
Q6	Alla Sheffer Hugues Hoppe
Q7	Brad Calder Timothy Sherwood
Q8	Yannis Papakonstantinou
Q9	Jagadish query
Q10	Stonebraker SIGMOD

Fig. 6. ER diagram of the ACM Digital Library dataset

Fig. 7. Queries used

of 1 because their search intentions are straightforward, and the relevant answers are ranked on the top by all the ranking schemes. Thus, we can see that `Path` is more effective than the other ranking schemes when the queries are ambiguous.

Table 1 shows two sample answers for query $Q3$ and their rankings by the different schemes. The keywords in this query match two paper titles. The first answer indicates that these two papers are related via 2 Paper-Cite relationships, while the second answer indicates that these two papers are related via a third paper published in the same conference and cites one of these papers. We observe that although the second answer is complex and not easy to understand, it is ranked higher by `Size` and `Prestige`. `Tf-idf` gives the similarly ranks to these two answers without considering the type of the semantic paths. In contrast, `Path` ranks the first answer much higher because it contains a recursive path while the second answer contains a complex path. Fig. 8 shows the corresponding annotated answers output by our approach to facilitate user understanding.

Table 1. Ranking of two sample answers for query $Q3$

	Sample query answers	Size[5]	Prestige[6]	Tf-idf[3,9]	Path[Ours]
(a)	p1: Resource usage analysis p2: Once upon a type p3: A call-by-need lambda calculus	9	7	10	4
(b)	c1: POPL'04 p4: Channel dependent types for higher-order... p1: Resource usage analysis p5: ... evaluation for typed lambda calculus...	2	5	11	21

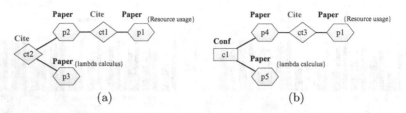

(a)   (b)

Fig. 8. Annotated answers for query $Q3$

6   Conclusion

In this paper, we have proposed a semantic approach to help users find informative answers in relational keyword search. This is achieved by capturing the semantics of objects and relationships in the database with the ORM data graph. We examined how objects are related and identified four types of semantic paths between object/mixed nodes. Based on these semantic paths, we have developed an algorithm to compute and rank the answers of keyword queries. We group the answers by the types of semantic paths and annotate them to facilitate user understanding. Experimental results on a real world dataset demonstrated the effectiveness of our path-based approach.

References

1. Agrawal, S., Chaudhuri, S., Das, G.: DBXplorer: A system for keyword-based search over relational databases. In: ICDE (2002)
2. Bergamaschi, S., Domnori, E., Guerra, F., Trillo Lado, R., Velegrakis, Y.: Keyword search over relational databases: a metadata approach. In: SIGMOD (2011)
3. Hristidis, V., Gravano, L., Papakonstantinou, Y.: Efficient IR-style keyword search over relational databases. In: VLDB (2003)
4. Hristidis, V., Hwang, H., Papakonstantinou, Y.: Authority-based keyword search in databases. ACM Trans. Database Syst. (2008)
5. Hristidis, V., Papakonstantinou, Y.: Discover: keyword search in relational databases. In: VLDB (2002)
6. Hulgeri, A., Nakhe, C.: Keyword searching and browsing in databases using BANKS. In: ICDE (2002)
7. Kacholia, V., Pandit, S., Chakrabarti, S.: Bidirectional expansion for keyword search on graph databases. In: VLDB (2005)

8. Liu, F., Yu, C., Meng, W., Chowdhury, A.: Effective keyword search in relational databases. In: SIGMOD (2006)
9. Luo, Y., Lin, X., Wang, W., Zhou, X.: Spark: top-k keyword query in relational databases. In: SIGMOD (2007)
10. Peng, Z., Zhang, J., Wang, S., Qin, L.: TreeCluster: Clustering results of keyword search over databases. In: Yu, J.X., Kitsuregawa, M., Leong, H.-V. (eds.) WAIM 2006. LNCS, vol. 4016, pp. 385–396. Springer, Heidelberg (2006)
11. Yu, X., Shi, H.: CI-Rank: Ranking keyword search results based on collective importance. In: ICDE (2012)
12. Zeng, Z., Bao, Z., Lee, M.L., Ling, T.W.: A semantic approach to keyword search over relational databases. In: Ng, W., Storey, V.C., Trujillo, J.C. (eds.) ER 2013. LNCS, vol. 8217, pp. 241–254. Springer, Heidelberg (2013)

Fraud Indicators Applied to Legal Entities: An Empirical Ranking Approach

Susan van den Braak[1], Mortaza S. Bargh[1,2], and Sunil Choenni[1,2]

[1] Research and Documentation Centre, Ministry of Security and Justice,
The Hague, The Netherlands
{s.w.van.den.braak,m.shoae.bargh,r.choenni}@minvenj.nl
[2] Creating 010, Rotterdam University of Applied Sciences,
Rotterdam, The Netherlands
{m.shoae.bargh,r.choenni}@hr.nl

Abstract. Legal persons (i.e., entities such as corporations, companies, partnerships, firms, associations, and foundations) may commit financial crimes or employ fraudulent activities like money laundering, tax fraud, or bankruptcy fraud. Therefore, in the Netherlands legal persons are automatically screened for misuse based on a set of so called risk indicators. These indicators, which are based on the data obtained from, among others, the Netherlands Chamber of Commerce, the Dutch police, and the Dutch tax authority, encompass information about certain suspicious behaviours and past insolvencies or convictions (criminal records). In order to determine whether there is an increased risk of fraud, we have devised a number of scoring functions to give a legal person a score on each risk indicator based on the registered information about the legal person and its representatives. These individual scores are subsequently combined and weighed into a total risk score that indicates whether a legal person is likely to commit fraud based on all risk indicators. This contribution reports on our two ranking approaches: one based on the empirical probabilities of the indicators and the other based on the information entropy rate of the empirical probabilities.

1 Introduction

A legal person allows one or more natural persons to act as a single entity for legal purposes. Examples of legal persons are corporations, companies, partnerships, firms, associations, and foundations. Each legal person must designate one or more representatives, such as a director, partner or manager, who is authorised by its articles of association or by power of attorney. Legal persons can in turn be represented by natural persons or other legal persons. This means that the network of persons related to a legal person may be several levels deep and may consist of relations between legal persons and natural persons (called 'LP-NP relations') or of relations amongst legal persons (called 'LP-LP relations').

Just like natural persons, legal persons may commit financial crimes or employ fraudulent activities such as money laundering, tax fraud, or bankruptcy fraud.

H. Decker et al. (Eds.): DEXA 2014, Part II, LNCS 8645, pp. 106–115, 2014.
© Springer International Publishing Switzerland 2014

Therefore, in the Netherlands legal persons are continuously being screened for misuse. In this automated process, based on a set of so-called *risk indicators*, it is determined whether a legal person is likely to commit fraud. To do so, legal persons and all persons holding key positions in the legal person are checked for certain suspicious behaviours and past insolvencies (bankruptcy) or convictions (e.g., criminal records). This analysis is based on the information obtained from the Netherlands Chamber of Commerce, the Dutch police, and the Dutch tax authority, among others. Most defined risk indicators refer to a legal person's network of representatives, the structure of this network, and the changes in it. For instance, an indicator may refer to the number of position changes in the previous year, the number of positions that the representatives hold simultaneously, and the number of positions with a short duration. Additionally there are indicators that take into account the criminal and financial past of the legal persons and their representatives.

In order to determine whether there is an increased risk of fraud, based on the registered information about the legal person and its representatives, a legal person is given a score on each risk indicator. These individual scores are combined (and weighed) into a total risk score, where a higher score means a high probability of fraudulent acts. Assigning scores to legal persons is not straightforward. Therefore, the Research and Documentation Centre of the Dutch Ministry of Security and Justice developed and tested several scoring functions based on a few rules of thumb from probability theory, information theory and the domain knowledge. This contribution reports on how the domain knowledge is used to devise two ranking functions based the empirical probabilities of the indicators. The first one is a linear function of the empirical probabilities and the other one is a logarithmic function (i.e., the information entropy rate) of the empirical probabilities. In order to evaluate whether these functions showed the desired behaviour they were applied to a test database [1]. This test database contained information on a large number of natural persons from the Chamber of Commerce that have been a representative of one or more legal persons in the Netherlands.

Constructing risk profiles to detect possible suspicious behaviours, such as fraud, is not new. Often profiles are generated based on statistical techniques such as data mining or logistic regression. However, in the case of fraud committed by legal persons there are very few examples of fraud, while there are many persons that are not suspicious. Additionally, due to privacy restrictions, we did not have any information about those persons in the datasets that were actually convicted of fraud. As a result, we did not have any positive examples of fraud. Having few examples of fraud and lack of actual positive examples of fraud led us to base the scoring functions on simple rules of thumb and domain knowledge. We were unable to use data mining or other statistical techniques to find the best possible functions. Moreover, we were also unable to test the performance of the our scoring functions with real fraudulent legal entities (i.e., to check whether there are not too many false positives or false negatives).

At the time that the bill on automatically screening legal persons for fraud became a law in the Netherlands [2], a set of risk indicators with their possible values had already been identified. However it was not yet determined how these could be translated into scores on individuals indicators and how they could be combined into a total risk score. Therefore, this contribution aims at addressing the following research questions: how can an appropriate score be defined for every risk indicator based on the available attribute(s)? How can the risk indicators be combined to obtain an overall risk indicator? In calculating these scores one needs to take into account the dependancy between the indicators and sometimes possible lack of information for some indicators.

2 Related Work

Fraud indicators are warning signals to predict/detect possible frauds so that some preventive actions can be taken against them. In [3] four types of fraud are investigated and their indicators are identified. In our study we use some indicators of just one type of fraud and try to combine their information as a total risk score of the fraud.

As mentioned already, constructing risk profiles to detect possible suspicious behaviours such as fraud is not new. These profiling approaches are generally based on statistical techniques (see [4]) or data mining techniques (see [5] and [6]). For example [4] presents a number of examples of risk profiles for social security fraud in the Netherlands using statistical analysis; while [5] studies a case to support planning of audit strategies in the fiscal and insurance domains, and [6] searches for the profiles of risky flights. In our case, i.e., fraud committed by legal persons, the number of unsuspicious legal persons was too high and we did not know which legal persons were fraudulent. Therefore, unlike the papers mentioned in this paragraph, we were unable to use data mining or other statistical techniques to find the best possible functions. Consequently, we used domain knowledge and simple rules of thumb to define our risk indicators.

In [7] a deterministic approach is proposed to combine information from different types (i.e., the way that the paper puts it: "combining apples and pears"). In this contribution, however, we use probabilistic approaches due to their strength to model data features in a large amount of complex and unlabelled data [8]. Probabilistic generative models [9] are used in [8] to propose some risk scoring schemes in order to rank the malicious behaviours of mobile apps. Similarly to [8], we present two scoring schemes that are inversely related to the empirical probabilities (in our study, those of legal persons being suspicious of fraud). The probabilistic generative model behind our scoring schemes is similar to the BNB model of [8], although our scoring schemes differ from those of [8]. Moreover, unlike [8] that compares the above mentioned probabilistic generative models, this paper investigates scoring schemes given a specific probabilistic generative model.

3   Scores of a Risk Indicator

There are $|I|$ risk indicators (or indicators in short): $I_1, I_2, \ldots, I_{|I|}$. Every indicator I takes a discrete value $m \in \{0, 1, \ldots, M\}$. Indicator I is observed N times, of which symbol m is observed N_m times and $\sum_{m=0}^{M} N_m = N$. We consider indicator I as an Independent and Identically Distributed (IID) random variable with an empirical Probability Density Function (PDF) of

$$p_m = Pr(I = m) = \frac{N_m}{N}, \qquad m \in \{0, 1, \ldots, M\}. \tag{1}$$

Investigating the available data closely we observe that the number of observations per indicator is a decreasing function of symbol $m \in \{0, 1, \ldots, M\}$. In other words, $N_{m-1} \geq N_m$, where $m \in \{1, \ldots, M\}$. As a result the following "monotonically decreasing property" holds:

$$p_0 \geq \cdots \geq p_m \geq \cdots \geq p_M. \tag{2}$$

In [8] three desiderata are proposed for a risk scoring function, namely: being monotonic, having high rise scores for dangerous apps, and being simple. These criteria lay down interesting characteristics of such risk scoring schemes. The monotonic property implies that when one indicator does not point to a risk, the total risk score corresponding to all indicators should be lower than the total score when the indicator does point to a risk (given the same values for the other indicators). The monologic property of (2) guarantees the monotonic property of our risk scoring schemes, to be described in the following subsections.

3.1   Heuristic Approach

Based on domain knowledge we know that the outcome of an indicator can determine the degree to which the corresponding data subject is suspicious of a fraud. For example, a zero outcome, i.e., $I = 0$, means that the data subject is unsuspicious from the viewpoint of indicator I. When $I = m \in \{1, \ldots, M\}$ the data subject is suspicious of a fraud from the viewpoint of indicator I and, moreover, the degree of suspiciousness increases as m increases. We define a **heuristic risk score** S^h to associate a degree of suspiciousness with a data subject, based on the value of indicator I. Specifically, when being unsuspicious (i.e., $I = 0$) the risk score is defined as: $S^h = 0$; when being suspicious (i.e., $I = m$, $m \in \{1, \ldots, M\}$) the risk score is defined as:

$$S^h = 1 - q_m = 1 - Pr(I = m | I \neq 0) = 1 - \frac{N_m}{\sum_{i=1}^{M} N_i}, \qquad m \in \{1, 2, \ldots, M\}, \tag{3}$$

where q_m in (3) is a PDF conditional on $I \neq 0$, i.e., $q_m = Pr(I = m | I \neq 0)$. Fig. 1 illustrates a typical PDF of an indicator obtained from our datasets and the corresponding conditional PDF q_m. The underlying assumption for this heuristic risk score is that the higher number of observations of these indicators the

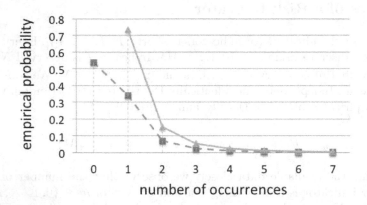

Fig. 1. PDF and conditional PDF (dashed) of an indicator, obtained from (1) and (3)

stronger indication of fraudulent behaviours. In other words, when an indicator is observed more often, there is a smaller possibility of fraud. Furthermore, the property in (2) implies that the entries of the defined PDF of q_m decrease monolithically as $m \in \{1, \ldots, M\}$ increases. Consequently, the heuristic risk score increases monolithically as m increases and has a maximum of 1.

3.2   Information Entropy Approach

Alternatively we define another risk score per indicator, which originates from Shannon entropy. Denoted by S^e, the information **entropy risk score** associates the amount of information conveyed by every outcome of an indicator with the degree to which the data subject is suspicious from the viewpoint of indicator I. In other words,

$$S^e = -\log_2 Pr(I = m) = -\log_2 p_m, \qquad m \in \{0, \ldots, M\}. \tag{4}$$

Unlike the heuristic risk score S^h, the entropy risk score S_n^e does not yield a zero value if an indicator's outcome is 0 (i.e., when $I = 0$). In this case, nevertheless, the S^h has a minimum value according to property (2). Like the heuristic risk score S^h, the entropy risk score S^e increases as m increases.

4   Combination Scores

The information from all indicators $I_1, I_2, \ldots, I_{|I|}$ should be integrated in order to provide a measure that indicates how much a data subject (i.e., legal person) is suspicious according to the indicators. This section starts with discussing the independency assumption among the indicators and then describes the way of information integration for the two scoring functions.

4.1 Independency Assumption

When risks scores are integrated, it is important to know whether or not the corresponding indicators are independent. This independency assumption was tested using both domain knowledge and statistical correlation. Using our domain knowledge, we examined the meaning of the indicators. Specifically, we investigated whether we were able to predict the value of one indicator based on the value of another indicator. We found that the predictability of indicator values was only the case for the records (i.e., individuals or organisations) that are registered on **blacklists**. These records receive a non-zero score on multiple indicators, which consist of some blacklist indicators and some other indicators referring to the reasons behind those blacklists. We note that the blacklist indicators are set to non-zero values only when the corresponding person has been convicted of fraud before (i.e., indicating the number of convictions). Therefore, the blacklist indicators overlap with and depend on the other indicators that refer to criminal records: a high score on one blacklist indicator automatically means a high score on the other indicator(s). This problem was solved by first checking whether a person is on one of the blacklists. For these persons who are on a blacklist, our scoring function is not applied and the risk that they are committing fraud is assessed manually. For those persons who are not on a blacklist, the blacklist indicators are discarded before calculating our risk scores.

Second, statistical techniques were used to determine whether certain indicators are correlated. We could not use a conventional analysis based on correlation coefficients because of the large number of zero outcomes. As an alternative, we examined the relative value of one indicator in relation to another indicator. This was done by calculating the mean value of one indicator for all possible values of another indicator. This analysis did not show any clear and definite correlations between indicators. Only some very weak one-way relations were found.

4.2 Heuristic Approach

The **total heuristic risk score** S_T^h is defined as a weighted average of individual scores S_i^h, where $i \in \{1, \ldots, |I|\}$. In other words,

$$S_T^h = \sum_{i=1}^{|I|} w_i \, S_i^h, \qquad \text{where} \sum_{i=1}^{|I|} w_i = 1. \qquad (5)$$

Using the weighted average of the risk scores of all indicators stems from the domain knowledge, based on which the smaller the number of nonzero outcomes that an indicator has the more information that the indicator provides about the data subject being fraudulent. Thus we define the weights of indicators proportional to the number of their zero outcomes $N_{i,0}$ for $i \in \{1, \ldots, |I|\}$, or

$$w_i = \frac{N_{i,0}}{\sum_{n=1}^{|I|} N_{n,0}}, \qquad i \in \{1, \ldots, |I|\}. \qquad (6)$$

From (6) we see that the larger the number of zero outcomes of an indicator the more significant the indicator is in calculating the total heuristic risk score. Moreover, in this way we allow participation of the number of zero outcomes in calculating the total risk value as we eliminated it in calculating the individual indicator S_i^h, see the definition of q_m in (3).

4.3 Information Entropy Approach

Using Shannon theory we simply add up all individual indicator scores to obtain the **total entropy risk score** for all indicators as:

$$S_T^e = \sum_{i=1}^{|I|} S_i^e. \tag{7}$$

This relation can be justified according to the definition of information entropy for a set of $|I|$ independent random variables or indicators $I_1, I_2, \ldots, I_{|I|}$ as $S_T^e = -\log_2(Pr(I_1 = m_1, I_2 = m_2, \ldots, I_{|I|} = m_{|I|}))$, resulting in (7). One can show that the average of total entropy risk score is the sum of the average of individual entropy rates, i.e., $\mathbb{E}[S_T^e] = H(I_1, I2, \ldots, I_{|I|}) = \sum_{i=1}^{|I|} H(I_i) = \sum_{i=1}^{|I|} \mathbb{E}[S_i^e]$ when its total number of observations N_i increases. Further, when all indicators point to an unsuspicious subject we have

$$S_T^e = \sum_{i=1}^{|I|} -\log_2(Pr(I_i = 0)) = \min_{p_{i,m}} S_T^e. \tag{8}$$

This total score in (8) is the minimum value possible for S_T^e according to the monotonically decreasing property mentioned in Section 3 stating that the PDFs of $p_{i,m}$ are monolithically decreasing function of outcomes m for all Indicators I_i. In conclusion, the subject is not suspicious when the total information entropy risk score is minimum (as derived from the domain knowledge). In other words, the minimum total information entropy risk score (8) is the *threshold value* above which a subject becomes suspicious of a fraud.

Because this information entropy approach considers both suspicious and non suspicious outcomes in calculating the risk score of individual indicators, we do not see the necessity of weighing the scores of individual indicators when calculating the total entropy risk score and we treat all indicators equally. Sometimes however, some indicators can be more critical than others (note that this was not the case for our problem at hand). To model this domain knowledge about unequal impacts of indicators on detecting a fraud, one can consider for example Naive Bayes with informative Priors (PNB) as proposed in [8].

5   Performance Evaluation

The NP dataset used contains $3,995,882$ records, consisting of three indicators: I_1 number of current positions taken by an NP, I_2 number of positions with a short duration taken by an NP in the last year, and I_3 the number of insolvencies in the past five years. Every record is a tuple $< i_1, i_2, i_3, n(i_1, i_2, i_3) >$ which defines how many NPs in the dataset, i.e., $n(i_1, i_2, i_3)$, have the same numbers of observations (i_1, i_2, i_3) for indicators (I_1, I_2, I_3). The dataset enabled us to define the joint empirical PDFs of these three indicators (I_1, I_2, I_3), which in turn led to three marginal empirical PDFs (i.e., one per each indicator) used in (1) and (3). Based on these PDFs we were able to derive the total heuristic risk score and the total entropy risk score for every data record of the dataset. The threshold values above which an NP is considered suspicious are 0 for the total heuristic risk score and 0.92 (obtained from (8)) for the total entropy risk score. Note that the dataset used for our evaluation does contain any information on whether a datarecord corresponds to an actual suspicious entity due to privacy concerns (thus we did not have any labelled datarecord to calculate false positive and false negative rates).

Fig. 2 illustrates how the total entropy risk score behaves with respect to the total heuristic risk score for all records of the dataset considered. As it can be seen from Fig. 2, the total entropy risk score increases on average as the total heuristic risk score increases. This implies that when the heuristic risk score becomes more indicative of a suspicious record (i.e., when it increases), the entropy risk score also emphasises the suspicion (due to its increase on average).

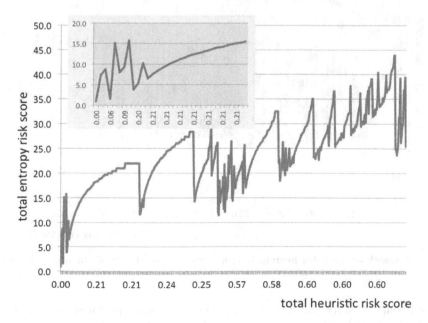

Fig. 2. A sketch of the total **entropy** risk score (versus the total heuristic risk score)

Similarly, Fig. 3 illustrates how the total heuristic risk score behaves with respect to the total entropy risk score. As it can be seen from Fig. 3, the total heuristic risk score oscillates around value 0.4 and gets saturated around value 0.6 as the total entropy risk score increases. This saturation is due to the fact that often just two out of three indicators are non zero (i.e., pointing to a suspicious legal person). These two heuristic risk scores are practically 1 when the corresponding indicators point to higher values, thus the total heuristic score becomes $\frac{1}{3}(1+1+0) = 0.66$. This implies that when the entropy risk score becomes more indicative of a suspicious record (i.e., when it increases), the heuristic risk score does not convey that much information about the severity of the suspicion (due to its saturation as explained). This behaviour can, on the one hand, be attributed to the linear versus logarithmic natures of the heuristic risk score and the entropy risk score, respectively. On the other hand, this behaviour can partly be attributed to the averaging versus the addition structures of the individual scores in the total heuristic risk score and the total entropy risk score, respectively.

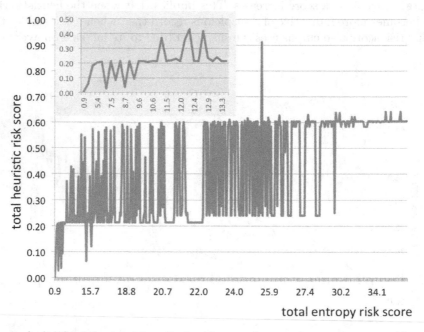

Fig. 3. A sketch of the total **heuristic** risk score (versus the total entropy risk score)

Both heuristic and entropy approaches show consistent performance in detecting whether a subject is suspicious or not. A subject is considered not suspicious when the total heuristic and entropy scores are 0 and 0.92, respectively (the later is derived from (8) for the dataset considered). One can conclude that the total entropy risk score behaves better than the total heuristic risk score when there are strong indications of a suspicious subject or the number of indicators increases. This is due to the logarithmic and additive structure of the total entropy risk score, as mentioned above.

6   Conclusion

In this paper we described two approaches on determining the probability of fraudulent behaviours by legal persons using risk indicators: one based on empirical probabilities of the indicators and the other based on the information entropy rate of the empirical probabilities. Both approaches are based on simple rules of thumb and incorporate the available domain knowledge. In both approaches, the first step is to assign scores to all individual risk indicators. Subsequently, these individual scores are combined into a total risk score. The evaluation of the two approaches on a test dataset containing natural persons showed that the approaches yield similar satisfying results. However, the information entropy approach is to be preferred over the heuristic approach as the total entropy risk score entails a confidence measure: it shows more confidence than the other approach does when more risk indicators are considered. To our knowledge, we are the first to devise and evaluate two separate approaches on combining risk scores in the judicial domain. Although not all relevant data were available to us due to privacy restrictions, we suspect our conclusions hold for the full set of indicators and LPs if the assumption of independence holds.

References

1. Verwer, S., van den Braak, S., Choenni, S.: Sharing confidential data for algorithm development by multiple imputation. In: Proc. of SSDBM 2013 (2013)
2. Herziening Toezicht Rechtspersonen, Verscherping van het toezicht op rechtspersonen, Ondernemingsstrafrecht en Compliance (February 2012),
   http://www.boekel.com/media/509294/nieuwsflits_februari_2012_-_verscherping_van_het_toezicht_op_rechtspersonen.pdf
3. Grobosky, P., Duffield, G.: Red flags of fraud. In: Trends and Issues in Crime and Criminal Justice, vol. (200). Australian Institute of Criminology (2001)
4. Freelik, N.F.: High-risk profiles and the detection of social security fraud. Journal of Social Intervention: Theory and Practice 19(1) (2010)
5. Bonchi, F., Giannotti, F., Mainetto, G., Pedreschi, D.: Using data mining techniques in fiscal fraud detection. In: Mohania, M., Tjoa, A.M. (eds.) DaWaK 1999. LNCS, vol. 1676, pp. 369–376. Springer, Heidelberg (1999)
6. Choenni, R.: Design and implementation of a genetic-based algorithm for data mining. In: Proc. VLDB 2000, pp. 33–42 (2000)
7. Taylor, C.: Composite indicators: reporting KRIs to senior management. RMA (Risk Management Association) Journal 88(8), 16–20 (2006)
8. Peng, H., Gates, C., Chris, S.B., Ninghui, L., Qi, Y., Potharaju, R., Cristina, N.R., Molloy, I.: Using probabilistic generative models for ranking risks of android apps. In: Proc. of Computer and Communications Security, pp. 241–252. ACM (2012)
9. Bishop, C.M.: it Pattern Recognition and Machine Learning. Information Science and Statistics, vol. 1, p. 740. Springer, New York (2006)

A Probabilistic Approach to Detect Local Dependencies in Streams

Qiyang Duan[1], MingXi Wu[2], Peng Wang[1], Wei Wang[1], and Yu Cao[3]

[1] Fudan University, No. 220, Handan Road, Shanghai, 200433, China
[2] Turn Inc., Redwood City, CA 94065, USA
[3] EMC Labs, Tsinghua Science Park, Beijing, China
{qduan,pengwang5,weiwang1}@fudan.edu.cn,
send2mingxiwu@gmail.com, yu.cao@emc.com

Abstract. Given m source streams $(X_1, X_2, ..., X_m)$ and one target data stream Y, at any time window w, we want to find out which source stream has the strongest dependency to the current target stream value. Existing solutions fail in several important dependency cases, such as the not-similar-but-frequent patterns, the signals with multiple lags, and the single point dependencies. To reveal these hard-to-detect local patterns in streams, a statistical model based framework is developed, together with an incremental update algorithm. Using the framework, a new scoring function based on the conditional probability is defined to effectively capture the local dependencies between any source stream and the target stream. Immediate real life applications include quickly identifying the causal streams with respect to a Key Performance Indicator (KPI) in a complex production system, and detecting locally correlated stocks for an interesting event in the financial system. We apply this framework to two real data sets to demonstrate its advantages compared with the Principal Component Analysis (PCA) based method [16] and the naive local Pearson implementation.

1 Introduction

1.1 The Motivating Problems

With modern technologies, people now can capture, digitalize, and archive almost every single signal in the real world. In the IT operations, startups like Splunk and Sumologic integrate clients' real time operation logs, and help them do almost real time analysis over any incident. When an anomaly is detected over a Key Performance Indicator (KPI) stream, the operation team must quickly identify which log streams are associated with the current KPI anomaly to understand the cause.

One KPI stream can be influenced by different streams at different occasions. For example, Fig 1 shows three interplaying database performance metrics. When the system is busy (687 to 721), 'SQL RSPNS Time' is heavily influenced by 'Consistent Read' (I/O). But when the system is idle (722 to 750), it mostly depends on 'CPU Usage Per Trx'. At the time of each high response time, the Database Administrator (DBA) must quickly evaluate all the related streams to pin down the current root cause. In this problem, users are more interested in finding those streams with *local dependencies*.

H. Decker et al. (Eds.): DEXA 2014, Part II, LNCS 8645, pp. 116–130, 2014.

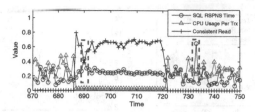

Fig. 1. The 'SQL Response Time' metric is influenced by different metrics at different occasions. Red dotted boxes indicate high response time incidents.

Many similar problems are seen in other applications as well. For example, (i) how to detect faulty components in a super computer when it fails [14]? (ii) how to find correlated stocks to a given stock in the last 5 minutes for arbitrage profits [21]? (iii) how to find problematic devices on a network when the traffic is abnormal [12]?

All these problems can be summarized into one: *from hundreds, or even thousands of monitored source streams, how can one identify the stream that has the strongest dependency on the target stream when an interested event happens?*

This problem has been treated as searching for the correlations among the monitored streams. Over static times series, various correlation scoring functions are proposed, e.g. Pearson coefficient, Cosine similarity, etc. [2,13,5]. Over the stream data, there are also some exploratory works on finding locally correlated streams over sliding windows, such as the local Pearson coefficient and the PCA based correlation [16,15]. Section 6 gives more details.

However, the correlation, in its most common form (e.g. Pearson), is only a measure of linear dependence [9,7]. The existing local correlation methods (Cosine, PCA, etc.) are also based on linear transformations. All these methods are essentially searching only for *linear dependence*. By these methods, the dependence between two streams is only judged by the similarity of their line shapes. In the aforementioned real life applications, the linear dependence based methods are facing several challenges.

1.2 Challenges to Existing Methods

Not Similar but Frequent Patterns. In many systems, certain patterns may occur frequently, though their shapes look very differently. In Fig 2a, all of the four dotted boxes illustrate one same type of patterns, in which one single point spike in the 'Transactions Initiated' (TRX) stream is always accompanied by a three-point plateau in the 'SQL Response Time' (SQL). This is a typical behavior of a database system. If one client initiates a large number of transactions, it will take the next three time points for the database server to finish them. This pattern is common in other modern complex systems as well, due to the limited processing power.

To test the performance of different correlation methods, we use a random stream (RAND) as the baseline. In Table 1, two dependency scores are recorded at each time window for each method. As an example, Pearson (TRX) column contains the scores between 'SQL' and 'TRX' at each time window by Pearson method. The next column Pearson (RAND) has the scores between 'SQL' and 'RAND'. At any time window,

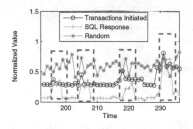 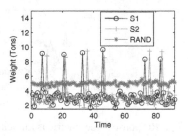

(a) Not-similar-but-frequent patterns.   (b) Signals with two different lags.

Fig. 2. Two challenges to existing linear dependence based methods

Table 1. The dependency scores for Fig 2a by four methods

Time Window	Pearson (TRX)	Pearson (RAND)	Cosine (TRX)	Cosine (RAND)	PCA (TRX)	PCA (RAND)	$\wp$ (TRX)	$\wp$ (RAND)
195-199	0.0602	**0.5793**	0.8235	**0.9799**	0.7168	**0.8973**	78261	1380
204-208	0.5815	**0.7731**	0.8924	**0.9946**	0.5618	**0.9864**	**52332**	328
218-222	**0.6095**	0.5805	0.8344	**0.9696**	0.8664	0.8528	**391.98**	33.48
230-234	0.5996	**0.6833**	0.9168	**0.9923**	0.7770	**0.9762**	**200.7**	9.93

if Pearson (TRX) is larger than Pearson (RAND), the pattern is correctly identified. According to Table 1, the Pearson and the PCA based methods can detect only one pattern correctly, while the Cosine based method misses all of the four patterns. Column $\wp$ is for the new method to be introduced.

Signals with Multiple Lags. Lagged signals are observed in many systems. For example, when a vehicle passes a bridge, the pressure sensors at different locations may generate a series of lagged events following the vehicle's movement.

Sakurai et al. [17] proposed a method to detect a global correlation lag. However, in the pressure streams, a global lag is not appropriate. Instead, many different lags exist because of different vehicle speed. For example, a truck usually passes the bridge at a speed of 80KM/H, while a sedan typically passes the bridge at 120KM/H. The analysis over these sensors' data can be effective only if we can correlate the events by different lags.

In Fig 2b, two sensor streams (S1, S2) are generated with two different lags, at the first and the third time points respectively. We apply several widely used correlation methods to detect these lagged signals against a random stream (RAND). According to Table 2, existing methods fail to detect most of them. For example, out of the six lagged signals, the Pearson coefficient can only detect two correctly.

Our Contribution: To uncover the aforementioned hard-to-detect patterns, we propose to use a probability based approach. We share the intuition that the historical streams can describe the entire system behaviors. The historical streams are fitted into a group of probabilistic models, which encapsulate how the system components interact with each other. We design an incremental model update algorithm to deal with the ever growing stream data, and show how it handles the concept drift problem [20]. Then based on the conditional probability, a scoring function is defined to measure the local dependency between any source stream and the target stream. Its advantages to traditional linear

Table 2. The dependency scores for Fig 2b by three different methods

Time Window	Pearson (S2)	Pearson (RAND)	PCA (S2)	PCA (RAND)	$\wp$ (S2)	$\wp$ (RAND)
6-11	0.2454	**0.3244**	0.4964	**0.5646**	**1.39e-3**	4.49e-4
18-23	0.2306	**0.4078**	0.4207	**0.6797**	**0.04**	9.5e-3
32-37	0.1709	**0.5449**	0.5366	**0.8768**	**0.0474**	0.0055
43-48	0.1909	**0.5054**	**0.6926**	0.6293	**0.0486**	0.0112
72-77	**0.1161**	0.0605	0.2992	**0.5688**	**0.0306**	0.0177
80-85	**0.3489**	0.2339	**0.8477**	0.5278	**0.0599**	0.0015

dependence based correlations are explained and then demonstrated on two applications over hundreds of real life streams.

The rest of this paper is organized as follows. Section 2 formalizes our problem and introduces the probabilistic dependency to attack this problem. Section 3 details our techniques of computing the probabilistic dependency. Section 4 elaborates the advantages of this new method and explains why our method can detect lagged signals. Section 5 gives experiment results. Section 6 discusses the related work. Finally Section 7 concludes the paper.

2   Problem Definition

2.1   Seeking the Local Dependency

Our problem is defined as follows. A stream X_i is a sequence of stream intervals $(x_{i1}, x_{i2}, ..., x_{iw}, ..., x_{iN})$. x_{iw} is the interval for the w-th time window and w is in $[0, N]$. All stream intervals are of the same length W, i.e. $x_{iw} = (x_{iw1}, x_{iw2}, ..., x_{iwW})$. Given m source data streams $(X_1, X_2, ..., X_m)$, one target data stream Y, and a user specified time window w, our goal is to find the source stream X_i that has the strongest dependency to the target stream Y at the time window w.

Definition 1: At a given time window w, the probabilistic dependency ($\wp$) between a source stream X_i and the target stream Y is defined as the conditional probability of $Y = y_w$ given $X_i = x_{iw}$, that is:

$$\wp(X_i, Y, w) = P(Y = y_w \mid X_i = x_{iw}) \tag{1}$$

In Equation 1, x_{iw}, y_w are the stream intervals of X_i, Y at the w-th time window. Following Definition 1, the source stream yielding the largest probabilistic dependency ($\wp$) is considered as having the strongest dependency with the target stream. The strongest dependency may not be unique.

To calculate the probabilistic dependency, we resort to the basic machine learning technique *Maximum Likelihood Estimation (MLE)*. In the MLE approach, choosing the right distribution is critical for the model accuracy. However, most real life streams do not conform to any well defined distribution. To make a generic solution, we choose the Gaussian Mixture Model (GMM). The reason is that with sufficient number of states, GMM can virtually approximate any distribution to any arbitrary accuracy [3,18].

2.2 Background: Gaussian Mixture Model

A GMM consists of multiple Gaussian distributions. If $x \in X$ is a d-dimensional vector, a Gaussian distribution can be written as[3]:

$$G(x \mid \mu, \Sigma) = \frac{1}{(2\pi)^{d/2}} \frac{1}{\Sigma^{1/2}} exp\{-\frac{1}{2}(x - \mu)^T \Sigma^{-1}(x - \mu)\} \tag{2}$$

In Equation 2, μ is the center (or the mean) of the Gaussian distribution, and Σ is the covariance matrix of x. If we partition the vector x into two parts, the Gaussian parameters can be partitioned in the same manner [3]:

$$x = \begin{pmatrix} x_a \\ x_b \end{pmatrix}, \mu = \begin{pmatrix} \mu_a \\ \mu_b \end{pmatrix}, \Sigma = \begin{pmatrix} \Sigma_{aa} & \Sigma_{ab} \\ \Sigma_{ba} & \Sigma_{bb} \end{pmatrix} \tag{3}$$

Then the marginal Gaussian distribution can be easily acquired by:

$$G(x_a) = G(x_a \mid \mu_a, \Sigma_{aa}) \tag{4}$$

The Gaussian distribution is a perfect symmetrical distribution, but most of our real life data does not look like that. To be able to model a complex distribution, a linear combination of multiple Gaussians is formed. This becomes a Gaussian Mixture Model (GMM), which is defined as:

$$GMM(x) = \sum_{k=1}^{K} \pi_k G(x \mid \mu_k, \Sigma_k) \tag{5}$$

In Equation 5, K is the number of Gaussian distributions (states) mixed into a single model and π_k is the mixing coefficient. These three parameters π, μ, Σ together determine a GMM. Given some training data, the Expectation Maximization (EM) algorithm can be applied to build the best fitting GMM [3].

3 Computing the Probabilistic Dependency

3.1 Using Paired GMMs to Compute Probabilistic Dependency

To capture the probabilistic dependencies between the source streams and the target stream, we learn a group of statistical models $\{GMM_i \mid i \in (1, 2, ..., m)\}$. Each model GMM_i describes the relationships between one pair of source stream X_i and target stream Y. We will first show how to learn the models over static time series, and then show how to update the models incrementally upon new data arrival.

To learn GMM_i, each stream interval x_{iw} from the source stream X_i is aligned with the corresponding target interval y_w. Then we have a set of $2W$ dimensional vectors:

$$< x_{iw}, y_w >=< x_{iw1}, x_{iw2}, ..., x_{iwW}, y_{w1}, y_{w2}, ..., y_{wW} > \tag{6}$$

Over the stream pair vectors $< x_{iw}, y_w >$, the EM algorithm [3] is applied to learn one GMM (i.e. GMM_i). Once the pairwise GMMs are ready, according to Definition 1, the probabilistic dependency can be computed as:

$$\wp(X_i, Y, w) = \frac{GMM_i(y_w, x_{iw})}{GMM_i(x_{iw})} \tag{7}$$

In Equation 7, $GMM_i(x_{iw})$ is the marginal probability on x_{iw} only. It can be directly computed from GMM_i by partitioning (π, μ, Σ) according to Equation 3 and Equation 4. We keep the division form in our implementation to avoid numerical issue arising in the matrix inversion.

3.2 Incremental Model Update

Considering the ever growing nature of the stream data, the EM algorithm over one fixed time period is insufficient. The models must be updated according to the non-stopping new data. Since most of the real life streams can be segmented on the time dimension (e.g. daily), we take a period by period approach. At the end of each time period, a new group of models are trained and then merged into the existing models. The model update frequency is determined by the stream velocity, usually hourly or daily.

The model update happens as an offline process, which is acceptable as long as the last model update finishes before the next time period starts. Only the dependency scoring will be done online and within a few seconds.

We start by merging two GMM models belonging to one stream pair. Then the models for all stream pairs can be processed in the same way. Let GMM_1 be the existing GMM model trained for all previous time periods. GMM_1 has K_1 states and C_1 stream intervals. At the end of a new time period, we receive C_2 stream intervals.

First, a new model GMM_2 with K_2 states can be acquired using a standard EM algorithm. To merge GMM_2 into GMM_1, we combine GMM_2 with GMM_1 and form a new GMM' with $(K_1 + K_2)$ states by equation 8.

$$GMM' = \sum_{k=1}^{K_1} \frac{C_1 \pi_{1k}}{C_1 + C_2} G(\mu_{1k}, \Sigma_{1k})$$
$$+ \sum_{k=1}^{K_2} \frac{C_2 \pi_{2k}}{C_1 + C_2} G(\mu_{2k}, \Sigma_{2k}) \tag{8}$$

Then we select two most similar Gaussian components $G_1(\pi_1, \mu_1, \Sigma_1)$, $G_2(\pi_2, \mu_2, \Sigma_2)$ from GMM' according to the Kullback-Leibler Divergence (KLD) [4] as defined in Equation 9.

$$KLD(G_1, G_2) = log(\frac{|\Sigma_2|}{|\Sigma_1|}) + tr(\Sigma_2^{-1} \Sigma_1)$$
$$+ (\mu_2 - \mu_1)^T \Sigma_1^{-1} (\mu_2 - \mu_1) - |w| \tag{9}$$

With the merging operations [4] defined by equations 10 - 14, the two selected components $G_1(\pi_1, \mu_1, \Sigma_1)$, $G_2(\pi_2, \mu_2, \Sigma_2)$ are merged into one new component $G'(\pi', \mu', \Sigma')$.

$$f_1 = \frac{\pi_1}{\pi_1 + \pi_2} \tag{10}$$

$$f_2 = \frac{\pi_2}{\pi_1 + \pi_2} \tag{11}$$

$$\pi' = \pi_1 + \pi_2 \tag{12}$$

$$\mu' = f_1\mu_1 + f_2\mu_2 \tag{13}$$

$$\Sigma' = f_1\Sigma_1 + f_2\Sigma_2 + f_1f_2(\mu_1 - \mu_2)(\mu_1 - \mu_2)^T \tag{14}$$

Iteratively, the number of states of GMM' is reduced to K_1 again. Algorithm 1 formally defines the incremental update process.

Algorithm 1. Updating the overall GMM_1 by the data from a new time period

---

Require: GMM_1 for all stream history, K_1 as the number of states for GMM_1, stream data for a new time period

Ensure: GMM' as the new model updated for the new time period data

1: Train a new GMM model GMM_2 by EM using the new time period data.
2: Merge GMM_1 and GMM_2 into GMM' according to equation 8.
3: Let K' be the number of state of GMM'.
4: **while** $K' > K_1$ **do**
5:　　Compute KLD for each pair of Gaussian components in GMM' according to equation 9.

6:　　Find the pair of Gaussian components (G_1, G_2) with the minimal KLD.
7:　　Merge G_1, G_2 into G' according to equations 10 - 14.
8:　　Remove components G_1, G_2 from GMM', and add G' into GMM'.
9:　　$K' = K' - 1$
10: **end while**
11: Output GMM' as the updated GMM_1.

Using the incremental update algorithm, new system states from a new time period can be incorporated into an existing model. Meanwhile, if a pattern disappears from the stream data, the corresponding system state will gradually be outweighed by other states. This solves the classical stream concept drift problem [20] . The concept drift phenomenon happened in our database experiment. Section 5.1 shows how our method handles it.

3.3　Complexity Evaluation

Our probabilistic dependency method has two phases: the offline training phase and the online evaluation phase. The offline training cost is all included in Algorithm 1:

1. Step 1 is the offline training process. For every stream pair in a new period, the computational cost is a standard EM cost, which is $O(N * K * W^2)$ for each iteration. In normal implementation, maximum number of iterations is a constant value. Therefore, the overall training cost for a new period is $O(N * K * W^2 * M)$.
2. The KLD computation in step 6 takes $O(W^2)$.
3. The Gaussian Merging in step 7 takes $O(K^2 * W^2)$.

Because the window length and the number of states are much smaller than the length of the full stream, we have $W << N$ and $K << N$. The overall computation cost for the offline training is $O(N * K * W^2 * M)$. Similarly, the online evaluation cost can be easily calculated as multiple Gaussian evaluation of $O(m * W^2 * K)$.

4 Benefits and Discussions

4.1 Detecting Not Similar but Frequent Patterns

Because the new method detects dependence by statistical co-occurrence, it can detect all chronic patterns, even if their shapes look very differently. The probabilistic dependency scores for Fig 2a are listed in the $\wp$ column of Table 1. One can easily verify that all the four patterns were correctly recognized by our new method with higher dependency scores.

The probabilistic dependency detection is also **insensitive to noises** because it evaluates streams over the whole history and keeps the scale information. Some background noises on one source stream may make it temporary look similar to the target stream in the current time window. This causes many false positives to traditional linear correlation methods. For example, in Section 1.2, linear dependence based methods had chosen 'RAND' because its shape is more similar than the real signal. Furthermore, most of linear dependence based methods imply a normalization process while searching for similarities. After eliminating the scale information, more streams shall look similar. These false positives can be avoided using the probabilistic method.

4.2 Detecting Lagged Signals

The probabilistic framework can detect lagged signals, even if multiple lags exist. This solves the second problem from Section 1.2. We first prove the ability of detecting lagged correlation by Lemma 1, and then verify it by experiments.

Lemma 1. *Assume that X_1 has an l-points lagged signal with dependency to Y and all the other time points are randomly distributed in X_1, X_2 and Y. When this lagged event happens at time window w and $W > l$, we will have :*

$$E(\wp(X_1, Y, w)) > E(\wp(X_2, Y, w))$$

Proof. In Lemma 1, $E(\wp(X_i, Y, w))$ is the expected value of $\wp(X_i, Y, w)$, which is defined as the weighted average of all possible scoring values at window w, i.e. as $\int_0^{+\infty} \wp * P(\wp)d\wp$. The reason of using expectation instead of actual $\wp(X_i, Y, w)$ is to *remove the noises. For easier presentation, we will use $\wp(x_{iw}, y_w)$ to denote $\wp(X_i, Y, w)$, and $GMM(x_{iw}, y)$ to denote $GMM(X_i, Y)$.*

Suppose that X_i contains an event which is l points ahead of the corresponding event in Y. Without loss of generality, we assume that the event is of length $W - l$, i.e. the time interval w consists of one event lasting $(W - l)$ points and l random points. Thus we can partition x_{iw} into two parts: the event part $x_{iw}^a = < x_{iw1}, x_{iw2}, ..., x_{iw(W-l)} >$ and the l-points random part $x_{iw}^b = < x_{iw(W-l+1)}, x_{iw(W-l+2)}, ..., x_{iwW} >$. Similarly y_w is partitioned into: $y_w^b = < y_{w1}, y_{w2}, ..., y_{wl} >$ and $y_w^a = < y_{w(l+1)}, y_{w(l+2)}, ..., y_{wW} >$.

Since all points from both x_{iw}^b and y_w^b are random, they are independent both to each other and to x_{iw}^a, y_w^a. With the learned pairwise GMM models over X_i, Y, according to Equation (4), we can slice the GMM models to be:

$$GMM(x_{iw}, y) = GMM(x_{iw}^a, y_w^a) \times GMM(x_{iw}^b, y_w^b) \tag{15}$$

Now assume that one dependent event happens between X_1 and Y, while nothing happens in another stream X_2 and Y. The event should fall in the $2(W - l)$ points in the (x_{1w}^a, y_w^a) subspace. Therefore, we should have:

$$E(\wp(x_{1w}^a, y_w^a)) > E(\wp(x_{2w}^a, y_w^a))$$

Since both (x_{1w}^b, y_w^b) and (x_{2w}^b, y_w^b) belong to random distributions, we have:

$$E(\wp(x_{1w}^b, y_w^b)) = E(\wp(x_{2w}^b, y_w^b))$$

Combining these two parts together, we can get:

$$E(\wp(x_{1w}, y_w)) > E(\wp(x_{2w}, y_w)). \qquad \square$$

Lemma 1 illustrates how one event with a single lag can be detected. When multiple lags exist, they are captured by different GMM states. Therefore, this proof can be easily extended to multiple lags.

To verify this ability, the dependency scores for Fig 2b is calculated in the $\wp$ column in Table 2. In Table 2, we can see that our probabilistic dependency always scores the stream 'S2' higher than the random signal 'RAND'. On the other hand, Pearson method and PCA based method succeed in only two time windows each. Section 5.1 will give another lagged example.

In the proof of Lemma 1, the lengths of the real signals x_{iw}^a and y_w^a are variable. It implies that our method may capture the interaction of any event length within the specified window length. This feature makes our methods very useful, because in certain realistic streams the event length may vary.

4.3 Detecting Single Point Dependency

Detecting dependency relationships over a single time point is an unique capability of this new method. It is required if the signal frequency is higher than the sampling rate. For example, the database metrics used in Section 5.1 are at the minute level. Since most system interactions finish in a few seconds, single point dependency can already reveal most of the meaningful relationships. Table 3 in the Experiment Section confirms that our method can recognize the right dependency by a single point ($W = 1$). In contrast, none of the linear dependence based methods [5,2,13,16,14] can work on single time point because the linear transformation cannot be performed on scalar values.

5   Experiment

Our method is applied on two real life data sets. The first data set is obtained by running the TPCC benchmark [1] in an Oracle database. The objective is to identify the dependent monitored stream at the time when a high 'SQL Response Time' is observed. The second data set contains all the China Security Index 300 (CSI 300) stock streams for five years. The target stream is set to be COFCO Property (000031.SZ), and the source streams are the rest stocks in the index. In the second data set, we want to detect the correlated stocks when the target stock's gain or loss exceeds 5%. The two data sets are published [6] for others to reproduce the results.

All the experiments are done via Matlab 7.5 on a machine equipped with Intel Core(TM)2 CPU U7600, 2GB Memory, and Windows XP operating system. For all the conducted experiments, we compare the result against the Pearson and the PCA based methods [16].

5.1   Experiment One: Database Metrics

Experiment Goal. In this experiment, the goal is to answer the following questions:

- Can our method *correctly* identify the metric streams with the strongest dependency when there is an exceptionally high value in the 'SQL Response Time'?
- Can we capture certain hard-to-detect patterns (such as lagged correlation) using our framework?
- Can we handle the concept drift in several days' data? Can we keep the correlation accuracy while merging GMM models?

Experiment Setup. To be able to record both busy and idle status of an Oracle database, a TPCC benchmark program was executed in an intermittent style. The program was executed for only 30 minutes in every hour. We ran the same test for totally 7 days (not consecutive), and collected the system metrics for each day. In each day, we extracted 210 metrics regarding the database operating status and then aggregated all metrics to the minute level.

Out of the 210 metrics, 'SQL Response Time' is chosen to be the target stream. Whenever there is an exceptionally high query response time, all the other metrics are inspected to understand which one caused the anomaly.

Result 1 (Accuracy). The top figure from Fig 3 shows the 'SQL Response Time'. By visual inspection, it is not hard to tell that there are four outstanding spikes occurred at time 53, 62, 665, and 733. Using our DBA knowledge, we first lay out the top correlated streams:

- At the second spike (time 62), the stream with the strongest dependency is the 'Average Active Sessions', which is shown in the second figure of Fig 3. It is because at time 62, a few testing clients were added, which led to the spikes in both the response time and the active sessions streams.
- Regarding the fourth spike (time 733), the stream with the strongest dependency is the 'Consistent Read Gets Per Txn'. It is because at time point 733, a single large background transaction incurred a large amount of IO and caused the fourth longer response time.

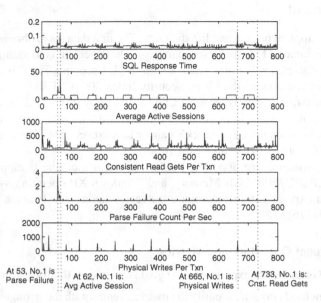

At 53, No.1 is: Parse Failure
At 62, No.1 is: Avg Active Session
At 665, No.1 is: Physical Writes
At 733, No.1 is: Cnst. Read Gets

Fig. 3. In all five figures, X-axis is time, and Y-axis is the database metric value. The vertical dotted lines help align time points, each indicating one anomaly.

Table 3. The accuracy of different window sizes and the Pearson/PCA based methods

Time Point	$W=1$	$W=3$	$W=6$	Pearson	PCA based
53	Y	Y	Y	N	N
62	Y	N	Y	N	N
665	Y	Y	Y	N	N
733	Y	N	Y	N	N

- Similarly, we conclude that the 'Parse Failure Count Per Sec' is accountable for the first spike (time 53), and the 'Physical Writes per Txn' is the most related to the third spike (time 665).

With the above findings as the standard, we compare the correlation accuracy between our method and the two existing ones. Three window sizes, respectively 1, 3, 6, are tested to evaluate all three methods on different window sizes.

Table 3 shows the accuracy results for three different window sizes. In the test, if the correct streams are ranked within top 10 out of the 209 source stream, we mark it as 'Y'. Otherwise, we mark 'N'. From Table 3, we see that both $W=1$ and $W=6$ achieved 100% accuracy. In contrast, the two existing alternatives both failed to capture the right dependent streams.

Result 2 (Lagged Correlation). While testing with window size 6, in the list of top ranking streams at time point 62, the top three dependent streams are: 'DBRM CPU Decisions Present', 'Average Active Sessions', 'Average Active Sessions'. Upon further examination, we found that Oracle database actually monitors two streams with the same name. The second 'Average Active Sessions' is mostly one time point behind the

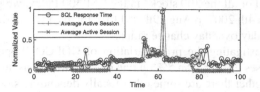

Fig. 4. 'SQL Response Time' is found to be correlated with two 'Avg Active Session'. One of them is one time point ahead of the 'SQL Response Time' .

first one, as shown in Fig 4. We do not know Oracle's reason of having two similar metrics, but it clearly demonstrates that our method can detect lagged signals.

Result 3 (Concept Drift). Over the 7 days' metrics, we trained one model for each day. The number of states for each model is still 5. Then all models are merged into a single GMM with 8 states using Algorithm 1. Fig 5a shows the data and the model from the first day, when the 'Executions Per Sec' is almost 6000. Fig 5b is from the seventh day, where the 'Executions Per Sec' dropped to around 1000. This is because the table size grows after each TPCC program execution, and the system is running slower and slower over time. Fig.5c shows the merged GMM for all 7 days over the first day's data. We can see that, the component centered at (900,0.2) from the second Fig.5b is added to Fig.5c, which reflects the new state of slower execution.

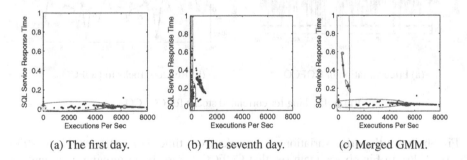

(a) The first day.   (b) The seventh day.   (c) Merged GMM.

Fig. 5. The GMMs for the first, seventh days, and the merged GMM

Using this merged model, we evaluated the top reasons for 4 anomalies again. The model still recognized the right metrics as No.1 for time points 53 and 665. For time points 62 and 733, the correct metrics were moved to ranking 100 and 30. It means the model still works, with minor accuracy loss.

5.2   Experiment Two: CSI Stocks

Experiment Goal. In this experiment, we want to verify if the GMM based method can identify local dependencies in the financial streams.

Experiment Setup. For all the 300 stocks in the CSI 300 index, we gathered the daily close prices from Jan 4th 2007 to Aug 29th 2011 for totally 1186 trading days, and then converted them into day over day change ratios.

We focus on investigating the price variation of COFCO Property (000031.SZ). When it gains or losses greater than 5% on one day, we treat it as an interesting event. We then analyze whether there are some other locally depending stocks. This can help the financial analysts understand the market situation in a much easier manner.

For all stock experiments, we set $W = 1$. We expect no lagged correlations since daily stock price is well known as un-predictable.

Result on COFCO Property. Fig.6a shows the price changes of COFCO Property (000031.SZ) between time 1(2007-01-04) and time 500(2008-11-08). All the points with over 8% loss (below the dotted line) are investigated.

COFCO Property is a real estate company, and globally its daily price variation should be highly correlated to other real estate companies. Indeed, the top 10 correlated streams from our method contained at least two real estate companies, except one time point at 95 (2007-05-30). After investigation, we found that Chinese government raised Stamp Tax from 0.1% to 0.3%, and all stocks dropped heavily. We believe COFCO actually descended together with all stocks, instead of with other real estate stocks. This information is very useful to assess the market situation.

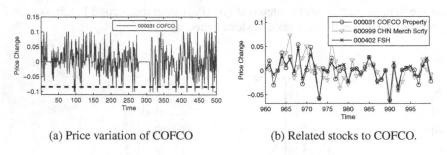

(a) Price variation of COFCO   (b) Related stocks to COFCO.

Fig. 6. Looking for correlated stocks for COFCO

Fig.6b shows the price variation of 3 stocks from time 940 (2010-08-24) to 1000 (2010-11-26). In Fig.6b we again see that COFCO is mostly co-moving with another real estate company FSH. But at the time 990 (2010-11-12), our method listed a few security trading and brokerage companies into top 10, including China Merchant Security (600999). When we look into COFCO's profile, we found that COFCO actually held about 5% share of the China Merchant Securities. On that day, the brokerage industry dropped more than real estate did. We can conclude that this caused the COFCO's loss at time 990. On the contrary, at time 973, China Merchant Security did not cause trouble to COFCO, and COFCO dropped along with FSH only.

Neither Pearson nor PCA based methods can identify the aforementioned relationships since they can not detect single point dependencies.

6   Related Work

Anomaly detection has been a hot topic in large scale systems, especially after the surge of big data. Kavulya et al. [11] used KL-Divergence to measure the distribution difference between the failed time and the normal time for each stream. The stream which yields higher difference is deemed as the most contributing to the failure. Wang et al. [19] proposed to use Tukey and Relative Entropy statistics to rank monitored streams for anomaly detection. Those two methods consider the anomaly for the individual stream only, but do not try to correlate them to find dependency.

There are some researches on identifying the local correlations among streams. One naive implementation could be calculating the Pearson coefficients in a sliding window[16]. Papadimitriou et al. proposed a PCA based method [16,15] and compared it with the naive Pearson one over a Currency Exchange Data. They gather the autocovariance data among m windows of length w, and form one $m * w$ matrix $\Gamma_t(X, w, m)$. Then the k largest Eigen vectors U_k from the PCA transformation over Γ are used to define the correlation score: $l_t(X, Y) = \frac{1}{2}(\| U_X^T u_Y \| + \| U_Y^T u_X \|)$. Oliner et al. also proposed a similar method using PCA together with the lag correlation [17] to analyze the system log of a super computer [14].

Guo et al. proposed a GMM based fault detection method [8] based on probabilistic relationship. They learn a GMM model for each pair of monitoring streams and then detect the outliers with extremely low probability density as anomalies. The difference from our work is that we use the model to find dependent streams for a target one, while they use the joint probability to detect anomalies.

Jiang et al. proposed a new Slope Duration Distance (SDD) function to detect local patterns in financial streams [10] and tested it on 1700 stocks. They are working on the same problem like us, but their approach is still based on linear dependence.

7   Conclusions and Future Work

In this paper, we proposed a GMM based framework to detect local dependencies between source streams and a target stream. Firstly we pair each source stream with the target stream and train a GMM model for each pair incrementally. Then for any given time window w, we compute the conditional probability of the target stream given a source stream at window w. The one with the largest conditional probability is returned as the most correlated source stream. We theoretically analyzed and experimentally demonstrated that the proposed framework can effectively capture lagged(or shifted) local correlation, single point local correlation, and the not-similar-but-frequent Patterns. As ongoing work, we are evaluating our methods on large scale system logs, e.g. Hadoop systems. In the future, we plan to explore correlations in longer windows. In the long run, the incremental GMM training process also requires a suitable decay scheme to handle the historical information more gracefully.

References

1. Transaction Processing Performance Council, TPC-C and TPC-H Benchmark, Online Document, http://www.tpc.org/tpcc/
2. Aggarwal, Yu: Finding generalized projected clusters in high dimensional spaces. ACM SIGMOD Record 29 (2000)
3. Bishop, C.M., Nasrabadi, N.M.: Pattern recognition and machine learning. J. Electronic Imaging, 78, 4, 14, 90, 111, 435 (2007)
4. Bouchachia, A., Vanaret, C.: Incremental learning based on growing gaussian mixture models. In: 10th International Conference on Machine Learning and Applications (2011)
5. Cha, S.-H.: Comprehensive survey on distance similarity measures between probability density functions. International Journal of Mathematical Models And Methods In Applied Sciences 1, 300 (2007)
6. Duan, Q.: Stream data for experiments. Online Data, https://www.dropbox.com/s/vuvnhl6n6mxompt/stream_data.zip
7. Embrechts, P., McNeil, A., Straumann, D.: Correlation and dependence in risk management: properties and pitfalls. In: Risk Management: Value at Risk and Beyond, pp. 176–223 (2002)
8. Guo, Z., Jiang, G., Chen, H., Yoshihira, K.: Tracking probabilistic correlation of monitoring data for fault detection in complex systems. In: DSN, pp. 259–268. IEEE Computer Society (2006)
9. Hrdle, W., Simar, L.: Applied multivariate statistical analysis. Springer (2007)
10. Jiang, T., Cai Feng, Y., Zhang, B., Sheng Cao, Z., Fu, G., Shi, J.: Monitoring correlative financial data streams by local pattern similarity. Journal of Zhejiang University SCIENCE (2009)
11. Kavulya, S.P., Daniels, S., Joshi, K., Hiltunen, M., Gandhi, R., Narasimhan, P.: Draco: Statistical diagnosis of chronic problems in large distributed systems. In: 2012 42nd Annual IEEE/IFIP International Conference on Dependable Systems and Networks (DSN), pp. 1–12. IEEE (2012)
12. Lakhina, A., Crovella, M., Diot, C.: Diagnosing network-wide traffic anomalies. In: SIGCOMM (2004)
13. Liao, T.W.: Clustering of time series data: A survey. Pattern Recognition 38(11), 1857–1874 (2005)
14. Oliner, A.J., Aiken, A.: Online detection of multi-component interactions in production systems, pp. 49–60. IEEE (2011)
15. Papadimitriou, S., Sun, J., Faloutsos, C.: Streaming pattern discovery in multiple time-series. In: Böhm, K., Jensen, C.S., Haas, L.M., Kersten, M.L., Larson, P.-Å., Ooi, B.C. (eds.) VLDB, pp. 697–708. ACM (2005)
16. Papadimitriou, S., Sun, J., Yu, P.S.: Local correlation tracking in time series. In: ICDM, pp. 456–465. IEEE Computer Society (2006)
17. Sakurai, Y., Papadimitriou, S., Faloutsos, C.: BRAID: Stream mining through group lag correlations. In: Özcan, F. (ed.) SIGMOD Conference, pp. 599–610. ACM (2005)
18. Sung, H.G.: Gaussian mixture regression and classification. PhD Thesis, Rice University, Houston, Texas (2004)
19. Wang, C., Viswanathan, K., Choudur, L., Talwar, V., Satterfield, W., Schwan, K.: Statistical techniques for online anomaly detection in data centers. In: 2011 IFIP/IEEE International Symposium on Integrated Network Management (IM), pp. 385–392. IEEE (2011)
20. Widmer, G., Kubat, M.: Learning in the presence of concept drift and hidden contexts. Machine Learning 23(1), 69–101 (1996)
21. Zhu, Y., Shasha, D.: Statstream: Statistical monitoring of thousands of data streams in real time. In: VLDB, pp. 358–369. Morgan Kaufmann (2002)

Inferring Topic-Level Influence from Network Data

Enliang Xu[1], Wynne Hsu[1], Mong Li Lee[1], and Dhaval Patel[2]

[1] School of Computing, National University of Singapore, Singapore
{xuenliang,whsu,leeml}@comp.nus.edu.sg
[2] Dept. of Electronics & Computer Engineering, IIT Roorkee, India
patelfec@iitr.ernet.in

Abstract. Existing influence analysis research has largely focused on studying the maximization of influence spread in the whole network, or inferring the "hidden" network from a list of observations. There is little work on topic-level specific influence analysis. Although some works try to address this problem, their methods depend on known social network structure, and do not consider temporal factor which plays an important role in determining the degree of influence. In this paper, we take into account the temporal factor to infer the influential strength between users at topic-level. Our approach does not require the underlying network structure to be known. We propose a guided hierarchical LDA approach to automatically identify topics without using any structural information. We then construct the topic-level influence network incorporating the temporal factor to infer the influential strength among the users for each topic. Experimental results on two real world datasets demonstrate the effectiveness of our method. Further, we show that the proposed topic-level influence network can improve the precision of user behavior prediction and is useful for influence maximization.

1 Introduction

Social networking sites such as Facebook, Twitter, Delicious and YouTube have provided a platform where user can express their ideas and share information. With the increasing popularity of these sites, social networks now play a significant role in the spread of information. Recognizing this, researchers have focused on influence analysis to discover influential nodes (users, entities) and influence relationships (who influences whom) among nodes in the network [4,5,2,1,3]. This will enable a company to target only a small number of influential users, thus leading to more effective online advertising and marketing campaigns. However, most often than not, influential users typically tweet on many topics and their followers generally follow them for different reasons. As a result, they may not be the ideal targets for targeted marketing.

Figure 1(a) shows 5 users and the tweets they make at different times. Based on their re-tweet information, we can construct the influence among the users as shown in Figure 1(b). We note that user u_1 is the most influential person as his/her tweets are re-tweeted by 3 other users. Yet, when we analyze the contents of the tweets, we discover that user u_1 only influences u_2 on the topic "iphone", whereas for the same topic "iphone", user u_4 influences users u_1 and u_5. Hence, if we wish to conduct a marketing campaign on "iphone", the most influential person, i.e. u_1, may not be the ideal target. Instead, we should target u_4.

H. Decker et al. (Eds.): DEXA 2014, Part II, LNCS 8645, pp. 131–146, 2014.

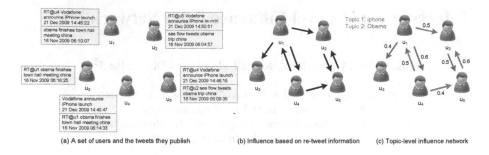

(a) A set of users and the tweets they publish   (b) Influence based on re-tweet information   (c) Topic-level influence network

Fig. 1. Example topic-level influence analysis

Further, temporal factor also plays an important role in differentiating the degree of influence among different users. For example, users u_1, u_3 and u_4 are connected to each other in Figure 1(b) and they have tweeted about "Obama" at time stamps 06:10, 06:16 and 06:14 respectively. Without utilizing the time information, the degree of influence from u_1 to u_3 and u_4 is the same. However, in real life, we observe that the influence is the greatest when the time lapse is the shortest [3]. In other words, the influence from u_1 to u_4 should be greater than that from u_1 to u_3.

To address this, the works in [8,7,9] have looked into capturing the micro-level mechanisms of influence, e.g. the influence relationship between two users on a specific topic. However, they require the connection among users to be explicitly modeled. In other words, suppose we wish to analyze the influence relationships among users on Twitter, these works can only report the topic-specific influence relationships among the followers where the follow relationships are explicitly modeled in Twitter. While this is useful for applications that concern only the explicitly modeled relationships, many applications need to go beyond the connected users.

Figure 2 shows the rumor "Two Explosions in the White House and Barack Obama is injured". The Twitter account of the Associated Press (@AP) was hacked and a tweet that reported a fake White House explosion caused the Dow Jones Index to drop more than 140 points within minutes. This tweet was retweeted by almost 1,500 Twitter users

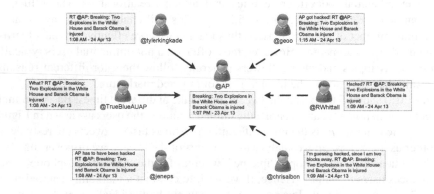

Fig. 2. "Two Explosions in the White House and Barack Obama is injured" rumor

within a short span of a few minutes and many of these users are not explicitly connected via the follow relationship in Twitter. Clearly, there is a need to capture topic-level influence among users that are not explicitly connected.

In this work, we introduce a guided topic modeling process that utilizes the knowledge of common representative words of popular topics to guide the topic extraction process. For each extracted topic, we infer topic-level influence among users who have posted documents on the extracted topic. We also take into account the temporal factor to infer the influential strength between users for this topic. We evaluate the proposed method on two real world datasets. Experimental results demonstrate the effectiveness of our method as well as the usefulness of the topic-level influence network in user behavior prediction and influence maximization.

The rest of the paper is organized as follows. Section 2 gives an overview of our approach. We describe the guided hierarchical LDA model in Section 3, and the topic-level influence network in Section 4. Experiment results are presented in Section 5. Section 6 discusses the related work before we conclude in Section 7.

2   Overview

A *topic-level influence network* is denoted as $G = (U, E)$, where U is the set of users and E is the set of labeled directed edges between users. An edge $e \in E$ from node u to node v with label (z, w) denotes user u influences user v on topic z with an influential strength of w, $w \geq 0$. Figure 3 gives an overview of our two-step approach to discover topic-level user influence network.

Given a collection of time-stamped documents D where a tuple $< u, d, t > \in D$ indicates that user u has published document d at time t, and a set of users who published these documents, the first step is to cluster the set of documents into various groups based on their topics. We propose a guided topic modeling approach based on the hierarchical LDA model [18]. The key idea is to utilize additional knowledge in the form of known popular topics to bias the path selection in the hierarchical LDA topic generation such that documents that belong to the same path are more similar than documents of another path. Since hierarchical LDA based topic generation allows infinite number of topics, we effectively remove the need to pre-determine the number of topics.

Once the documents are clustered, the second step is to compute the KL-divergence based similarity between pairs of documents in each cluster. Then we utilize user and temporal information of each document to obtain the influential strength among the users for each topic and construct the topic-level user influence network.

3   Guided Hierarchical LDA

In this section, we briefly review the original hierarchical LDA model [18] and then describe our proposed guided hierarchical LDA topic model. In the original hierarchical LDA model, a document is generated by choosing a path from the root to a leaf, and as it moves along the path, it repeatedly samples topics along that path, and then samples the words from the selected topics. The path selection is based on the nested Chinese

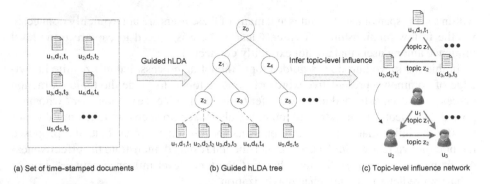

(a) Set of time-stamped documents   (b) Guided hLDA tree   (c) Topic-level influence network

Fig. 3. Overview of proposed solution

Restaurant Process (nCRP) which is a stochastic process that assigns probability distribution to an infinitely branched tree. In nCRP, the first customer sits at the first table, and the nth subsequent customer sits at a table drawn from the following distribution:

$$p(occupied\ table|previous\ customers) = \frac{n_i}{\gamma + n - 1}$$

$$p(next\ unoccupied\ table|previous\ customers) = \frac{\gamma}{\gamma + n - 1} \tag{1}$$

where n_i is the number of customers currently at table i, and γ is a real-valued parameter which controls the probability of choosing new tables.

Careful observation of this distribution shows that the probability of choosing a table depends on the number of customers already assigned to the table at that level. Thus, tables with more customers will have a higher probability to be selected. However, this does not consider the similarity of the customers at the table. For short documents such as tweets and microblogs, the length of each path is short and hence it is vital to ensure the similar documents are assigned to the same table as early on the path as possible.

Fortunately, in real life social networks, we often have some rough ideas what are the hot topics being discussed and the commonly used words associated with these topics. Taking advantage of such knowledge, we propose to guide the topic generation of hierarchical LDA model by biasing the path selection at the beginning of each path by favoring the table (the preferred table) whose customers are most similar to the incoming customer. This is achieved by changing the probability distribution of path selection at level 2 as follows:

$$p(preferred\ table|previous\ customers) = \frac{n_i + \delta}{\gamma + n + \delta}$$

$$p(next\ occupied\ table|previous\ customers) = \frac{n_i}{\gamma + n + \delta}$$

$$p(next\ unoccupied\ table|previous\ customers) = \frac{\gamma}{\gamma + n + \delta} \tag{2}$$

where δ adds an increment to the table where the most similar customers are seated.

More specifically, in our guided hierarchical LDA model, a document is drawn by first choosing a L-level path and then drawing the words from the L topics which are associated with the nodes along that path. The generative process is as follows:

(1) For each table k in the infinite tree,
    (a) Draw a topic $\beta_k \sim \text{Dirichlet}(\eta)$.
(2) For each document, $d \in \{1, 2, \ldots, D\}$,
    (a) Let c_1 be the root node.
    (b) Let *hot* be the most similar hot topics to d.
    (c) Mark the table corresponding to the *hot* as "preferred table".
        i. Draw a table from c_1 using Equation (2).
        ii. Set c_2 to be the restaurant referred to by that table.
    (d) For each level $l \in \{3, \ldots, L\}$,
        i. Draw a table from c_{l-1} using Equation (1).
        ii. Set c_l to be the restaurant referred to by that table.
    (e) Draw a distribution over levels in the tree, $\psi_d \mid \{m, \pi\} \sim \text{GEM}(m, \pi)$.
    (f) For each word,
        i. Choose level $z_{d,n} \mid \psi_d \sim \text{Discrete}(\psi_d)$.
        ii. Choose word $w_{d,n} \mid \{z_{d,n}, c_d, \beta\} \sim \text{Discrete}(\beta_{c_d}[z_{d,n}])$, which is parameterized by the topic in position $z_{d,n}$ on the path c_d.

where $z_{d,n}$ denotes the topic assignments of the nth word in the dth document over L topics, $w_{d,n}$ denotes the nth word in the dth document, and m, π, γ and η are the same hyperparameters used in hierarchical LDA [18].

Having defined guided hierarchical LDA model, the next step is to learn the model from data. We adopt the Gibbs sampling approach and iteratively sample each variable conditioned on the rest. First, we sample a path c_d for each document conditioned on the path assignment of the rest documents in the corpus and the observed words:

$$p(c_d|w, c_{-d}, z, \eta, \gamma, \delta)$$
$$\propto p(c_d|c_{-d}, \gamma, \delta)p(w_d|c, w_{-d}, z, \eta) \tag{3}$$

where c_{-d} and w_{-d} denote the vectors of path allocation and observed words leaving out c_d and w_d respectively. $p(w_d|c, w_{-d}, z, \eta)$ is the probability of the data given a particular choice of path and $p(c_d|c_{-d}, \gamma, \delta)$ is the prior on paths implied by the modified nested Chinese Restaurant Process.

Given the path assignment, we sample the level allocation variable $z_{d,n}$ for word n in document d conditioned on all the other variables:

$$p(z_{d,n}|z_{-(d,n)}, c, w, m, \pi, \eta)$$
$$\propto p(z_{d,n}|z_{d,-n}, m, \pi)p(w_{d,n}|z, c, w_{-(d,n)}, \eta) \tag{4}$$

where $z_{-(d,n)}$ and $w_{-(d,n)}$ denote the vectors of level allocation and observed words leaving out $z_{d,n}$ and $w_{d,n}$ respectively. The first term in Eq. 4 is a distribution over levels and the second term is the probability of a given word based on the topic assignment.

4 Topic-Level Influence Network

Having organized the documents into topic-specific groups, our next task is to determine the influential strength among the users on each topic. From the proposed guided hLDA model, we find the topic-specific documents by following each path in the model. Let

d_u be the document published by user u and d_v be the document published by user v. Suppose d_u and d_v share the same path that corresponds to topic z.

We say user u influences user v on topic z if the time associated with d_v is greater than d_u. Furthermore, we realize the degree of influence is greater when the time lapse between documents is less. We model this effect using a time decay function $g(d_u, d_v)$. Let t_u and t_v be the times at which users u and v post documents d_u and d_v respectively. Then, we have

$$g(d_u, d_v) = \begin{cases} e^{-\frac{\Delta}{\alpha}}, & \text{if } t_u < t_v \\ 0, & \text{otherwise} \end{cases} \tag{5}$$

where $\Delta = t_v - t_u$ and $\alpha > 0$.

The parameter α controls the time window to compute $g(d_u, d_v)$. Note that for a fixed α, $e^{-\frac{\Delta}{\alpha}} \to 1$ when $\Delta \to 0^+$ and $e^{-\frac{\Delta}{\alpha}} \to 0$ when $\Delta \to +\infty$. This implies that if user v posts a document just after u then u may have an influence on v. On the other hand, if v posts a document after a long elapse time, then u has little influence on v.

Another factor determining the strength of influence between user u and user v on topic z is the degree of similarity among the documents published by u and v on topic z. Let D_u and D_v be the sets of documents published on topic z by users u and v respectively.

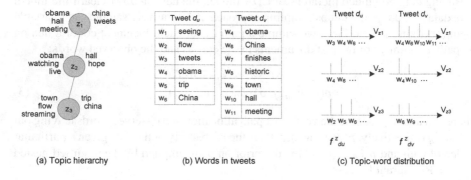

(a) Topic hierarchy   (b) Words in tweets   (c) Topic-word distribution

Fig. 4. (a) Topic hierarchy for tweet d_u and d_v. (b) Words in tweet d_u and d_v. (c) Topic-word distribution for tweet d_u and d_v at each level. Distribution of words in tweet d_u and d_v at each topic w.r.t all the words assigned to that topic.

For each pair of documents (d_u, d_v) where $d_u \in D_u$ and $d_v \in D_v$, we obtain the normalized topic-word distributions of d_u and d_v on topic z from guided hLDA model, denoted as $f_{d_u}^z$ and $f_{d_v}^z$ respectively (see Figure 4). The similarity of these two documents on topic z is evaluated based on the commonly used measure $S(f_{d_u}^z, f_{d_v}^z)$ [19]:

$$sim(d_u, d_v) = 10^{-S(f_{d_u}^z, f_{d_v}^z)}$$

$$= 10^{-[KL(f_{d_u}^z \| \frac{f_{d_u}^z + f_{d_v}^z}{2}) + KL(f_{d_v}^z \| \frac{f_{d_u}^z + f_{d_v}^z}{2})]}$$

where $KL(P\|Q) = \Sigma_i P(i) \log \frac{P(i)}{Q(i)}$ defines the divergence from distribution Q to P.

With this, we define the influential strength between u and v on topic z as follows.

$$strength(u,v) = max_{d_u \in D_u, d_v \in D_v}[g(d_u, d_v) * sim(d_u, d_v)] \qquad (6)$$

Using max function reflects the scenario whereby a user may publish many documents on a topic. As long as one of his published document has large overlapped with another user, we may conclude that this user has influenced the other user.

Algorithm 1 shows the details of our **T**opic-level **I**nfluence **N**etwork **D**iscovery algorithm (TIND). The input is a guided hLDA tree T, time threshold τ, and similarity threshold σ. The output is topic-level influence network G. For each path in the tree T, we obtain the set of documents D associated with the path (Line 3). For each pair of documents in D, we check if their time difference is within the threshold τ (Line 7). If yes, we calculate their similarity for each topic along the path (Line 8). If the similarity for a topic exceeds the threshold σ, we add an edge (u,v) or (v,u) to G with weights w denoting the maximum influential strength between u and v on topic z (Lines 9-18). Finally, in Line 19, we return the constructed topic-level influence network.

Algorithm 1. TIND(T, τ, σ)

Require: guided hLDA tree T, time threshold τ, and similarity threshold σ
Ensure: topic-level influence network G
1: Initialize $G = \emptyset$
2: **for** each path p in T **do**
3:      let D be the set of documents associated with path p
4:      **for** for each pair of documents in D **do**
5:          let d_u be the document published by user u at time t_u
6:          let d_v be the document published by user v at time t_v
7:          **if** $|t_u - t_v| \leq \tau$ **then**
8:              compute sim(d_u, d_v) for each topic z along p
9:              **if** sim(d_u,d_v) $\geq \sigma$ **then**
10:                 compute strength(u,v)
11:                 **if** edge between u and v does not exist **then**
12:                     **if** $t_u < t_v$ **then**
13:                         $G = G \cup (u,v)$ with label (z,strength(u,v))
14:                     **else**
15:                         $G = G \cup (v,u)$ with label (z,strength(u,v))
16:                 **if** strength(u,v) $> max\_strength_{uv}$ **then**
17:                     $max\_strength_{uv}$ = strength(u,v)
18:                     update the label for edge (u,v) to (z,max_strength$_{uv}$)
19: return G

---

5   Experimental Evaluation

In this section, we present the results of experiments conducted to evaluate our proposed method. We implemented the proposed algorithm in C#. The experiments are carried out on an Intel Core 2 Quad CPU 2.83 GHz system with 3GB RAM running Windows.

We use two real world datasets in our experiments. The first is the Twitter dataset [20,21], which consists of 64,451 tweets published by 880 users over a 7 month period from June 1 2009 to December 31 2009. Each tweet has the following information: user, time and content. We preprocess the tweets by stemming and removing stopwords. The tweets are then manually categorized into 6 hot topics and each topic is described by top-5 representative words as shown in Table 1.

We generate the ground truth as follows. A user u is said to be influenced by v on topic z if there is a "follow" relationship from u to v and both u and v have published tweets on topic z with the tweets published by v on z being earlier than that by u. The last column of Table 1 gives the number of influence relationships among the users for each topic.

Table 1. Characteristics of Twitter dataset

Topic	Top-5 representative words	# of tweets	# of ground truth
freeiran	iran, khamenei, tehran, regime, islamic	10,469	992
litchat	litchat, good, think, literature, books	13,511	940
lovestories	karma, forgive, love, lovestories, get	12,502	406
ObamaCN	obama, china, watch, town, hall	1,706	154
supernatural	supernatural, de, dean, que, assistir	13,504	550
Yahoo	yahoo, search, content, site, fav	12,759	326

For our second dataset, we extract from the MemeTracker dataset [6] the quotes, phrases, and hyperlinks of articles/blogposts that appear in prominent online news sites from August to September 2008. Each post contains a URL, time stamp, and all of the URLs of the posts it cites. A site publishes a piece of information and uses hyperlinks to refer to the same or closely related pieces of information published by other sites.

We use the hyperlink information to obtain the ground truth for this dataset. A site u is influenced by another site v on topic z if there exists a hyperlink from u to v on topic z. Table 2 shows the characteristics of this MemeTracker dataset. The default values for the time threshold τ and similarity threshold σ are 20 hours and 0.5 respectively.

Table 2. Characteristics of MemeTracker dataset

Top-5 topics	Top-5 representative words	# of documents	# of ground truth
election	obama, mccain, campaign, vote, political	14,846	2,228
social media	blog, social, media, twitter, post	32,962	5,453
Iraq war	government, military, iraq, security, troop	15,379	1,080
finance	financial, market, credit, money, banks	10,293	2,033
apple	apple, iphone, store, macbook, ipod	11,668	2,059

5.1 Effectiveness Experiments

We carried out two sets of experiments to evaluate the effectiveness of our approach. In the first set of experiments, we evaluate the effectiveness of the guided hierarchical LDA model for grouping the documents into topic-specific clusters. The second set of experiments compare our TIND algorithm with the TAP method [8] which requires the network structure to be known for inferring topic-level influence relationships.

Guided hLDA vs Clustering. We first evaluate the effectiveness of guided hierarchical LDA model for grouping documents into topic-specific clusters. We compare the guided hierarchical LDA model with the original hierarchical LDA model and a clustering based method. The clustering based method compares each tweet with the 6 known hot topics using cosine similarity and groups the tweet under the most similar topic.

For each topic cluster, we determine the influence relationships among the users whose tweets are in the cluster. Let E_{truth} be the set of influence relationships in the ground truth for a topic, and E_θ be the set of influence relationships obtained at various cut-off thresholds, θ. Then the precision and recall of the models are defined as follows:

$$precision = \frac{|E_{truth} \cap E_\theta|}{|E_\theta|} \qquad recall = \frac{|E_{truth} \cap E_\theta|}{|E_{truth}|} \qquad (7)$$

Figure 5 shows the average precision and recall on Twitter data for all the 6 topics in the Twitter dataset as we vary θ from 0.1 to 0.8. We observe that the precision of guided hLDA outperforms that of the original hLDA and the clustering based method. Further, the gaps in precision widen as θ increases. The recall for all three models decreases as θ increases. This is because all the models predict only the influence relationships with influential strength greater than θ. As a result, the number of influence relationships decreases, leading to lower recall. Guided hLDA and clustering based method outperform hLDA in both precision and recall measures, because both methods utilize the hot topics to do clustering, while hLDA does not utilize any additional information.

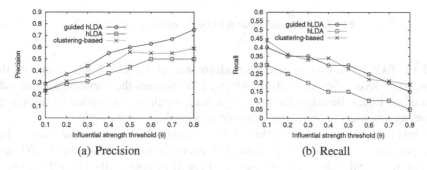

(a) Precision   (b) Recall

Fig. 5. Guided hLDA vs. clustering for varying θ on Twitter data

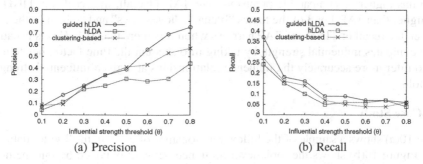

(a) Precision   (b) Recall

Fig. 6. Guided hLDA vs. clustering for varying θ on MemeTracker data

Figure 6 shows the average precision and recall on MemeTracker dataset as we vary θ from 0.1 to 0.8. Once again, we observe that guided hLDA outperforms both clustering based method and hLDA especially when θ is large.

Figure 7 shows the hierarchical topic tree generated by guided hLDA and hLDA as well as example tweets assigned to each path. We observe that hLDA may assign tweets with different topics into the same branch, while guided hLDA can correctly assign tweets into the appropriate branch based on their topics.

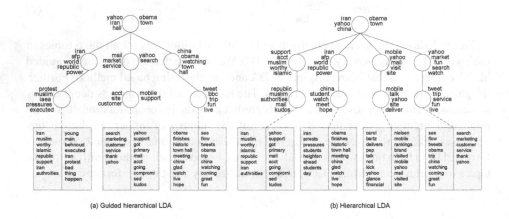

Fig. 7. Guided hierarchical LDA vs. hierarchical LDA

TIND vs. TAP. Next, we compare the performance of our TIND algorithm with the existing topic-level influence method TAP [8]. TAP assumes the documents are already grouped into topics. Based on the groupings, it then utilizes the explicit modeled connections among users to derive the influence relationships for the topic.

We first apply the guided hierarchical LDA to obtain the topic-specific clusters. For each topic cluster, we generate topic-level influence relationships using both TIND and TAP. Figures 8 and 9 show the precision and recall of both methods on the 6 topics in the Twitter dataset as we vary θ from 0.1 to 0.8. We observe that in all the topics, TIND has higher or comparable precision than TAP. Overall, the recall for TIND is also higher than TAP. For 3 of the topics "litchat", "lovestories" and "Obama", the gap between the recall of TIND and TAP narrows when θ is more than 0.4. This is because TIND computes influential strength by taking into account the time factor, hence it is able to infer more accurately the influence relationships at a given influential strength threshold.

5.2 Case Study

Figure 10(a) shows a sample of the follow relationships of users in the Twitter dataset, while Figure 10(b) shows the topic-level influence network obtained by our method. We see that when there is a following relationship from users u to v in Figure 10(a), our method will correctly infer that v influences u on the associated topic. For example,

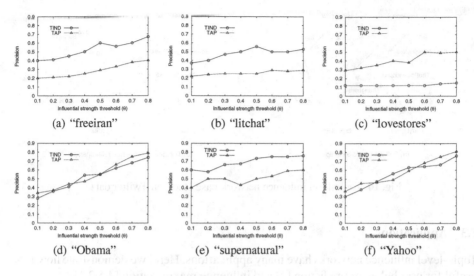

Fig. 8. Precision of TIND vs. TAP for varying θ

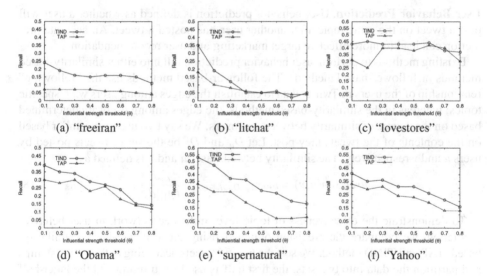

Fig. 9. Recall of TIND vs. TAP for varying θ

user *Indexma* is following user *SearchEngineNow*, and our network shows that user *Indexma* is influenced by user *SearchEngineNow* on the topic "Yahoo".

In addition, our method can also infer influence relationship between two users although they are not following each other. For example, there is no edge between user *CaryHooper* and *cspodium* in Figure 10(a), indicating that *CaryHooper* is not following *cspodium*. However, our topic-level influence network discovers that *cspodium* influences *CaryHooper* on topic "Obama". When examining the tweets of *CaryHooper*, we realize that his tweets are very similar to *cspodium*'s and have been posted soon after *cspodium*'s tweets, indicating that *cspodium* could have some influence on *CaryHooper*.

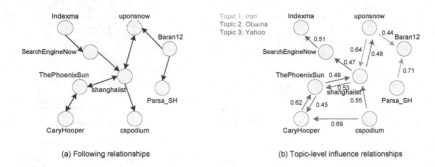

(a) Following relationships   (b) Topic-level influence relationships

Fig. 10. Topic-level influence network case study on Twitter data

5.3   Applications

Topic-level influence networks have many applications. Here, we demonstrate how it is useful for user behavior prediction [7] and influence maximization [4,5,2,1].

User Behavior Prediction. User behavior prediction is defined as whether a user will post a tweet on the same topic after another user has posted a tweet. An accurate prediction can lead to more effective target marketing and user recommendation.

Existing methods to perform user behavior prediction fall into either similarity-based methods or follower-based methods. The follower-based methods use the "following" relationship of the users in Twitter data to establish the edges among users w.r.t. specific topics; whereas in the similarity-based methods, the edges among users are determined based on the degree of similarity between two users. We say two users are similar based on the contents of the tweets they post. Let D_u and D_v be the set of tweets posted by users u and v respectively. The similarity between user u and v is defined as:

$$user\_sim(u,v) = max_{d_u \in D_u, d_v \in D_v} \left[\frac{d_u \cdot d_v}{||d_u|| ||d_v||} \right] \tag{8}$$

To demonstrate the effectiveness of topic-level influence network in user behavior prediction, we compare the precision obtained using the follower-based, similarity-based, TAP and TIND method. We sort the 64,451 tweets according to their time stamps and partition the data into two sets: the first half is used for training and the latter half is used for testing. Note that if the time difference between the two tweets posted by u and v is larger than a given time threshold, we consider the two tweets to be unrelated and there will be no edge between u and v. In this experiment, we set the time threshold τ to 30 for all the 4 methods.

We use the training dataset to construct four networks using the four methods. Based on the constructed networks, we perform user behaviour prediction as follows: Let M be the constructed network. For each user u in the training set, let Y_u^M be the set of users that are connected to u in M. Let X_u be the set of users who have posted a tweet on the same topic within the time threshold after u's tweet in the test dataset. Then the precision of model M is given as:

$$precision(M) = \frac{\sum_u |X_u \cap Y_u^M|}{\sum_u |Y_u^M|} \qquad (9)$$

We use the influential strength threshold of 0.6 and similarity threshold of 0.6 as determined in an empirical study. We plot the precision of the four models as we vary the number of users involved in the training and testing datasets. We repeat the experiment for two topics, i.e. "Obama" and "Yahoo". Figure 11 shows the results.

We observe that as the number of users increases, the precision for similarity-based and follower-based models decreases whereas the topic-level influence network is more stable. This is because similarity-based and follower-based models simply predict the most popular users without taking into consideration the topic. On the other hand, topic-level influence network predicts accurately users that are interested in the specific topic. Note that both TIND and TAP consider topic information, however, TIND outperforms TAP as TAP relies on the network structure. This demonstrates that topic-level influence can indeed improve the performance of user behavior prediction.

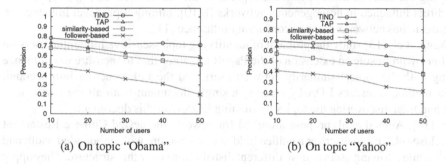

(a) On topic "Obama"   (b) On topic "Yahoo"

Fig. 11. User behavior prediction

Influence Maximization. The problem of influence maximization in a social network is to find k nodes in the network such that the expected number of nodes influenced by this k nodes is maximized. The work in [4] proposed a greedy algorithm to identify the k nodes. At each iteration, it selects a node that leads to the largest increase in the number of nodes influenced. The algorithm stops when k nodes are selected.

For topic-specific influence maximization, we define the influence spread of the k nodes as the set of nodes influenced by these k nodes on a given topic. A node u is said to be influenced by another node v on a topic z if the tweets posted by u and v contain topic z. In our experiments, we use the 64,451 tweets to construct four networks using the follower-based, similarity-based, TAP and TIND method. We run the greedy algorithm to find k nodes in each of the four networks. We select two topics, namely "Obama" and "Yahoo", and determine the influence spread of the k nodes on each topic.

Figure 12 shows the influence spread as we vary k from 10 to 50. We observe that the influence spread of all the four methods increases as k increases with TIND clearly in the lead. This demonstrates that topic-level influence network is effective for influence maximization.

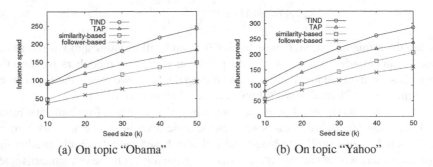

(a) On topic "Obama"   (b) On topic "Yahoo"

Fig. 12. Influence maximization

6 Related Work

Research on influence analysis has focused on validating the existence of influence [12], studying the maximization of influence spread in the whole network [4,5,2,1], modeling direct influence in homogeneous networks [8,10], mining topic-level influence on heterogeneous networks [7], and conformity influence [11].

Weng et al. [13] study the problem of identifying topic-sensitive influential users on Twitter by proposing an extension of the PageRank algorithm to measure the influence taking both the topical similarity between users and the link structure into account. Their method leverages LDA by creating a single document from all the tweets of a user and then discovering the topics by running LDA over this document.

In [14], Ahmed et al. propose a unified framework, the nested Chinese Restaurant Franchise (nCRF), to discover a unified hidden tree structure with unbounded width and depth while allowing users to have different distributions over this structure. They apply the framework to organize tweets into a hierarchical structure and show that this tree structure can predict locations of unlabeled messages, resulting in significant improvements to state-of-the-art approaches, as well as revealing interesting hidden patterns.

Tang et al. [8] introduce topic-based influence analysis and present a method to quantify the influential strength in social networks. Given a social network and a topic distribution for each user, the problem is to find topic-specific sub-networks, and topic-specific influence weights between members of the sub-networks. They propose a Topical Affinity Propagation (TAP) model to model social influence in a network for different topics, which are extracted by using topic modeling methods [17]. Later, Wang et al. [9] extend the TAP model further by considering the dynamic social influence.

In [7], Liu et al. introduce a probabilistic model for mining direct and indirect influence between nodes of heterogeneous networks. They measure influence based on clearly observable "following" behaviors and study how influence varies with number of hops in the network. All these works assume that the network connections are known.

Temporal factor plays an important role in determining social influence [12,15]. Kossinets et al. [16] take into account the temporal factor in information network and show that considering only the topology is not sufficient to understand how information diffuses through a network. Gomez et al. [3] study the diffusion of information among blogs and online news sources. They assume that connections between nodes are not known and use the observed cascades to infer the "hidden" network of diffusion and influence. They do not model topics and ignore the influential strength at topic level.

7 Conclusion

In this paper, instead of focusing on macro-level social influence, we have investigated the micro-level mechanisms of social influence and have taken into account the temporal factor in social influence to infer the influential strength between users at topic-level. Our approach does not require the underlying network structure to be known. To achieve this, we have proposed a guided hierarchical LDA model to automatically identify topics without using any structural information. Experimental results on two real world datasets have demonstrated the effectiveness of our method. Further, we have shown that the proposed topic-level influence network can improve the precision of user behavior prediction and is useful for influence maximization.

References

1. Chen, W., Wang, C., Wang, Y.: Scalable Influence Maximization for Prevalent Viral Marketing in Large-Scale Social Networks. In: KDD, pp. 1029–1038 (2010)
2. Chen, W., Wang, Y., Yang, S.: Efficient Influence Maximization in Social Networks. In: KDD, pp. 199–208 (2009)
3. Gomez-Rodriguez, M., Leskovec, J., Krause, A.: Inferring Networks of Diffusion and Influence. In: KDD, pp. 1019–1028 (2010)
4. Kempe, D., Kleinberg, J., Tardos, É.: Maximizing the Spread of Influence through a Social Network. In: KDD, pp. 137–146 (2003)
5. Leskovec, J., Krause, A., Guestrin, C., Faloutsos, C., VanBriesen, J., Glance, N.: Cost-effective Outbreak Detection in Networks. In: KDD, pp. 420–429 (2007)
6. Leskovec, J., Backstrom, L., Kleinberg, J.: Meme-tracking and the Dynamics of the News Cycle. In: KDD, pp. 497–506 (2009)
7. Liu, L., Tang, J., Han, J., Jiang, M., Yang, S.: Mining Topic-level Influence in Heterogeneous Networks. In: CIKM, pp. 199–208 (2010)
8. Tang, J., Sun, J., Wang, C., Yang, Z.: Social Influence Analysis in Large-scale Networks. In: KDD, pp. 807–816 (2009)
9. Wang, C., Tang, J., Sun, J., Han, J.: Dynamic Social Influence Analysis through Time-Dependent Factor Graphs. In: ASONAM, pp. 239–246 (2011)
10. Xiang, R., Neville, J., Rogati, M.: Modeling Relationship Strength in Online Social Networks. In: WWW, pp. 981–990 (2010)
11. Tang, J., Wu, S., Sun, J.: Confluence: Conformity Influence in Large Social Networks. In: KDD, pp. 347–355 (2013)
12. Anagnostopoulos, A., Kumar, R., Mahdian, M.: Influence and Correlation in Social Networks. In: KDD, pp. 7–15 (2008)
13. Weng, J., Lim, E.P., Jiang, J., He, Q.: TwitterRank: Finding Topic-sensitive Influential Twitterers. In: WSDM, pp. 261–270 (2010)
14. Ahmed, A., Hong, L., Smola, A.J.: Hierarchical Geographical Modeling of User Locations from Social Media Posts. In: WWW, pp. 25–35 (2013)
15. Saez-Trumper, D., Comarela, G., Almeida, V., Baeza-Yates, R., Benevenuto, F.: Finding Trendsetters in Information Networks. In: KDD, pp. 1014–1022 (2012)

16. Kossinets, G., Kleinberg, J., Watts, D.: The Structure of Information Pathways in a Social Communication Network. In: KDD, pp. 435–443 (2008)
17. Blei, D.M., Ng, A.Y., Jordan, M.I.: Latent Dirichlet Allocation. J. Mach. Learn. Res. 3, 993–1022 (2003)
18. Blei, D.M., Griffiths, T.L., Jordan, M.I.: The Nested Chinese Restaurant Process and Bayesian Nonparametric Inference of Topic Hierarchies. J. ACM 57(2), 1–30 (2010)
19. Manning, C.D., Schütze, H.: Foundations of Statistical Natural Language Processing. MIT Press (1999)
20. Yang, J., Leskovec, J.: Patterns of Temporal Variation in Online Media. In: WSDM, pp. 177–186 (2011)
21. Kwak, H., Lee, C., Park, H., Moon, S.: What is Twitter, a Social Network or a News Media? In: WWW, pp. 591–600 (2010)

A Synergy of Artificial Bee Colony and Genetic Algorithms to Determine the Parameters of the Σ-Gram Distance

Muhammad Marwan Muhammad Fuad

Forskningsparken 3, Institutt for kjemi, NorStruct
The University of Tromsø, The Arctic University of Norway
NO-9037 Tromsø, Norway
marwan.fuad@uit.no

Abstract. In a previous work we presented the Σ-gram distance that computes the similarity between two sequences. This distance includes parameters that we calculated by means of an optimization process using artificial bee colony. In another work we showed how population-based bio-inspired algorithms can be sped up by applying a method that utilizes a pre-initialization stage to yield an optimal initial population. In this paper we use this pre-initialization method on the artificial bee colony algorithm to calculate the parameters of the Σ-gram distance. We show through experiments how this pre-initialization method can substantially speed up the optimization process.

Keywords: Artificial Bee Colony, Bio-inspired Optimization, Genetic Algorithms, Pre-initialization, Σ-gram.

1   Introduction

Optimization is a rich domain of research and application in computer science and applied mathematics. An optimization problem can be defined as follows: Given a function $f : U \subseteq \mathbf{R}^{nbp} \to \mathbf{R}$ (*nbp* is the number of parameters), find the solution $\overrightarrow{X^*} = \left[x_1^*, x_2^*, ..., x_{nbp}^* \right]$ which satisfies: $f\left(\overrightarrow{X^*} \right) \le f\left(\overrightarrow{X} \right), \forall \overrightarrow{X} \in U$. The function f is called the *fitness function*, or the *objective function*. Informally, the purpose of an optimization process is to find the best-suited solution of a problem subject to given constraints.

Bio-inspired, also called *nature-inspired*, optimization algorithms have gained popularity in many applications because they handle a variety of optimization problems. These algorithms are inspired by natural processes, natural phenomena, or by the collective intelligence of natural agents.

One of the main bio-inspired optimization families is *Evolutionary Algorithms* (*EA*). *EA* are population-based metaheuristics that use the mechanisms of Darwinian evolution. The *Genetic Algorithm* (*GA*) is the main member of *EA*. GA is an

H. Decker et al. (Eds.): DEXA 2014, Part II, LNCS 8645, pp. 147–154, 2014.
© Springer International Publishing Switzerland 2014

optimization and search technique based on the principles of genetics and natural selection [1]. GA has the following elements: a population of individuals, selection according to fitness, crossover to produce new offspring, and random mutation of new offspring [5].

Data mining is a field of computer science which handles several tasks such as classification, clustering, anomaly detection, and others. Processing these tasks usually requires extensive computing. As with other fields of computer science, different papers have proposed applying bio-inspired optimization to data mining tasks [8] [9] [10].

In [6] we presented a new distance metric, the *Sigma Gram* distance (*SG*) that is applied to sequences. *SG* uses parameters which we computed using an optimization algorithm called *Artificial Bee Colony* (*ABC*); one of the bio-inspired optimization algorithms.

Applying *ABC*, and other bio-inspired algorithms, to the data mining problems requires recruiting extensive computing resources and long computational time. This is a part of what is called *expensive optimization*. In [7] we presented a new technique to handle such optimization problems. In this work we re-visit the work presented in [6] and apply the technique we introduced in [7] to speed up the optimization process.

The rest of this paper is organized as follows: Section 2 is a background section, in Section 3 we explain the pre-initialization method and we show how it can be used to speed up the optimization process, we test this pre-initialization method to compute the parameters of the Σ-gram in Section 4. Section 5 is a concluding section.

2   Background

Let Σ be a finite alphabet of a set of characters. A *string* is an ordered set of this alphabet. Strings appear in a variety of domains in computer science and bioinformatics. The *Edit Distance* is the main distance used to compare two strings. In a previous work [6] we presented an extension of the edit distance, which is based on the sum of *n*-grams. The proposed distance Σ-gram (which we refer to in this paper as *SG*) is defined as follows:

Let Σ^* be the set of strings on Σ. Given a positive integer n, let $f_{a_n}^{(S)}$ be the frequency of the *n-gram* a_n in S, and $f_{a_n}^{(T)}$ be the frequency of the *n-gram* a_n in T, where S, T are two strings in Σ^*. Let $\mathbf{N}$ be the set of integers, and $\mathbf{N}^+$ the set of positive integers.

Let $g : \mathbf{N}^+ \times \Sigma^* \to \mathbf{N}$

$$g(n,S) = n \qquad \text{if} \quad 1 \le n \le |S|$$

$$g(n,S) = |S| + 1 \qquad \text{if} \qquad |S| < n$$

Then SG is defined as:

$$SG(S,T) = \sum_{n=1}^{max(|S|,|T|)} \lambda_n \left[|S| + |T| - g(n,S) - g(n,T) + 2 - 2 \cdot \sum_{a_n \in \Sigma^n} min\left(f_{a_n}^{(S)}, f_{a_n}^{(T)}\right) \right] \quad (1)$$

where $|S|, |T|$ are the lengths of the two strings S, T respectively, and where $\lambda_n \in \mathbf{R}^+ \cup \{0\}$.

Determining the values of the parameters λ_n is not a trivial task. In [6] these values were obtained as the outcome of an optimization problem. The optimization algorithm we used was artificial bee colony (*ABC*).

2.1 Artificial Bee Colony (*ABC*)

Artificial Bee Colony (*ABC*) [2] is an optimization algorithm inspired by the foraging behavior of bees. In *ABC* each food source represents a potential solution to the optimization problem and the quality of the food represents the value of the objective function to be optimized. Artificial bees explore and exploit the search space. These bees communicate and share information about the location and quality of food sources. Bees exchange of information by performing a *waggle dance* which takes place in the dancing area in the hive. In *ABC* there are three kinds of bees; *employed bees*: these are the bees that search in the neighborhood of a food source. They perform a dance with a probability that is proportional to the quality of the food source, *onlooker bees*: these bees are found on the dance floor, and *scouts*: these bees explore the search space randomly.

The first step of *ABC* is generating a randomly distributed population of size (*pop_size*) of food sources which correspond to potential solutions. Each solution $\vec{x}_i, i \in \{1,..,pop\_size\}$ is a vector whose dimension is (*nr_par*) which is equal to the number of parameters of the function f to be optimized. The population is subject to change for a number of cycles (*nr_cycles*). In each cycle every employed bee perturbs the current solution using a local search procedure. The perturbation produces a new solution:

$$\vec{x}_i^* = \vec{x}_i + rand(-1,1)(\vec{x}_i - \vec{x}_k) \quad , i \neq k \quad (2)$$

The above relation is not applied to all parameters but only to a certain number of them. The parameters to be altered are chosen randomly. The algorithm uses a greedy selection to decide if the new solution should be kept or discarded, i.e.:

$$\vec{x}_i = \begin{cases} \vec{x}_i^* & if \quad f(\vec{x}_i^*) < f(\vec{x}_i) \\ \vec{x}_i & otherwise \end{cases} \quad (3)$$

After all employed bees have modified their positions the onlooker bees choose one of the current solutions depending on a probability that corresponds to the fitness value of that solution according to the following rule:

$$p_i = \frac{f(\vec{x}_i)}{\sum_{k=1}^{pop\_size} f(\vec{x}_k)} \tag{4}$$

After that the onlooker bees try to improve the solution using the same mechanism that was described in (4). The number of trials the algorithm attempts to improves the same solution is limited by a maximum number (*max_nr*) after which the solution is abandoned and the bees employed by that food source become scouts. The abandoned solution is replaced by a new solution found by the scouts.

3   A Pre-initialized Artificial Bee Colony Algorithm

The optimization problem we presented in Section 2 requires extensive computing. This type of optimization problems is called expensive optimization. In [7] we introduced a new method that can be applied to any population-based optimization algorithm to speed up the optimization process. The principle of this method is to use an "optimal" initial population by adding to the main problem, which we call *MainOptim*, an artificial optimization problem to optimize the initial population. We call this latter problem *SecOptim*. As a fitness function of *SecOptim* we choose one that gives as much information as possible about the search space of *MainOptim* since this initial population will eventually be used to optimize *MainOptim*. The fitness function for *SecOptim* will be the one that maximizes the average distance of the chromosomes of the population, i.e.:

$$f_{secOptim} = \frac{2}{secPopSize\,(secPopSize\,-1)} \sum_{i=1}^{secPopSize-1} \sum_{j=i+1}^{secPopSize} d(ch_i, ch_j) \tag{5}$$

where *secPopSize* is the population size of *SecOptim*, *ch* is the chromosome. *d* is a distance, which we choose to be the Euclidean distance. Notice that $d(ch_i, ch_j) = d(ch_j, ch_i)$ so we only need to take half of the summation in (5).

The other component of *SecOptim* is the search space. As indicated earlier, *SecOptim* is a separate optimization problem from *MainOptim* with its own search space. The search space of *SecOptim* is a discrete one whose points are feasible solutions of *MainOptim*. In other words, the search space of *SecOptim* is a *pool* of solutions of *MainOptim*. The cardinality of this pool is denoted by *poolSize*.

Now all the elements of *SecOptim* are defined. *poolSize* is a new element that is particular to our method. In the experimental section we discuss this element further.

As we can see, *MainOptim* and *SecOptim* are two independent problems, so we can use two different optimization algorithms. One of the optimization algorithms that we can use for *SecOptim* is the Genetic Algorithms for its exploiting ability.

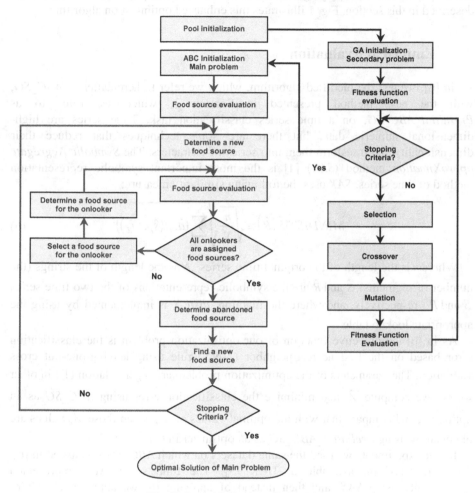

Fig. 1. Pre-initialized *ABC* algorithm

The Genetic Algorithm (*GA*): *GA* is a widely-known bio-inspired optimization algorithm. *GA* starts by randomly generating a number of chromosomes. This step is called *initialization*. The fitness function of each chromosome is evaluated. The next step is *selection*. The purpose of this procedure is to determine which chromosomes are fit enough to survive. *Crossover* is the next step in which offspring of two parents are produced to enrich the population with fitter chromosomes. The last element is *Mutation* of a certain percentage of chromosomes.

The basis of our work we present here is that instead of applying *ABC* directly, which was the case in [6], to obtain the optimal values of λ_n , we use an optimal initial population by applying the method we presented in [7], and which we described in this section. Fig. 1 illustrates this enhanced optimization algorithm.

4   Empirical Evaluation

As in [6], we test the modified algorithm, which we refer to hereinafter as *ABC_SG*, with the new method presented in Section 3, which we refer to as *PreInitial_ABC_SG*, on a time series classification task. Time series are high-dimensional numeric data, but there are some techniques that reduce their dimensionality and transform them into series of characters. The *Symbolic Aggregate approXimation* method (*SAX*) [4] is the most important symbolic representation method of time series. *SAX* uses the following similarity measure:

$$MINDIST(\hat{S},\hat{R}) \equiv \sqrt{\frac{n}{N}} \sqrt{\sum_{i=1}^{N} (dist(\hat{s}_i, \ \hat{r}_i))^2} \qquad (6)$$

Where n is the length of the original time series, N is the length of the strings (the number of segments), $\hat{S}$ and $\hat{R}$ are the symbolic representations of the two time series S and R , respectively, and where the function *dist*( ) is implemented by using the appropriate lookup table.

As in [6] the objective function of our optimization problem is the classification error based on the first nearest-neighbor (*1*-NN) rule using leaving-one-out cross validation. The parameters of the optimization problem are λ_n in relation (1), in other words, we compute λ_n that minimize the classification error using *ABC_SG* as an optimizer, and compare that with the optimal values of λ_n when those λ_n values are obtained by using *PreInitial_ABC_SG* as an optimizer in (1)

In our experiments we used the same datasets on which *ABC_SG* was tested in [6]. These datasets are available at UCR [3]. The time series were represented symbolically using *SAX*, and then instead of applying (6) we apply *ABC_SG* (or *PreInitial_ABC_SG*).

We used 3 different values of the alphabet size in *SAX* : 3,10, 20. As for n in (1) we used $n \in \{1,2,3\}$. When testing a method, we first apply it to the training sets to obtain the optimal λ_n then these values are used on the testing sets.

The aim of the experiments is to show that by using *PreInitial_ABC_SG* , we can get classification errors close to those obtained by *ABC_SG* in a shorter time. This is achieved practically by running *PreInitial_ABC_SG* for a smaller number of generations; *NrGen*=20 (*MainOptim*), and running *ABC_SG* for *NrGen*=100. We then compare the results in term of classification error and

In Table 1 we show the classification error of *PreInitial_ABC_SG* and *ABC_SG* (because of space limitation we only show a part of the tested datasets).

Table 1. Comparison between the classification error of *PreInitial_ABC_SG* and that of *ABC_SG* for different values of the alphabet size and for different *n*-grams

Dataset	Method		*n*-gram		
			n=1	*n*=2	*n*=3
ECG	*PreInitial_ABC_SG*	α= 3	0.210	0.210	0.220
		α=10	0.200	0.210	0.220
		α=20	0.220	0.240	0.250
	ABS_SG	α= 3	0.190	0.210	0.240
		α=10	0.200	0.220	0.220
		α=20	0.230	0.230	0.260
Gun_Point	*PreInitial_ABC_SG*	α= 3	0.180	0.193	0.180
		α=10	0.133	0.133	0.127
		α=20	0.073	0.073	0.067
	ABS_SG	α= 3	0.193	0.193	0.180
		α=10	0.146	0.127	0.133
		α=20	0.087	0.073	0.073
FaceFour	*PreInitial_ABC_SG*	α= 3	0.057	0.057	0.045
		α=10	0.045	0.045	0.102
		α=20	0.090	0.114	0.102
	ABS_SG	α= 3	0.057	0.057	0.057
		α=10	0.045	0.057	0.114
		α=20	0.114	0.114	0.102
OSULeaf	*PreInitial_ABC_SG*	α= 3	0.331	0.343	0.322
		α=10	0.298	0.298	0.298
		α=20	0.306	0.343	0.322
	ABS_SG	α= 3	0.351	0.343	0.331
		α=10	0.298	0.306	0.298
		α=20	0.322	0.331	0.331

In Table 2 we present the wall clock time comparison between *PreInitial_ABC_SG* and *ABC_SG* for the datasets presented in Table 1. The experiments were conducted on Intel Core 2 Duo CPU with 3G memory.

Table 2. Run time comparison between *PreInitial_ABC_SG* and *ABC_SG*

Dataset	Method		*n*-gram		
			n=1	*n*=2	*n*=3
ECG	*PreInitial_ABC_SG*	α= 3	01h 38m 43s	02h 00m 51s	03h 46m 25s
		α=10	01h 42m 08s	02h 08m 03s	03h 59m 35s
		α=20	01h 47m 26s	02h 53m 52s	34h 37m 22s
	ABS_SG	α= 3	08h 16m 49s	10h 06m 15s	18h 54m 17s
		α=10	08h 30m 24s	10h 30m 42s	19h 53m 23s
		α=20	08h 49m 34s	14h 23m 12s	154h 07m 20s
Gun_Point	*PreInitial_ABC_SG*	α= 3	01h 54m 06s	02h 22m 51s	04h 15m 34s
		α=10	02h 12m 17s	02h 38m 54s	04h 48m 22s
		α=20	02h 56m 48s	03h 34m 16s	42h 52m 32s
	ABS_SG	α= 3	09h 58m 34s	13h 16m 35s	21h 37m 18s
		α=10	11h 12m 41s	14h 52m 42s	24h 27m 52s
		α=20	14h 21m 26s	16h 46m 47s	138h 36m 62s
FaceFour	*PreInitial_ABC_SG*	α= 3	01h 22m 31s	01h 55m 24s	03h 25m 46s
		α=10	01h 38m 08s	02h 01m 45s	03h 14m 51
		α=20	01h 42m 52s	02h 43m 18s	29h 26m 26s
	ABS_SG	α= 3	07h 26m 52s	09h 57m 32s	17h 17m 52s
		α=10	08h 04m 28s	10h 03m 52s	19h 54m 26s
		α=20	08h 36m 26s	13h 52m 52s	148h 17m 53s
OSULeaf	*PreInitial_ABC_SG*	α= 3	12h 14m 52s	15h 45m 31s	25h 24m 35s
		α=10	13h 46m 26s	16h 52m 03s	23h 35m 43s
		α=20	15h 25m 26s	22h 12m 04s	184h 26m 54s
	ABS_SG	α= 3	62h 51m 02s	72h 26m 01s	81h 19m 31s
		α=10	64h 56m 41s	74h 55m 10s	88h 01m 47s
		α=20	67h 42m 45s	81h 57m 29s	543h 47m 29s

As we can see from Tables 1 and 2, *PreInitial_ABC_SG* is on average almost 5 times faster than *ABC_SG* although they both give quite comparable classification errors.

5   Conclusion

In this paper we applied a method from a previous work that speeds up population-based bio-inspired algorithms by initializing the optimization process using an optimal population. We applied this method to compute the parameters λ_n of the Σ-gram distance and we showed experimentally how by using an optimal initial population the optimization process can be sped up substantially.

References

1. Haupt, R.L., Haupt, S. E.: Practical Genetic Algorithms with CD-ROM. Wiley-Interscience (2004)
2. Karaboga, D.: An idea based on honey bee swarm for numerical optimization. Technical Report TR06, Erciyes University, Engineering Faculty, Computer Engineering Department (2005)
3. Keogh, E., Zhu, Q., Hu, B., Hao, Y., Xi, X., Wei, L., Ratanamahatana: The UCR Time Series Classification/Clustering Homepage, C. A. (2011),
   http://www.cs.ucr.edu/~eamonn/time_series_data/
4. Lin, J., Keogh, E., Lonardi, S., Chiu, B.Y.: A Symbolic Representation of Time Series, with Implications for Streaming Algorithms. In: DMKD 2003, pp. 2–11 (2003)
5. Mitchell, M.: An Introduction to Genetic Algorithms. MIT Press, Cambridge (1996)
6. Muhammad Fuad, M.M.: ABC-SG: A New Artificial Bee Colony Algorithm-Based Distance of Sequential Data Using Sigma Grams. In: The Tenth Australasian Data Mining Conference, AusDM 2012, Sydney, Australia, December 5-7 (2012)
7. Muhammad Fuad, M.M.: A Pre-initialization Stage of Population-Based Bio-inspired Metaheuristics for Handling Expensive Optimization Problems. In: Motoda, H., Wu, Z., Cao, L., Zaiane, O., Yao, M., Wang, W. (eds.) ADMA 2013, Part II. LNCS, vol. 8347, pp. 396–403. Springer, Heidelberg (2013)
8. Muhammad Fuad, M.M.: Differential Evolution versus Genetic Algorithms: Towards Symbolic Aggregate Approximation of Non-normalized Time Series. In: Sixteenth International Database Engineering & Applications Symposium, IDEAS 2012, Prague, Czech Republic, August 8-10. BytePress/ACM (2012)
9. Muhammad Fuad, M.M.: Towards Normalizing the Edit Distance Using a Genetic Algorithms–Based Scheme. In: Zhou, S., Zhang, S., Karypis, G. (eds.) ADMA 2012. LNCS, vol. 7713, pp. 477–487. Springer, Heidelberg (2012)
10. Muhammad Fuad, M.M.: Using Differential Evolution to Set Weights to Segments with Different Information Content in the Piecewise Aggregate Approximation. In: 16th International Conference on Knowledge-Based and Intelligent Information & Engineering Systems, KES 2012, San Sebastian, Spain, September 10-12. Frontiers of Artificial Intelligence and Applications (FAIA). IOS Press (2012)

Web Application Relations Mapping

Radek Mařík[1], Zdeněk Kouba[2], and Michal Pantůček[3]

[1] FEE CTU, Department of Telecommunications, Technicka 2
166 27 Praha 6, Czech Republic
marikr@fel.cvut.cz
http://www.comtel.cz/
[2] FEE CTU, Department of Cybernetics, Technicka 2
166 27 Praha 6, Czech Republic
kouba@fel.cvut.cz
http://cyber.felk.cvut.cz/
[3] Neoware s.r.o., Prague, Czech Republic
michal.pantucek@neoware.cz
http://www.neoware.cz/

Abstract. Web applications are developed and modified often so fast
that architects, developers, and managers lose their control over quality of such software products. Reverse engineering techniques focused on
different aspects of software implementation might help in keeping comprehension to implementation of web applications at appropriate level
required for fulfilling change requests. We focus our effort on a reconstruction of implicit associations among GUI data items and back-end
database entities. Our approach is based on an association analysis and
mining using a synchronized log of GUI events and related counterparts
of SQL statements. Recovered dependencies between GUI and back-end
databases can be utilized advantageously in an automated design of web
application data flow testing.

Keywords: data flow, reverse engineering, GUI events, SQL statements,
association mining, software comprehension.

1 Introduction

Existing web applications are modified so often, so quickly, and by so many
different people so that responsible developers, architects, and their managers
lose their in-depth view and understanding to the implementation structure.
Although the appropriate software engineering principles are well known, their
use is usually neglected or not applied at all. Consequently, the architecture of
software and a related documentation deteriorate through modifications. Thus,
application analysis, comprehension and modifications require considerable effort.

A typical architecture of web applications follows the Model-View-Controller
(MVC) pattern that is known since 1979 [13] and followed by a number of frameworks including Java EE [8]. GUI widgets belonging to the view part are well

H. Decker et al. (Eds.): DEXA 2014, Part II, LNCS 8645, pp. 155–163, 2014.

separated from the model entities that are implemented usually using a relational database. If such web applications utilize advanced features of complex implementation frameworks, such as Java EE, then interactions and relationships between widgets of graphical user interface mutually (GUI) and between GUI widgets and DB entities might be hidden through a number of procedural implementation layers. We do not expect that even parsing of source code based on state of art post-processing techniques can recover such relations.

To make justifiable decisions on implementations of required modifications one needs to comprehend such relationships easily. In this paper, we propose a method of relationship reconstruction that is based on observations of activities occurring on both application ends, i.e. events happening at the graphical user interface and triggered operations performed by the DB engine. Our association mining method focused on static aspects of application is based on a comparison of synchronized traces logged at both GUI and DB layers. Input values entered by the user and activities triggered by the user are assumed to be passed from the GUI layer through a complex business logic layer to the back-end DB layer processing. The actual input values serve as binding links between the GUI widgets and DB entities. Although the basic idea of the method is rather simple, we have not found any reference that exploits such an approach.

The paper is organized as follows. We provide a general overview of state of art methods in Section 2. A detailed description of the problem is specified in Section 3. We propose the new method of association mining in Section 4. Experimental results are presented in Section 5. Finally, Section 6 concludes and presents future work.

2   State of Art

Data mining and association mining of software applications has become a standard way how to ensure appropriate comprehension to their implementation. Results are also utilized for evaluation of implementation quality, for automated test suite generation, and implementation refactoring. There are many aspects of software implementation that can be retrieved [14]. In this paper we focus on **data value propagation**. Nevertheless, identification of related graphical user interface widgets, DB entities and their relations, and behavioral aspects of software implementation cannot be avoided in this case. We provide a basic overview of published techniques in the following paragraphs.

A general overview of reverse engineering methodology applied to web applications can be found in [12,7]. The approach aims to produce a subset of UML diagrams such as class, use case, and sequence diagrams. Benslimane et al [5] present an approach analyzing HTML web pages to derive a first cut version of the conceptual schema modeling the web application based on domain ontology. A method transforming the GUI into class diagram with processes based on the interpreted Petri nets models is provided in [11]. Starting from Data Flow Diagrams Canfora et al [6] synthesize semantic abstractions called Dynamic Data Flow Diagrams which can be used for a production of executable models of a

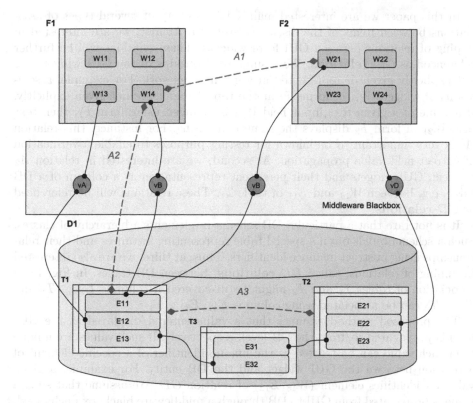

Fig. 1. Items and their relations of the association mining problem

software system. The authors of [4] proposed the use of DB reverse engineering in order to extract the data structures from the legacy systems and integrate them with the new technology systems. Mihancea [10] presents the framework MEMBRAIN that is a dataflow analysis infrastructure for reverse engineering.

The authors of this paper are not aware of any publication that covers recovering of mutual dependencies between GUI widgets, and between GUI widgets and DB entities. In the following sections we draw a framework capable to recover at least a subset of such relations.

3   Problem Model

In this section we provide a formal specification of the problem. We will use an exemplar diagram in Fig. 1 to guide the reader through our proposal. There are two GUI forms F_1 and F_2 at the top of the diagram. The forms consist of several widgets W_{ij} which mutual relationships are not known. Database D_1 consists of a number of tables T_k. Let E_{kl} is a subset of column names of a given table T_k. A business logic layer remains unknown, it is depicted in Fig. 1 in the middle of the diagram as a middleware blackbox. We assume that all items F_i, W_{ij}, T_k, and E_{kl} can be identified uniquely.

In this paper we are interested mainly in discovery of several types of associations between items of the discussed problem. At first, we are interested in mining of relations between GUI form widgets. These relations will be further referenced as GG **relations**. We assume relationships of the GUI widgets are not explicitly given or represented in a given framework. For example, associations of widgets W_{ij} of forms F_i at the top of Fig. 1 are not given explicitly. If a value of a given text input field W_{14} is changed using form F_1, then text field W_{21} of form F_2 displays the same text string. For instance, this relation A_1 is very important to be known for testing purposes to enable a verification of correct field value propagation. At second, we are interested in relation A_2 between GUI widgets and their persistent representation as a column of a DB table, e.g. between W_{14} and E_{11} of table T_1. These relations will be referenced as GD **relations**.

It is not rare that a particular DB schema implements a hierarchy of classes. Such a schema builds often a special table representing instances and their relationships using abstract instance identifiers. Thus, at third, we are also interested in mining of relations, called DD **relations**, between DB tables. In Fig. 1, an association of tables T_1 and T_2 might be discovered if columns $E_{13} \in T_1$ and $E_{23} \in T_2$ can be associated using columns $E_{31}, E_{32} \in T_3$.

The proposed method assumes that a value entered/displayed at the GUI level is propagated to/from the DB level, respectively. If such values are unique then each value can be treated as the unique identifier of a specific element of a relation between the GUI widget and the DB entity. For example, a given value v_A identifies element (W_{13}, E_{12}) of relation GD. We assume that such a value is propagated from GUI to DB through a middleware blackbox unchanged. Furthermore, element (W_{ij}, E_{kl}) can be decomposed as two tuples (W_{ij}, v_{ijkl}) and (v_{ijkl}, E_{kl}) using the unique value v_{ijkl}. Thus, it is sufficient to generate pairs (W_{ij}, v_{ijkl}) at the GUI level and pairs (v_{ijkl}, E_{kl}) at the DB level. In fact, it can be shown that values v_{ijkl} need not to be unique. However, a mining procedure is more complex leading to a constraint satisfaction method. In this paper we will assume that the values v_{ijkl} can be set uniquely as it is easy to ensure such a condition by a test driver.

Having all tuples $(W_{ij}, E_{kl}) \in R$ discovered, we can also compose relation GG. Let all W_{ij} and E_{kl} are unique. Then we can create equivalence classes W^{kl} of all related GUI widgets

$$W^{kl} = \{W_{ij} | (W_{ij}, E_{kl}) \in R\}$$

Any pair of $(W_{i_1 j_1}, W_{i_2 j_2})$, where $W_{i_1 j_1}, W_{i_2 j_2} \in W^{kl}$, determines an element of GG relation.

Relation DD between the DB entities can be discovered through an analysis of SQL statements with JOIN operation. The related details are out of scope of this paper.

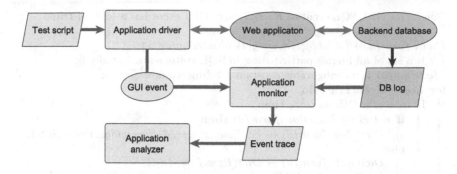

Fig. 2. A general data relationship mining architecture

4   Proposed Method

The proposed architecture of the entire reverse-engineering system is depicted on Figure 2. The aim of the tool is to bring the DB log records into a context of the application activity represented by GUI events. To simplify our explanation, the figure shows the back-end DB as a standalone block separated from the web application. Our method makes use of the fact that all relevant DB management systems keep track of all DB operations in their respective DB log.

The *Application driver* is an off-the-shelf component providing any infrastructure necessary for an efficient implementation of test scripts and model-based testing based on a finite state machine. The *Application driver* executes the *Test script* and controls the *Web application* accordingly by generating respective GUI events (e.g. *push the particular button*). In parallel, the *Application driver* sends a *GUI event notification* to the server-side process called *Application monitor* any time when the current GUI form was instructed by the *Application driver* to change its status.

The *Application monitor* is the core of the reverse engineering tool. It implements a web service aimed at listening to above mentioned *GUI event notifications*. Simultaneously, the *Application monitor* listens to all changes of the *DB log* (e.g. using the `tail -f` comand on Unix based platforms). Any time, when the *Application monitor* receives a *GUI event notification*, it records it to the *Event trace* file (see Figure 3) and appends all subsequent SQL command traces that appeared in the *DB log* afterwards. When the *DB log* has not changed for a specified amount of time, the *Application monitor* assumes that all DB activity invoked by the respective GUI event has been completed and sends a *GO-ON* message to the *Application driver* to allow him to execute the next instruction of the *Test script*.

The *GUI event notifications* are expected to cover both identification of activated widgets and their input/output values if there are any. In this paper, we consider only widgets enabling entering information and their input values. If both traces are created in a synchronized manner, we may expect that events triggered at the GUI level will be followed by operations at the DB level in the

Row ... an GUI/SQL event of *EventTrace*, GUI event has a form of triple
  $< formId, widgetId, widgetVal >$;
Dict : $formId \mapsto \{< widgetId, widgetValue, counter >\}$;
Lit is a set of all literals participating in SQL statements, initially $\emptyset$;
MaxCompar is a configurable constant (sliding window depth);
for *Row* $\in$ *EventTrace* **do**
  if *Row is a GUI message* **then**
    if *not exists Dict(Row.formId)* **then**
      | $Dict(Row.formId) \leftarrow \{< Row.widgetId, Row.widgetValue, 0 >\}$;
    else
      $Dict(Row.formId) \leftarrow Dict(Row.formId) \cup$
      $\{< Row.widgetId, Row.widgetValue, 0 >\}$;
    end
  else
    /* *Row* is an SQL message */;
    Lit $\leftarrow$ all literals from INSERT/UPDATE assignment section and/or
    from all WHERE/JOIN conditions;
    for *lit* $\in$ *Lit* **do**
      for *formId* $\in$ *Dict* **do**
        ElemSet $= Dict(formId)$;
        for *Elem* $\in$ *ElemSet* **do**
          if *Lit* $==$ *Elem.widgetValue* **then**
            Remove *Elem* from $Dict(formId)$;
            Record the new discovered *GD* association;
          else
            Elem.counter $=$ *Elem.counter* $+ 1$;
            if *Elem.counter* $>$ *MaxCompar* **then**
              /* Sliding window */;
              Remove *Elem* and olders from $Dict(formId)$;
            end
          end
        end
      end
    end
  end
end

Algorithm 1. *GD* relation discovery algorithm. It processes trace events sequentially. A GUI event is a triple specifying a form *fromId*, a widget *widgetId*, and its value *widgetVal*. Structure *Dict* maintains a set of active GUI events composing a sliding window characterized by history depth *MaxCompar*. In fact, *MaxCompar* is the maximum number of unsuccessful trials to match a given GUI event against an SQL event. Each active GUI event keeps the current number of such trials in *counter*. If the number of the trials exceeds the limit, the GUI event and all older ones are abandoned as there is a very low chance to find a match in future. The process of matching tests detected SQL literals against widget values of all active GUI events.

```
GUIREPORT:FORM:state=NewServiceProvider
GUIREPORT:STEP:state=NewServiceProvider,
   elementName=OrganisationName,elementData='OrganisationName1'
GUIREPORT:STEP:state=NewServiceProvider,
   elementName=BusPhoneNumber,elementData='+420 8484848'
...
CET LOG: execute <unnamed>: INSERT INTO service_provider
   (business_postal_code, business_house_number,
   business_phone_number, organisation_name, business_email, comm_id)
   VALUES ($1, $2, $6, $7, $8, $9, $10)
CET DETAIL: parameters: $1 = '23456789', $2 = '234567',
   $6 = 'BusinessStreet1', $7 = '+420 8484848', $8 = 'OrganisationName1',
   $9 = 'bus@email.com', $10 = 'spfirstname1.spsurname181'
```

Fig. 3. A fragment of the *Event trace*

monitor log. Then, relationships between the GUI widgets and the DB entities can be recovered by processing of matched *Event trace* entries. The *Application analyser* analyses the *Event trace* off-line to discover the correspondences introduced in Section 3. The algorithm for discovery of GD associations is introduced in Algorithm 1.

5   Experimental Results

Our association reconstruction method has been tested with a web-based application developed within the MONDIS project using the standard Java EE Glassfish framework with some additional extensions. The middleware implemented through Jave EE and its extensions does not provide any explicit mapping among GUI elements and the DB entities. The middle size DB consists of over 80 tables.

The behavioral model of the application in a form of finite state X-machine [9] was created manually. We use the Selenium extension for the Mozilla Firefox browser that allows driving interactions with a web based applications by scripts. In our case we selected Python scripting language [3]. SQL commands are parsed by a parser generated by the JavaCC (Java Compiler compiler) tool [2] using a slightly modified SQL grammar originating in the JSqlParser project [1].

By a manual investigation of a log file we reconstructed some relations among tables. Later on we received similar results through processing using the prototype implementation of our method. The results differed as the software processing were more precise and discovered even relations that were missed during the human based processing. Fig. 4 depicts a result of the automated entity matching. The middle (blue/shaded) record specifies entry fields in the form *NewServiceProvider*. The left and right white blocks reflect columns of two DB tables *person* and *service_provider*. The line connectors between columns and fields identify discovered relationships. Even this example is very promising, the overall efficiency of the method does not appear so high. For the example application, we reconstructed 38% of entity pairs as the finite state X-machine was

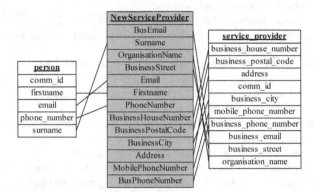

Fig. 4. An example of automatically matched entities. The GUI widgets are grouped in the middle block, while the DB entities are in the left and right tables. The straight connectors present discovered relationships.

not complete. This coverage is encouraging as it corresponds to the functional coverage of the MONDIS application by its simplified model. Furthermore, the applied method does not explore read-only widgets. Nevertheless, we are preparing extensions allowing to include read-only widgets and to reach better coverage, too. We should notice that otherwise huge DB logs contain important invaluable information for testers/maintainers/developers that cannot be utilized normally so easily. As a side effect, our method enables better comprehension of log records.

6   Conclusions

In this paper we deal with a reconstruction of association software implementation entities. We have focused on web applications with the traditional MVC pattern built around a complex implementation framework such as Java EE. More specifically, relationship GD between active GUI widgets and the related DB schema entities is explored. Then, GG relationship between active GUI widgets can be reconstructed from relationship GD. It was demonstrated that at least a part of such relationships can be reverse engineered through tracking activities at both GUI and DB layer as the unique input values are passed through the logic complex layer. Key names and values passed from the GUI are synchronized with DB entities receiving the values. Furthermore, implicit associations between DB entities can be reconstructed by analysis of the structure of SQL statements. Although the basic method principle is very simple, one needs to deal with asynchronous nature of the trace and its processing. We provided an algorithm based a sliding window.

Acknowledgments. This work was supported by the grant No. DF11P01OVV002 of the NAKI programme funded by the Ministry of Culture of Czech Republic.

References

1. JSqlParser, http://jsqlparser.sourceforge.net/ (retrieved March 10, 2012)
2. Java Compiler Compiler (JavaCC) - the java parser generator (2012), http://javacc.java.net/ (retrieved March 20, 2012)
3. Python programming language, official website (March 2012), http://www.python.org/ (retrieved March 20, 2012)
4. Abbasifard, M.R., Rahgozar, M., Bayati, A., Pournemati, P.: Using automated database reverse engineering for database integration. In: Proc. of the World Academy of Science, Engineering, and Technology, Budapest, Hungary, vol. 19, pp. 338–342 (2006)
5. Benslimane, S.M., Malki, M., Bouchiha, D.: Deriving conceptual schema from domain ontology: A web application reverse engineering approach. Int. Arab J. Inf. Technol. 7(2), 167–176 (2010)
6. Canfora, G., Sansone, L., Visaggio, G.: Data flow diagrams: reverse engineering production and animation. In: Proceedings of the Conference on Software Maintenance, pp. 366–375 (November 1992)
7. Di Lucca, G.A., Fasolino, A.R., Tramontana, P.: Reverse engineering web applications: the WARE approach. J. Softw. Maint. Evol. 16(1-2), 71–101 (2004)
8. Gulzar, N.: Fast track to struts: What it does and how (November 2012), http://www.theserverside.com/news/1364986/Fast-Track-to-Struts-What-it-Does-and-How (retrieved March 10, 2012)
9. Holcombe, M., Ipate, F.: Correct Systems - Building a Business Process Solution. Applied Computing Series. Springer (1998)
10. Mihancea, P.F.: Towards a reverse engineering dataflow analysis framework for Java and C++. In: Proceedings of the 2008 10th International Symposium on Symbolic and Numeric Algorithms for Scientific Computing, SYNASC 2008, pp. 285–288. IEEE Computer Society, Washington, DC (2008)
11. Muhairat, M.I., Al-Qutaish, R.E., Athamena, B.M.: From graphical user interface to domain class diagram: A reverse engineering approach. Journal of Theoretical and Applied Information Technology 24(1), 28–40 (2011)
12. Patel, R., Coenen, F., Martin, R., Archer, L.: Reverse engineering of web applications: A technical review. Technical Report ULCS-07-017, University of Liverpool, Department of Computer Science (2007)
13. Reenskaug, T.: THING-MODEL-VIEW-EDITOR, an example from a planning system, technical note. Technical report, Xerox PARC (May 1979), http://heim.ifi.uio.no/~trygver/themes/mvc/mvc-index.html (retrieved March 10, 2012)
14. Roscoe, J.F.: Looking Forwards to Going Backwards: An Assessment of Current Reverse Engineering. Technical report, Current Issues in Software Engineering, 2010-2011 Aberystwyth University, UK (2011)

Towards Practical Anomaly-Based Intrusion Detection by Outlier Mining on TCP Packets

Prajowal Manandhar and Zeyar Aung

Institute Center for Smart and Sustainable Systems (iSmart)
Masdar Institute of Science and Technology, Abu Dhabi, UAE
{pmanandhar,zaung}@masdar.ac.ae

Abstract. Intrusion detection System (IDS) is an important part of the security of large networks like the Internet. With increasing number of data being transmitted day by day from one subnetwork to another, the system needs to identify intrusion in such large datasets in an effectively and timely manner. So the application of knowledge discovery comes handy to identify unusual accesses or attacks. Improving an IDS's performance and accuracy is one of the major challenges network security research today. In this paper, we propose a practical anomaly-based IDS using outlier mining of the readily available basic Transmission Control Protocol (TCP) header information as well as other easily derivable attributes. We use a two-step approach of k-means clustering and one-class support vector machine (SVM) to model the normal sessions presented in MIT DARPA '99 dataset. We then feed the testing set to the resultant model to predict the attacks sessions.

Keywords: IDS, outlier mining, k-means clustering, one-class SVM, TCP.

1 Introduction

Intrusion detection is a network security mechanism to protect the computer network system from invasions or attacks. As technologies evolve, it increases the risks of new threats evolving along with them. Thus, when a new type of attack emerges, an intrusion detection system (IDS) needs to be able to respond in an effectively and timely fashion in order to avoid hazardous effects. Intrusion detection has been one of the core areas of computer security whose objective is to identify these malicious activities in network traffic, and importantly, to protect the resources from threats. In today's context, the biggest challenge could be dealing with "big data", i.e., a huge volume of network traffic data which gets collected dynamically in the network's communications [1].

The common use of TCP means that it is likely to be exploited for misuse and various forms of attacks. Thus, there is a possibility of malicious behavior that can be executed through the TCP/IP protocols without being noticed/blocked by a firewall or a traditional signature-based IDS as they do not recognize new variations of attacks. So, there is a pressing need for a practical and easy-to-deploy IDS for TCP based on anomaly detection. Such an anomaly-based IDS

H. Decker et al. (Eds.): DEXA 2014, Part II, LNCS 8645, pp. 164–173, 2014.

will be able to identify new attack variations and thus supplement the existing firewalls and signature-based IDSs. In fact a number of anomaly-based IDSs has been proposed by researchers throughout the years [2]. However, none of those become widely used in practise till now, and signature-based IDSs, both commercial and freeware, are still dominating the market [3].

One of the main reasons behind the underdeployment of anomaly-based IDSs is that they cannot be easily deployed. A majority of anomaly-based IDSs use data mining and machine learning algorithms, which take feature vectors containing some complicated features as their inputs. However, extracting such complicated features from raw TCP network traffic data is not a straight-forward task. The most notable example is the widely-used KDD '99 dataset [4] where, 13 out of 41 features are extracted based on domain knowledge. Unfortunately, to our best knowledge, there is no publicly available tools to automatically extract those features from the raw TCP data. As such, it is virtually impossible for an average network administrator to deploy an anomaly-based IDS which relies on such complicated feature vectors. In order to address this, instead of complicated feature vectors, we opt to use the raw TCP data, which can be easily obtained using the tcpdump tool. Many researchers have pointed out that different patterns of TCP flags associated with anomalies. In our approach, we inspect the TCP headers information from the TCP/IP packets which allows for high-speed network status detection.

Although a number anomaly-based IDS using raw TCP data has been proposed in the past, some of them like PHAD/ALAD [5] are packet-based. They mainly rely on the information of individual TCP packets in detecting anomalies and hence, it is not able to capture higher-level anomalous events that may occur in a "session" involving multiple packets, even though the individual packets in that particular session seem innocent. In order to address this issue, some session-based approach such as NATE [6], have been proposed. Our method in this paper also employs a session-based approach. We use MIT's DARPA '99 tcpdump dataset [7] in order to evaluate method's effectiveness. Although the DARPA dataset is known to have some weaknesses [7,8], we use it for a proof of concept of our proposed IDS method, which can be readily applied to any tcpdump files. Our objective is to provide network security professionals with a practical anomaly-based IDS that can be easily deployed as a supplementary tool to the existing firewalls and and signature-based IDSs. To this end, we make the scripts/soure code and the running intrusions of our IDS program available in http://www.aungz.com/IDS/.

2 Background

2.1 Transmission Control Protocol (TCP)

The TCP protocol is a connection-oriented protocol that enables the reliable and ordered transmission of traffic in Internet application. The TCP header information is important for the establishment of a trusted connection. The TCP sequence and acknowledgement numbers provide unique identifiers of the connection to the client and server, and also provide confirmation that data has been received. TCP flags are used to control the state of the TCP connection. The client and server must complete a three-way handshake before any data exchange takes place. However, either party can initiate termination of the connection by resetting with a TCP RST packets. Thus, we have focused on areas where, intruders look to exploit this weakness in the protocol and can come up with various attacks.

2.2 Intrusion Detection Systems

An intrusion detection system (IDS) is a device or tool that monitors network activities for malicious activities and generates alarms and logs. IDS can be categorized into two types based on how it works:

1. **Signature-based IDS:** A rule-based approach, where pattern matching on known signatures leads to high accuracy for detecting threats. But it cannot detect novel attacks as novel attack signatures are not available for pattern matching. The limitation with this IDS is their downfall to detect new attacks and also neglect minor variations of known patterns as well as an overhead costs in order to maintain signature databases.
2. **Anomaly-based IDS:** A statistical approach which analyzes the deviation from the normal activities by mining information from available data. It can detect novel attacks by comparing suspicious ones with normal traffics but has high false alarm rate due to difficulty of generating practical normal behavior profiles for protected systems [9]. More recently, different statistical techniques have been applied for IDSs, especially for anomaly-based ones. These techniques are used when there is no structured knowledge about the pattern of data. Many data mining/machine learning algorithms are considered good for attack detection, in which decision tree is considered one of the most powerful and effective ways of detecting attacks in anomaly detection [10,3]. In addition, k-nearest neighbor (KNN), Naive Bayes, Artificial neural network (ANN), support vector machine (SVM), random forest, and fuzzy logic are also known to be used for learning and classifying the network traffic data as normals or attacks [10]. Ensemble learning, which is a combination of multiple machine learning approaches, also is being used to improve the result of intrusion detection as compared to the result of a single classifier.

2.3 Anomaly-Based IDS Using Outlier Mining

An outlier can be regarded as an observation which varies from the other observations as to arouse suspicions that it was generated by a different mechanism. It was also observed that researchers have tried two approaches either to model only normal data or both normal and abnormal data for finding intruders. Modeling both the normal and abnormal data allows the system to be tight on false positive and false negatives as both the normal and abnormal data needs to be modeled; but it puts the limitation in modeling the abnormal patterns. Similarly, using only normal patterns allows the system to model the boundaries of normal data but due to unknown nature of the boundaries of abnormal data, it gives a possibility of false positive as well if it comes to overlap with the normal data. Thus, a proper tuning needs to be done for defining the threshold of the system to balance the number of true positives and false positives. Outlier mining has been a popular technique for finding intrusions [6,11].

Network Analysis of Anomalous Traffic Events (NATE) [6] is a good example of an anomaly-based IDS using outlier mining. It is a low cost network IDS, which measures only TCP packets' headers information. It generates several concise descriptions using clustering to model normal network traffic behaviors. The measure of detection is the deviations of new TCP header data from the existing clusters, which allows for high speed network traffic monitoring once the normal behavior base is built. They performed a cluster analysis and found 7 such clusters that represent their normal TCP sessions. And Chebyshev's inequality was used to perform the boundary analysis. After that, Mahalanobis distance or Euclidean distance was used to detect intrusions in their test sessions. They have identified that false positives could be a potential problem with this approach, and hence, they have used significance level as a tuning parameter in response to decrease the false positives.

3 Proposed Method

Our proposed method's objective is to model the normal network data and then predict the abnormality based on the deviation from the normal patterns. Normal data may be of various types and may show various types of behaviors. Therefore, it is important to understand and analyze the behavior for each of these normal patterns and try to distinguish one another from abnormal data. In this end, we try to explore the TCP protocol in the network data and traverse through the TCP headers to understand the normal patterns. Exploring the TCP headers allows for high speed network traffic monitoring because it can be performed in real time and once the normal behavior base is built, it greatly reduces analysis complexity. Therefore, our design objective is to identify suitable tools to model the normal network data, and then implement the same to achieve an improved output in identifying the anomalous (attack) data. Our proposed method is partially inspired by NATE in defining TCP sessions, outlier mining based on TCP header information, and sample clustering. However, we have added substantial improvements like clustering quality analysis, one-class SVM, and model validation and updating. As a result, our method offers better results than those by NATE (as discussed later in Section 4). Our approach is to use basic TCP headers which can be used to detect the anomaly intrusions in real time. We tried modeling using a different model like clustering and one-class classifier and tried to vary over a range of different model parameters in order to achieve a stable and accurate model. Finally we came up with a model that has a high accuracy and a perfect recall along with a high degree of consistency.

The proposed architecture of the system is depicted in Figure 1(i). We take the tcpdump data files as input, analyze and process the TCP headers, and then finally predict the intrusions.The four steps involving in our method, namely (i) preprocessing, (ii) initial model creation, (iii) model validation and updating, and (iv) anomaly prediction are briefly descried below.

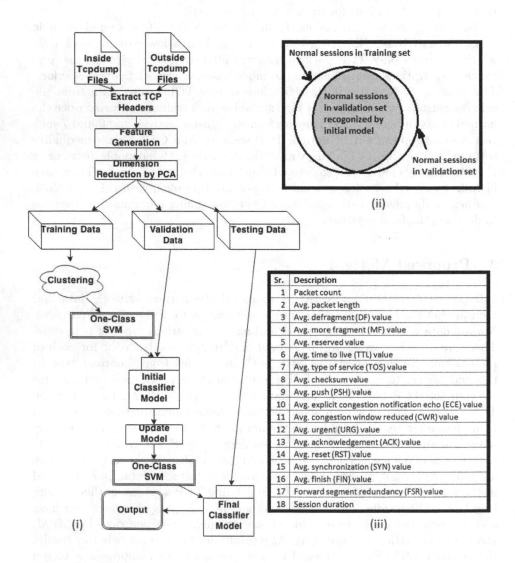

Fig. 1. (i) Overall work flow of the proposed intrusion detection system using outlier mining. (ii) Modeling normal sessions using both training and validation sets. (iii) Attributes used to represent a TCP session.

3.1 Preprocessing

From the input tcpdump data, we extract TCP headers of individual packets. Then, following the assumptions by Taylor and Alves-Foss [6], we define 'session', a group of packets having a unique combination of source IP and port number to destination IP and port number. This definition varies from the standard one practised in network traffic, which is one complete TCP connection beginning with SYN flags from source, response of SYN/ACK flags from destination and ending along with FIN flag from source, FIN/ACK flags from destination. The reason for defining our session is due to the fact that it is very difficult to extract such independent connections as many connection do not complete until more than one FIN flags are sent out. Since it is technically difficult and time consuming to capture such activities, instead, we collect and analyze the aggregated traffic between the unique pairs of senders and receivers. This allows for a quick and real-time analysis of the network traffic. The important attributes in an individual TCP packet's header include its packet length, DF, MF, and reserved flags, time-to-live (TTL), type of service (TOS), checksum, and all 8 TCP flags. Once a session is defined, the number of packets in it is counted and the average values of the above attributes from the session's packets are computed. In addition, the attribute FSR (forward segment redundancy) is derived as below. FSR shows the percentage of connection control flags in total packets of the session and observed to have high values for abnormal traffics.

$$FSR = \left(\sum FIN + \sum SYN + \sum RST\right) / \text{packet count} \qquad (1)$$

The resultant feature vector containing 18 attributes, shown in Figure 1(iii), represents a session's behavior. Since the values of attributes vary significantly from one attribute to another, we use a natural logarithmic transformation as a form of normalization. In order to avoid negative infinite values caused by logarithm on zeros, we modify transformation function as: $normalize(x) = \log(1+x)$. After that, we apply principal component analysis (PCA) technique to reduce the dimensionality of our feature vectors. The first c number of principal components are taken. (c is an empirically determined integer.)

3.2 Initial Model Creation

We split the whole dataset into 3 portions as follows:

1. **Training set:** contains normal sessions only; used for initial model creation.
2. **Validation set:** contains both normal and attack sessions; used for model update after removing attack zone from the validation set.
3. **Testing set:** containing both normal and attack sessions; for prediction.

The normal sessions, represented by c-dimensional feature vectors, in the training set are divided into clusters by means of k-means clustering in order to model the overall behaviors of the normal sessions. Different values of k are tried, and the one that gives the highest average silhouette value is taken. Since the k-means algorithm is affected by first initialized center values, prior to deciding with k, we make sure we get the reliable value of k, by executing the clustering process multiple times for each value of k. Then, Euclidean distance is used to measure the distances from k cluster centers to each training data point. From this, new k-dimensional feature vectors, each containing k distance values, are constructed. Then, these new feature vectors in the training set are fed to one-class SVM, resulting in the "initial classifier model".

3.3 Model Validation and Updating

Now, a similar k-dimensional feature vector is also constructed for each of the normal sessions in the validation set by calculating its Euclidean distances to the k cluster centers obtained before. Then, in order to validate the initial model, each validation set's feature vector is sent to the initial classifier model for classification. Those which are misclassified (as attacks) are removed from the data pool, as they do not agree with the initial model defined by the normal sessions in the training data. Finally, the original c-dimensional feature vectors of the correctly classified normal session in the validation set (depicted as the gray area in Figure 1(ii)) used to build another one-class SVM model. This results in the "final classifier model", which reflects both the training and validation sets.

3.4 Anomaly Prediction

When a new unknown TCP session is to be classified, we first need to construct its c-dimensional feature vector. Then, it is sent to the one-class SVM of the final classifier model in order to predict whether it is a normal or an attack session.

4 Experimental Results

4.1 Dataset Used

We use the publicly available MIT DARPA '99 dataset [7]. It has been one of the well-known benchmarks for IDS for a long time. One drawback with DARPA dataset concerns with its simulated nature as pointed out by McHugh [8]. However, we chose to proceed with this dataset because our results would be comparable with others when we use a standard and recognized dataset. We use the three weeks of data starting from third, fourth and fifth. The third week of dataset comprises of only normal packets and hence is used for training purpose. While fourth and fifth week's datasets, containing both normal and attack packets, are used for validation and testing (evaluation) purposes respectively. The different types of TCP attacks included in DARPA '99 dataset are Portsweep, Satan, Neptune, Mailbomb, nmap, etc.

In our experiment, we look at the distributions of of the numbers of sessions over the three weeks' data. Based on the assumption presented by Taylor and Alves-Foss [6], we similarly define our session. The number of normal and attack sessions in training, validation, and testing sets are shown in Table 1.

Table 1. Statistics of training, validation, and testing sets

Dataset	DARPA'99	#Files	Total Size (GB)	#Normal Sessions	#Attack Sessions
Training	week 3	16	6.46	1,318,656	0
Validation	week 5	10	4.85	1,232,356	159,929
Testing	week 4	9	2.80	639,326	2,304

4.2 Results and Discussions

Our first objective is to model the normal sessions, for which we begin with performing PCA on the features of all three sets in order to reduce the dimensionality of our feature set. Here, we choose to proceed with the first $c = 8$ principal components out of 18 because those 8 covers above 99.6% of variance of the data. Now, in order to model the normal sessions in the training set, we apply k-means clustering to the resultant 8-dimensional feature vector set. The best k is chosen based on the highest average silhouette value obtained, which is 0.805 when $k = 3$. Once clustering of the normal sessions in the training set is done, we construct its 3-dimensional feature vectors to build the initial model using one-class SVM. Then, we feed the similar 3-dimensional feature vectors from the validation set to the initial model in order to perform a validation. We would like to achieve a clear distinction between the normal sessions and the attack ones. Thus, we select the one-class SVM parameters that allows us to achieve the maximum number of true negatives (normal sessions). After that, we use only the true negative in the validation set in order to build the final model using one-class SVM. Figure 2 shows the distribution of the validation set containing normal and attack sessions.

Finally, we feed the testing set, to the final model for prediction. We tried building our final model using different kernel types in one-class SVM. Among different kernel used, radial basis kernel (RBF) gave us the best results as shown in Table 2, with a perfect recall of 100% and a high accuracy of 91.85%. A relatively high false positive rate incurred (8.18%) can be attributed to the underlying similarities between the normal and the attack sessions. The reason for RBF being stand out amongst other kernels is that, RBF kernel in support vector algorithm automatically determines centers, weights and threshold such as to minimize an upper bound on the expected test error.

Table 2. Results using different kernel types in one-class SVM. The best results, which are obtained with radial basis kernel, are highlighted.

	Kernel Type			
	Linear	Polynomial (degree=3)	Sigmoid	**RBF**
True Positive (TP)	2,205	2,166	2,285	**2,304**
True Negative (TN)	502,094	515,047	532,681	**587,033**
False Positive (FP)	137,232	124,279	106,645	**52,293**
False Negative (FN)	99	138	19	**0**

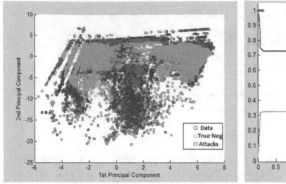

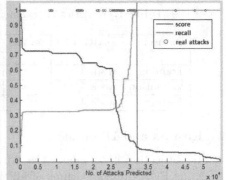

Fig. 2. Modeling true negatives detected in the validation set

Fig. 3. Positions of 'real attacks' among all predicted attacks ordered by their 'prediction scores' in descending order with their corresponding 'recall' values

Evaluation Criteria: Using radial basis kernel (RBF), we achieve the best results with a perfect recall of 100% (because there are no false negatives or dismissals) and a high accuracy of 91.85%. A relatively high false positive rate incurred (8.18%) can be attributed to the underlying similarities between the normal and the attack sessions.

$$\text{recall} = \text{TP}/(\text{TP} + \text{FN}) \tag{2}$$

$$\text{face positive rate} = \text{FP}/(\text{TN} + \text{FP}) \tag{3}$$

$$\text{accuracy} = (\text{TP} + \text{TN})/(\text{TP} + \text{TN} + \text{FP} + \text{FN}) \tag{4}$$

Comparison with NATE: In the NATE method presented by Taylor and Alves-Foss [6], among the 4 selected types of attacks explored, they were able to detect the Portsweep, Satan, and Neptune attacks, but not Mailbomb (ref. Tables 4 and 5 in [6]). That means their recall is less than 100%. In our case, we are able to detect all attacks presented in the 4th week of DARPA '99 dataset with a perfect recall of 100%. Our model overcomes the problem of NATE, which requires absolutely normal packets for normal session modeling. Since we have used the validation set to cross-check our normal session modeling, our final model reflects both the normal session and the attack session (by means of using the negative space).

Reducing False Positives: Apart from taking the default labels given by the one-class SVM classifier implemented with LIBSVM, we can also take the probabilistic or decision values (prediction scores) as output. Figure 3 shows the positions of real attacks (true positives) among all predicted attacks (predicted positives) ordered by their prediction scores, normalized into the range of 0 to 1, in descending order. It can be observed that among the 54,597 predicted positives, after ~31,700 highest scoring ones (the region after the black vertical line), the recall is already 99.87%, with only 3 true positives remain undetected. That point corresponds to the prediction score of ~0.1. Thus, by sacrificing a very small fraction in recall value, we can significantly reduce the false positive rate and substantially improve the accuracy. By setting the score threshold as 0.1 and taking only the sessions whose scores are ≥ 0.1, we can achieve: TP=2,301; TN=609,913; FP=29,413; FN=3; recall=99.87%; false positive rate=4.60%; and accuracy=95.42%.

5   Conclusion and Future Work

In this research work, we have been able to model the normal sessions and detect the attack sessions derived from '99 DARPA dataset. Our work provides a practical solution for IDS based on outlier mining method on TCP headers. In today's world of increased network-based security threats, it enables network security professionals to analyze the TCP header information and perform anomaly detection in fast incoming traffic in a real-time manner.

For future work, apart from using basic TCP header information, we look forward to using derived information from the connection analysis that might help us to further reduce the false positive rate. And we also look to compare our results with ALAD/PHAD's.

References

1. Suthaharan, S., Panchagnula, T.: Relevance feature selection with data cleaning for intrusion detection system. In: Proc. 2012 IEEE SECon., pp. 1–6 (2012)
2. Zhang, X., Jia, L., Shi, H., Tang, Z., Wang, X.: The application of machine learning methods to intrusion detection. In: Proc. 2012 S-CET, pp. 1–4 (2012)
3. Kumar, M., Hanumanthappa, M., Kumar, T.V.S.: Intrusion detection system using decision tree algorithm. In: Proc. 14th IEEE ICCT, pp. 629–634 (2012)
4. Tavallaee, M., Bagheri, E., Lu, W., Ghorbani, A.A.: A detailed analysis of the KDD CUP 99 data set. In: Proc. 2nd IEEE CISDA, pp. 53–58 (2009)
5. Mahoney, M.V., Chan, P.K.: Learning nonstationary models of normal network traffic for detecting novel attacks. In: Proc. 8th ACM KDD, pp. 376–385 (2002)
6. Taylor, C., Alves-Foss, J.: NATE – network analysis of anomalous traffic events, a low-cost approach. In: Proc. 2001 NSPW, pp. 89–96 (2001)
7. Mahoney, M.V., Chan, P.K.: An analysis of the 1999 DARPA/Lincoln Laboratory evaluation data for network anomaly detection. In: Vigna, G., Kruegel, C., Jonsson, E. (eds.) RAID 2003. LNCS, vol. 2820, pp. 220–237. Springer, Heidelberg (2003)
8. McHugh, J.: Testing intrusion detection systems: A critique of the 1998 and 1999 DARPA intrusion detection system evaluations as performed by Lincoln Laboratory. ACM Transactions on Information System Security 3, 262–294 (2000)
9. Gharibian, F., Ghorbani, A.: Comparative study of supervised machine learning techniques for intrusion detection. In: Proc. 5th CNSR, pp. 350–358 (2007)
10. Sarvari, H., Keikha, M.M.: Improving the accuracy of intrusion detection systems by using the combination of machine learning approaches. In: Proc. 2010 SoCPaR, pp. 334–337 (2010)
11. Hofmeyr, S.A., Forrest, S.A.: Architecture for an artificial immune system. Evolutionary Computation 8, 443–473 (2000)

Survey on Big Data and Decision Support Benchmarks

Melyssa Barata[2], Jorge Bernardino[1,2], and Pedro Furtado[1]

[1] CISUC – Centre of Informatics and Systems of the University of Coimbra
FCTUC – University of Coimbra, 3030-290 Coimbra, Portugal
[2] ISEC – Superior Institute of Engineering of Coimbra
Polytechnic Institute of Coimbra, 3030-190 Coimbra, Portugal
melyssa_ox@hotmail.com, jorge@isec.pt, pnf@dei.uc.pt

Abstract. Benchmarking is a common practice for the evaluation of database computer systems. By executing certain benchmarks, manufacturers and researchers are able to highlight the characteristics of a certain system and are also able to rank the system against the rest. On the other hand, at the moment, BigData is a hot topic. It concerns dealing efficiently with information that is challenging to handle, due to volume, velocity or variety. As more and more platforms are proposed to deal with BigData, it becomes important to have benchmarks that can be used to evaluate performance characteristics of such platforms. At the same time, Decision Support applications are related to BigData, as they need to efficiently deal with huge datasets. In this paper we describe benchmarks representing Decision Support Systems (TPC-H, SSB, TPC-DS), and benchmarks for the Big Data class (YCSB, BigBench, and BigFrame), in order to help users to choose the most appropriate one for their needs. We also characterize the relationship between Big Data benchmarks and Decision Support benchmarks.

Keywords: Benchmarking, Decision Support Systems, Big Data, YCSB, BigBench, BigFrame, TPC-H, SSB, TPC-DS.

1   Introduction

Benchmarking the performance of a DBMS-Database Management System consists of performing a set of tests in order to measure the system response under certain conditions. Benchmarks are used to illustrate the advantages of one system or another in a given situation, as well as comparing the global performances of DBMS, or to determine an optimal hardware configuration for a given DBMS and/or application. A database benchmark is a standard set of executable instructions that are used to measure and compare the relative and quantitative performance of two or more database management systems through the execution of controlled experiments.

A Decision Support System (DSS) is a computer-based application, developed to collect, organize and analyze business data in order to facilitate quality business decision-making for tasks such as management, planning and operations. Decision support systems are interactive, computer-based systems that aid users in judgment and choice activities. They provide data storage and retrieval but enhance the

H. Decker et al. (Eds.): DEXA 2014, Part II, LNCS 8645, pp. 174–182, 2014.

traditional information access and retrieval functions with support for model building and model-based reasoning.

Big Data refers to datasets whose size and complexity is beyond the ability of typical database software tools to capture, manage, and analyze. While earlier DBMSs focused on modeling operational characteristics of enterprises, big data systems are now expected to model user behaviors by analyzing vast amounts of user interaction logs [11]. In order to make the idea of Big Data clearer, some experts began to summarize the subject in ways that can satisfactorily describe the basis of the concept: the four 'Vs': Volume, Velocity, Variety and Veracity.

The purpose of this paper is to overview the main properties of different benchmarks for Decision Support and Big Data systems. We will explain their characteristics in order to understand the main purpose and functionality of each class, and to help users decide what specific benchmark they comply for their wants or needs. We analyze three Big Data benchmarks: YCSB, BigBench, BigFrame, and three Decision Support benchmarks: TPC-H, SSB, and TPC-DS.

The remainder of this paper is organized as follows. Section 2 gives a short introduction to Big Data benchmarks, followed by detailed information of YCSB, BigBench, and BigFrame. Section 3 defines Decision Support, followed by explaining thoroughly the following benchmarks: TPC-H, SSB, and TPC-DS. Section 4 presents the relationship between Big Data and Decision Support System benchmarks. Finally section 5 presents our main conclusions and future work.

2   Big Data Benchmarks

Data handling in the order of terabytes involves the requirement of high computational processing, data storage and manipulation power. A lot has been written about Big Data and how it can serve as a basis for innovation, growth and differentiation.

The purpose of Big Data benchmarks is to evaluate Big Data systems. In this section YCSB, BigBench, and BigFrame will be explained in thorough detail.

2.1   YCSB – Yahoo! Cloud Serving Benchmark

The Yahoo! Cloud Serving Benchmark (YCSB) is one of the most used benchmarks, and provides benchmarking for the bases of comparison between NoSQL systems. The YCSB Client can be used to benchmark new database systems by writing a new class to implement the following methods [6]: read, insert, update, delete and scan. These operations represent the standard CRUD operations: Create, Read, Update, and Delete.

A cloud service testing client consists of two parts: workload generator and the set of scenarios. Those scenarios are known as workloads, which perform reads, writes and updates. The YCSB predefined workloads are [17]:

- Workload A: Update heavy. Mix of 50% reads and 50% updates.
- Workload B: Read heavy. 95% of reads and 5% of updates.

- Workload C: Read only. 100% read.
- Workload D: Read latest. 95% of reads and 5% of inserts.
- Workload E: Short ranges. 95% of scans and a 5% of inserts.
- Workload F: Read-modify-write. In this workload, the client will read a record, modify it, and write back the changes.

YCSB will measure throughput in operations per second and record the latency in performing these operations [1].

The main goal that the Yahoo! Cloud Serving Benchmark developers make is to increase the variety of cloud serving systems. In the future, this benchmark wants to build tiers to evaluate performance, scalability, elasticity, availability, and replication [17].

2.2   BigBench

BigBench is an end-to-end Big Data benchmark proposal. The underlying business model of BigBench is a product retailer [2]. Figure 1 illustrates the data model for BigBench. The structured part of this model is derived from the TPC-DS data model. The semi-structured data focuses on click-streams contained in web log files, and the unstructured data resembles written text that is associated with items and product reviews offered by the retailer [3].

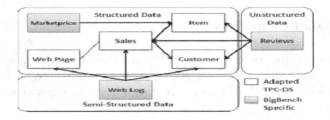

Fig. 1. Data Model for BigBench [3]

The model contains a Marketprice table which was added to the structured data because it contains competitor's names and prices for each item so that price comparisons performed by online users who are interested in a specific item can also be captured [3].

The unstructured data in the data model is composed of a table named Reviews that is used for customers or guest users to submit product reviews.

The table named "Web Log" in the semi-structured data part focuses on click-streams, which are contained in the web log files.

BigBench has a total of 30 queries designed along one business dimension and three technical dimensions, which aim to cover different business cases and technical perspectives.

2.3 BigFrame

Unlike all other benchmarks which are either micro benchmarks or benchmarks for very specific domains, BigFrame is a benchmark generator, which captures, volume, variety and velocity emphasized in Big Data environments. With BigFrame, a user can generate their own benchmark fit to their special needs.

For instance, BigFrame generates a benchmark specification containing relational data and SQL queries. It consists of large volumes of relational data that has mostly aggregations and few joins. It can also generate a benchmark specification that contains a specified number of concurrent query streams with class labels for the queries. These tend to be Big Data deployments and are increasingly multi-tenant.

Since BigFrame provides a variety of workflows involving different combinations of data types. The current version implements a Business Intelligence app domain, which provides relational data, graph data and nested textual data [8]. It is based on the TPC-DS benchmark.

Figure 2 illustrates BigFrame's data model, which reflects the analytics in a retail company that has carried out a promotion on a number of items and its main goal is to analyze the effectiveness of a promotion.

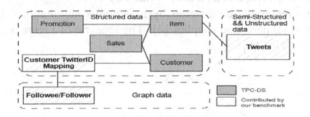

Fig. 2. Data Model of BigFrame

BigFrame distinguishes two types of workloads based on the data model: (1) Historical Query and (2) Continuous Query. In the workload case of Historical Query, data is refreshed at a rate as in traditional data warehouses. In comparison, the Continuous Query is a standing query that continuously crawls the latest tweets [9].

3 Decision Support Benchmarks

All the three benchmarks (TPC-H, SSB, TPC-DS), analyzed in this section are intended to provide a fair and honest comparison of various vendor implementations to accomplish an identical, controlled and repeatable task in evaluating the performance of DSS systems. The defined workloads are expected to test hardware system performance in the areas of CPU utilization, memory utilization, I/O subsystem utilization and the ability of the operating system and database software to perform various complex functions important to DSS - examine large volumes of data, compute and execute the best execution plan for queries with a high degree of complexity, schedule efficiently a large number of user sessions, and give answers to critical business questions.

3.1   TPC-H

TPC-H is a benchmark that simulates a Decision Support System database environment. The transactions in such environments are characterized by business intelligence intensive complex data mining queries and concurrent data modifications. The performance of a system can only be measured when it is providing means for a business analyses on a dataset.

The read-intensive Decision Support System (DSS) database is designed to accurately simulate for DSS repository for commercial order-processing Online Transaction Processing databases, obtaining realistic results for the business analysis. These can be performed in very large scale factor datasets [5].

TPC-H consists of a suite of business oriented queries. Both the queries and the data that populate the database were carefully chosen in order to keep the benchmark's implementation easy but at the same time to be relevant. The benchmark is composed of 22 read-only queries and 2 update queries. These queries are performed on considerably large amounts of data, have a high degree of complexity, and were chosen to give answers to critical business questions. The queries represent the activity of a wholesale supplier that manages, sells, or distributes a product worldwide. Both the queries and data are provided by TPC.

TPC-H schema represents data stored by a fictitious multinational importer and retailer of an industrial parts and supplies [4]. The scenario of Figure 3 illustrates the schema diagram for TPC-H [16].

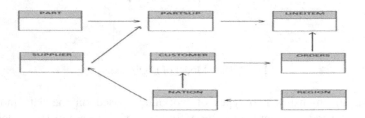

Fig. 3. TPC-H database schema

Customers and suppliers for this retail business can come from different parts of the worlds (regions) and different countries (nations) within those regions. Customers place several orders and each order can contain many different part purchases from different suppliers at different prices (Lineitem). The list of parts and list of suppliers are connected by a table (Partsup) to indicate the specific parts supplied each supplier.

3.2   Star Schema Benchmark

The Star Schema Benchmark (SSB) is designed to measure performance of database products in support of classical data warehousing applications, and is a variation of the TPC-H benchmark. TPC-H models the data in 3rd normal form, while SSB implements the same logical data in a traditional star schema. SSB models the data warehouse of a wholesale supplier and its queries are simplified versions of TPC-H queries. They are organized in four flights of three to four queries each [13]. Each flight consists of a sequence of queries that someone working with data warehouse system would ask.

The SSB benchmark, as the name itself says, has a star formation. Figure 4 shows the SSB database schema. This schema registers the sales that are made [13].

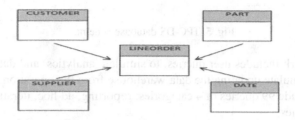

Fig. 4. Star Schema Database Schema

Performance measurement will result in a report with the following information: the processor model, memory space, disk setup, number of processors being used in the test with breakdown of schema by processor, and any other parameter of the system that interferes on performance.

3.3   TPC-DS

TPC-DS models the decision support functions of a retail product supplier. The supporting schema contains vital business information such as customer, order, and product data. It models the two most important components of any mature decision support system: user queries and data maintenance.

TPC-DS's workloads test the upward boundaries of hardware system performance in the areas of CPU utilization, memory utilization, I/O subsystem utilization and the ability of the operating system and database software to perform various complex functions important to DSS - examine large volumes of data, compute and execute the best execution plan for queries with a high degree of complexity.

This benchmark, models the challenges of business intelligence systems where operational data is used to support the making of sound business decisions in near real time and to direct long-range planning and exploration [15].

TPC-DS models a data warehouse. This database schema, whose fact tables are represented in Figure 6 models the decision-support functions of a retail product supplier including customer, order, and product data. Catalog and web sales and returns are interrelated, while store management is independent. Its database schema consists of 7 fact tables and 17 dimension tables that are shared among the fact tables [2]. Thus, the whole model is a constellation schema, and only a reduced version is shown in Figure 5 [14].

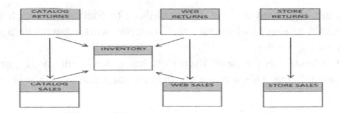

Fig. 5. TPC-DS database schema

This benchmark includes user queries, to simulate analytics, and data maintenance operations, to simulate updating the data warehouse from the operational sources. The user queries include 99 queries of 4 categories: reporting, ad-hoc, iterative OLAP, and data mining queries.

Lastly, the TPC-DS benchmark includes primary metrics, in which its performance metric is measured in queries per hour.

TPC-DS is widely used by vendors to demonstrate their capabilities to support complex decision support systems, by customers as a key factor in purchasing servers and software, and by the database community for research and development of optimization techniques. The TPC-DS benchmark is expected to be the next generation industry standard decision support benchmark [12].

4   Relationship Between Big Data and Decision Support Benchmarks

There are some important differences between the main objectives of Big Data and Decision Support. Decision Support Systems help organizations analyze data about current situations, compare available options and assess probable outcomes. Data sources in this system are mostly structured [10].

Big Data includes more flexible information, adds procedural analysis like MapReduce to examine data in parallel, is associated with large volumes of data, and is highly scalable [7].

The real use for Big Data comes in when it is used for business outcomes in decision-making or in operation management. As it stands today, Big Data demands the maturity of diverse technologies to deliver insights as well as an evolved consumer or corporate to handle big decisions out of it.

BigData systems are frequently oriented towards scalable storage and retrieval of key-value organized data, offering as main pillars put and get primitives to store and retrieve individual records based on an indexed key value, and simple scans. This is why benchmarks such as YCSB stress these kinds of operations. On the other hand, DSS systems are more oriented towards complex analytical processing of the data in order to compute meaningful multidimensional analysis and exploration of business data. For that reason, DSS benchmarks such as TPC-H or SSB are more oriented towards elaborate queries scanning and joining big datasets. In contrast, many NoSQL engines do not even process joins or group by, which are main pillars in DSS.

Another important aspect tested in some DSS benchmarks, in particular in TPC-DS, is a set of operations and structures related to DSS. These include the extract-transform-load (ETL) operations used to migrate data, and several auxiliary structures, such as data staging area, materialized views or indexes.

However, Big Data is related to DSS, since decision support over big amounts of data can be considered a Big Data issue. This is why the BigBench benchmark, which is identified as a Big Data benchmark, is based on TPC-DS. The BigBench data model is based on the TPC-DS model, adding new semi-structured and unstructured data sources to it. Likewise, the query workload of BigBench is an augmented version of the TPC-DS workload, adding data mining analysis and other Big Data processing.

Since these research areas overlap in some cases, solutions targeted at DSS can also benefit research in Big Data and vice-versa. Successfully managing big data and using them for decision making and business analytics include the improvement of overall efficiency, the improvement of speed and accuracy of decision making, the ability to forecast, the identification of business opportunities and a greater understanding of citizens' and customers' needs. Examples of some commonly useful contributions are: DSS processing of complex queries over huge datasets; automatic parallelization and scalability of processing in Big Data systems. In the future, these systems will probably partially converge, and there will be common benchmarks.

5   Conclusions and Future Work

In this paper we reviewed a set of benchmarks. The information we provided eases the job of anyone who needs to choose the most adequate benchmark for their evaluation. Both Big Data and DSS systems have gained a lot of popularity in the last years and have become successful in many different ways.

This paper helps the reader to have a better understanding of different Big Data and Decision Support System Benchmarks that exist and their major characteristics, which allows us to differentiate one from another, making it easier to choose which one we want for different purposes. It also compares Big Data and DSS concepts and benchmarks, concluding what is similar and different between them. We concluded that Big Data and DSS are very much related to each other, and in the future we should see more and more common benchmarks and systems.

Acknowledgments. This work was partially financed by iCIS – Intelligent Computing in the Internet Services (CENTRO-07- ST24 – FEDER – 002003), Portugal.

References

1. 5 Steps To Benchmarking Managed NoSQL - DynamoDB Vs Cassandra,
   http://highscalability.com/blog/2013/4/3/
   5-steps-to-benchmarking-managed-nosql-dynamodb-vs-
   cassandra.html (accessed November 2, 2013)

2. Bog, A.: Benchmarking Transaction and Analytical Processing Systems. In: The Creation of a Mixed Workload Benchmark and its Application. Springer (2009) ISBN: 978-3-642-38069-3

3. Ghazal, A., Rabl, T., Hu, M., Raab, F., Poess, M., Croletteand, A., Jacobsen, H.A.: BigBench: Towards an Industry Standard Benchmark for Big Data Analytics. In: SIGMOD Conference, pp. 1197–1208 (2013)

4. Thanopoulou, A., Carreira, P., Galhardas, H.: Benchmarking with TPC-H on Off-the-Shelf Hardware. In: ICEIS (1), pp. 205–208 (2012)

5. An overview of the TPC-H benchmark on HP ProLiant servers and server blades, ftp://ftp.hp.com/pub/c-products/servers/benchmarks/HP_ProLiant_tpcc_Overview.pdf (accessed October 14, 2013)

6. Cooper, B.F., Silberstein, A., Tam, E., Ramakrishnan, R., Sears, R.: Benchmarking Cloud Serving Systems with YCSB. In: SoCC, pp. 143–154 6 (2010)

7. Big Data Insights, http://www.sas.com/big-data/ (accessed November 1, 2013)

8. Big Data, http://wikibon.org/wiki/v/Enterprise_Big-data (accessed December 10, 2013)

9. BigFrame User Guide, https://github.com/bigframeteam/BigFrame/wiki/BigFrame-User-Guide (accessed November 1, 2013)

10. Decision Support, http://www.slideshare.net/sursayantan92/decision-support-systemdss (accessed December 10, 2013)

11. Li, F., Ooi, B.C., Ozsu, M.T., Wu, S.: Distributed Data Management Using MapReduce. ACM Comput. Surv. 46(3), 31 (2014)

12. Poess, M., Nambiar, R.O., Walrath, D.: Why You Should Run TPC-DS:A Workload Analysis. In: VLDB, pp. 1138–1149 (2007)

13. O'Neil, P., O'Neil, E., Chen, X.: The star schema benchmark (SSB). University of Massachusetts, Boston, Technical Report (2007)

14. TPC-DS, http://www.tpc.org/information/sessions/tpc_ds.pdf (accessed October 12, 2013)

15. TPC-DS, http://www.tpc.org/tpcds/default.asp (accessed October 12, 2013), 4

16. TPC-H Benchmark, http://cs338-s13.cs.uwaterloo.ca/ (accessed October 18, 2013)

17. YCSB, https://github.com/brianfrankcooper/YCSB/wiki/core-workloads (accessed November 2, 2013)

Parallelizing Structural Joins to Process Queries over Big XML Data Using MapReduce

Huayu Wu

Institute for Infocomm Research, A*STAR, Singapore
huwu@i2r.a-star.edu.sg

Abstract. Processing XML queries over big XML data using MapReduce has been studied in recent years. However, the existing works focus on partitioning XML documents and distributing XML fragments into different compute nodes. This attempt may introduce high overhead in XML fragment transferring from one node to another during MapReduce execution. Motivated by the structural join based XML query processing approach, which uses only related inverted lists to process queries in order to reduce I/O cost, we propose a novel technique to use MapReduce to distribute labels in inverted lists in a computing cluster, so that structural joins can be parallelly performed to process queries. We also propose an optimization technique to reduce the computing space in our framework, to improve the performance of query processing. Last, we conduct experiment to validate our algorithms.

1 Introduction

The increasing amount of data generated by different applications and the increasing attention to the value of the data marks the beginning of the era of big data. It is no doubt that to effectively manage big data is the first step for any further analysis and utilization of big data. How to manage such data poses a new challenge to the database community.

Gradually, many research attempts converge to a distributed data processing framework, MapReduce [9]. This programming model simplifies parallel data processing by offering two interfaces: map and reduce. With a system-level support on computational resource management, a user only needs to implement the two functions to process underlying data, without caring about the extendability and reliability of the system. There are extensive works to implement database operators [5][12], and database systems [3][11] on top of MapReduce.

Recently, researchers started looking into the possibility of managing big XML data in a more elastic distributed environment, such as Hadoop [1], using MapReduce. Inspired by the XML-enabled relational database system, big XML data can be stored and processed by relational storage and operators. However, shredding XML data in big size into relational tables is extremely expensive. Furthermore, with relational storage, each XML query must be processed by several θ-joins among tables. The cost for joins is still the bottleneck for Hadoop-based database systems.

H. Decker et al. (Eds.): DEXA 2014, Part II, LNCS 8645, pp. 183–190, 2014.

Most recent research attempts [8][7][4] leverage on the idea of XML partitioning and query decomposition adopted from distributed XML databases [13][10]. Similar to the join operation in relational database, an XML query may require linking two or more arbitrary elements across the whole XML document. Thus to process XML queries in a distributed system, transferring fragmented data from one node to another is unavoidable. In a static environment like a distributed XML database system, proper indexing techniques can help to optimally distribute data fragments and the workload. However, for an elastic distributed environment such as Hadoop, each copy of XML fragment will probably be transferred to undeterminable different nodes for processing. In other words, it is difficult to optimize data distribution in a MapReduce framework, thus the existing approach may suffer from high I/O and network transmission cost.

Actually different approaches for centralized XML query processing have been study for over a decade. One highlight is the popularity of the structural join based approach (e.g., [6]). Compared to other native approaches, such as navigational approach and subsequence matching approach, one main advantage of the structural join approach is the saving on I/O cost. In particular, in the structural join approach, only a few inverted lists corresponding to the query nodes are read from the disk, rather than going through all the nodes in the document. It will be beneficial if we can adapt such an approach in the MapReduce framework, so that the disk I/O and network cost can be reduced.

In this paper, we study the parallelization of the structural join based XML query processing algorithms using MapReduce. Instead of distributing a whole big XML document to a computer cluster, we distribute inverted lists for each type of document node to be queried. Since the size of the inverted lists that are used to process a query is much smaller than the size of the whole XML document, our approach potentially reduces the cost on cross-node data transfer.

2   Related Work

Recently, there are several works proposed to implement native XML query processing algorithms using MapReduce. In [8], the authors proposed a distributed algorithm for Boolean XPath query evaluation using MapReduce. By collecting the Boolean evaluation result from a computer cluster, they proposed a centralized algorithm to finally process a general XPath query. Actually, they did not use the distributed computing environment to generate the final result. It is still unclear whether the centralized step would be the bottleneck of the algorithm when the data is huge. In [7], a Hadoop-based system was designed to process XML queries using the structural join approach. There are two phases of MapReduce jobs in the system. In the first phase, an XML document was shred into blocks and scanned against input queries. Then the path solutions are sent to the second MapReduce job to merge to final answers. The first problem of this approach is the loss of the "holistic" way for generating path solutions. The second problem is that the simple path filtering in the first MapReduce phase is not suitable for processing "//"-axis queries over complex structured XML data with recursive nodes. A recent demo [4] built an XML query/update system on top of MapReduce framework. The technical details were not thoroughly presented. From the system architecture, it shreds an XML document and asks each mapper to process queries against each XML fragment.

In fact, most existing works are based on XML document shredding. Distributing large XML fragments will lead high I/O and network transmission cost. On contrast, our approach distributes inverted lists rather than a raw XML document, so that the size of the fragmented data for I/O and network transmission can be greatly reduced.

3   Framework

3.1   Framework Overview

In our approach, we implement the two functions, map and reduce in a MapReduce framework, and leverage on underlying system, e.g., Hadoop for program

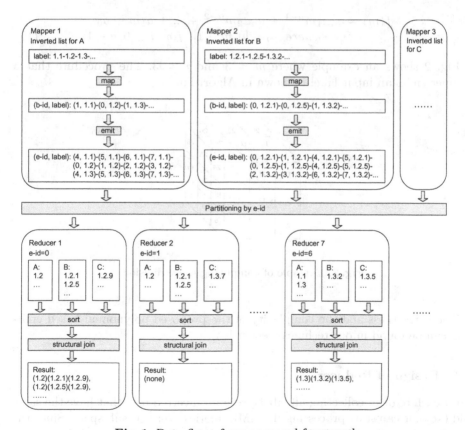

Fig. 1. Data flow of our proposed framework

execution. The basic idea is to equally (or nearly equally) divide the whole computing space for a set of structural joins into a number of sub-spaces, and each sub-space will be handled by one reducer to perform structural joins.

Each mapper will take a set of labels in an inverted list as input, and emit each label with the ID of the associated sub-space (called e-id, standing for emit id). The reducers will take the grouped labels for each inverted list, re-sort it, and apply holistic structural join algorithms to find answers. The whole process is shown in Fig. 1, and the details will be explained in the following sections.

3.2 Design of Mapper

The main task of a mapper is to assign a key to each incoming label, so that the overall labels from each inverted list are nearly equally distributed in a given number of sub-spaces for the reducers to process. To achieve this goal, we adopt a polynomial-based emit id assignment.

For a query using n inverted lists and each inverted list is divided into m sub-lists, the total number of sub-spaces is m^n. We construct a polynomial function f of m with the highest degree of $n-1$, to determine the emit id of each sub-space.

$$f(m) = a_{n-1}m^{n-1} + a_{n-2}m^{n-2} + ... + a_1 m + a_0$$
$$where \quad a_i \in [0, m-1] \quad for \quad i \in [0, n-1] \qquad (1)$$

Fig. 2 shows an example where $m = 3$ and $n = 3$. The procedure that a mapper emits an input label is shown in Algorithm 1.

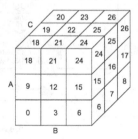

Fig. 2. Example of computing space division

The correctness and the complexity of the proposed polynomial-based emitting can be found in our technical report [14].

3.3 Design of Reducer

After each reducer collecting all labels from all inverted lists that have the same e-id (key), it can start processing the XML queries over this sub-space. Since the map function only splits the computing space into small sub-spaces without other

Algorithm 1. Map Function

Input: an empty key, a label l from an inverted list I as the value; variables m for the
   number of partitions in each inverted list, n for the number of total inverted lists,
   r a random integer between 0 and $n - 1$

1: identify the coefficient corresponding to I, i.e., a_i where i is the index
2: initiate an empty list L
3: toEmit(L, i) /*m, n, r, l are globally viewed by Function 1 and 2*/

Function 1. toEmit(List L, index i)

1: **if** L.length == number of inverted lists **then**
2:    Emit(l)
3: **else**
4:    **if** L.length == i **then**
5:       toEmit($L.append(r), i$)
6:    **else**
7:       **for all** $j \in [0, n - 1]$ **do**
8:          toEmit($L.append(j), i$)

Function 2. Emit(List L)

1: initiate the polynomial function $f(m)$ as defined in (1)
2: set the coefficients of f as the integers in L, by order
3: calculate the value of f given the value of m, as the emit key e-id
4: emit(e-id, l)

---

operations on data, in the reduce function, any structural join based algorithm
can be implemented to process queries.

In the example in Fig. 1, for each reducer, after getting a subset of labels
in each inverted list, the reducer will sort each list and then perform holistic
structural join on them, to find answers.

4   Optimization

Let us start with a motivating example. Suppose an algorithm tries to process a
query //A[//B]//C. In the sorted inverted list for A, the first label is 1.3, while
in the sorted inverted list for B the first label is 1.1.2. Obviously, performing a
structural join A//B between 1.3 and 1.1.2 will not return an answer.

Recall that in Algorithm 1 when a map function emits a label, it randomizes
a local partition and considers all possible partitions for other inverted lists for
the label to emit. In our optimization, to emit a label from an inverted list I, we
(1) set the local partition in I for the label according to an index pre-defined
based on node statistics, i.e., the cut-off index, and (2) selectively choose the
coefficients (i.e., the partitions) for all the ancestor inverted lists of I, such that
the current label can join with the labels in those partitions and return answers.
The *toEmit* function for an optimized mapper is presented in Function 3.

Function 3. toEmit^O(List L, index i)

Input: the partition cut-off index cutoff[x][y] for inverted list I_x and partition y; the
current inverted list I_u with numerical id u; other variables are inherited from
Algorithm 1

1: **if** L.length == number of inverted lists **then**
2:    Emit(l)
3: **else**
4:    **if** L.length == i **then**
5:      initiate $v = 0$
6:      **while** cutoff[x][y].precede(l) && v<m **do**
7:        v++
8:      toEmit($L.append(v), i$)
9:    **else**
10:     **if** the query node for $I_{L.length}$ is an ancestor of the query node for I_u **then**
11:       initiate $v = 0$
12:       **while** cutoff[x][y].precede(l) && v<m **do**
13:         v++
14:       **for all** $k \in [0, v-1]$ **do**
15:         toEmit($L.append(k), i$)
16:     **else**
17:       **for all** $j \in [0, n-1]$ **do**
18:         toEmit($L.append(j), i$)

The intuition of the optimization is to prune a label emitting to certain reducers in which it will not produce structural join result. The details of defining
the cut-off index and some running examples can be found in [14].

5 Experiment

5.1 Settings

We use a small Hadoop cluster with 5 slave nodes to run experiments. Each
slave node has a dual core 2.93GHz CPU and a 12G shared memory. We keep
all default parameters of Hadoop in execution.

We generated a synthetic XML dataset with the size of 10GB, based on
the XMark [2] schema. The document is labeled with the containment labeling scheme so that the size of each label is fixed. We randomly compose 10
twig pattern queries with the number of query nodes varying from 2 to 5, for
evaluation. The result presented in this section is based on the average running
statistics.

5.2 Result

Since the main query processing algorithm is executed in the Reduce function, we
first vary the number of reducers to check the impact on the performance for both

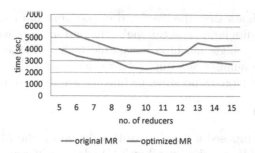

Fig. 3. Performance under different number of reducers

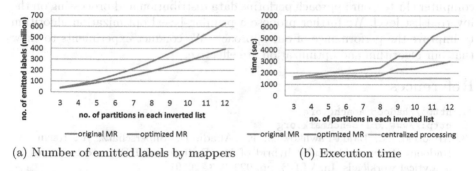

(a) Number of emitted labels by mappers   (b) Execution time

Fig. 4. Performance comparison between the original and the optimized algorithms

the original MapReduce algorithm and the optimized MapReduce algorithm. We set the number of mappers to be 10, and try different numbers of reducers.

From the result in Fig. 3 we can see that the performance for both the original algorithm and the optimized algorithm are improved as the number of reducers increases, until it reaches 10 or 11. After that the performance is not stable. The result accords with the hardware settings. There are 5 dual core processors, which can support 10 tasks to run parallelly. When the processors are fully utilized, the performance may be significantly affected by other issues, such as network transmission overhead.

Also, the optimized algorithm is always better than the original algorithm without optimization. We further show this point in the second experiment.

We keep the number of mappers and the number of reducers to be 10. We vary the number of partitions in each inverted list, which determines the number of reduce jobs. Fig. 4(a) shows the number of labels emitted by the mappers for the two algorithms under different partition numbers for inverted lists. It clearly tells that the optimized algorithm prunes more labels as the sub-spaces are more fine-grained. As a consequence, the performance of the optimized algorithm is better than the original algorithm, as shown in Fig. 4(b).

We can also see from Fig. 4(b) that as the number of partitions in each inverted list increases, the overall performance will drop. We also managed to run the structural join algorithm in a computer with a 3.2GHz CPU and a 8GB memory, and show the average execution time for queries in Fig. 4(b) as well.

This comparison shows that the overhead on network data transmission for a Hadoop cluster is quite large, and can make the performance worse than a single machine (suppose a single machine is capable to handle the program). Thus, we should limit the number of reduce jobs so that each reducer will take over a computing space as large as possible, to fully utilize its resource.

6   Conclusion

In this paper, we proposed a novel algorithm based on the MapReduce framework to process XML queries over big XML data. Different from the exiting approaches that shred and distribute XML document into different nodes in a computer cluster, our approach performs data distribution and processing on the inverted list level. We further propose a pruning-based optimization algorithm to improve the performance of our approach. We conduct experiments to show that our algorithm and optimization are effective.

References

1. http://hadoop.apache.org
2. http://www.xml-benchmark.org
3. Abouzeid, A., Bajda-Pawlikowski, K., Abadi, D., Silberschatz, A., Rasin, A.: HadoopDB: an architectural hybrid of MapReduce and DBMS technologies for analytical workloads. In: VLDB, pp. 922–933 (2009)
4. Bidoit, N., Colazzo, D., Malla, N., Ulliana, F., Nole, M., Sartiani, C.: Processing XML queries and updates on map/reduce clusters. In: EDBT, pp. 745–748 (2013)
5. Blanas, S., Patel, J.M., Ercegovac, V., Rao, J., Shekita, E.J., Tian, Y.: A comparison of join algorithms for log processing in MapReduce. In: SIGMOD, pp. 975–986 (2010)
6. Bruno, N., Koudas, N., Srivastava, D.: Holistic twig joins: optimal XML pattern matching. In: SIGMOD, pp. 310–321 (2002)
7. Choi, H., Lee, K., Kim, S., Lee, Y., Moon, B.: HadoopXML: a suite for parallel processing of massive XML data with multiple twig pattern queries. In: CIKM, pp. 2737–2739 (2012)
8. Cong, G., Fan, W., Kementsietsidis, A., Li, J., Liu, X.: Partial evaluation for distributed XPath query processing and beyond. ACM Trans. Database Syst. 37(4), 32 (2012)
9. Dean, J., Ghemawat, S.: MapReduce: simplified data processing on large clusters. In: USENIX Symp. on Operating System Design and Implementation, pp. 137–150 (2004)
10. Kling, P., Ozsu, M.T., Daudjee, K.: Generating efficient excution plans for vertically partitioned XML databases. PVLDB 4(1), 1–11 (2010)
11. Lin, Y., Agrawa, D., Chen, C., Ooi, B.C., Wu, S.: Llama: leveraging columnar starage for scalable join processing in the MapReduce framework. In: SIGMOD, pp. 961–972 (2011)
12. Okcan, A., Riedewald, M.: Pricessing theta-joins using MapReduce. In: SIGMOD, pp. 949–960 (2011)
13. Suciu, D.: Distributed query evaluation on semistricutred data. ACM Trans. Database Syst. 27(1), 1–62 (2002)
14. Wu, H.: Parallelizing Structural Joins to Process Queries over Big XML Data Using MapReduce. Tech Report, http://www1.i2r.a-star.edu.sg/~huwu/paraSJ.pdf

Optimization Strategies for Column Materialization in Parallel Execution of Queries

Chi Ku[1], Yanchen Liu[1,2], Masood Mortazavi[1], Fang Cao[1],
Mengmeng Chen[1], and Guangyu Shi[1]

[1] Huawei Innovation Centeri, Santa Clara, CA, USA
[2] City College, CUNY, New York, NY, USA
{chi.young.ku,yanchen.liu,masood.mortazavi,
fang.cao,mengmeng.chen,shiguangyu}@huawei.com

Abstract. All parallel query processing frameworks need to determine the optimality norms for column materialization. We investigate performance trade-off of alternative column materialization strategies. We propose a common parallel query processing approach that encapsulates varying column materialization strategies within *exchange nodes* in query execution plans. Our experimental observations confirm the theoretically deduced trade-offs that suggest optimality norms to be dependent on the scale of the cluster, data transmissions required for a query, and the predicate selectivities involved. Lastly, we have applied a probit statistical model to the experimental data in order to establish a system-dependent adhoc performance estimation method that can be used to select the optimal materialization strategy at runtime.

Keywords: query optimization, databases, materialization.

1 Introduction

Parallel processing of queries is a common feature of modern database systems [1]. As such, optimization techniques need to account for data transfer costs, whether among cores [2] or among nodes in a distributed system [3]. As an optimization strategy, materializing of intermediate values affects the efficiency of column-oriented analytic data processing [4]. In the course of evaluating a query, intermediate columns are produced from input columns. Intermediate columns may need to be reshuffled among nodes for further processing. An execution plan often involves columns of *relative* (implicit) or *absolute* (explicit) row identifiers produced as a result of some selection predicate. In this paper, we seek to determine the optimal point in a parallel query plan where one should materialize column values corresponding to column row identifiers.

2 Framework

Materialization builds an attribute needed by subsequent processing. We use the query in Listing 1.1 as our running example. After evaluating the predicate, $l\_quantity > 0$, the attribute vector for column $l\_partkey$ needs to be built

H. Decker et al. (Eds.): DEXA 2014, Part II, LNCS 8645, pp. 191–198, 2014.
© Springer International Publishing Switzerland 2014

so that it could be used to compute $l\_partkey = p\_partkey$ predicate to find rows that satisfy it. Here, $l\_suppkey$ is char(30), and table *lineitem* is randomly horizontally partitioned. Table *part* is horizontally partitioned on $p\_partkey$.

Listing 1.1. Example Query 1

```
1  SELECT  l_suppkey
2  FROM    lineitem, part
3  WHERE   l_partkey = p_partkey
4  AND     l_quantity > 0;
```

All the materialization strategies we consider operate within the context of an execution plan expressed as a Directed Acyclic Graph (DAG). At operators in this DAG, the materialization methods can be organized into two categories: 1. *Relative Row IDs*: Using relative (implicit) row IDs, which are the row IDs of intermediate results. 2. *Absolute Row IDs*: Using absolute (explicit) row IDs, which are the row IDs of base tables. Given these two basic methods, materialization strategies involve, but are not limited to: 1. *Early Materialization* (EM): For one column only, with *relative row IDs* being used for the operators. 2. *Late Materialization* (LM): For one column only, with *absolute row IDs* being used for the operators. 3. *Mixed Materialization* (MM): Some columns could use EM while others use LM. MM strategy takes advantage of both EM and LM. In *parallel processing*, exchange nodes (used to implement data exchange operators) are included. The choice of materialization strategy affects both memory usage and communication cost. Therefore, the materialization cost (at an operator with an exchange node) involves the memory to hold the attribute, as well as the communication bandwith to re-shuffles the attributes (that are not immediately needed after the current operation), while the benefit is the savings from not reading the attribute columns from disks until they are absolutely needed and not shipping the selected data through the communication bandwith.

Most commercial databases seem to use early materialization strategies. Although there are some differences in implementation, Vertica [5], SQL Server [6], IBM DB2 [7], Oracle 12C [8], Teradata [9], and InfoBright [10] all seem to use early materialization strategies. MonetDB [11], with which we have done most of our own research work, uses a just-in-time late materialization: Columns are accessed when needed as input to a BAT operation. [12] demonstrates how early materialization combined with sideways information passing allows them to get the benefits of late materialization. However, their approach is not proposed for distributed cluster environments. To the best of our knowledge, none of the existing systems optimize materialization strategies, especially for distributed database systems with parallel query processing. That has been our focus.

3   Materialization Strategies

In what follows, we will discuss three materialization strategies: early, late and mixed. We have elided the details of the processing cost model we developed for these strategies due to space limitations.

Early materialization (EM) uses *relative materialization* to materialize each column needed by an operator. Each exchange node materializes all columns needed by an ancestor operator. Take the query in Listing 1.1 (in Section 2) as an example. In EM, the runtime processing steps are as follows: 1. On each node where a partition of *lineitem* table resides, scan *l_quantity* to evaluate the predicate, *l_quantity* > 0. 2. On each node where a partition of *lineitem* table resides, materialize *l_partkey* and *l_suppkey* and re-shuffle them according to *l_partkey*. 3. On each node where a *part* table partition resides, perform the join, *l_partkey* = *p_parkey*, and materialize *l_suppkey*. Early materialization integrates into the common parallel query processing framework for column store by utilizing exchange nodes that materialize and re-shuffle all columns needed by their ancestor relational operators [13][14].

Late materialization (LM) uses *absolute materialization*. An exchange node only materializes the column which is needed by its ancestor operators. Again, taking the query 1.1 (Section 2) as an example, runtime processing steps for LM are: 1. On each node where a partition of *lineitem* table resides, scan *l_quantity* to evaluate the predicate, *l_quantity* > 0. 2. On each node where a partition of *lineitem* table resides, materialize *l_partkey* and re-shuffle the row ID and *l_partkey* according to *l_partkey*. 3. On each node where a *part* table partition resides, perform the join, *l_partkey* = *p_parkey*, and materialize *l_suppkey* by fetching using *lineitem* table row ID from the join output. In late materialization, each exchange node re-shuffles only the absolute row IDs. When a column is needed, it is materialized by fetching it using absolute row IDs from nodes where the column resides. In the parallel query processing of LM, the parallel Rel DAG is transformed into a DAG of functions calls and data re-shuffling actions. The cost model for both early and late materialization the CPU cost, $cost_{cpu}(Q)$, the disk I/O cost, $cost_{io}(Q)$, and the column values transmission costs $cost_{comm}(Q)$. As noted earlier, we elide the details of both cost models here, due to space limitations.

Mixed materialization (MM) materializes columns depending on their access patterns. Based on the trade-offs captured in a cost model, columns that have clustered access patterns are materialized early while columns that have scattered access patterns are materialized late. The early-materialized column is materialized/re-shuffled at every exchange node that has ancestor relational operators needing the column, while the late-materialized column is materialized (by means of row IDs) as needed. Take the following query as an example:

Listing 1.2. Example Query 2

```
1  SELECT  l_suppkey
2  FROM    lineitem, part a, part b
3  WHERE   l_partkey = a.p_partkey
4  AND     l_quantity = b.p_partkey
5  AND     l_quantity > 0;
```

Let's assume *lineitem* is randomly partitioned, while *part* is partitioned according to *p_partkey*. Here, *l_quantity* is materialized early, and *l_suppkey* is materialized late. In MM, the runtime processing steps are: 1. On each node where a partition of *lineitem* table resides, scan $l\_quantity$ to evaluate the predicate, $l\_quantity > 0$. 2. On each node where a partition of *lineitem* table resides, reshuffles *lineitem* row IDs, $l\_partkey$ and $l\_quantity$ according to $l\_partkey$. 3. On each node where a *part* table partition resides, perform the join, $l\_partkey = a.p\_parkey$. We, then, re-shuffles *lineitem* table row ID and $l\_quantity$. 4. On each node where a *part* table partition resides, perform the join, $l\_quantity = b.p\_partkeyi$. We, then, materialize $l\_suppkey$ by fetching it from *lineitem* table row IDs that are the output of the join. Mixed materialization integrates into the common parallel query processing framework for column store by utilizing each exchange node that materializes and re-shuffles row IDs and all early materialization columns. The mixed materialization specific part of parallel query processing strategy is summarized in the following: *(i)* A depth-first traversal of the parallel Rel DAG is utilized to analyze the access pattern of each column for computing estimates of cardinality of output of each relational operator. *(ii)* Based on the theoretical cost model developed for early and late materialization, an experimental Probit model for the probability that LM outperforms EM is built. This is then exploited to determine whether a column is materialized by EM or LM.

4   Performance Modeling

In order to validate our theoretical cost model and develop a suitable cost model for deciding which materialization strategy to use, we firstly create and measure various experiments based on the strategies proposed in the previous sections. Then, we use the GRETL econometric analysis software to deduce the best Probit model based on our experimental results.

According to our previous analysis, the testable predictions are: (i) Late materialization is better when access pattern is scattered. (ii) Late materialization is better when the selectivity of operators is low. (iii) Late materialization is better when the number of nodes in the system increases.

Experiments. Physical compute resources, within which we run our experiments, have the following configuration: Ubuntu Linux 3.5.0-17-generic x86_64, 188 GB of RAM, 2x Intel(R) Xeon(R) CPU, E5-2680 @ 2.70GHz, 32 effective cores @ 1.2GHz, cache size: 20480 KB. Our laboratory environment for distributed query processing experimentation and analysis consists of clusters of Linux containers (also known as LXCs [15]) installed on physical machines with separate system administration and data exchange networks. We implemented and compared our strategies in MonetDB, an open-source, column oriented database management systems [16].

In order to discover how the trade-offs among EM, LM and MM depend on the amount of data transmitted, the number of processing nodes, the number of columns in the results, and the number of join operations, etc., we performed several experiments. These experiments were based on the queries listed in Table 1. Here, *lineitem* table is randomly partitioned, and the *part* table is partitioned

Table 1. Experimental Queries

1-column-1-join (A)	2-column-1-join (B)	1-column-2-join (C)
SELECT l_suppkey FROM lineitem, part WHERE l_partkey = p_partkey AND l_quantity > 0;	SELECT l_suppkey, l_suppkey_2 FROM lineitem, part WHERE l_partkey = p_partkey AND l_quantity > 0;	SELECT l_suppkey FROM lineitem, part a, part b WHERE l_partkey = a.p_partkey AND l_quantity = b.p_partkey AND l_quantity > 0;

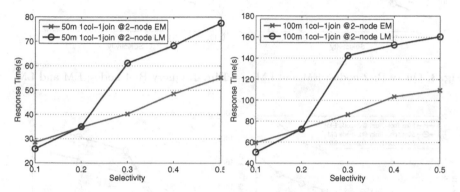

Fig. 1. Query A: 2 nodes, 50 million rows; EM and LM

Fig. 2. Query A: 2 nodes, 100 million rows; EM and LM

on *p_partkey*. The tables are loaded with 50, 100 and 200 million rows. And, the join selectivity varies from 0.1 to 0.9. In Query A, *l_suppkey* is either early or late materialized. In Query B, *l_suppkey*, *l_suppkey_2* and *l_quantity* are either early or late materialized. In Query C, *l_quantity* is either early, mixed or late materialized, and *l_suppkey* is either early or late materialized.

Figure 1 and Figure 2 present the time latency of materialization strategies for executing the query A in Table 1 with 50 million and 100 million row input, respectively. It is indicated that LM is favored when join selectivity descreases because smaller join selectivity decreases the amount of data transmitted in LM. On the other hand, the amount of data transmitted in EM does not change with join selectivity, and, thus, the benefit for EM brought in by the lower selectivity is smaller than that for LM.

The impacts of the number of returned columns and the cluster scale on the query latency are presented in Figure 3 and Figure 4. It is indicated that the trade-off between EM and LM is magnified as the number of columns in the select list increases. Since the ratio of data transmitted in EM and LM is constant, the time latency is proportionally increased for both EM and LM. In Figure 4, LM is favored when the number of nodes in the system increases because the amount of data re-shuffled in EM increases with the number of nodes; while, the amount of data transmitted in LM does not change with the number of nodes.

Figure 5 and Figure 6 present the time latency for query C with 50 million and 100 million row input, respectively. In the 50 million input case, the selectivity of the second join is fixed to 0.1, and the one for the first join is varied. On the

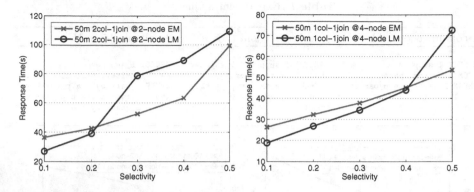

Fig. 3. Query B: 2 nodes; EM and LM   **Fig. 4.** Query B: 4 nodes; EM and LM

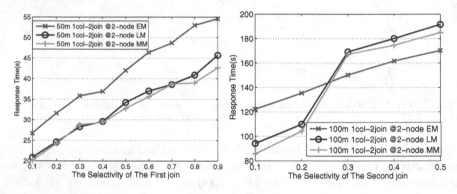

Fig. 5. Query C: 2 nodes, 50 million rows, second join selectivity of 0.9; EM, LM and MM

Fig. 6. Query C: 2 nodes, 100 million rows, first join selectivity of 0.1; EM, LM and MM

other hand, in the 100 million input case, the selectvity of the first join is fixed to 0.9, and the one for the second join is varied. It is shown that LM outperforms EM when the number of join increases because the amount of data transmitted in LM decreases (generally) with the increasing number of joins (as the number of output rows from each join is progressively smaller in the experiment); On the other hand, the data transmitted in EM does not decrease with the increase of the number of joins. It is also shown that MM outperforms LM when the selectivity of the first join is high because $l\_quantity$ is early materialized in MM, and the higher join selectivity favors EM.

Model. We fit a Probit model to the experimental data. The Probit model from the GRETL software indicates that the number of nodes, the number of columns re-shuffled, the number of joines, the amount of data transmitted in EM, and the amount of data transmitted in LM are indeed very good determinants

Table 2. Probit model fitting

	Coefficient	Std. Error	z-stat	p-value
const	10.6210	4.65447	2.2819	0.0225
l_Data_EM	2.04058	0.874308	2.3339	0.0196
l_Data_LM	−3.66815	1.10964	−3.3057	0.0009
Joins	−2.69016	1.11177	−2.4197	0.0155

Mean dependent var	0.645833	S.D. dependent var	0.091623
McFadden R^2	0.623500	Adjusted R^2	0.495292
Log-likelihood	−11.74659	Akaike criterion	31.49318
Schwarz criterion	38.97798	Hannan–Quinn	34.32170

Number of cases 'correctly predicted' = 44 (91.7 percent)
Likelihood ratio test: $\chi^2(3) = 38.906\ [0.0000]$

for whether one should use LM or EM for a shorter query response time. The discriminator function for a given query is then given as:

$$E(Q) = f\left(log(\frac{NOB_{EM}}{NUM_n}) \times SPB_{EM} - log(\frac{NOB_{LM}}{NUM_n}) \times SPB_{LM} + NUM_j \times J + K\right) \quad (1)$$

where $E(Q)$ selects materialization strategy for a given query Q, $f()$ is the Normal CDF function, NOB_{EM} and NOB_{LM} are the numbers of transfered bytes in EM and LM, respectively. NUM_n represents the number of nodes, and NUM_j represents the number of joins in the query. SPB_{EM} and SPB_{LM} represent second per byte for early and late materializations, respectively. J is the join specific overhead, while K is a constant. All coefficients of independent variables of this Probit Model are statistically significant and the in-sample error rate is less than 10%. A similar approach can result in a system specific adhoc cost model for deciding when to use EM or LM for a column.

5 Conclusion

We have investigated how a column can be materialized in parallel query processing for column store databases. Three different column materialization approaches are implemented within a common parallel query processing framework. To investigate how our theoretical prediction coincides with the actual data, we ran a number of queries consisting of commonly used relational operators against data with different join selectivities and systems with different number of nodes. The experimental results confirm all the theoretical predictions: (i) The relative performance of different materialization strategies is a function of the amount data transmitted, the number of nodes in the system, and the number of join operators in the query. (ii) Early Materialization outperforms Late Materialization when the column access is clustered, while Late Materialization outperforms Early Materialization when the access is scattered. (iii) Late Materialization outperforms Early Materialization for the relational operator with low selectivity. (iv) Higher query complexities favor Early Materialization, while larger system scales favor Late Meterialization. Our experimental result confirms our theoretical model firmly. The resulting Probit model for selecting the optimal materialization stratery enjoys a 91% in-sample prediction accuracy.

References

1. DeWitt, D., Dray, J.: Parallel database systems: the future of high performance database systems. Comm. ACM 35, 85–98 (1992)
2. Anikiej, K.: Multi-core parallelization of vectorized queries. Master Thesis, University of Warsaw and VU University of Amsterdam (2010)
3. Thomson, A., et al.: Calvin: Fast Distributed Transactions for Partitioned Database Systems. In: Proceedings of the 2012 ACM SIGMOD International Conference on Management of Data, Scottsdale, Arizona, USA (2012)
4. Abadi, D., Myers, D.S., DeWitt, D.J., Samuel, R.M.: Materialization strategies in a column-oriented DBMS. In: IEEE 23rd International Conference on Data Engineering, pp. 466–475 (2007)
5. Lamb, A., Fuller, M., Varadarajan, R., Tran, N., Vandiver, B., Doshi, L., Bear, C.: The vertica analytic database C-store 7 years later. In: Proceedings of the 38th International Conference on Very Large Data Bases, pp. 1790–1801. VLDB Endowment (2012)
6. Larson, P., Hanson, E.N., Price, S.L.: Columnar Storage in SQL Server 2012. IEEE Data Eng. Bull. 35, 15–20 (2012)
7. IBM DB2, http://pic.dhe.ibm.com/infocenter/db2luw/v10r5/index.jsp?topic=%2Fcom.ibm.db2.luw.admin.dbobj.doc%2Fdoc%2Fc0060592.html
8. Oracle, http://www.oracle.com/us/corporate/features/database-in-memory-option/index.html
9. Teradata, https://www.teradata.com/white-papers/Teradata-14-Hybrid-Columnar/
10. Infobright, https://www.infobright.com/index.php/Products/MySQL-Integration/
11. Boncz, P.A., Marcin, Z., Niels, N.: MonetDB/X100: Hyper-Pipelining Query Execution. In: CIDR, vol. 5, pp. 225–237 (2005)
12. Shrinivas, L., et al.: Materialization Strategies in the Vertica Analytic Database: Lessones Learned. In: IEEE 29th International Conference on Data Engineering, pp. 1196–1207. IEEE, Brisbane (2013)
13. MonetDB Kernel Modules, http://www.monetdb.org/Documentation/Manuals/MonetDB/Kernel/Modules
14. MonetDB MAL reference, http://www.monetdb.org/Documentation/Manuals/MonetDB/MALreferenceMonetDBstatement
15. Linux Container, http://lxc.sourceforge.net/
16. Idreos, S., et al.: MonetDB: Two Decades of Research In Column-Oriented Database Architectures. In: IEEE Data Engineering Bulletin, vol. 35, pp. 40–45 (2012)

Evaluating Cassandra Scalability with YCSB

Veronika Abramova[1], Jorge Bernardino[1,2], and Pedro Furtado[1]

[1] CISUC – Centre of Informatics and Systems of the University of Coimbra
FCTUC – University of Coimbra, 3030-290 Coimbra, Portugal
[2] ISEC – Superior Institute of Engineering of Coimbra
Polytechnic Institute of Coimbra, 3030-190 Coimbra, Portugal
veronika@student.dei.uc.pt, jorge@isec.pt, pnf@dei.uc.pt

Abstract. NoSQL data stores appeared to fill a gap in the database market: that
of highly scalable data storage that can be used for simple storage and retrieval
of key-indexed data while allowing easy data distribution over a possibly large
number of servers. Cassandra has been pinpointed as one of the most efficient
and scalable among currently existing NoSQL engines. Scalability of these engines means that, by adding nodes, we could have more served requests with
the same performance and more nodes could result in reduced execution time of
requests. However, we will see that adding nodes not always results in performance increase and we investigate how the workload, database size and the level of concurrency are related to the achieved scaling level. We will overview
Cassandra data store engine, and then we evaluate experimentally how it behaves concerning scaling and request time speedup. We use the YCSB – Yahoo! Cloud Serving Benchmark for these experiments.

Keywords: NoSQL, YCSB, Cassandra, Scalability.

1 Introduction

NoSQL databases represent the new technology developed as the new solution that
provides a set of additional characteristics that defer this database model from relational model. Mostly, all the concerns are focused on the high data volumes as well as
gown in application use. That means that used technology must be capable of fast
execution of requests as well as efficient data management, when it comes to storage
and extraction. Over past years Web data has been gowning exponent, and has highly
contributed to the technological evolution. The 3V concept characterized by the data
volume, velocity and variety has affected the entire data industry and showed that
there was a need for the evolution and improvement in used management tools, concerning data storage [8]. The data storage itself was solved by data distribution, meaning that NoSQL databases are capable of easily place parts of the same database on
different servers that are distributed geographically [2, 3]. However, this distribution
should provide performance increase and overcome communication delays, providing
fast query execution. After improvements in horizontal scalability of NoSQL databases, the companies started to defend that scaling your data highly affects the number of

H. Decker et al. (Eds.): DEXA 2014, Part II, LNCS 8645, pp. 199–207, 2014.
© Springer International Publishing Switzerland 2014

the served clients and shows good execution speed. First, with parallel execution of the requests, more operations are performed per time unit and more requests are executed. Secondly, adding more and more servers is expected to continue improving this performance and directly contribute to the overall request serving. This idea characterizes the data scalability that is tightly connected to the system speedup. This is the concept that corresponds to the reduced execution time of the requests that was obtained by the database scaling.

Companies started noticing that there is a need in efficiently manage client's requests and reduce waiting time to the lowest point, since performance problems and badly implemented system may directly affect companies' profit [10, 11]. From that point of view there are two questions. How many clients may be served? And how fast those can be served? Those aspects are closely related and both are connected to the database scaling and data distribution. The capability of serving high amount of load is managed by the amount of the servers and is highly important when we assume that the response time is acceptable. But it may not be and, therefore, it has to be reduced and put under a certain limit that would be considered reasonable. Scaling may also directly affect and improve execution time, lower execution time per client results in bigger number of served clients. Increased interest in non-relational technology, shown by the researchers, has contributed to the evaluations and studies of the NoSQL databases and, consequently, into big amount of the published papers. But, those studies try to simulate high corporate environment that does not correctly translates into smaller companies.

The main goal of this paper is to study scalability characteristics and architecture of the one of the most popular NoSQL databases, Cassandra. We test the capability of this database to manage and distribute stored data as well as the capability of efficiently manage offered load that represents higher number of simultaneous requests. We start by reviewing the architecture of Cassandra and some of its more important mechanisms that directly affect overall performance. We compare different execution times, obtained by varying the number of used servers as well as database size. As the second part of our experimental evaluation we increase the number of threads and test if Cassandra is able to manage those efficiently. During our evaluation we used the standard YCSB - Yahoo! Cloud Serving Benchmark that is one of the most popular benchmarks used to test NoSQL databases, and that provides data generator and executable workloads.

The remainder of this paper is organized as follows: section 2 discusses Cassandra's internal mechanisms that are highly related to the overall performance. Section describes the experimental setup and section 4 discusses the experiments. In that section we present our results and analyze Cassandra's performance concerning scalability and load management. Section 5 reviews related work and finally, section 6 presents the conclusions and future work.

2   Cassandra's Architecture

Cassandra is a Column Family NoSQL database that was originally developed by Facebook and currently is developed by the Apache Software [6]. As a Column Family database, the data in Cassandra is stored as sets of vertically oriented tables and this database is highly suited for heavy write systems, such as logging. This database has different features that contribute for performance increase such as, indexing and memory mapping. In this section we will describe some of the Cassandra's mechanisms that, in our view, cause high performance impact.

Internally, each request, received by Cassandra, is divided into stages that are part of the SEDA – Staged Event-Driven Architecture, which allows Cassandra being able to handle and serve high number of simultaneous requests [7]. That mechanism provides necessary system variations, meaning that it increases the number of running threads accordingly to the demand, allowing Cassandra to execute higher number of operations. The maximum number of allowed threads is limited by the operating system, based on the hardware limitations. However, as long as there is a possibility, Cassandra will keep showing better performance during execution of requests. As previously described, memory usage does highly affects the performance and allows to considerably reduce execution time. Since volatile memory is much faster, compared to standard disk, Cassandra uses memory mapping [10]. This part of the architecture consists of the two similar mechanisms: Row cache and Key cache (Figure 1). As the name suggests, Key Cache is used to map, in memory, all the stored keys. It is simply responsible for put in RAM primary index that allows fast key-value data retrieval. On the other hand, Row Cache enables memory mapping of the entire records. This behavior is similar to the in-memory databases such as, Redis or Memcached.

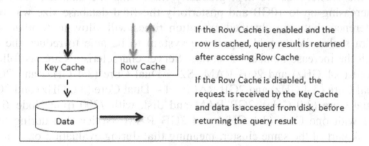

Fig. 1. Cassandra caching

As we have seen, besides memory mapping mechanisms, indexing takes an important role and provides better performance. In Cassandra, primary (key) index is generated automatically and is important, allowing accessing all the stored records. Since all the hashed keys are unique, when new data is added, the index is refreshed and all the records may be retrieved. The data retrieval used by YCSB is performed by generating a random record key and accessing its value that represent the data itself. While working in a cluster environment, when the system receives the request it should create node communication and serve the client. For the data distribution, each record is assigned to a specific server, according to its hashed key. Each node becomes responsible for all the records that it manages and is also aware of the hash ranges stored by the other servers [9]. This approach allows efficient request management. Since any of Cassandra's nodes may receive the request, it redirects the request to the node that stores the requested data.

The next section describes the experimental environment that was used during our evaluation by presenting the benchmark, hardware summary and performed operations.

3   Experimental Environment

As described in section 2, during our evaluation we choose to use the standard Yahoo! Cloud Serving Benchmark [1].This benchmark consists of two components: data generator and a set of tests that perform specific operations over the databases. The generated data is synthetic and is represented by random characters organized into records. Each record has 10 fields plus key, with a total of 1KB per record. Each test scenario is called workload and is defined by a set of parameters such as a percentage of read and write operations, the total number of operations performed, the number of used records, etc. We executed two YCSB workloads: workload C – 100% read and E – 100% scan. We modified the original workload E that it 95% scan and 5%insert up to 100% scan. We choose those workloads because we focused our evaluation on the read performance, even though Cassandra is faster performing writes. We consider that after the bigger part of the data is loaded it is important to study if the database is capable of serving read requests and verify its performance during execution of scans.

For the evaluation of Cassandra's scalability we created different database size and cluster size scenarios. We varied the number of servers from 1 to 3 and up to 6. While varying the cluster size we kept gradually increasing the database size, starting at 1GB, increasing up to 10GB and posteriorly the final database size was of 100GB. The differences between obtained execution times will allow us to understand how Cassandra scales. We would expect the system to be able to reduce the execution time with the increase number of servers. Server characteristics are, as follows: S1 – Dual Core (3.4 GHz) and 2GB RAM; S2 – Dual Core (3.4 GHz) and 2GB RAM; S3 – Dual Core (3.4 GHz) and 2GB RAM; S4 – Dual Core (3.0 GHz) and 2GB RAM; S5 – Dual Core (3.0 GHz), 2GB RAM and disk with 7200 rpm; Node_6 – Virtual Machine with one Core (3.4 GHz) and 2GB RAM. Notice that during tests, same nodes took part of the same cluster, meaning that during evaluation on 3 node cluster were always used S1, S2 and S3 and S1 was assigned for execution of the single server tests. During evaluation of the request management, we varied the number of client's threads, option provided by the YCSB. We started at 1 thread and kept increasing up to maximum of 6000 threads. Those tests showed Cassandra's efficiency and capability of managing high load.

Next section presents our experimental results. First, we evaluated Cassandra's scalability while executing workloads C and E over 1, 3 and 6 servers and database size variation: 1GB, 10GB and 100GB. Secondly, we present performance displayed by Cassandra while serving different numbers of requests (thread variation): 1, 6, 60, 600, 1200, and 6000 threads.

4   Experimental Results

In this section we present obtained results concerning Cassandra's scalability and load management. The execution time, in seconds, is obtained by executing a total of 10.000 operations.

4.1   Scalability

First, we evaluated Cassandra's scalability by executing workloads C (100% read) and E (100% scan) in different cluster environments (1, 3 and 6 servers) and while varying the database size. Figure 2 presents the execution time of workload C.

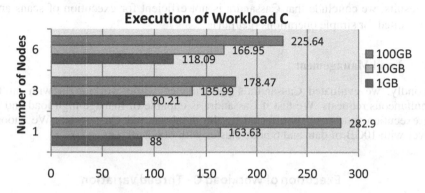

Fig. 2. Workload C (100% read) - Scalability

While running the tests over smaller database size used (1GB of data) we observed constant grown in execution time. However, as the database size kept increasing, 3 node cluster was capable of showing a better performance for both 10GB and 100GB of data, while results in single server environment continued getting worse. These results may be explained by the database sizes that we used, meaning that 3 node cluster started showing much better results with larger database size but 100GB database was not big enough to see high performance of 6 node cluster.

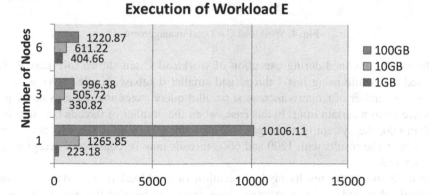

Fig. 3. Workload E (100% scan) - Scalability

Figure 3 presents the results of execution of workload E that executes only scans. We observed some similarities while comparing these results with the results for workload C. Overall, execution time in single server environment showed to be worse than expected. Due to the workload type, 1 server with 100GB of data was not capable of efficiently execute all the requests. System got overloaded and this test had very high execution time. Also, high execution time is explained by the operation type. Previously, during execution of the workload C, we performed a total of 10.000 read operations but scans are slightly different. While 1 read operation corresponds to a simple get operation, 1 scan reads between 1 and 100 records. While observing all the results, we conclude that Cassandra is not efficient for execution of scans and is more suited for simple operations (get/put).

4.2   Load Management

Secondly, we evaluated Cassandra's performance while working in with multiple simultaneous requests. We test if Cassandra is capable of manage high load and how the execution time itself is affected by the database and cluster size. We choose 1 server with 10GB of data and 6 node cluster with 60GB of data.

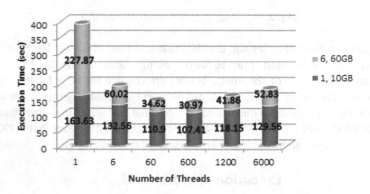

Fig. 4. Workload C – Load management

The results obtained during execution of workload C are shown in Figure 4. We observed that while using just 1 thread and smaller database, single server is faster. But, as the number of requests increases, parallel query execution presents better performance up to a certain limit. In this case, when the number of threads becomes bigger than 600, the system is no longer able to reduce even more the execution time. Even though the results with 1200 and 6000 threads may be considered acceptable, it has worsened.

Figure 5 shows the results of the execution of workload E. The system showed good scalability and load management since execution time of the workload continued decreasing. However, we were not able to execute workload E in single server environment with 60 threads. The system was not capable of processing all the demand. Since the query execution is not parallelized, the response time was that big that system would announce timeout.

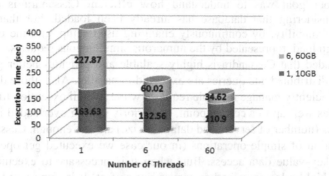

Fig. 5. Workload E - Load management

5   Related Work

A literature study of Cassandra, HBase and Riak was also presented by Konstantinou et al. [4]. Authors presented throughput results and discussed the scalability as well as the use of hardware, shown during execution of workloads. Pirzadeh et al. [5] evaluated performance of several NoSQL databases: Cassandra, HBase and Project Voldemort, using YCSB [1]. These authors focused their analysis mainly on execution of scan operations that allowed understanding scan performance of those NoSQL databases. During our evaluation we executed scan workload but our focus was to understand how it is affected by the scalability. Fukuda et al. [12] propose a method for reducing Cassandra's request execution time of simple operations as well as range queries [12]. Another performance increasing method was evaluated by Feng et al. [13]. For decrease of execution times were used new indexes that provide faster execution of requests. Authors defend that Cassandra should have index optimization and the use of CCIndexes offers high performance increase. Differently from previous authors, we focus our study on Cassandra's standard performance that would be expected without modifications. Our approach shows how Cassandra would behave in real enterprise environment, using standard version of Cassandra. Alternative performance improvement of Cassandra was projected by Elif et al. [14]. In this case the strategy was combining Cassandra with MapReduce engine, originally used by Hadoop framework [15]. Executed tests proved that Cassandra may be combined with other mechanisms in order to increase overall performance.

In our paper we evaluate scalability and how execution time is affected by database and cluster sizes, which is not done in the referenced works.

6   Conclusions and Future Work

In this paper we have evaluated one of most popular NoSQL databases, Cassandra. We studied Cassandra's internal architecture that allowed us to better understand its functioning and used mechanisms. During experimental evaluation we used YCSB and executed simple get operations and, posteriorly, scans, We choose those operations since our goal was to understand how efficient Cassandra is during data querying, considering that database has already been loaded. For that, we tested Cassandra's scalability, by continuously changing used setup, and the capability of managing high load, represented by the numerous simultaneous requests.

We concluded that Cassandra is highly scalable and provides better performance while data is distributed and queries are executed in parallel. Also, those databases are capable of efficiently manage multiple requests while overall execution time becomes lower. It scales well up to a certain point, posteriorly getting overloaded if the cluster characteristics (number of servers and database size) are not enough. Cassandra is fast during execution of simple operations (in our case we executed get operations) and while using key-value data access. But, when it is necessary to execute scans, the performance highly decreases and system is not as fast. It is important to remember that regardless operation type the system must be scaled adequately, according to the database size. Distributing considerably small amounts of data will create unnecessary network overhead and lower the overall performance. As future work we plan to study the scalability of another popular NoSQL database, MongoDB.

Acknowledgment. This work was partially financed by iCIS – Intelligent Computing in the Internet Services (CENTRO-07- ST24 – FEDER – 002003), Portugal.

References

1. Cooper, B.F., Silberstein, A., Tam, E., Ramakrishnan, R., Sears, R.: Benchmarking cloud serving systems with YCSB. In: Proceedings of the 1st ACM Symposium on Cloud Computing (SoCC 2010), pp. 143–154. ACM, New York (2010)
2. A community white paper developed by leading researchers across the United States. Challenges and Opportunities with Big Data
3. Cattell, R.: Scalable SQL and NoSQL data stores. SIGMOD Rec. 39(4), 12–27 (2011)
4. Konstantinou, I., Angelou, E., Boumpouka, C., Tsoumakos, D., Koziris, N.: On the elasticity of nosql databases over cloud management platforms. In: CIKM, pp. 2385–2388 (2011)
5. Pirzadeh, P., Tatemura, J., Hacigumus, H.: Performance evaluation of range queries in key value stores. In: IPDPSW, pp. 1092–1101 (2011)
6. http://cassandra.apache.org/
7. Welsh, M., Culler, D., Brewer, E.: SEDA: an architecture for well-conditioned, scalable internet services. In: Proceedings of the Eighteenth ACM Symposium on Operating Systems Principles (SOSP 2001), pp. 230–243. ACM, New York (2001)
8. Talia, D.: Clouds for Scalable Big Data Analytics. IEEE Computer (COMPUTER) 46(5), 98–101 (2013)
9. Garefalakis, P., Papadopoulos, P., Manousakis, I., Magoutis, K.: Strengthening Consistency in the Cassandra Distributed Key-Value Store. In: Dowling, J., Taïani, F. (eds.) DAIS 2013. LNCS, vol. 7891, pp. 193–198. Springer, Heidelberg (2013)
10. Hewitt, E.: Cassandra - The Definitive Guide: Distributed Data at Web Scale. Springer (2011)

11. Beernaert, L., Gomes, P., Matos, M., Vilaça, R., Oliveira, R.: Evaluating Cassandra as a manager of large file sets. In: Proceedings of the 3rd International Workshop on Cloud Data and Platforms (CloudDP 2013), pp. 25–30. ACM, New York (2013)
12. Fukuda, S., Kawashima, R., Saito, S., Matsuo, H.: Improving Response Time for Cassandra with Query Scheduling. In: Proceedings of the 2013 First International Symposium on Computing and Networking (CANDAR 2013), pp. 128–133. IEEE ComputerSociety, Washington, DC (2013)
13. Feng, C., Zouand, Y., Xu, Z.: CCIndex for Cassandra: A Novel Scheme for Multi-dimensional Range Queriesin Cassandra. In: SKG 2011, pp. 130–136 (2011)
14. Dede, E., Sendir, B., Kuzlu, P., Hartog, J., Govindaraju, M.: An Evaluation of Cassandra for Hadoop. In: IEEE CLOUD 2013, pp. 494–501 (2013), Dean, J., Ghemawat, S.: MapReduce: a flexible data processing tool. Commun. ACM 53(1), 72–77 (2010)
15. Dean, J., Ghemawat, S.: MapReduce: a flexible data processing tool. Commun. ACM 53(1), 72–77 (2010)

An Improved Method for Efficient PageRank Estimation

Yuta Sakakura[1], Yuto Yamaguchi[1], Toshiyuki Amagasa[2], and Hiroyuki Kitagawa[2]

[1] Graduate School of Systems and Information Engineering,
University of Tsukuba, Tsukuba, Japan
[2] Faculty of Engineering, Information and Systems,
University of Tsukuba, Tsukuba, Japan
{yuuta,yuto_ymgc}@kde.cs.tsukuba.ac.jp,
{amagasa,kitagawa}@cs.tsukuba.ac.jp

Abstract. PageRank is a link analysis method to estimate the importance of nodes in a graph, and has been successfully applied in wide range of applications. However, its computational complexity is known to be high. Besides, in many applications, only a small number of nodes are of interest. To address this problem, several methods for estimating PageRank score of a target node without accessing whole graph have been proposed. In particular, Chen et al. proposed an approach where, given a target node, subgraph containing the target is induced to locally compute PageRank score. Nevertheless, its computation is still time consuming due to the fact that a number of iterative processes are required when constructing a subgraph for subsequent PageRank estimation. To make it more efficient, we propose an improved approach in which a subgraph is recursively expanded by solving a linear system without any iterative computation. To assess the efficiency of the proposed scheme, we conduct a set of experimental evaluations. The results reveal that our proposed scheme can estimate PageRank score more efficiently than the existing approach while maintaining the estimation accuracy.

Keywords: PageRank, Local Estimation, Link Structure Analysis.

1 Introduction

PageRank [13] is one of the most popular link analysis method, whereby one can estimate the importance of nodes in a graph, and has widely been applied in a wide spectrum of applications where a graph is used as the primary data model, such as web search [6], social data analysis [17], and bioinformatics [12]. It should be noticed that the size of graphs tend to be large, e.g., large web crawl datasets may contain millions of web pages. Hence, it becomes hard to apply PageRank to such huge data due to its computational complexity. For this reason, so far, many research have been done to accelerate PageRank computation [11,2,10].

An important point that should be noticed is that it is not always possible to access the whole graph, in particular when the data is huge. Taking the example of web crawl data, in many web search engines, the crawled data are maintained in distant data centers. It is thus difficult to make analysis of whole data due to its excessive volume as well as the cost for data transmission. Moreover, in many cases, users are not always

H. Decker et al. (Eds.): DEXA 2014, Part II, LNCS 8645, pp. 208–222, 2014.

interested in knowing the importance of all nodes, i.e., it is sufficient if we can estimate the importance of some selected nodes. Some application scenarios are 1) one wishes to observe the temporal variation of a particular node's PageRank score, or 2) one wishes to identify the most important node within a community. In these cases, the PageRank scores of nodes other than the target(s) are not of interest. It is desirable if we can compute the PageRank scores of some specific nodes very quickly.

To meet these demands, several methods have been proposed to estimate PageRank scores without accessing whole data [8,4,5]. Chen et al. [8] is one of the primary works in this research domain. They assume that the graph data is stored in a database, and want to minimize the cost for accessing the database. More precisely, their idea is to construct a local subgraph containing the target node on which we would like to estimate the PageRank score. To induce the local subgraph, they introduce the concept of *influence* which quantifies the effect of a node to the target nodes' PageRank score. A local subgraph including nodes with high influence on the target node approximates PageRank score more accurately. Having constructed a local subgraph, the approximated PageRank score is calculated by applying power-iteration over the (small) local subgraph, thereby reducing the processing cost. Nevertheless, it can be said that it is still time consuming due to the fact that it requires a lot of iterative processing to estimate the influence of each node during expanding the local subgraph. Consequently, the time for local subgraph construction is significantly affected by the graph size, though the estimation accuracy is directly influenced by the graph size.

To address this problem, in this paper, we propose a novel approach to estimate the PageRank score of a specific node in a more efficient way. Specifically, we are based on the Chen's approach [8]. Some readers may think that it is too old, but, in fact, the subgraph-based method has not been so well-studied as we will review in Section 2. Consequently, [8] can be considered as one of the latest work in this direction. It should be also noticed that the subgraph-based method has a strong advantage that one can use existing software for computing PageRank almost with little modifications. From the users' perspective, this beneficial to lower the cost to introduce the improved method.

In this work, we attempt to improve the performance by making the local subgraph construction phase more efficient. More precisely, we exploit the property that, for a node, its influence to the target being investigated can be recursively determined by solving a linear system describing the influences of its connected neighbors. Owing to this property, we can expand the local subgraph in a stratified manner without conducting costly iterative processing, thereby allowing us to estimate the PageRank score of a specific node more efficiently while maintaining the estimation accuracy.

To evaluate the effectiveness of the proposed scheme, we conduct a set of experimental evaluations using different kinds of datasets, including Twitter social graph and web crawl data. The experimental results show that the proposed scheme can reduce the time for constructing local subgraphs, while maintaining the estimation accuracy in terms of ranking by estimated PageRank scores. Specifically, then the subgraph size is the same, the proposed scheme is about 2 to 3 times faster then the original Chen's method, but the estimation accuracy remains the same.

The rest of this paper is organized as follows. Section 2 briefly overviews the related work. Then, we introduce the local PageRank estimation method by Chen et al. [8] in

Section 3. We proposed an improved method in Section 5. In Section 6, we evaluate the performance of the proposed scheme through a set of experiments. Finally, Section 7 concludes this paper.

2 Related Work

There have been a lot of attempts to approximately compute PageRank scores due to the popularity of PageRank and the emerging large graph data in various application domains.

One of the major ways to approximate PageRank score computation is that, given a *target* node which we want to estimate the importance, we induce local subgraph containing the target, whereby we can estimate PageRank. Chen et al. [8] is one of the primary works. They extend a local subgraph containing the target in a step-by-step manner by estimating each node's *influence* to the target. After constructing a subgraph, they apply power-iteration over it to estimate the target's PageRank. One of the major advantages of this approach is that one can use the software for PageRank computation almost as it is. For this reason, we exploit this method as our basis, and will give a detailed explanation in Section 3.

Bar-Yossef et al. [4] also propose a similar, but different method for local PageRank estimation. In their approach, they do not explicitly construct a local subgraph for subsequent power-iteration process. Instead, they use the property that the PageRank score of the target can directly be computed by the influence of the connected nodes. Specifically, from the target, they recursively follow the incoming edges while estimating the nodes' influence and the target's PageRank score. In this way, unlike Chen's method, they avoid power-iteration for PageRank estimation. However, they do not provide any evaluation in terms of estimation accuracy.

Bressan et al. [5] investigate the problem of ranking k distinct nodes according to the PageRank scores, and devise the accuracy at different costs. Specifically, they employ a modified version of [4] in their experiments. However, they do not compare the accuracy with other related work.

Another direction of PageRank approximation is to estimate the PageRank scores of the nodes in a given subgraph. In Davis et al. [9], for a given subgraph, they extend the subgraph by adding edges linking to it, and apply power-iteration over the subgraph for estimating PageRank scores. In Xue et al. [19], for a target subgraph, they create a new subgraph by using user access patterns, and estimate the PageRank scores using the created subgraph. In Wu et al. [18], they first transform the graph containing the target subgraph in such a way that the nodes outside of the target subgraph is represented as a node for subsequent PageRank estimation. In Vattani et al. [16], for a given set of nodes, they induce a subgraph based on it in such a way that the personalized PageRank score is preserved. These works are different from our proposal in the sense that they deal with a subgraph as the target, whereas our target is to estimate the PageRank of a node in a whole graph.

Broder et al. [7] propose a method for estimating PageRank scores by graph aggregation. Their method constructs a host graph which is a summary of entire graph by aggregating nodes corresponding to web pages to an aggregated node if they are of the same host. Then, they estimate PageRank scores of all nodes using the host graph. Their method is specifically dedicated for web crawl data where each node is assigned with a URL as its identifier, thereby they can sum up those nodes in a same site.

In addition, there is a work [14] that deals with ObjectRank [3], rather than PageRank, by applying local computation.

3 Chen's Method for Local PageRank Estimation

This section overviews the method of local PageRank estimation by Chen et al. [8] on which our proposed method is based. Given a unlabeled-directed graph and a node in the graph, called *target node*, it estimates the PageRank score of the target node only accessing the local subgraph containing the target. One basic assumption is that the graph is stored in a database, and *fetch* operation is used to access the data. The fetch operation takes a node ID as its argument, and returns the IDs of the nodes, each of which is connected to the target with either an incoming or an outgoing edge. Notice that, since the database is assumed to be stored in a secondary storage, the number of disk accesses is one of the major cost factors. Hence, reducing the local subgraph size is important to reduce the time required.

The process is divided into two phases, namely, local subgraph construction, followed by power-iteration. In the following, we elaborate these steps in detail.

3.1 Local Subgraph Construction

Starting from the target node, the algorithm iteratively *expands* the local subgraph by following the incoming edges. Initially, the local subgraph is consisting only of the target node. Then, the algorithm collects the node IDs that are connected to the target by incoming edges using fetch operation. Having expanded the local subgraph, in the next, it checks whether or not each newly added node should further be expanded according to the criteria described below. In this way, the process continues until there is no node to expand.

Figure 1 shows an example of local subgraph construction. In this example, the target node is denoted as node 0. Initially, the local subgraph consists only of node 0 (Figure 1 (1)). First, by fetching and expanding target node 0, the algorithm adds nodes 1, 2, and 3 (Figure 1 (2)). Next, we fetch nodes 1, 2, and 3 and check whether these nodes should further be expanded according to the criteria. Let us assume that node 2 and 3 are selected for further expansion, while node 1 is not (Figure 1 (3)). Then, by following the incoming edges from nodes 2 and 3, nodes 5 and 6 are added (Figure 1 (4)). Likewise, we check nodes 5 and 6 for further expansion and assume that both nodes are not selected. (Figure 1 (5)). We stop expanding the local subgraph because there are no selected nodes. Finally, we get Figure 1 (6) as the induced subgraph of the expanded nodes.

Let us take a closer look at a local subgraph. Regarding the edges, we call such edges that connect two nodes within the local subgraph *internal edges*. In addition, we refer to such edges that link from outside (inside) nodes to inside (outside) nodes of the local subgraph *incoming edges* (*outgoing edges*, respectively). As for the nodes, *boundary*

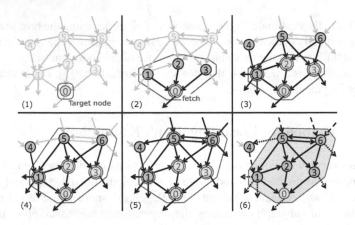

Fig. 1. An example of local subgraph construction

nodes are those nodes that are pointed by the incoming edges. The rest of the nodes, other than the target node, are called *internal nodes*. In Figure 1 (6), there are seven outgoing edges (dotted lines), four incoming edges (dashed lines), and eleven internal edges (solid lines). Besides, there are one target node, two internal nodes, and three boundary nodes.

Influence. To judge whether a node should be expanded or not, the *influence* of the node to the target is evaluated. The basic idea of influence is to evaluate a node in terms of how much of its evaluation score will eventually arrive at the target node, i.e., if a node has higher influence, it will contribute to improve the accuracy of the PageRank estimation. It is also assumed that the nodes with edges directed to the nodes with high influence have large influence, too. Another characteristic that needs to be taken into account is the cost for fetching data from the disk. Specifically, if we fetch a node with high degree, it causes a large number of fetch operations for further expanding the node. Consequently, to judge whether a node is to be further expanded, its influence is divided by its indegree and the result is checked to see whether it is larger than or equal to the user-specified threshold.

However, in fact, computing exact influence requires the computation of exact PageRank scores, which is unacceptable. For this reason, they employ OPIC approach [1] to approximately compute the influence, which is outlined below:

1. Assign score 1.0 to node v which we want to estimate the influence.
2. Transfer v's score to the neighboring nodes by following the outgoing edges. At this step, we just ignore the score that goes out from the local subgraph.
3. Select a node having the largest score except for the target node, and transfer its score to the neighbors by following its outgoing edges.
4. Repeat step (3) until the sum of the scores of internal and boundary nodes becomes less than the user-specified threshold.
5. Output the score of the target node as the estimated influence of node v.

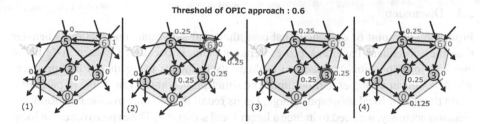

Fig. 2. An example of OPIC approach

Note that the user-defined threshold used in the termination condition affects the estimation accuracy, as well as the execution time; the smaller the threshold is, the better estimation accuracy is achieved, but it requires a longer execution time.

Figure 2 shows an example of OPIC approach where the influence of node 6 is being evaluated. Assume that the threshold for termination condition is 0.6. First, the initial score 1.0 is given to node 6 (Figure 2 (1)), and the score is transferred to its neighbors by following the outgoing edges. There are four outgoing edges, nodes 2, 3, and 5 receive 0.25, but the rest of the score is ignored, because it goes out of the local subgraph (Figure 2 (2)). Then we select a node with the largest score. In this case, nodes 2, 3, and 5 have the same score (0.25). Hence, we randomly choose one of them (node 3) (Figure 2 (3)). In the next, we transfer its score to the neighbors, and node 0 receives $0.125 = 0.25/2$ (Figure 2 (4)). After this step, we check that the sum of the scores of internal and boundary nodes ($0.5 = 0.25 + 0.25$) is less than the threshold (0.6). As a result, we terminate the process, and take the score of target node (0.125) as the estimated influence of node 6.

3.2 PageRank Estimation

Before estimating the target's PageRank score, we need to estimate the scores of the boundary nodes. This is because, the scores flowing into the local subgraph is determined by them. Consequently, it is quite important to estimate the scores of boundary nodes as accurate as possible in order to estimate the target's PageRank score accurately. In Chen's approach, they exploit some heuristics for efficiency; they assume that, for each boundary node, d/E is transferred from the outside by following each incoming edge, where E and d are the total number of edges in the entire graph and the PageRank's dumping factor (0.85), respectively. More than that, by taking into account the score flowing into the boundary node from the connected inside nodes and random jump, we can estimate the PageRank score of the boundary nodes.

Finally, we estimate the PageRank score of the target node by conducting a power-iteration within the local subgraph. The difference with the ordinary PageRank computation is the boundary nodes. For each iteration, as explained above, we assume that d/E flows into the boundary noes by following incoming edges, while those flowing out of the local subgraph are just ignored.

3.3   Discussion

From the viewpoint of computational cost, the local subgraph construction is heavier than the subsequent PageRank power-iteration. This is because, for each node being checked for expansion, we need to evaluate its influence using the OPIC approach, which results in a long execution time. The situation becomes even worse in particular when the size of local subgraph is large. Let us recall that, to improve the PageRank estimation accuracy, we need to induce a larger local subgraph. The expensiveness in local subgraph construction has a dampening effect on the performance of Chen's approach.

Another important observation is that, in OPIC approach, the influence of internal nodes are already estimated in the process of expanding the local subgraph. However, such previously estimated influence values are not used and are just ignored. We think that it is reasonable to reuse those (intermediate) results to speed up the process.

4   An Improved Method for Efficient PageRank Estimation

4.1   Redefinition of Influence

To improve the subgraph construction phase, in this work, we redefine the influence $inf(v, u)$ of node v to node u as follows:

$$
inf(v, u) = \begin{cases} \dfrac{1}{outdeg(v)} \displaystyle\sum_{w \in N_v} inf(w, u) & \text{if there is the path to } u \text{ from } v \\ 0 & \text{if there is not the path to } u \text{ from } v \\ 1 & u = v \end{cases} \tag{1}
$$

where $outdeg(v)$ is the outdegree of node v and N_v is the set of out-neighbor nodes of node v. In other words, the influence of node v to node u is equal to the average of the influence of out-neighbor nodes $w \in N_v$ of node v to node u. However, if there is no path to u from v, the influence of node v to node u is 0, because the score of node v never reaches to node u.

5   The Proposed Method

According to the definition, we estimate the influence of node v recursively by using those of out-neighbor nodes of node v. However, in the process of the local subgraph construction, the current local subgraph does not always include complete set of out-neighbor nodes. For this reason, we estimate the influence of node v only using the currently available out-neighbors. For example, let us consider Figure 1 (5). At this step, the influence of nodes 1, 2, and 3 is already decided. Therefore, it is possible to compute the influence of nodes 5 and 6 using those of 1, 2, and 3, as well as the information how they are connected by edges.

The problem is that, in general, there are two or more nodes being expanded and they may have edges in between. Then, the influence of such nodes are not yet decided. In the above example, to compute the influence of 5 (6), that of 6 (5) is needed. Our idea is to represent the relationship using a linear system where the undecided influence values are represented using variables, and solve it.

5.1 Efficient Influence Estimation

Now, we explain in detail how to estimate the influence of nodes efficiently. Let $G = (V,E)$ and v_t denote a graph and the target node such that $v_t \in V$, respectively. Let V' be the nodes in the current local subgraph, such that $v_t \in V'$. Then, we classify the set of nodes V' into two categories: 1) let $C = \{v_i \mid v_i \in V'$ and v_i is a boundary node being expanded$\}$ be the set of boundary nodes, whose influence is not yet decided, and 2) $W = V' - C$ be the rest of the nodes whose influence is already decided. Then, the influence α_i of node $v_i \in C$ to the target node $v_t \in W$ can be represented as the following equations.

$$\alpha_i = \frac{1}{outdeg(v_i)} \left(\sum_{v_j \in C_i} \alpha_j + \sum_{v_k \in W_i} \beta_k \right) \tag{2}$$

$$\beta_t = 1.0 \tag{3}$$

where $outdeg(v_i)$ is the outdegree of node v_i; C_i is the set of nodes in C that receive edges from v_i; W_i is the set of nodes in W that receive edges from v_i; and α_j (β_k) is the influence of node $v_j \in C$ ($v_k \in W$, resp.) to the target node v_t. Notice that β_k is already determined, while α_j is not. Basically, given the node v_i, Equation (2) represents the influence of v_i to the target v_t by summing up the influence of out-neighbor nodes, and divide it by v_i's outdegree. Therefore, except for ignoring the influence of nodes outside of the current local subgraph, Equation (2) is equivalent to the definition of the influence (Equation (1)).

An important point is that, if v_i has links to some of the boundary nodes (C_i), the equation contains some undecided variables (α_j). Consequently, we need to determine those values in order to determine the influence of c_i (α_i). In fact, if we look at other nodes in the boundary C, we have the same number of equations as the number of undetermined variables, and they are linearly independent due to the fact that they reflect the graph topology. Therefore, by solving the linear system comprised of the equations corresponding to the boundary nodes (C), we can compute α_i ($i = 1, 2, \ldots, |C|$) as the answer.

Those having discussed, we can construct a local subgraph in the following way:

1. Let target v_t's score be 1.0, and identify the boundary nodes C.
2. Solve the linear system consisting of the equations corresponding to each boundary node in C, and get the influence of each boundary node.
3. For each boundary node, check if the influence is larger than the predefined threshold. If so, we expand it in the next step; otherwise, we do not.
4. Repeat 2 to 3 until there is no node to expand.

Algorithm 1 shows the detailed algorithm. Note that we modify the condition to expand a node from the original Chen's method (Section 3). The reason is that, in the experiments, it turned out that the performance was not good when we employ Chen's original condition. Note also that, even with the same target node, the local subgraph induced by the proposed method and those by other methods are not the same due to the difference in the algorithm and in the above condition, as well. We observe that the estimation accuracy is still maintained even though the subgraphs are not the same.

Algorithm 1. subgraphConst(G, v_t, γ)

Input: Graph $G = (V, E)$, target node v_t, threshold γ
Output: Local subgraph H
1: $H \leftarrow \{v_t\}; C \leftarrow \emptyset; W \leftarrow \{v_t\};$
2: $\beta_t \leftarrow 1.0;$
3: $V_{in} \leftarrow$ get in-neighbors of target node v_t;
4: $C \leftarrow V_{in};$
5: $H \leftarrow H \cup V_{in};$
6: **while** $C \neq \emptyset$ **do**
7:     $C_{next} \leftarrow \emptyset;$
8:     **for** v_i in C **do**
9:         $\alpha_i \leftarrow$ estimate the influence of node v_i;
10:        **if** $\alpha_i \geq \gamma$ **then**
11:            $V_{in} \leftarrow$ get in-neighbors of node v_i;
12:            $C_{next} \leftarrow C_{next} \cup Vin$;
13:            $H \leftarrow H \cup V_{in};$
14:        **end if**
15:        $\beta_i \leftarrow \alpha_i;$
16:     **end for**
17:     $W \leftarrow W \cup C;$
18:     $C \leftarrow C_{next};$
19: **end while**
20: Return H;

5.2 A Running Example

Figure 3 shows a running example of local subgraph construction by the proposed scheme. In this example, again, node 0 is the target node, and we assume that the threshold for expansion is 0.5 (Figure 3 (1)).

First, we fetch the target node 0, and add nodes with edges directed to the target node (i.e., 1, 2, and 3) to the local graph (Figure 3 (2)). Then we fetch nodes 1, 2, and 3, and calculate the influence of these nodes by solving a linear system. From $C = \{1, 2, 3\}$ and $W = \{0\}$, we get the following information:

Node	$outdeg(i)$	C_i	W_i
1	5	$\{2\}$	$\{0\}$
2	1	$\emptyset$	$\{0\}$
3	2	$\emptyset$	$\{0\}$

According to this, we can generate the following linear system:

$$\alpha_1 = \frac{1.0}{5}(\alpha_2 + 1.0) \tag{4}$$

$$\alpha_2 = 1.0 \cdot 1.0 \tag{5}$$

$$\alpha_3 = \frac{1.0}{2} \cdot 1.0 \tag{6}$$

By solving this, we get $\alpha_1 = 0.4(= \beta_1)$, $\alpha_2 = 1.0(= \beta_2)$, and $\alpha_3 = 0.5(= \beta_3)$. Then, we compare them with the threshold, and it turns out that nodes 2 and 3 are expanded in

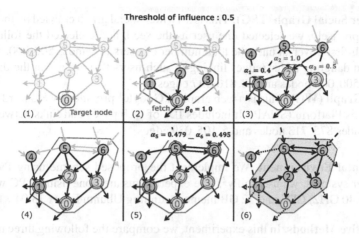

Fig. 3. A running example of local subgraph construction

the next step, whereas node 1 is not, because the influence is smaller than the threshold (0.5) (Figure 3 (3)).

Next, by expanding nodes 2 and 3, we add nodes 5 and 6 to the local graph (Figure 3 (4)). Then, we fetch nodes 5 and 6, and compute the influence of these nodes. Now, $C = \{5, 6\}$ and $W = \{0, 1, 2, 3\}$, and we get the following information:

Node	$outdeg(i)$	C_i	W_i
5	5	$\{6\}$	$\{1, 2, 3\}$
6	4	$\{5\}$	$\{2, 3\}$

According to this, we can generate the following linear system:

$$\alpha_5 = \frac{1.0}{5}(\alpha_6 + 0.4 + 1.0 + 0.5) \tag{7}$$

$$\alpha_6 = \frac{1.0}{4}(\alpha_5 + 1.0 + 0.5) \tag{8}$$

By solving this, we get $\alpha_5 = 0.479(= \beta_5)$ and $\alpha_6 = 0.495(= \beta_6)$ (Figure 3 (5)). We stop the local subgraph expansion, because the influence is all less than the threshold. Finally, we get the local subgraph as shown in Figure 3 (6).

6 Experimental Evaluation

In this section, we evaluate the proposed scheme through some experiments, and show its efficiency.

6.1 Experimental Setting

Dataset. In this work, we use a couple of real datasets:

- **Twitter Social Graph (TSG):** This is a Twitter social graph crawled by the authors. More precisely, we selected one user as the seed, and collected the followees by breadth-first traversal until the total number of users became $2,500,000$. Then, we deleted dangling edges (edges linking to such users that are not in the dataset). It has $2,500,000$ nodes and $28,810,947$ edges.
- **Web Graph (WG):** This is a web graph of Google[1] provided by Stanford Network Analysis Platform (SNAP). It includes IDs of web pages and links between them. It includes $875,713$ nodes and $5,105,039$ edges.

Experimental Environment. We implement the proposed scheme using Python. To solve linear systems, we use SciPy[2]. The experiments are done using a PC with Core i7-3770 (3.40 GHz) CPU and 32 GB memory running Ubuntu Linux 13.04.

Comparative Methods. In this experiment, we compare the following three methods:

- *Proposed:* the proposed method described in Section 5.
- *Chen*: Chen's method described in Section 3.
- *Naive*: It computes the local subgraph by taking the nodes within the distance s from the target. It is regarded as the method where all boundary nodes are always expanded.

To make a fair comparison, we apply parameter s (maximum distance to the target) not only to *Naive*, but also to *Proposed* and *Chen*, i.e., the local subgraph expansion continues if 1) the condition regarding influence is satisfied and 2) the distance to the target is less than s.

Experimental Methodology. We compare the methods in both accuracy and efficiency. As for the accuracy, we focus on the ranking accuracy according to the estimated PageRank scores, rather than directly comparing (estimated) PageRank scores. This is because, in real applications, it is more important to preserve relative order among the nodes according to the scores.

To compare the accuracy from that perspective, we first apply PageRank to the entire graph to get the exact PageRank scores. Next, we sample the target nodes by reference to Chen et al.'s study. Chen et al. randomly take 100 nodes out of all the nodes as a sample. However, we focus on the upper rank of the ranking, because many people pay more attention to the upper rank of the ranking than the low rank of the ranking. Specifically, we take top 1,000 nodes according to the true PageRank scores, and randomly take 100 nodes out of the 1,000 nodes as a sample. We then estimate PageRank scores of the 100 nodes using the comparative methods (*Proposed*, *Chen*, and *Naive*), and sort the nodes according to the estimated scores in descending order. Finally, the accuracy is evaluated as Spearman's rank correlation coefficient [15] between the ranking by the true PageRank scores and that by one of the methods being compared. Spearman's rank correlation coefficient is defined as follows:

$$\rho = 1 - \frac{6 \Sigma D^2}{N^3 - N} \tag{9}$$

---

[1] https://snap.stanford.edu/data/web-Google.html
[2] http://www.scipy.org/

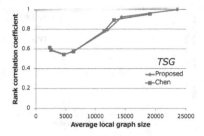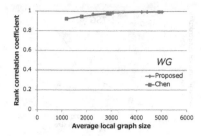

Fig. 4. Comparison of the rank correlation coefficient and the average local graph size by *Proposed* and *Chen* (Left: *TSG*, Right *WG*)

where N is the number of nodes (100 in the experiments) and D is the difference of the ranks of the same node in different orderings. If the rank correlation coefficient ρ is close to 1, the ranking based on the estimated scores is more accurate.

To evaluate the efficiency, we exploit the local subgraph size as well as the processing time. As mentioned earlier, the major cost factor is the number of disk accesses, and that can be approximated by the size of subgraphs. For this reason, we compare the execution time when the subgraph size is the same.

6.2 Performance Comparison with *Chen*

We conducted the preliminary experiments, and set s (maximum distance to the target) to 4 and 3 for *TSG* and *WG*, respectively. Similarly, we set the thresholds for OPIC to 0.5 and 0.9 for *TSG* and *WG*, respectively, so that *Chen* performs the best on each dataset. Furthermore, we set the following threshold of the influence in each method and conduct the experiments.

– 0.1, 0.05, 0.01, 0.005, 0.001, 0.0005, 0.0001

In the following graphs, each point corresponds the result of each threshold of the influence.

Accuracy. Figure 4 shows the ranking accuracy in terms of rank correlation coefficient versus the average local graph size by *Proposed* and *Chen* on *TSG* (left) and *WG* (right). The vertical axis and horizontal axis show the rank correlation coefficient and the average local graph size, respectively.

According to Figure 4, when the average local graph size is almost the same, *Proposed* shows almost the same rank correlation coefficient as *Chen*. This means that, even though we are based on a different schemes of generating local subgraphs, the generated local subgraphs show similar characteristics for estimating PageRank scores. Moreover, in the case of *TSG* (left), the graph size contributes to improve the accuracy, while it does not in the case of *WG* (right). This is due to the nature of graph being analyzed.

Efficiency. Figures 5–7 show the comparison of the execution time versus the average local graph size by *Proposed* and *Chen* on *TSG* (left) and *WG* (right). The vertical axis

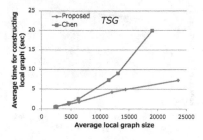 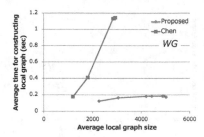

Fig. 5. Comparison of the average time for local graph construction and the average local graph size by *Proposed* and *Chen* (Left: *TSG*, Right: *WG*)

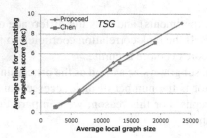 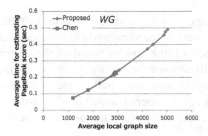

Fig. 6. Comparison of the average time for PageRank estimation and the average local graph size by *Proposed* and *Chen* (Left: *TSG*, Right: *WG*)

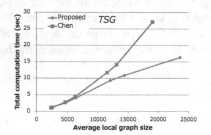 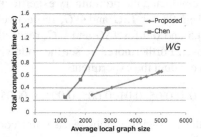

Fig. 7. Comparison of the total computation time and the average local graph size by *Proposed* and *Chen* (Left: *TSG*, Right: *WG*)

shows the average time for local subgraph construction, PageRank estimation, and total computation, respectively. The horizontal axis shows the average local subgraph size.

According to Figure 5, *Proposed* requires significantly smaller amount of time to construct local subgraphs if the subgraph size is the same. If we look at Figure 7, the time for PageRank estimation is almost the same when the graph size is similar. In total, our scheme is successful in estimating PageRank scores more quickly (Figure 6). Recall that, if the subgraph size is the same, we can maintain the same level of estimation accuracy.

6.3   Performance Comparison with *Naive*

Finally, to show the effectiveness of the proposed scheme, we compare the performance of the proposed scheme with a naive method (*Naive*). To do this, for *Proposed*, we tried several combinations of parameters to find the best parameters. Table 1 shows the result. As can be seen, *Proposed* achieves almost the same accuracy with smaller local subgraph sizes. The reason of this result is that *Proposed* successfully prunes the insignificant nodes for PageRank estimation.

Table 1. Comparison of the rank correlation coefficient and the average local graph size by *Proposed* and *Naive*

	TSG		WG	
	Naive	Proposed	Naive	Proposed
Rank correlation coefficient	1.000	0.998	0.993	0.983
Average local graph size	57247	23578	5029	3055

7   Conclusion

In this paper, we proposed a novel method to estimate the PageRank scores of a given specific node efficiently without accessing the whole graph. Our method computes the influence of a node in a stratified manner by solving linear system generated according to the boundary nodes and their neighbors. As a result, we can avoid iterative process to estimate the influence of a node, which is quite time consuming. The experimental results suggest that the proposed scheme is more efficient than the original Chen's method, while maintaining the estimation accuracy. Our future work includes applying the approach to other ranking methods, such as ObjectRank [3]. Another future work is to decide the threshold of the influence automatically because the threshold is an important factor for the estimation accuracy. The appropriate threshold differs according to the property of dataset. An automatic way to decide the appropriate threshold improves the estimation accuracy. Furthermore, we plan to make theoretical analysis over the proposed scheme.

Acknowledgments.   This work was partly supported by JSPS KAKENHI Grant Number 25330124.

References

1. Abiteboul, S., Preda, M., Cobena, G.: Adaptive on-line page importance computation. In: World Wide Web Conference Series, pp. 280–290 (2003)
2. Arasu, A., Novak, J., Tomkins, A., Tomlin, J.: PageRank Computation and the Structure of the Web: Experiments and Algorithms. In: World Wide Web Conference Series (2002)
3. Balmin, A., Hristidis, V., Papakonstantinou, Y.: ObjectRank: Authority-Based Keyword Search in Databases. In: Very Large Data Bases, pp. 564–575 (2004)

4. Bar-yossef, Z., Mashiach, L.T.: Local approximation of pagerank and reverse pagerank. In: International Conference on Information and Knowledge Management, pp. 279–288 (2008)
5. Bressan, M., Pretto, L.: Local computation of PageRank: the ranking side. In: International Conference on Information and Knowledge Management, pp. 631–640 (2011)
6. Brin, S., Page, L.: The anatomy of a large-scale hypertextual Web search engine. Computer Networks and ISDN Systems 30, 107–117 (1998)
7. Broder, A.Z., Lempel, R., Maghoul, F., Pedersen, J.O.: Efficient PageRank approximation via graph aggregation. Information Retrieval 9, 123–138 (2006)
8. Chen, Y.-Y., Gan, Q., Suel, T.: Local methods for estimating pagerank values. In: International Conference on Information and Knowledge Management, pp. 381–389 (2004)
9. Davis, J.V., Dhillon, I.S.: Estimating the global pagerank of web communities. In: Knowledge Discovery and Data Mining, pp. 116–125 (2006)
10. Haveliwala, T.H.: Efficient computation of pagerank. Technical report, Stanford InfoLab (October 1999)
11. Kamvar, S.D., Haveliwala, T.H., Manning, C.D., Golub, G.H.: Extrapolation methods for accelerating PageRank computations. In: World Wide Web Conference Series, pp. 261–270 (2003)
12. Morrison, J.L., Breitling, R., Higham, D.J., Gilbert, D.R.: GeneRank: Using search engine technology for the analysis of microarray experiments. BMC Bioinformatics 6 (2005)
13. Page, L., Brin, S., Motwani, R., Winograd, T.: The pagerank citation ranking: Bringing order to the web. Technical Report 1999-66, Stanford InfoLab (November 1999)
14. Sakakura, Y., Yamaguchi, Y., Amagasa, T., Kitagawa, H.: A Local Method for ObjectRank Estimation. In: International Conference on Information Integration and Web-Based Applications & Services, pp. 92–101 (2013)
15. Spearman, C.: Footrule for Measuring Correlation. British Journal of Psychology 2, 89–108 (1906)
16. Vattani, A., Chakrabarti, D., Gurevich, M.: Preserving personalized pagerank in subgraphs. In: International Conference on Machine Learning, pp. 793–800 (2011)
17. Weng, J., Lim, E.-P., Jiang, J., He, Q.: TwitterRank: finding topic-sensitive influential twitterers. In: Web Search and Data Mining, pp. 261–270 (2010)
18. Wu, Y., Raschid, L.: ApproxRank: Estimating Rank for a Subgraph. In: International Conference on Data Engineering, pp. 54–65 (2009)
19. Xue, G.-R., Zeng, H.-J., Chen, Z., Ma, W.-Y., Zhang, H.-J., Lu, C.-J.: Implicit link analysis for small web search. In: Research and Development in Information Retrieval, pp. 56–63 (2003)

An Enhanced Genetic Algorithm for Web Service Location-Allocation

Hai Huang, Hui Ma, and Mengjie Zhang

Victoria University of Wellington, New Zealand
simonhuang211@gmail.com, {hui.ma,mengjie.zhang}@ecs.vuw.ac.nz

Abstract. Network latency has a significant impact on the response time of web services. Thus, the proper choice of network locations for the deployment of web services is of major importance for the performance of web services. In this paper, we present an enhanced genetic algorithm with self-adaptive feature and memory filter to solve the location-allocation problem for web services. A simulated experiment is conducted using the WS-DREAM dataset with 8 different complexities. The results show that our approach is able to efficiently compute good solutions for the location-allocation problem.

1 Introduction

A web service is a software that can be discovered via the Internet-based protocols [19]. Their convenient service registration, search and discovery processes enables Web service receivers (WSRs) to rapidly find web service providers (WSPs) and their services. In most situations, functions of existing web services cannot fulfill the customer requirements. Web service intermediaries (WSIs) integrate the different services from WSPs and deliver a bundle of services to the WSRs. WSIs choose the most suitable service composition that meets requirements of the WSRs. However, the evaluation of all possible service compositions can lead to a combinatorial explosion of the search space [15].

Web service composition aims to meet functional requirements and non-functional requirements. Non-functional requirements, known as Quality of Service (QoS), is the degree to which a service meets specified requirements or user needs [19], such as response time, security, or latency. The factors that can affect QoS are various, such as composition processes, selection processes, or physical locations of users and services. However, some researches on service composition [4,5,9,11,16,17] conducted in recent years excessively concentrate on composition and selection, without considering the influence of service locations. Some others [1,7,8,14] take account of location factor of web service, assuming that the locations of web services have already been decided. Studies [7,8] investigate the influence of component service locations on the QoS of composite services and show that the locations of the component services have big impact on the performance of composite services. Very few researches [1,14] study the service location-allocation problem. Both studies try to solve the the problem by integer

H. Decker et al. (Eds.): DEXA 2014, Part II, LNCS 8645, pp. 223–230, 2014.

programming techniques. However, integer programming techniques do not scale well, so that no satisfactory results can be obtained for large-scale datasets.

Response time is regarded as paramount concern in time-sensitive applications, like those used in financial markets [1]. Response time consists of local processing time and network response time that consists of three components including transmit time, queuing delays, and latency. [1] states network latency can only be improved by moving servers nearer to their clients. That is to say, the location of web service and service requester can impact the service response time enormously. Although some studies [10,15] study the location-allocation problem using a classical genetic algorithm (GA). Its shortcomings, i.e. poor convergence rate and premature peculiarity, however, have imposed a bottleneck on its efficiency. In this paper, we propose an enhanced GA-based approach to efficiently obtain nearly-optimal web service location-allocation. The main objectives are:

- To model the web service location-allocation problem so that it can be tackled with genetic algorithms.
- To develop an enhanced GA-based approach for the web service location-allocation problem.
- To evaluate our approach by comparing it to a standard GA-based approach using public test datasets.

2   Problem Description

The aim of service location-allocation is to find optimal locations for deploying web service instances such that the overall response time becomes minimal for a target user group. This problem needs to be tackled with care. To fulfil response time requirements of users, it is tempting to use replication, i.e., to deploy multiple instances of atomic web services to different network locations. This idea, however, is likely to boost the operational costs for the service providers. Therefore, service location-allocation typically asks to the satisfy given response time requirements while at the same time restricting the service costs.

Objective. The objective of service location-allocation is to minimize the aggregated response time when a group of users invokes the services that are deployed across the network and, simultaneously, restrict the overall service costs.

Response Time Calculation. To model the service location-allocation problem we use three matrices: the *server response time matrix* $T = [t_{ij}]$, the *service location matrix* $A = [y_{sj}]$ and the *user response time matrix* $R = [r_{is}]$. The server response time matrix T is used to record estimated response times from users to servers with different locations, where t_{ij} is the estimated response time from user i to server j. The service allocation matrix $A = [y_{sj}]$ represents the actual service location-allocation, where y_{sj} is a binary value (i.e., 1 or 0) that shows whether an atomic service s is deployed in server j or not. The user response time matrix $R = [r_{is}]$ is calculated using the matrices T and A as follows (with n being the number of locations):

$$r_{is} = MIN\{t_{ij} \mid j \in \{1, \ldots, n\} \text{ and } y_{sj} = 1\} \tag{1}$$

In Fig. 1, for example, the server response time matrix T_1 records the estimated response times from 4 user locations to 4 server locations. A_1 is an example of a service allocation matrix that records allocation information of two atomic services to server locations. Note that for service s_1 two instances, $s_{1,1}$ and $s_{1,3}$, are deployed to two servers, 1 and 3, respectively. When there are multiple instances of the same service available for a user i, an instance with the shortest response time is chosen. For example, user i_1 chooses service $s_{1,1}$.

$$
T_1 = \begin{array}{c} \\ i_1 \\ i_2 \\ i_3 \\ i_4 \end{array}
\begin{array}{cccc} j_1 & j_2 & j_3 & j_4 \end{array}
\begin{bmatrix} 0.2 & 0.3 & 0.5 & 0.7 \\ 0.4 & 1.2 & 0.9 & 0.6 \\ 0.7 & 0.6 & 1 & 0.8 \\ 1.4 & 0.9 & 0.6 & 1.1 \end{bmatrix}
\quad
A_1 = \begin{array}{c} \\ s_1 \\ s_2 \end{array}
\begin{array}{cccc} j_1 & j_2 & j_3 & j_4 \end{array}
\begin{bmatrix} 1 & 0 & 1 & 0 \\ 0 & 1 & 0 & 0 \end{bmatrix}
\quad
R_1 = \begin{array}{c} \\ i_1 \\ i_2 \\ i_3 \\ i_4 \end{array}
\begin{array}{cc} s_1 & s_2 \end{array}
\begin{bmatrix} 0.2 & 0.3 \\ 0.4 & 1.2 \\ 0.7 & 0.6 \\ 0.6 & 0.9 \end{bmatrix}
$$

Fig. 1. Response Time Calculation Example

Service Demand. Atomic services are composed fulfil the functional requirements of users. Finding a best service composition, however, is known to be NP-hard thus leading to an exponential increase of the search space. To improve the tractability the problem, we assume that service requests are extracted from historical statistical data. For example, consider two composite service A and B that are composed of atomic services from a set $[a, b, c, d]$ with different workflow patterns. Say, service A is a sequential pattern with services $[a, b, c]$, while service B is a choice pattern with atomic service b or services $[c, d]$.

The demand of an atomic service s depends on the frequency of composite services c having s as a component and on its position in the composite service c. Taking the popular 20/80 rule that assumes 20% of service requests count for 80% of overall service requests [6], we focus on the most frequent 20% service requests when deciding service location-allocation. To measure the importance of an atomic service we use *atomic service demand* $\alpha_{sc} = \omega_{sc} \times \beta_c$, where β_c is the *composite service demand*, and ω_{sc} is the *service position weight* of a component service s in the workflow structure of composite service c.

Composite service demand β_c of a composite service c is the ratio of frequency $freq_c$ of service c over total number of service demands $freq_{all}$ within a period of time, i.e., $\beta_c = freq_c / freq_{all}$. Service position weight ω_{sc} is a proportional variable which relates to the position of component service s in the workflow pattern of a composite service c.

For example, assuming we collect 100 records of service requests, where composite service A and service B were invoked 46 and 54 times, respectively. Composite service demand of service A and B are $\beta_A = 0.46$ and $\beta_B = 0.54$. As a valid invocation of composite service B should contain a service b and one of optional service c or d. Service b take a more important position in this workflow pattern than c and d. Service position weight $\omega_{bB} = 0.5$ and for services $\omega_{cB} = \omega_{dB} = 0.25$ in composite service B. Atomic service b is a component service of both composite service A and B. The atomic service demand of b is $(0.46 \times 0.33) + (0.54 \times 0.5) = 0.4218$. Service a is not involved in composite service B, the atomic service demand of a should be $(0.46 \times 0.33) + (0.54 \times 0) = 0.1518$.

3   Enhanced Genetic Algorithm

GA is a powerful tool to solve combinatorial optimizing problems based on Darwinian natural selection theory [13]. In general, GA contains a fixed-size population of potential solutions over the search space. Potential solution is also called *chromosome* representing a string of genes. Gene is considered to be part of the potential solution. Current population generates a new population based on genetic operations, *crossover*, *mutation* and *selection*. The process will stop when termination condition is satisfied.

Standard GA is known to suffer from two weaknesses, poor convergence and premature phenomenon. A common countermeasure that we will apply, too, is to use adaptive parameters to dynamically adjust the genetic operation possibility in order to avoid being trapped in some local optimum during evolution [2,12]. In addition, we will employ a memory scheme to remove duplicates from the population that are likely to occur during evolution when the average fitness deacrases. To continue with, we will define the variables used in our approach, and then present our Memory Filter Genetic Algorithm (MFGA) for service location-allocation.

Chromosome. In our approach, the chromosome models a service location-allocation matrix $A = [y_{sj}]$, see Section 2. Since service providers expect the total number of service instances not exceed a particular amount, an integer number L is applied to satisfy the constraint in Eq. 2. Furthermore, an atomic service has at least one instance in a potential solution. This yields the constraint in Eq. 3. To satisfy the properties of feasible solutions, the initial chromosome is created under these two constraints:

$$\sum_{s=1}^{l}\sum_{j=1}^{n} y_{sj} \leq L \tag{2}$$

$$\sum_{j=1}^{n} y_{sj} \geq 1 \qquad s \in \{1 \ldots l\} \tag{3}$$

Crossover and Mutation. A *crossover operation* mates pairs of chromosomes to create offspring, which is done by a random selection of a pair of chromosomes from the generation probability P_c [10]. The chromosomes are selected on the basis of fitness values. Due to the matrix structure of the chromosome, the crossover operation is a multi-point crossover. As shown in Fig. 2, p_1 and p_2 are two selected chromosomes to apply the crossover operation. A random position *randomPos* is selected from 1 to the maximum row number of two parent matrix. Supposed *ramdomPos* = 1, the entire first row of p_1 and p_2 is exchanged to reproduce children c_1 and c_2.

Mutation is another vital operation in the GA algorithm for exploring new chromosomes. A mutation operation replace a gene with a random selected number within the boundaries of the parameter [10]. In this study, a chromosome in a population will be selected based on mutation possibility P_m. As shown in Fig. 2, p_3 is a chromosome selected to be mutated. A random position will be selected to replace the gene according to original value. Supposed that the random number is 5, namely position (s_2, j_2). The original value of (s_2, j_2) is 0 which replaced by 1 in c_3.

$$
p_1 = \begin{array}{c} \\ s_1 \\ s_2 \\ s_3 \end{array} \begin{array}{c} j_1\ j_2\ j_3 \\ \left[\begin{array}{ccc} 0 & 0 & 1 \\ 0 & 1 & 0 \\ 1 & 0 & 0 \end{array}\right] \end{array}
\quad
p_2 = \begin{array}{c} \\ s_1 \\ s_2 \\ s_3 \end{array} \begin{array}{c} j_1\ j_2\ j_3 \\ \left[\begin{array}{ccc} 1 & 0 & 0 \\ 0 & 1 & 0 \\ 0 & 0 & 1 \end{array}\right] \end{array}
\quad
p_3 = \begin{array}{c} \\ s_1 \\ s_2 \\ s_3 \end{array} \begin{array}{c} j_1\ j_2\ j_3 \\ \left[\begin{array}{ccc} 1 & 0 & 0 \\ 0 & 0 & 1 \\ 1 & 0 & 0 \end{array}\right] \end{array}
$$

$$
c_1 = \begin{array}{c} \\ s_1 \\ s_2 \\ s_3 \end{array} \begin{array}{c} j_1\ j_2\ j_3 \\ \left[\begin{array}{ccc} 1 & 0 & 0 \\ 0 & 1 & 0 \\ 1 & 0 & 0 \end{array}\right] \end{array}
\quad
c_2 = \begin{array}{c} \\ s_1 \\ s_2 \\ s_3 \end{array} \begin{array}{c} j_1\ j_2\ j_3 \\ \left[\begin{array}{ccc} 0 & 0 & 1 \\ 0 & 1 & 0 \\ 0 & 0 & 1 \end{array}\right] \end{array}
\quad
c_3 = \begin{array}{c} \\ s_1 \\ s_2 \\ s_3 \end{array} \begin{array}{c} j_1\ j_2\ j_3 \\ \left[\begin{array}{ccc} 1 & 0 & 0 \\ 0 & 1 & 1 \\ 1 & 0 & 0 \end{array}\right] \end{array}
$$

Fig. 2. Crossover and Mutation Example

Evaluation. A fitness function is required to evaluate the quality of a solution. Two constraints (Eq. 2 and Eq. 3) are defined to the feasibility of solutions. In this study, an appropriate penalty function [3] is employed to reduce the fitness value of infeasible solutions proportionally. The penalty function $P(x)$ and fitness function for the individual x is shown in below:

$$P(x) = p_1 * p_2 \tag{4}$$

$$p_1 = \frac{1}{L} \sum_{s=1}^{l} \sum_{j=1}^{n} y_{sj} \qquad p_2 = \begin{cases} 1 & \text{if } \sum_{j=1}^{n} y_{sj} \geq 1, \forall s \in \{1 \ldots l\} \\ 0 & \text{otherwise.} \end{cases} \tag{5}$$

$$fitness(x) = \frac{1}{m * l} \sum_{i=1}^{m} \sum_{s=1}^{l} \alpha_s r_{is} * P(x) \tag{6}$$

where p_1 and p_2 are two decimal number stand for the satisfaction degree of the constraints, with p_1 in interval $(0, 2)$ representing the ratio of actual concrete service number over required concrete service number and p_2 representing whether all atomic services have at least one concrete instance. Variables α_s and r_{is} are the user response time and service demand (see Section 2), while m and l are the total number of users and atomic services, respectively.

Algorithm. In Algorithm 1, let P be a population with a constant size. M is a memory chunk that stores P_m, P_c and m, where P_m and P_c are possibility values for mutation and crossover, respectively, and m is a list of individuals that is sorted by the fitness values. The size of m should be much smaller than population size since maintaining a sorted list will increase the overhead of the algorithm. 10 to 20 percent of the population size are normal. The algorithm terminates when the maximum generation number is reached or a satisfactory result is found.

Algorithm 1. MFGA for web service location-allocation

Initialize random population P
Evaluate each individual i of population P
Initialize memory filter $M \leftarrow (P_m, P_c, m)$
while Termination condition is not met **do**
   Perform crossover operation with P_c
   Perform mutation operation with P_m
   Generate new population P'
   if Memory filter M contains individual i **then**
     $i.fitness \leftarrow M.getVal(i)$
   else
     $i.fitness \leftarrow FitnessFuction(i)$
     $M.update(m)$
   end if
   $M.update(P_m)$
   $M.update(P_c)$
end while

Table 1. Hypothetical Web Service Location-allocation Problems

Problem ID	Users	Potential Server Location	Composite Services	Atomic Services	Constrains
1	20	100	1	5	None
2	40	250	10	30	None
3	80	500	20	100	None
4	160	1000	40	200	None
5	200	2000	80	400	None
6	80	500	20	100	150
7	160	1000	40	200	300
8	200	2000	80	400	800

4   Experimental Evaluation

To investigate the efficiency and effectiveness of our approach, we have conducted a full experimental evaluation to compare our proposed MFGA algorithm with a traditional GA-based approach, using an existing test dataset, WS-DREAM dataset [18]. The WS-DREAM dataset is a historical dataset on user-dependent QoS of Web services from different locations. It contains the data of response time and throughput from 339 different users invoked 5824 web services scattered all over the Internet. The response time value is normalized into interval [0,1], the higher value means the faster response time. The location information is realistic data. We randomly generate 80 different composite web services with different workflows and calculate the service demands for the atomic services. Table 1 outlines the hypothetical web service location-allocation problems used to evaluate our proposed approach. Each experiment has different users number, potential server number, composite services, atomic services and constraints. The complexity of problems rises as the number of the problem ID increase.

The population size of GA increases with the complexity of the problem. For problems 1 to 5, the population size is 30, 50, 100, 200, 300, respectively. The maximum number of generations is 50, 100, 500, 1000, 1500 for problems 1 to

5, respectively. For the standard GA, the predefined crossover and mutation possibility are set to 0.6 and 0.2, respectively, as this parameter combination achieves relatively good and stable results. The memory size m in MFGA is set to 20 percent of the population.

The experiments are conducted on a laptop with 2.3GHz CPU and 4.0 GB RAM. As GA is non-deterministic, 30 independent runs are performed for each experiment. Table 2 displays the comparison of standard GA and MFGA to solving the hypothetical problems in Table 1. For fitness values, searching time and evaluation time, a statistical significance test has done to examine whether the comparison result is significant different.

Table 2. Experimental Results for the MFGA and GA

ID	Fitness Value		Search Time(s)		Evaluation Time(ms)		Avg. No. Generation	
	GA	MFGA	GA	MFGA	GA	MFGA	GA	MFGA
1	0.809 ± 0.03	0.803 ± 0.03	4.3 ± 0.3	3.4 ± 0.3	≤ 10	≤ 10	46	39
2	0.794 ± 0.09	0.802 ± 0.08	5.6 ± 0.8	5.0 ± 0.7	≤ 10	≤ 10	89	72
3	0.722 ± 0.10	0.733 ± 0.10	10.1 ± 1.1	7.9 ± 1.2	15	≤ 10 *	438	389
4	0.703 ± 0.13	0.719 ± 0.09	22 ± 1.5	17.2 ± 1.5*	23	15 ± 8 *	866	632
5	0.694 ± 0.18	0.702 ± 0.11	48.2 ± 1.8	38.6 ± 1.6*	37	25 ± 8 *	1212	893
6	0.679 ± 0.14	0.708 ± 0.11	12.8 ± 1.8	9.6 ± 1.3 *	15	≤ 10 *	467	421
7	0.656 ± 0.12	0.687 ± 0.11 *	23.9 ± 2.4	20.5 ± 1.6	24	17 ± 7 *	902	777
8	0.638 ± 0.18	0.672 ± 0.16 *	50.9 ± 2.7	39.6 ± 1.8 *	40	28 ± 10 *	1291	927

[a] * sign means significant better in comparison   [b] GA: standard genetic algorithm
[c] MFGA: memory filter genetic algorithm

Table 2 shows that for small and medium-scale of problems, except for problem 1, the fitness value of MFGA is slightly better than standard GA for problems 2 to 6. For larger problems, problem 7 and 8, MFGA produces significantly better fitness values than standard GA. This is because the solutions of MFGA are not easily trapped into local optima due its self-adaptive feature. The search time of MFGA is always smaller than that of standard GA. For problem 3 to 8, the evaluation time of MFGA is significantly smaller than for standard GA. The self-adaptive feature of MFGA also reduces the search time because it converges faster. The experimental evaluation results indicates that our proposed MFGA is more efficient and effective than standard GA.

5 Conclusion

In this paper, we presented a new GA-based approach for web service location-allocation problem. Our approach utilizes a self-adaptive strategy to dynamically adjust genetic operation possibility, so that better convergence of the evolutionary process can be achieved. Moreover, our approach employs a memory filter that greatly reduces the evaluation time while the diversity of the population gradually decreases. We have conducted a full experimental evaluation using the public WS-DREAM dataset to compare our approach to standard GA. The experimental results show that our approach is efficient and effective in computing near-optimal solutions for the web service location-allocation problem.

References

1. Aboolian, R., Sun, Y., Koehler, G.J.: A location-allocation problem for a web services provider in a competitive market. Europ. J. Operat. Research 194, 64–77 (2009)
2. Cheng, J., Chen, W., Chen, L., Ma, Y.: The improvement of genetic algorithm searching performance. In: Int. Conf. on Machine Learning and Cybernetics, pp. 947–951 (2002)
3. Gen, M., Cheng, R.: Foundations of genetic algorithms. In: Genetic Alg. Eng. Design, pp. 1–41 (1997)
4. Heydarnoori, A., Mavaddat, F., Arbab, F.: Towards an automated deployment planner for composition of web services as software components. Electr. Notes Theor. Comput. Sci. 160, 239–253 (2006)
5. Hoffmann, J., Bertoli, P., Pistore, M.: Web service composition as planning, revisited: In between background theories and initial state uncertainty. In: Nat. Conf. on Artificial Intelligence, vol. 22, p. 1013. AAAI Press (2007)
6. Huang, L., Nie, J.: Using pareto principle to improve efficiency for selection of QoS web services. In: 7th IEEE Conf. on Consumer Communicat. Networking, pp. 1–2. IEEE (2010)
7. Liu, T., Liu, Z., Lu, T.: A location & time related web service distributed selection approach for composition. In: 9th Int. Conf. on Grid and Cooperative Computing, pp. 296–301. IEEE (2010)
8. Liu, Z., Liu, T., Cai, L., Yang, G.: Quality evaluation and selection framework of service composition based on distributed agents. In: 5th Int. Conf. on Next Generation Web Services Practices, pp. 68–75. IEEE (2009)
9. Martin, D., et al.: OWL-S: Semantic markup for web services. W3C (2004)
10. Pasandideh, S.H.R., Niaki, S.T.A.: Genetic application in a facility location problem with random demand within queuing framework. J. Intell. Manufact. 23, 651–659 (2012)
11. Rodríguez-Mier, P., Mucientes, M., Lama, M., Couto, M.I.: Composition of web services through genetic programming. Evolut. Intell. 3, 171–186 (2010)
12. Srinivas, M., Patnaik, L.: Adaptive probabilities of crossover and mutation in genetic algorithms. IEEE Trans. Systems, Man and Cybernetics 24, 656–667 (1994)
13. Srinivas, M., Patnaik, L.: Genetic algorithms. Computer 27, 17–26 (1994)
14. Sun, Y., Koehler, G.J.: A location model for a web service intermediary. Decision Support Systems 42, 221–236 (2006)
15. Vanrompay, Y., Rigole, P., Berbers, Y.: Genetic algorithm-based optimization of service composition and deployment. In: 3rd Int. Workshop on Services Integration in Pervasive Environments, pp. 13–18. ACM (2008)
16. Yu, Y., Ma, H., Zhang, M.: An adaptive genetic programming approach to QoS-aware web services composition. In: IEEE Congress on Evolutionary Computation, pp. 1740–1747. IEEE (2013)
17. Zeng, L., Benatallah, B., Dumas, M., Kalagnanam, J., Sheng, Q.Z.: Quality-driven web services composition. In: 12th int. Conf. on World Wide Web, pp. 411–421. ACM (2003)
18. Zheng, Z., Zhang, Y., Lyu, M.R.: Distributed QoS evaluation for real-world web services. In: IEEE Int. Conf. on Web Services, pp. 83–90. IEEE (2010)
19. Zhou, J., Niemela, E.: Toward semantic QoS-aware web services: Issues, related studies and experience. In: IEEE/WIC/ACM Int. Conf. on Web Intelligence, pp. 553–557. IEEE (2006)

Ascertaining Spam Web Pages
Based on Ant Colony Optimization Algorithm

Shou-Hong Tang, Yan Zhu, Fan Yang, and Qing Xu

School of Information Science and Technology, Southwest Jiaotong University, China
{magicdreams,yangfan57319}@163.com,
yzhu@home.swjtu.edu.cn, qqwdgdx@126.com

Abstract. Web spam is troubling both internet users and search engine companies, because it seriously damages the reliability of search engine and the benefit of Web users, degrades the Web information quality. This paper discusses a Web spam detection method inspired by Ant Colony Optimization (ACO) algorithm. The approach consists of two stages: preprocessing and Web spam detection. On preprocessing stage, the class-imbalance problem is solved by using a clustering technique and an optimal feature subset is culled by Chi-square statistics. The dataset is also discretized based on the information entropy method. These works make the spam detection at the second stage more efficient and easier. On next stage, spam detection model is built based on the ant colony optimization algorithm. Experimental results on the WEBSPAM-UK2006 reveal that our approach can achieve the same or even better results with less number of features.

Keywords: Web spam detection, ACO, Class-imbalance, Feature selection, Discretization.

1   Introduction

Nowadays Web spam is troubling both internet users and search engine companies, because Web spammers play tricks to obtain a high ranking in the search engine results, which is not equivalent to the true quality of their pages. Generally speaking, there are three types of Web spam: link spam, content spam, and cloaking. All these spamming tricks bring a great damage to Web and Web applications. Therefore, it is urgent to ascertain fraudulent pages and punish them by some legislative measures.

Many researchers have worked on detecting fraudulent pages via different methods, such as tag-based methods [1] and cross-language Web content quality assessment method [2]. However, new spamming hoaxes have also emerged in various forms, which ceaselessly force us to explore new detection techniques. To this end, we design a framework of Web spam detection based on Ant Colony Optimization (WSD-ACO) inspired by the idea of [3, 4].

The contributions of this paper are as follows:

1. Applying K-means to solve the class-imbalance problem. The experiment results prove such a preprocessing makes spam detection more efficient.

H. Decker et al. (Eds.): DEXA 2014, Part II, LNCS 8645, pp. 231–239, 2014.

2. Using Chi-square statistics to obtain an optimal feature subset from the high dimensional feature space.

3. Discretizing the dataset through an information entropy method. This can further improve the efficiency and predictive accuracy of WSD-ACO.

The rest of this paper is organized as follows: Section 2 gives a brief review of related work; Section 3 discusses the proposed WSD-ACO approach; Section 4 presents the experimentation carried out on a public dataset. Finally, Section 5 concludes this paper and proposes further work in the area.

2   Related Work

Many studies have addressed different Web spam detection methods. For example, Araujo et al. [5] proposed to combine new link-based features with language model-based features to form an efficient spam detection system. Niu et al. [6] improved the performance of Web spam detection by training a discrimination model using genetic algorithm. Liu et al. [7] added user behavior analysis to a learning scheme to detect spam pages.

Taweesiriwate et.al [8] proposed a LSD-ACO algorithm to generate rules following the TrustRank assumption for spam detection. Our work differs from theirs on the detection model construction and preprocessing. Rungsawang et.al [9] combined content and link features to build a learning model by ant colony optimization. The differences between our method and theirs are: (1) we introduce a clustering method to solve biases of machine learning algorithms caused by the imbalance dataset. The experiment results show that this work can improve the detection performance greatly. They did not consider this issue. (2) In our framework, Chi-square statistics is used to search an optimal feature subset. The different feature subsets (e.g. 10, 20... 139 features) are experimented and 20 features are proven to obtain the best F-measure value. They adopted IG method and selected just 10 features, which may not be enough for learning a model. (3) In our WSD-ACO approach, we have used a different pheromone function $\tau_{ij}(t)$. Besides, when updating pheromone, we increased the pheromone of the terms that occurring in the rule antecedent and decreased the pheromone of the terms that does not occurring in the rule antecedent, by which the process of extracting detection rules can converge more quickly. However, they did not adopt such a pheromone evaporation mechanism in their work.

3   Web Spam Detection Based on ACO

3.1   The Framework of WSD-ACO

The key idea of WSD-ACO is to build a detection model, which contains lots of detection rules extracted by ACO. Discovering an optimal rule based on the features of Web pages is similar to the process of searching an optimal path to the food by a colony of ants. The framework of Web spam detection is described in Fig. 1. In this frame work, some difficulties including class-imbalance, the high dimensionality of

data, and high interdependence among some features, are solved during the preprocessing stages. The core parts of WSD-ACO approach including rule extraction and
rule pruning. Once the detection model has been built, we use the testing dataset to
validate its effectiveness.

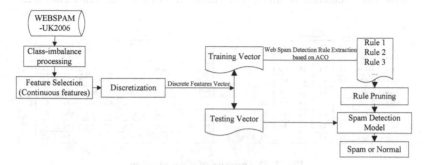

Fig. 1. The framework of WSD-ACO

3.2   The Data Preprocessing Stage

Class Imbalance Processing

The class-imbalance may lead to the biased problems of machine learning. In this
paper, K-means is used to solve the class-imbalance problem. Assuming that a dataset
contains k_1 samples with class label C_1 and k_2 samples with class labels C_2 ($k_1>>k_2$).
Samples that labeled C_1 are firstly clustered to m groups by K-means, cluster Cl_i ($i=1$,
$2... m$) corresponds to cluster center w_i. Supposing that cluster Cl_i contains n_i samples, So we need cull [n_i*k_2/k_1] samples from cluster Cl_i according to the distance
between sample and cluster center w_i. The more the samples close to the cluster center, the more representative. Finally, we obtain N samples closed to the number of
samples with class label C_2. Note that the operator [·] stands for a method of integer.

$$N = \sum_{i=1}^{m}\left[n_i * \frac{m_2}{m_1}\right]$$

(1)

Feature Selection and Discretization

In this section, a feature selection method based on Chi-square statistics is adopted to
filter redundant features. Chi-square statistical value of each feature is calculated by
Eq. (2). The larger the Chi-square value is, the more correlative the given feature
associates with the class.

$$\chi^2 = \sum_{i=1}^{m}\sum_{j=1}^{n}\frac{\left(A_{ij}-E_{ij}\right)^2}{E_{ij}}$$

(2)

Where m,n is the number of the values of feature A and of classes ($n=2$ in this paper,
normal or spam); A_{ij} is the number of samples with the feature value x_i that belong to
class C_j; E_{ij} is the expected frequencies of A_{ij} and can be calculated by $E_{ij}=(r_i \times c_j)/M$,
r_i is the amount of the samples with value x_i of   feature A and c_j is the number of the
samples belonging to class label C_j.

In discretization stage, a set of continuous features is divided into a smaller number of intervals to improve the efficiency and predictive accuracy of machine learning. We use the information entropy method proposed by Fayyad et.al in 1993 to discretize these features. Assuming that a dataset contains 8 samples and two class labels, each sample has 5 features. A binary discretization process is described in Fig.2.

Features

	F1	F2	F3	F4	F5	Class
Sample 1:	0.07	2.3	3.4	4.3	8.4	C2
Sample 2:	0.03	1.7	7.1	3.8	10.3	C2
Sample 3:	0.01	3.4	6.2	2.9	9.7	C2
Sample 4:	0.04	2.1	4.3	1.8	9.4	C1
Sample 5:	0.02	4.5	5.7	3.4	8.8	C2
Sample 6:	0.05	2.4	9.2	2.3	9.1	C1
Sample 7:	0.06	1.8	2.6	3.1	8.5	C2
Sample 8:	0.08	3.6	8.7	2.8	7.6	C1

Discretization $\nearrow$

Features

F1	F2	F3	F4	F5	Class
[0.05,0.07)	[2.1,3.6)	$(-\infty,3.4)$	[2.8,+∞)	[7.6,8.8)	C2
$(-\infty,0.03)$	$(-\infty,1.8)$	[4.3,7.1)	[2.8,+∞)	[9.4,+∞)	C2
$(-\infty,0.03)$	[2.1,3.6)	[4.3,7.1)	[2.8,+∞)	[9.4,+∞)	C2
[0.03,0.05)	[1.8,2.1)	[3.4,4.3)	$(-\infty,1.8)$	[8.8,9.4)	C1
[0.03,0.05)	[3.6,+∞)	[4.3,7.1)	[2.8,+∞)	[7.6,8.8)	C2
[0.03,0.05)	[2.1,3.6)	[7.1,+∞)	$(-\infty,1.8)$	[8.8,9.4)	C1
[0.05,0.07)	$(-\infty,1.8)$	$(-\infty,3.4)$	[2.8,+∞)	[7.6,8.8)	C2
[0.07,+∞)	[2.1,3.6)	[7.1,+∞)	$(-\infty,2.8)$	$(-\infty,7.6)$	C1

Fig. 2. The process of discretization

3.3   Classification Rule Construction Based on ACO

A path from the start node to the goal node (normal or spam) via multiple feature nodes can be viewed as a detection rule discovered by an ant. Each detection rule is expressed as If <feature1_value and feature2_value and ...> then <class>. Each feature has many values, but just one of them can be selected to associate with one class label in a rule. Fig. 3 describes the process of rule discovering.

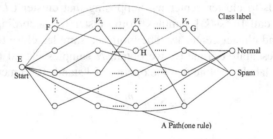

Fig. 3. Paths corresponding to Web spam detection rules found by ants

The process of rule extraction is as follows. At first, the RuleList is empty and the training set contains all the training samples. Each iteration of the outer WHILE loop in Algorithm I obtains a best detection rule that covers several samples in the training set and then adds the best detection rule to RuleList and finally updates the training set by moving the samples that covered by this best detection rule. The outer WHILE loop will be continuously executed until a stop criterion is reached. In the inner WHILE loop, the current ant adds one feature node at a time to its current detection rule to form its local optimal detection rule and then updates the pheromone to guide the following ants to discover their local optimal detection rules.

WSD-ACO Algorithm : *Web Spam Detection Rule Extraction Based on ACO*

Initialize the *TrainingSet* with all training samples;

Initialize the *RuleList* with an emptylist;

Initialize the *TempRuleList* with an emptylist;

WHILE (*TraningSet* > *Max_uncovered_training* samples)

    t = 1; /* ant index beginning */

    j = 1; /* convergence test index   */

       Initialize all nodes of features with the same amount of pheromone;

      WHILE (t <*Num_of_ants* and j < *Num_rules_converg*)

          Ant$_t$ starts with an empty detection rule and gradually constructs a Web spam detec-

          tion rule R_t by adding a node of features to the current detection rule R_t at a time;

             Prune rule R_t;

             IF (R_t is equal to R_{t-1})

                  j = j + 1;

             ELSE

                  j = 1;

             ENDIF

             Adding the current detection rule to *TempRuleList*;

             Update the pheromone of all nodes of features;

             t = t + 1;

      END WHILE

          Choose the best detection rule R_{best} from *TempRuleList*   and add it to *RuleList*;

          Move all the samples that covered by the detection rule R_{best} from the *TrainingSet*;

          Update the *TempRuleList* with an emptylist;

  END WHILE

Let *feature$_{ij}$* be a rule condition in the form $A_i=V_{ij}$, where A_i represents the ith feature and V_{ij} denotes its jth value. Each ant chooses *feature$_{ij}$* and adds it to the current partial rule according to the following probability:

$$P_{ij} = \frac{\eta_{ij} \bullet \tau_{ij}(t)}{\sum\limits_{i=1}^{a} x_i \sum\limits_{j=1}^{b_i} \left(\eta_{ij} \bullet \tau_{ij}(t) \right)} \qquad (3)$$

Where η_{ij} and $\tau_{ij}(t)$ denote a problem-dependent heuristic function and a pheromone information function obtained at the tth iteration respectively, which can be obtained by Eq. (4) and Eq. (6) respectively. *a* represents the total amount of features and b_i stands for the number of values of the ith feature. x_i is an indicator to avoid features chosen repeatedly in the current rule. If the feature A_i is not added to the current rule, then x_i is set to 1, otherwise 0. The larger the value of η_{ij} and $\tau_{ij}(t)$ are, the more possible the *feature$_{ij}$* is added to the current partial rule.

$$\eta_{ij} = \frac{\log_2 k - H\left(W \mid A_i = V_{ij}\right)}{\sum\limits_{i=1}^{a} x_i \bullet \sum\limits_{j=1}^{b_i} \left(\log_2 k - H\left(W \mid A_i = V_{ij}\right) \right)} \qquad (4)$$

$$H\left(W \mid A_i = V_{ij}\right) = -\sum\limits_{w=1}^{k} \left(P\left(w \mid A_i = V_{ij}\right) \bullet \log_2 P\left(w \mid A_i = V_{ij}\right) \right) \qquad (5)$$

Note that $H(W / A_i = V_{ij})$ in Eq. (4) denotes the entropy of the *feature_{ij}* which can be derived from the Eq. (5), where W is the class label, $P(w/A_i=V_{ij})$ is the empirical probability of observing class w conditional on having observed $A_i=V_{ij}$, and k is the number of classes. Since the optimal detection rule is found iteratively, the pheromone in Eq. (3) will be updated for controlling the movement of ants at each round. The pheromone update at the iteration time $t = 1, 2, 3...$ is calculated by Eq. (6) and Eq. (7). If *feature_{ij}* occurring in the rule found by an ant, the pheromone of *feature_{ij}* will be increased by Eq. (6), otherwise, the pheromone will be decreased by Eq. (7) to simulate pheromone evaporation in real ant colonies.

$$\tau_{ij}(t) = \tau_{ij}(t-1) + \tau_{ij}(t-1) \bullet \frac{q}{1+q} \tag{6}$$

$$\tau_{ij}(t) = \frac{\tau_{ij}(t-1)}{\sum_{i=1}^{a}\sum_{j=1}^{b_i}\tau_{ij}(t-1)} \tag{7}$$

The quality of rules over the training dataset (q in Eq. (6)) is measured by:

$$q = \frac{TP}{TP+FN} \bullet \frac{TN}{FP+TN} \tag{8}$$

Where TP is the number of samples covered by the rule that have the class predicted by the rule, FN is the number of samples that are not covered by the rule but that have the class predicted by the rule, FP is the number of samples that covered by the rule but have a different class predicted by the rule, and TN is the number of samples that not covered by the rule and that do not have the class predicted by the rule.

At each iteration of the outer WHILE loop in Algorithm, the initial pheromone of all *feature_{ij}* are equally set to:

$$\tau_{ij}(t = 0) = \frac{1}{\sum_{i=1}^{a}b_i} \tag{9}$$

3.4 Rule Pruning

The local optimal detection rule cannot be added to TempRuleList directly, because some rules may contain redundant feature nodes. In Fig. 3, a Web page that satisfies the rule E-F-G or the rule E-F-H-G can be predicted as a spam page. Obviously, H in E-F-H-G is redundant and decreases the efficiency. Besides, some features may make little or even negative contribution to the detection result for certain categories. Overfitting may occur during the process of extracting detection rules from the training data. Therefore, rule pruning is necessary. A simple method is to iteratively remove one feature node from a rule at a time which can greatly improve the quality of this rule. This process is repeated until only one feature remains in the antecedent part of the rule or until any further removal of a feature will degrade the quality of the rule.

3.5   Web Spam Filtering

Finally, a spam detection model has been built. If the first rule's antecedent matches with the new test sample, it will be assigned to the consequent of the first rule. Otherwise, the sample will be detected by other rules in turn. If the test sample is not detected by any rules, it will be predicted by a default rule, which simply assign it to the majority class in the set of uncovered training samples in Algorithm I.

4   Experiments and Results

4.1   Dataset Preparation

We used the public Web spam collection WEBSPAM-UK2006 [10] for evaluating WSD-ACO and combined the content features (e.g. number of words in the title, fraction of anchor text), the linked features (e.g. TrustRank of home page, PageRank of home page), and two direct features (the number of pages in home page, the length of host name) together. The prepared dataset contains 6511 "normal" pages and 1978 "spam" pages. To solve the imbalance dataset problem, we clustered "normal" sites into three clusters (several experiments show that three clusters can obtain the best detection results). The proportion of normal pages of cluster one (3661 samples), two (1110 samples), three (1740 samples) in total normal pages is 56%, 17%, 27% respectively. Finally we chosen 1108, 336, 534 samples from cluster one, two, three respectively. After this process, we obtain 1978 "normal" and 1978 "spam" samples.

4.2   Experimental Results

We firstly culled feature subsets from 139 features on by Chi-square statistics and Table 1 lists the top 6 features. The selected features are then discretized according to the top 10, 20, ..., 130, and 139 features to obtain 14 groups of dataset. WSD-ACO is conducted on these 14 data subsets respectively. The results of our framework comparing with NaiveBayes and RandomTree are presented at Fig. 4 and Table 2 (each method achieve its highest F-Measure on 20, 139, 20 features respectively). Each point on the curves is obtained by ten-fold cross validation.

Table 1. The top ten features selected by Chi-square statistics

Rank	Chi-square value	Feature
1	2766.24837	Number of words in the title (average value for all pages in the host)
2	1876.93377	Fraction of anchor text
3	1702.45330	TrustRank of home page
4	1641.43114	Top 1000 corpus precision
5	1637.70541	Top 1000 queries recall
6	1635.41410	Number of words in the title (Standard deviation for all pages in the host)

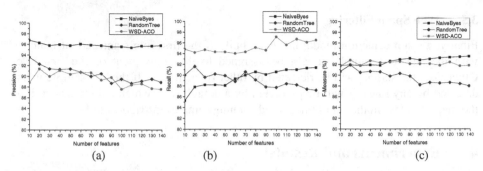

Fig. 4. The performance of different methods

Table 2. Number of features, Precision, Recall, F-Measure for different methods

Algorithm	Number of Features	Precision	Recall	F-Measure
NaiveBayes	139	0.958	0.914	0.935
RanomTree	20	0.923	0.916	0.919
WSD-ACO	20	0.917	0.943	0.930

5   Conclusions

The WSD-ACO approach exploits the idea of the optimal paths searching by ant colony for finding the Web spam detection rules. The WSD-ACO framework has two
notable merits: (1) it can solve class-imbalance problem, search for optimal feature
subsets and discretize feature subsets which can boost the detection model. (2) It can
build optimal detection model owing to the optimal detection rules discovery by ants.
WSD-ACO has a better performance in recall, which illustrates that WSD-ACO can
detect more spam pages than the other detection models. Although it may not be as
effective as some anti-spam algorithms for filtering spam pages, it can help search
engines find the most troublesome spam pages with fewer features. However, the
framework of WSD-ACO is very time consuming due to the preprocessing and rule
pruning. In the near future, we plan to deal with this issue and add the logical operator
OR in the conditions of the rules for enhancing the prediction performance of efficient
spam detection rules.

Acknowledgment. This work has been supported by the academic and technological
leadership training foundation of Sichuan province, China (X8000912371309).

References

[1] Lin, J.L.: Detection of cloaked Web spam by using tag-based methods. Expert Systems
    with Applications 36(4), 7493–7499 (2009)
[2] Geng, G.G., Wang, L.M., Wang, W., Hu, A.L., Shen, S.: Statistical cross-language Web
    content quality assessment. Knowledge-Based Systems 35, 312–319 (2012)

[3] Parpinelli, R.S., Lopes, H.S., Freitas, A.A.: Data mining with an ant colony optimization algorithm. IEEE Transactions on Evolutionary Computation 6(4), 321–332 (2002)

[4] Liu, X.P., Li, X., Liu, L., He, J.Q., Ai, B.: An innovative method to classify remote-sensing images using ant colony optimization. IEEE Transactions on Geoscience and Remote Sensing 46(12), 4198–4208 (2008)

[5] Araujo, L., Martinez-Romo, J.: Web spam detection: new classification features based on qualified link analysis and language models. IEEE Transactions on Information Forensics and Security 5(3), 581–590 (2010)

[6] Niu, X., Ma, J., He, Q., Wang, S., Zhang, D.: Learning to detect web spam by genetic programming. In: Chen, L., Tang, C., Yang, J., Gao, Y. (eds.) WAIM 2010. LNCS, vol. 6184, pp. 18–27. Springer, Heidelberg (2010)

[7] Liu, Y., Chen, F., Kong, W., Yu, H., Zhang, M., Ma, S., Ru, L.: Identifying Web Spam with the Wisdom of the Crowds. ACM Transactions on the Web 6(1) (2012)

[8] Taweesiriwate, A., Manaskasemsak, B., Rungsawang, A.: Web spam detection using link-based ant colony optimization. In: Processings of 26th International Conference on Advanced Information Networking and Applications (AINA) (2012)

[9] Rungsawang, A., Taweesiriwate, A., Manaskasemsak, B.: Spam host detection using ant colony optimization. In: Park, J.J., Arabnia, H., Chang, H.-B., Shon, T. (eds.) IT Convergence and Services. LNEE, vol. 107, pp. 13–21. Springer, Heidelberg (2012)

[10] Castillo, C., Donato, D., Becchetti, L., Boldi, P., Leonardi, S., Santini, M., Vigna, S.: A reference collection for Web spam. ACM SIGIR Forum 40(2), 11–24 (2006)

Reducing Redundant Information in Search Results Employing Approximation Algorithms[*]

Christos Makris[1], Yannis Plegas[1], Yannis C. Stamatiou[2],
Elias C. Stavropoulos[1,3], and Athanasios K. Tsakalidis[1]

[1] Computer Engineering and Informatics Department, University of Patras, Greece
{makri,plegas,estavrop,tsak}@ceid.upatras.gr
[2] Business Administration Department, University of Patras, Greece
stamatiu@ceid.upatras.gr
[3] Business Administration Department, Technological Educational Institute of Patras, Greece
estavrop@teipat.gr

Abstract. It is widely accepted that there are many Web documents that contain identical or near-identical information. Modern search engines have developed duplicate detection algorithms to eliminate this problem in the search results, but difficulties still remain, mainly because the structure and the content of the results could not be changed. In this work we propose an effective methodology for removing redundant information from search results. Using previous methodologies, we extract from the search results a set of composite documents called *SuperTexts* and then, by applying novel approximation algorithms, we select the SuperTexts that better reduce the redundant information. The final results are next ranked according to their relevance to the initial query. We give some complexity results and experimentally evaluate the proposed algorithms.

Keywords: Web search, redundant information, approximation algorithms, SuperTexts, ranking, shingling, semantics.

1   Introduction

Reducing redundant information from web documents is as well-studied problem [1], [14], [15], [16], [21], [22]. Initial approaches identify documents that are duplicates according to their contextual content [3], [17, 18]. The improvement of the previous algorithms led to the creation of algorithms that identify near-duplicate documents, i.e. documents that are not the same but they share common content. The most representative algorithms of this category represent the content of the documents as vectors like n-grams and find the percentage of the overlapping information between the documents in order to determine where the documents are near-duplicate [11], [25].

[*] This research has been co-financed by the European Union (European Social Fund – ESF) and Greek national funds through the Operational Program "Education and Lifelong Learning" of the National Strategic Reference Framework (NSRF) - Research Funding Program: Thales; Investing in knowledge society through the European Social Fund.

H. Decker et al. (Eds.): DEXA 2014, Part II, LNCS 8645, pp. 240–247, 2014.

Another important approach, called w-shingling, proposed in [2], [31], with several variations in [3], [15], uses the percentage of the overlapping sequences of w tokens between two documents to determine the redundant information. An interesting work in [11] contains a probabilistic algorithm that proposes near-duplicates documents with high accuracy.

A recent study connects the problem of reducing redundant information with the diversification problem, and a diversity-aware algorithm is given in [4]. The most up to date approaches use both the lexical and the semantic content of the documents. An interesting approach that follows these principles is given in [21]. This work initially identifies the duplicates and near duplicates documents of the search results and next maintains a new document called *SuperText*. SupertTexts contain the overlapping parts of the documents only one time in conjunction with the non-overlapping parts. Consequently, the overlapping information appears only once, in a SuperText.

Based on the notion of the SuperText, in this work we propose a methodology for effectively selecting proper SuperTexts that replace the initial list of search results. The selection of SuperTexts is based on (i) the elimination of redundant information, (ii) the fully coverage of initial information, and (iii) the semantic cohesion of the final SuperTexts that may represent different domains of knowledge in order to help users to browse fast and accurate to the information they seek. The whole procedure consists of multiple steps. Initially the proposed system takes an input the results obtained from a user query. Next it creates an initial set of SuperTexts getting all possible combinations of the search results that contain replicated contextual information more than a predefined threshold. The system adopts the approach of [21] to connect a number of search results to a SuperText and to create the final SuperTexts. These act as the input of the proposed algorithms which select the final Supertexts that eliminate redundancy in the search results. Compared with the work in [21], the novelty of our approach is that we model the problem of selecting the suitable SuperTexts as a variant of a well-studied combinatorial optimization problem called *maximum k-intersection* (MAXI), where given a collection of subsets over a universe, we are asked to find a subcollection that its cardinality exceeds a given threshold, with maximum intersection. The decision version of MAXI belongs to **NP**, while it was proved by Vinterbo in [26, 27, 28] and, independently by Xavier in [29], that it is **NP**-hard, thus MAXI is **NP**-complete. For a variety of areas of application of the MAXI problem, and variants of it, see, for example, [10], [20], [24], [27], [29].

We also performed experiments that show that the proposed algorithms remove redundancy without loss of the initial information and compared with the approach in [21], improve significantly the total process and enhance the use of SuperTexts. We analytically describe the whole process in the sequel.

2   Operating Model

The whole process constitutes a post processing of the results of a search engine query. When the user submits a query, the search engine retrieves a predefined number, say N, of the top pages, which act as the input for our methods. More specific, the system downloads the contents of the top-N pages which are fetched for the query, removes the mark-up and apply tokenization to their textual body. Finally, it lexically analyzes the tokenized text into canonical sequences of tokens giving as input to the proposed methods for each result (page) a document with the respective tokens.

Next, to reduce the information redundancy in search results, our approach proceeds to the following steps: **Creation of SuperTexts:** The duplicates and near duplicates documents are determined. Duplicate documents are removed and near duplicate documents are processed for the creation of the SuperTexts. **Selection of the Supertexts:** The final SuperTexts are effectively selected by employing approximation algorithms for solving variants of MAXI. **Ranking of the Supertexts:** The resulting SuperTexts are ranked according to their similarity with the user preferences.

For the first and the last step, we adopt the approaches in [21] with slight improvements. We next describe the way we select the final SuperTexts.

3   Selection of the Final Supertexts

The most innovative part of this work is the selection of the final SuperTexts. The usage of approximation algorithms optimizes the whole process and produces a faster and more accurate way for the reduction of the information redundancy in the search results. In the previous section we created an extended set of SuperTexts covering the most important combinations of the search results. We now tackle the problem of reducing information redundancy by selecting the appropriate set of SuperTexts.

For every SuperText, consider the corresponding set of tokens (the final set of the words), defined over a universal set of tokens. By this representation, our problem is modeled as a well-defined combinatorial optimization problem where given a family of subsets (the SuperTexts) over a universe of items (the tokens), and a constant number, say k, we want to select a subfamily having at least k subsets that maximizes the number of common items, i.e. the cardinality of the intersection of subsets in the subfamily. We want the solution to consist of at least k subsets, so that the coverage of the information will be satisfactory, and also the cardinality of their intersection to be maximized, so that to reduce the redundancy and at the same time to increase the semantic cohesion of the SuperTexts returned to user. However, one would also ask to maximize the cardinality of union of the solution, so that to increase the coverage of the information. An alternative request would be a solution with the greatest *importance*, where the importance of a SuperText can be defined as the number of tokens it contains, even multiplied by a constant *weight*.

All these thoughts are modeled by different variants of the well-known *Maximum k-Intersection* problem (MAXI), formally defined as follows: Given a collection $C = [C_j]_{j \in J}, J = \{1, 2, \ldots, m\}$ of m subsets over a universe $U = \{e_1, e_2, \ldots, e_n\}$ of n elements, and a positive integer $k \leq m$, find a subset $I \subseteq J$ with $|I| \geq k$ that maximizes $|\bigcap_{i \in I} C_i|$. For the decision version of MAXI, the existence of a subset $I \subseteq J$ with $|I| \geq k$, that satisfies $|\bigcap_{i \in I} C_i| \geq l$ for a given positive integer l, is asked.

We next define a variant of MAXI, named *Maximum k-Intersection-Maximum k-Union* problem (MAXI–MAXU): Given a collection $C = [C_j]_{j \in J}, J = \{1, 2, \ldots, m\}$ of m subsets over a universe $U = \{e_1, e_2, \ldots, e_n\}$ of n elements, and a positive integer $k \leq m$, find a subset $I \subseteq J$ with $|I| \geq k$ that maximizes both $|\bigcap_{i \in I} C_i|$ and $|\bigcup_{i \in I} C_i|$. For the decision version of MAXI–MAXU, the existence of a subset $I \subseteq J$ with $|I| \geq k$, that satisfies $|\bigcap_{i \in I} C_i| \geq l$ and $|\bigcup_{i \in I} C_i| \geq u$ for the given positive integers l, u, is asked. This problem obviously also belongs to **NP**, while the next proposition states its hardness (we omit proofs due to lack of space).

Proposition. The decision version of the MAXI–MAXU problem is **NP**-hard.

Lemma. Unless **P=NP**, there is no polynomial time algorithm for MAXI–MAXU.

In the Template Algorithm, we give a greedy approach for solving variants of the MAXI problem. A similar algorithm was given for the MAXI problem in [27, 28] for the MAXI-MAXU version. By setting condition X to be the maximization of the cardinality of the intersection and condition Y to be the maximization of the cardinality of the union of the returned sets, the template algorithm is self-explained.

Input: A collection $C = [C_j]_{j \in J}, J = \{1, 2, \ldots, m\}$ of m subsets over a universe $U = \{e_1, e_2, \ldots, e_n\}$ of n elements, and a positive integer $k \leq m$

Output: A subset $I \subseteq J$ with $|I| \geq k$ that satisfies X and Y.

1.　　Initialize $I = \emptyset$, $H = \emptyset$, iteration = 1;
2.　　Compute the frequencies f_i, for every element $e_i \in U$;
3.　　Set $H = \{e_i \mid f_i \geq k\}$;
4.　　**while** { $H \neq \emptyset$ and iteration $\leq k$ } **do**
5.　　　　Select the subset, say C_0, that satisfies condition X. If this is not unique, among all subsets that satisfies X, pick one that satisfies condition Y;
6.　　　　Set $I = I \bigcup C_0$, $H = H \bigcap C_0$, $C = C - C_0$, iteration = iteration + 1;
7.　　Return I

Template Algorithm: A greedy approach for solving variants of the MAXI problem.

As already described, the above approach aims to maximize first the cardinality of the intersection of the subsets of the solution, and next the cardinality of their union. An alternative solution for our problem can obtained if we swap these priorities: let condition X be maximization of the cardinality of the union and condition Y be the maximization of the cardinality of the intersection of the sets returned by the algorithm. As we describe in Section 4, experimental evaluation of this approach shows that this approach exhibits better performance in practice.

We next formalize another version of MAXI, the *weighed maximum k-intersection* problem (WMAXI), where instead of maximizing the union of the selected subsets, we aim on maximizing the importance of them: Given a collection $C = [C_j]_{j \in J}, J = \{1, 2, \ldots, m\}$ of m subsets over a universe $U = \{e_1, e_2, \ldots, e_n\}$ of n elements, a positive integer $k \leq m$, and positive numbers $w_j, w_j \geq 1, j = 1, \ldots, m$, find a subset $I \subseteq J$ with $|I| \geq k$ that maximizes both $|\bigcap_{i \in I} C_i|$ and $\sum_{i \in I} w_i |C_i|$. For the decision version of WMAXI, the existence of a subset $I \subseteq J$ with $|I| \geq k$, that satisfies $|\bigcap_{i \in I} C_i| \geq l$ and $\sum_{i \in I} w_i |C_i| \geq w$ for the given positive integers l, w, is asked.

Proposition. The decision version of the WMAXI problem is **NP**-hard.

Lemma. Unless **P=NP**, no polynomial time algorithm for WMAXI does exist.

We can suitable define conditions X and Y in the Template algorithm to obtain two more variants of the MAXI problem: Let condition X be the maximization of the cardinality of the intersection and condition Y be the maximization of the importance of the solution, where the importance is defined by a linear modular function defined as the sum of terms $w_i |C_i|$. One variant aims firstly to maximize the cardinality of the intersection and next the importance of the solution. An alternative approach is obtained if we swap the priority of these conditions. For the experimental evaluation of these approaches, we set $w_i = 1$, for every Supertext.

More variants can be obtained if we aim to *minimize* the cardinality of intersection of the sets of the solution. Intuitively, this condition will radically reduce the redundant information, but also it may reduce the quality of the solution, since the results will be loosely cohesive. As an obvious choice, no theoretical issues are discussed for these variants, but only for the more theoretical ones. We next summarize the variants of the *Maximum k-Intersection* problem, according to the definition of conditions X and Y in the Template Algorithm, and experimental evaluate them in Section 4:

Algorithm 1: *Maximize Intersection* (cond. X) and *Maximize Importance* (cond. Y).
Algorithm 2: *Maximize Importance* (cond. X) and *Maximize Intersection* (cond. Y).
Algorithm 3: *Maximize Intersection* (cond. X) and *Maximize Union* (cond. Y).
Algorithm 4: *Maximize Union* (cond. X) and *Maximize Intersection* (cond. Y).
Algorithm 5: *Minimize Intersection* (cond. X) and *Maximize Importance* (cond. Y).
Algorithm 6: *Maximize Importance* (cond. X) and *Minimize Intersection* (cond. Y).
Algorithm 7: *Minimize Intersection* (cond. X) and *Maximize Union* (cond. Y).
Algorithm 8: *Maximize Union* (cond. X) and *Minimize Intersection* (cond. Y).

4   Experimental Evaluation

The evaluation of the performance of our approaches was carried out on data query testing datasets, obtained by exploring 200 queries for the *TREC WebTrack 2009 – 2012* [6, 7, 8, 9]. The *TREC Web Tracks* use the 1 billion page ClueWeb09 dataset (see http://lemurproject.org/clueweb09/) and has a diversity task that contains for every query a ranked list of results that provide a complete coverage for the query topic, while avoiding excessive redundancy in the query results. According to [21], despite the content duplication algorithms that were performed, there is information redundancy in the search results justifying the need a solution like the proposed one.

For our experiments, we retain for each query the first $N=50$ results. We explored three different aspects of the problem, the redundancy among the final SuperTexts, the coverage of the initial information (when we set limits in the number of Super-Texts), and the ranking of the final SuperTexts. The redundancy among the final *SuperTexts* was measured using the *Containment* values between the SuperTexts. The selection of the SuperTexts was stopped when the initially information was covered, after all the w-shingles of the initial search results were contained in the final Super-Texts. In our experiments, we set the threshold *thresh* = 0.95.

The coverage of the initial information is measured by computing the percentage of the w-shingles that contained in the SuperTexts in relation to the w-shingles that existed initially in the search results. For this measure we set the number of the Super-Texts equals to $k=10$, assuming that search engines initially shows the first 10 results.

Finally, we ranked the SuperTexts according to Section 6 and we compare the *a-nDCG* (α -*Normalized Discounted Cumulative Gain*) [5] values with α=0.5, which quantifies the usefulness of a SuperText based on its position in the final result list. The *a-nDCG* measure extends the *nDCG* metric [13] rewarding both diversity and novelty. Also comparing the results of the Algorithms 4, 5, 7 and 8 with the results from the rank method in [21], clearly these Algorithms improve the total procedure.

4.1 Experimental Results

In Figure 1 we depict the average Redundancy values for all our Algorithms. As it can be seen, Algorithm 7 performs better while Algorithms 4, 5 and 8 perform quite well. In Figure 2 we give the average Coverage values. We can see that Algorithms 4 and 8 outperform the other ones, while Algorithms 5 and 7 perform quite well.

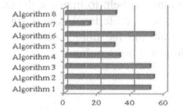
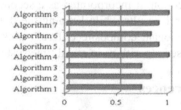

Fig. 1. Redundancy in the final SuperTexts   **Fig. 2.** Coverage in the final SuperTexts

Finally we compute the average *α-nDCG* values and compare them with the respective values from the methods presented in [21]. We give the results in Figure 3. The values for the initial search list are depicted with blue, the values from the methods in [21] with red, and the values for each of our Algorithms with green color. The values for the initial list and for the method in [21] are always the same – we repeat their values to clarify the performance of each Algorithm. It's obvious that Algorithms 7 and 4 perform better from the others. Also Algorithm 5 has a slightly better performance than the method in [21] while the rest of the Algorithms perform worst. We compare out Algorithms with the method in [21], and not only with the values of the initial list, because the creation of the SuperTexts reduces the number of the final results, increasing the α-$nDCG$ values automatically (as analytically described in [21]). To sum up, experimental evaluation shows that Algorithms 4, 5, 7 and 8 outperform the other ones and the method in [21], too. It is quite interesting that Algorithm 4 that maximizes union and then (counter intuitively) maximizes intersection performs competitively well with the obvious alternatives that minimizes intersection in order to remove redundancy. Hence a more refined approach is needed that will simultaneously reduce redundancy and guarantee an acceptable level of cohesion.

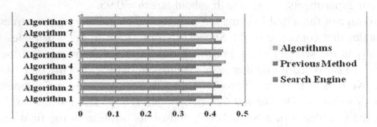

Fig. 3. Average α-$nDCG$ values

5   Conclusions and Future Work

We presented a collection of approximation algorithms that are applied to the framework of SuperText, to improve the quality of the returned results. The selection of SuperTexts that will reduce the information redundancy in the search results reduces in turn, to a number of variants of the *Maximum k-Intersection* problem. We gave some theoretical results for the problem and experimentally evaluate a number of greedy approaches, to conclude that the counter intuitive approach of maximizing intersection, if incorporated as a secondary criterion, can help to improve the quality of search results while keeping the redundancy in acceptable level. The design of a more elegant composite criterion ranking algorithm and the implementation of the proposed algorithms in a real time system need further consideration.

References

1. Broder, A.Z., Glassman, S.G., Manasse, M.S., Zweig, G.Z.: Syntactic Clustering on the Web. Computer Networks 29(8-13), 1157–1166 (1997)
2. Broder, A.Z.: Identifying and Filtering Near-Duplicate Documents. In: Giancarlo, R., Sankoff, D. (eds.) CPM 2000. LNCS, vol. 1848, pp. 1–10. Springer, Heidelberg (2000)
3. Charikar, M.: Similarity Estimation Techniques form Rounding Algorithms. In: 34th Annual Symposium on Theory of Computing, pp. 380–388 (2002)
4. Chen, H., Karger, D.R.: Less is More: Probabilistic Models for retrieving Fewer Relevant Documents. In: 29th Intern. ACM SIGIR Conf., pp. 429–436 (2006)
5. Clarke, C., Kolla, M., Cormack, G., Vechtomova, O., Ashkan, A., Büttcher, S., MacKinnon, I.: Novelty and diversity in information retrieval evaluation. In: 31st Annual International ACM SIGIR Conf., Singapore, July 20-24 (2008)

6. Clarke, C.L.A., Craswell, N., Soboroff, I.: Overview of the TREC 2009 Web Track. In: 18th TREC Conference (2009)
7. Clarke, C.L.A., Craswell, N., Soboroff, I., Cormack, G.: Overview of the TREC 2010 Web Track. In: 19th TREC Conference (2010)
8. Clarke, C.L.A., Craswell, N., Soboroff, I.: Overview of the TREC 2011 Web track. In: 20th TREC Conference (2011)
9. Clarke, C.L.A., Craswell, N., Voorhees, E.M.: Overview of the TREC 2012 Web track. In: 21th TREC Conference (2012)
10. Clifford, R., Popa, A.: Maximum subset intersection. Information Processing Letters 111, 323–325 (2011)
11. Deng, F., Rafiei, D.: Estimating the Number of Near Duplicate Document Pairs for Massive Data Sets using Small Space. University of Alberta, Canada (2007)
12. Fellbaum, C.: WordNet: An Electronic Lexical Database. MIT Press (1998)
13. Jarvelin, K., Kekalainen, J.: IR Evaluation Methods for Retrieving Highly Relevant Documents. In: 23rd International ACM SIGIR Conference, pp. 41–48 (2000)
14. Hajishirzi, H., Wen-Tau, Y., Kolcz, A.: Adaptive Near-Duplicate Detection via Similarity Learning. In: 23rd International ACM SIGIR Conference, pp. 10–17 (2000)
15. Henzinger, M.R.: Finding Near-Duplicate Web Pages: A Large-Scale Evaluation of Algorithms. In: 29th Intern. ACM SIGIR Conference, pp. 284–291 (2006)
16. Huffman, S., Lehman, A., Stolboushkin, A., Wong-Toi, H., Yang, F., Roehrig, H.: Multiple-Signal Duplicate Detection for Search Evaluation. In: 30th International ACM SIGIR Conference, pp. 223–230 (2007)
17. Kumar, J.P., Govindarajulu, P.: Duplicate and Near Duplicate Documents Detection: A Review. European Journal of Scientific Research 32(4), 514–527 (2009)
18. Manning, C.D., Raghavan, P., Schutze, H.: Introduction to Information Retrieval. Cambridge University Press (2008)
19. Navigli, R.: Word Sense Disambiguation. ACM Computing 41(2), 10:1–10:69 (2003)
20. Peeters, R.: The maximum edge biclique problem is NP-complete. Discrete Applied Mathematics 131, 651–654 (2003)
21. Plegas, Y., Stamou, S.: Reducing information redundancy in search results. In: SAC 2013, pp. 886–893 (2013)
22. Radlinski, F., Bennett, P.N., Yilmaz, E.: Detecting Duplicate Web Documents using Clickthrough Data. In: 4th Intern. Conf. on WSDM, pp. 147–156 (2011)
23. Salton, G., McGill, M.: Introduction to Modern Information Retrieval. McGraw-Hill (1998)
24. Shieh, M.-Z., Tsai, S.-C., Yang, M.-C.: On the inapproximability of maximum intersection problems. Information Processing Letters 112, 723–727 (2012)
25. Theobald, M., Siddharth, J., Paepcke, A.: Spotsigs: Robust and Efficient Near Duplicate Detection in Large Web Collections. In: 31st ACM SIGIR C., pp. 563–570 (2008)
26. Vinterbo, S.A.: A note on the hardness of the κ-ambiguity problem. Technical Report DSG TR 2002/06 (2002)
27. Vinterbo, S.A.: Maximum κ-intersection, edge labeled multigraph max capacity κ-path, and max factor κ-gcd are all NP-hard. Technical Report DSG TR (2002)
28. Vinterbo, S.A.: Privacy: A machine learning view. IEEE Transactions on Knowledge and Data Engineering 16(8), 939–948 (2002)
29. Xavier, E.C.: A note on a maximum κ-subset intersection problem. Information Processing Letters 112, 471–472 (2012)
30. Wu, Z., Palmer, M.: Web Semantics and Lexical Selection. In: ACL Meeting (1998)
31. Zhang, Y., Callan, J., Minka, T.: Novelty and Redundancy Detection in Adaptive Filtering. In: 25th International ACM SIGIR Conference (2002)

Dynamic Privacy Policy Management in Services-Based Interactions

Nariman Ammar[1], Zaki Malik[1], Elisa Bertino[2], and Abdelmounaam Rezgui[3]

[1] Department of Computer Science, Wayne State University Detroit, Michigan, USA
{nammar,zaki}@wayne.edu
[2] Department of Computer Science, Purdue University, West Lafayette, IN, USA
bertino@cerias.purdue.edu
[3] Department of Computer Science and Engineering,
New Mexico Tech, Socorro, New Mexico, USA
rezgui@cs.nmt.edu

Abstract. Technology advancements have enabled the distribution and sharing of patient personal health data over several data sources. Each data source is potentially managed by a different organization, which expose its data as a Web service. Using such Web services, dynamic composition of atomic data type properties coupled with the context in which the data is accessed may breach sensitive data that may not comply with the users preference at the time of data collection. Thus, providing uniform access policies to such data can lead to privacy problems. Some fairly recent research has focused on providing solutions for dynamic privacy policy management. This paper advances these techniques, and fills some gaps in the existing works. In particular, dynamically incorporating user access context into the privacy policy decision, and its enforcement. We provide a formal model definition of the proposed approach and a preliminary evaluation of the model.

1 Introduction

Technology advancements enable the online collection and publication of vast amount of data about individuals. Atomically, these data sources may not reveal personally identifiable information for individuals (e.g., HIPAA regulating EHRs), but linking a number of distributed sources may lead to unintended breach of privacy. We discuss privacy risks associated with personal health data collection and subsequent sharing in service-oriented environments, which may lead to breaching sensitive data by linking health data to other data in publicly available records. Specifically, we target genomic data, which is potentially transformative for public health. Gene sequencing data-collection has been on a consistent (often exponential) rise for the past few years. The success of Gene sequencing as a tool for diagnosis has resulted in a recent announcement of the Genomic Sequencing and Newborn Screening Disorders (GSNSD) program; which collects Genomic data of new-borns and keeps the data as hospital records. Data sharing not only provides the ability to view data but also to further analyze

H. Decker et al. (Eds.): DEXA 2014, Part II, LNCS 8645, pp. 248–262, 2014.

it by linking it to other publicly available data sets. In service-oriented environments, dynamic composition enables the specification of composite services without knowing a priori which Web services will be actually used at run-time. Such dynamic composition of different data items (retrieved through participating Web services), coupled with the context in which these data items are accessed may reveal sensitive information, which was not deemed as such by the user at the time of data collection. Thus, one challenge that service-oriented environments pose is *context-sensitivity*. Context has been defined in the literature in terms of client trust, affiliation, query history, query type, temporal or spatial relationships, among others. A context-aware system consists of an infrastructure to capture context and a set of rules to govern how the system should respond to context changes. Most context-aware systems do not take the dynamicity at the rule level when they make their decisions. Researchers who have proposed privacy policy management systems often used policy definitions that are statically chosen by data owners at the time of data collection. However, context-sensitivity implies that data owner's consent may not be enough for data disclosure. Recently, few researchers proposed solutions to dynamically handle context [6,9,11,3]. However, the dynamicity of these solutions is only at the decision level, not at the rule level. The few approaches that have dealt with rule-level dynamicity still predetermine the rule and the context types based on a set of activities, states, and contexts in which the user could be [19]. Moreover, these rules are not defined in semantic terms and do not govern what is potentially sensitive data. Thus, we need a mechanism to dynamically identify what is considered as potentially sensitive data and make a decision regarding data disclosure not only based on the current context, but also based on the previous context. We present an approach that defines privacy policy rules in terms of concepts and relations from domain ontologies and dynamically handles context by updating policy rules at the time of data access based on previous contexts.

Outline. The paper is organized as follows: First, we motivate the problem in Section 2 by defining a scenario involving potential adversary actions. This is followed by a formal definition of the proposed solution in Section 3. We then provide our model evaluation in Section 4. Finally, we discuss some related works (Section 5) and conclude.

2   Adversary Model Definition

Consider a collaborative health data sharing environment in which data is distributed among several organizations, each of which manages data access and usage through a Web service WS_i. Through each service interface, a requester can perform a set of operations. Assume also that all the concepts that can be searched for are stored in a generic ontology, which has a taxonomy for purposes P and another for data type properties D. Each operation Op_j exposed by WS_i queries an ontology-based repository and returns a set of data type properties D_j. WS_i defines a privacy policy for each patient instance in it's repository. Together with every instance, WS_i records the patient's predefined disclosure

Table 1. Sample requests that reflect the scenario

R_j	WS_i/Op_j	Q_j	
		P_j	D_j
1	WS1(GenomicWS)/getGeneInfo()	Research	G,GL
2	WS1(GenomicWS)/getVariantInfo()	Research	RS,G,GL
3	WS1(GenomicWS)/getGenPhenAsso()	Research	PH,RS,G,VL
4	WS2(DrugWS)/getDrugInfo()	Research	D,DT,IG
5	WS3(ClinicalWS)/getPatientInfo()	Research	G,A,GE
6	WS3(ClinicalWS)/getPatientDiagnosis()	Diagnosis	PH,G,V,D,DT,IG,VD
7	WS3(ClinicalWS)/getPatientDiagnosis()	Marketing	PH,VL,D,A,GE,AD
8	WS3(ClinicalWS)/getPatientDiagnosis()	Marketing	PH,VL,DT,A,GE,AD
9	WS3(ClinicalWS)/getPatientDiagnosis()	Marketing	PH,G,V,VL,DT,A,GE,AD
10	WS3(ClinicalWS)/getPatientDiagnosis()	Marketing	PH,RS,G,GL,V,VL,D,DT,A,GE,AD

preferences over his data type properties and the purpose of disclosure. An adversary submits a set of requests, each of which invokes an operation. A request R_j eventually gets translated into a SPARQL query Q_j.

Assume that the purposes from which a requester can choose are represented by the set $P = \{Research, Diagnosis, Marketing\}$, and the data type properties that a requester can ask for are *Phenotype, RsNo., GeneSymbol, GeneLocation, Variant, VariantLocation, Drug, DrugTarget, InteractingGene, VisitDate, Age, Gender, Name, Address.* For simplicity, we refer to these data type properties as the set $D = \{P, RS, G, GL, V, VL, D, DT, IG, VD, A, GE, N, AD\}$. Assume that the adversary submits several requests to a number of Web services (Table 1) to perform *Genetic Variation Detection.* In each request, the adversary claims a different purpose asking for one of these data type properties. For example, in R_1 he asks for $D_1=\{G, GL\}$ and indicates the purpose as $P_1=\{Research\}$. Then, in R_2 he asks for $D_2=\{RS, G, GL\}$ and indicates the purpose as $P_2=\{Research\}$. In R_3 he asks for $D_3=\{PH, RS, G, VL\}$ and also indicates the purpose as $P_3=\{Research\}$. Later, in R_6 he asks for $D_6=\{PH, G, V, D, DT,...\}$, but he indicates the purpose as $P_6=\{Diagnosis\}$. The adversary asks for some of the data type properties in more than one request to be able to link the data at later stages. For example, the G and GL data type properties appear in the first three requests. Later, in requests R_7 through R_{10} the adversary links those data type properties to other sensitive data type properties (e.g. A, GE, AD).

Moreover, the adversary may ask for data in subsequent phases seeking more sensitive data in each phase. In an initial phase, he may submit initial exploratory queries that do not explicitly ask for sensitive data. The purpose of those queries is to probably get an overall view of the data. For example, he can first ask for the gene (G) data of all patients that are within some age (A) range for *Research* purposes. In later phases, he may look for patients who have been diagnosed for depression for *Diagnosis* purpose. To this effect, he changes the purpose of the query.

We would like to monitor different aspects of user access context and return data based on the inferred context. We wish to preserve privacy whenever values of data type properties retrieved from different Web services are combined in a way that violates privacy policy rules tied to those data type properties. We achieve that by making WS_i context-sensitive. Next, we present our solution by defining our notion of context and we explain how we incorporate context into dynamic privacy policy rule enforcement.

3   Dynamic Privacy Policy Management Solution

In our proposed architecture (Fig. 1), a requester enters the purpose P_j and the Web service operation WS_i/Op_j. WS_i has a Policy Enforcement Point PEP agent client, which interprets the request as a SPARQL query Q_j and forwards it to the *Semantic Handler* (**SH**). The **SH** component runs Q_J by an ontology-based repository and passes the set of instances I that match the query together with the query Q_j to the *Context Handler* (**CH**). The *Context Handler* consists of two sub components. The *Classifier*, which dynamically classifies a query as being potentially *malicious* or *legitimate*, and the *Sensitive Data Detector*, which dynamically determines the subset of data type properties in a query that could potentially be sensitive. WS_i uses the context $CTXT$ inferred by it's sub components to update the context of each instance in I. The PEP uses $CTXT$ to make the final decision by performing *Dynamic Rule Check* (**DRC**) and *Query Rewriting* (**QR**), which will be explained in Section 3. The PEP then looks up the updated policy rule context and sends the response back to the user.

For computational purposes, we express a query Q_j that pertains to a request R_j as a vector that consists of the purpose P_j and the set of data type properties D_j. For example, Table 2 shows the requests in Table 1 as query vectors. We represent each purpose by a numerical value, and we represent each data type property by a binary value, where 1 indicates that a data type property d_k

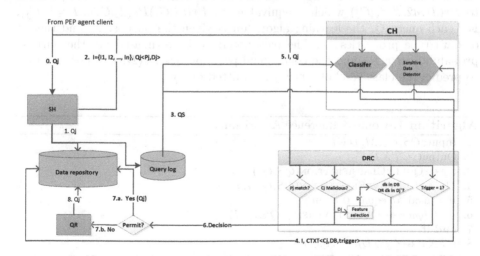

Fig. 1. Dynamic privacy policy management solution

Table 2. Queries as vectors of purposes and data type properties

Qj	Pj	PH	RS	G	GL	V	VL	D	DT	IG	VD	A	GE	N	AD
1	1.0	0	0	1	1	0	0	0	0	0	0	0	0	0	0
2	1.0	0	1	1	1	0	0	0	0	0	0	0	0	0	0
3	1.0	1	1	1	0	0	1	0	0	0	0	0	0	0	0
4	1.0	0	0	0	0	0	0	1	1	1	0	0	0	0	0
5	1.0	0	0	1	0	0	0	0	0	0	0	1	1	0	0
6	2.0	1	0	1	0	1	0	1	1	1	1	0	0	0	0
7	3.0	1	0	0	0	0	1	1	0	0	0	1	1	0	1
8	3.0	1	0	0	0	0	1	0	1	0	0	1	1	0	1
9	3.0	1	0	1	0	1	1	0	1	0	0	1	1	0	1
10	3.0	1	1	1	1	1	1	1	1	0	0	1	1	0	1

appears in the query and 0 to indicate that d_k does not appear in the query. Algorithm 1 summarizes our context inference algorithm. We break our solution to context inference into the following two sub problems:

Query Classification: For classifying queries we use the Naive Bayesian learning algorithm. The input to the learning algorithm is the query space QS and the output is a classification C_j. We assume that the presence of one data type property in a classification is conditionally independent of another data type property. We also assume that the data type properties asked for in a query are dependent on a query's purpose. Based on that, we construct a *Naive Bayesian Classification* model by converting a query Q_i into a *Bayesian Network*, where the root node represents a query's purpose P_i and the children represent data type properties $d_1, ..., d_k$. In the rest of the paper, we treat the data type properties in a query as a set D_i. So, based on our definition, $Pr(D_i|C_j)$ is equivalent to $Pr(d1, d2, .., d_k|C_j)$ which is equivalent to $Pr(d_1|C_j)Pr(d_2|C_j)...Pr(d_k|C_j)$. For each query Q_i the learning algorithm is given the purpose P_i and the set of data type properties D_i. The parameters to be estimated are the purpose probabilities $Pr(P_i)$ and the conditional probabilities $Pr(D_i|P_i)$. For example, to predict the class label of a newly submitted query

---

Algorithm 1. Context Inference Algorithm

1. Input: QS, Q_j, M, D_A, t
2. Output: $CTXT$
3. $C_j \leftarrow$ QUERYCLASSIFICATION(QS, Q_j)
4. $D_B \leftarrow$ RELATIVESENSITIVITY(QS, Q_j, t, D_A)
5. **if** i mod M equals 0 **then**
6. $trigger \leftarrow$ QUERYDIVERSITY(QS, t, M)
7. **end if**
8. $CTXT \leftarrow C_j \cup D_B \cup trigger$

$$Q_{i+1} = [C_{j+1} =?, P_{i+1} = Research, D_{i+1}], \text{ where } D_{i+1} = \{PH, G, GL, RS, A, ...\}$$

the algorithm computes $Pr(Q_{i+1}|C_j)Pr(C_j)$, for $j = \{malicious, legitimate\}$ based on the estimated parameters from the training data.

Algorithm 2. Relative Sensitivity Algorithm

---
1. input: QS, Q_j, D_j, t, D_A
2. output: D_B
3. $D_B \leftarrow \phi$
4. **if** $I(D_j, D_A) \geq t$ **then**
5. $CPM \leftarrow$ CONDITIONALPROBABILITYMATRIX(D_j, D_A)
6. $SVD \leftarrow$ SINGULARVALUEDECOMPOSITION(CPM)
7. $D_B \leftarrow D_B \cup$ FINDMAX(SVD)
8. **end if**
9. **for each** Q_k in QS **do**
10.   **if** $I(D_j, D_k) \geq t$ **then**
11. $CPM \leftarrow$ CONDITIONALPROBABILITYMATRIX(D_j, D_k)
12. $SVD \leftarrow$ SINGULARVALUEDECOMPOSITION(CPM)
13. $D_B \leftarrow D_B \cup$ FINDMAX(SVD)
14.   **end if**
15. **end for**

Sensitive Data Detection: For sensitive data detection, our goal is to determine, for a set of queries, a set of data type properties that could potentially be sensitive even though the data has not been deemed sensitive at the time of data collection. This problem reduces to two sub-problems:

Sensitivity of a Set of Data Type Properties: We apply conditional entropy to measure the relative sensitivity of a subset D_B of the set of data type properties D_i that is asked for in a newly submitted query with respect to two things. First, users are often asked to make privacy decisions regarding their sensitive data (e.g., *Name*) at the time of data collection. Let D_A be the set of predetermined sensitive data type properties. We apply conditional entropy to measure the relative sensitivity of D_i with respect to D_A. Second, we measure the relative sensitivity of D_i with respect to all sets of data type properties $D_1, ..., D_k$ in the previously submitted queries in QS. In both cases, we use the notion of information gain as a measure of the mutual information between two random variables. We define the information gain $I(D_A, D_i)$ for D_i with respect to D_A, as the reduction in uncertainty about the value of D_A when the requester knows the value of D_i. Formally:

$$I(D_A, D_i) = H(D_A) - H(D_A|D_i)$$

$$H(D_A) = - \sum_{d_a \in D_A} Pr(d_a)log_2 Pr(d_a)$$

$$H(D_A|D_i) = - \sum_{d_a, d_i \in D_A, D_i} Pr(d_a, d_i)log_2 Pr(d_a|d_i)$$

We apply the same formulas above to measure $I(D_k, D_i)$. The relative sensitivity algorithm (Algorithm. 2) first computes the information gain between the sets D_i and D_A. It then computes the information gain between D_i and each set of data type properties $D_1, ..., D_k$ in the set of previously submitted queries QS. If either case results in an information gain that is higher than a threshold t, the algorithm distills the data type properties in D_i that caused the highest information gain. The algorithm does that by calculating the Singular Vector Decomposition (SVD) of the conditional probability matrix (CPM) of D_i and D_k or D_A. The singular vectors with the largest values indicate which subset of D_i and D_k or D_A interact the most with each other. The resulting data type properties are then added to the set of relatively sensitive data D_B.

Data diversity for a set of queries. We use the notion of joint entropy as a measure of data diversity. The diversity of a set of data depends on the number of homogeneous groups of data and the proportion of attributes in each group. The data set in our case is a set of submitted queries QS. Our desired metric shares some properties that Shannon sought in his measure of information uncertainty. First, if there are multiple possible options which are equally likely, there is more uncertainty. Thus, the smaller the entropy, the fewer the number of different queries or the more regular the queries are. Second, if a data set is defined as the combination of several disjoint data sets, the entropy for them combined should be at least the weighted sum of the individual entropy values for the individual sets. In our case, for a query set QS composed of query subsets QS_I and QS_{II} submitted in two phases, the overall entropy should be higher, or at least equal to the weighted entropy of the query sets involved. Formally:

$$H(QS) = H(x,y) + xH(QS_I) + yH(QS_{II}), \text{ where } x = \frac{QS_I}{QS}, y = \frac{QS_{II}}{QS}$$

We use the above formula to measure the change in diversity among a set of queries by determining the constant and varying attributes of QS assuming all queries are submitted by the same source. The attributes in our case are the purpose P_i and the set D_i. In each phase, some query attributes are expected to change abnormally. To measure this change we track the entropy for both query attributes. Formally, for a query set QS we calculate the entropy H for each group of homogeneous queries. We determine the homogeneity of a group of queries based on the following criteria:

Algorithm 3. Query Diversity Algorithm

1. input QS, t, M
2. output: *trigger*
3. $QS_{selected} \leftarrow \phi, QS_{previous} \leftarrow \phi, H_{phase} \leftarrow \phi, count \leftarrow 0, phase \leftarrow 1$
4. **while** $count < size(QS)$ **do**
5.   **for each** Q_k in QS such that $k \leq M$ **do**
6. $QS_{selected} \leftarrow QS_{selected} \cup Q_k$
7. $k \leftarrow k + 1$
8.   **end for**
9.   Calculate diversity maps $H_P, H_{PD}, H_{P|D}$
10.   Update phase diversity map H_{phase}
11.   Update $QS_{previous}$
12. $phase \leftarrow phase + 1$
13. $count \leftarrow count + M$
14. **end while**
15. **if** DIVERSITYCHANGEDETECTION$(H_{phase}) \geq t$ **then**
16. $trigger \leftarrow 1$
17. **end if**

- The purpose regardless of the data (P);
- The data given a purpose ($D|P$);
- Both attributes combined (PD).

We calculate the entropy for each criteria a as:

$$H_a(QS) = -\sum_{i=0}^{n} Pr(a_i) log_2 Pr(a_i), \text{ where } a = \{P, D|P, PD\}$$

The diversity detection part of the sensitive data detector provides an extra check for data sensitivity. The query diversity algorithm (Algorithm. 3) calculates the entropy for each criterion in a and creates a map of entropy values for each phase. It then uses the resulting phase diversity map to monitor the change in diversity between phases by comparing the change to a threshold t. The algorithm takes the query space QS, the threshold t, and the number of queries M to consider in each phase as inputs and returns a boolean value to the context inference algorithm (Algorithm. 1) indicating whether there is an attempt to breach sensitive data in the recently submitted query.

Dynamic Rule Check and Query Rewriting. After inferring the context, the Web service updates the context block each instance in the set of matching instances I in it's repository (Fig. 1). The inferred context is a combination of the classification C_j, the set D_B, and the diversity *trigger*. In addition to the context information, the query id Q_i, the data D_i, and purpose P_i are stored. Listing 1.1 demonstrates an instance with recently stored context.

Listing 1.1. An instance with recently stored context

```
<rdf:RDF xmlns:mc="http://www.michcare.com/michcare.owl#">
<mc:Patient rdf:about="patient1">
 <mc:hasGene>HLA-B<mc:hasGene>
 <mc:hasContext>
   <mc:hasQi>1</mc:hasQi>
   <mc:hasP>Research</mc:hasP>
   <mc:hasD>Gene,Location,Drug,Age</mc:hasD>
   <mc:hasDB>Gene,Age</mc:hasDB>
   <mc:hasC>malicious<mc:hasC>
<mc:hasContext>
</rdf>
```

The Web service then uses the set I and the query Q_j to perform *Dynamic Rule Check* and *Query Rewriting*. For each instance in I, the policy rules that govern each of the data type properties in D_j of that instance are checked to see if the purpose P_j of the query matches the purpose indicated in each rule. In the case of a purpose mismatch, the data type property will not be disclosed. However, if a rule permits the disclosure of a data type property d_k, further check is performed to investigate if there are contradictions between the permissions that a rule states regarding d_k and the previously stored contexts of the matching instances. First, if the query is classified as malicious, feature selection[1] is used to filter out the subset of data type properties D'_j in the query that resulted in such a classification. Second, a check is made to see if any of the data type properties d_k of D_i is included in a previously detected relatively sensitive data set D_B. Finally, the diversity trigger is checked to determine if there has been irregularity in the query sets due to the newly submitted query. If any data type property d_k requested in the query is either sensitive, relatively sensitive, resulted in a malicious classification, or caused irregularity in the query set, the query Q_j is rewritten[2] to exclude d_k and the new query Q'_j is run by the repository.

4   Evaluation

We implemented our context handler in Java using the Weka API [12] for the classifier component and JavaMI API [7] for the query diversity and relative sensitivity components. For relative sensitivity we used the Chi-Squared test to measure the significance of the mutual information between two sets of data type properties with an alpha level of 0.05. For query diversity, we chose an M value of 5. To build our classification model we needed a set of labeled queries. Since we did not have labeled queries we applied clustering to initial sets of queries. Since our query data consists mostly of binary attributes, and since we want to measure similarities between queries based on the '1' value of the query attributes then hierarchical clustering is most suitable in our case. To cluster the first set of

---

[1] We use feature selection techniques implemented in Weka.

[2] For query rewriting, we use the query rewriting technique proposed in [5].

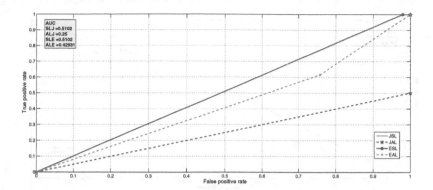

Fig. 2. ROC curves of different clusterers

queries we applied four configurations of Agglomerative Hierarchical Clustering as implemented in Matlab: Single linkage using Jaccard Coefficient, Average linkage using Jaccard Coefficient, Single linkage using Euclidean distance, and Average linkage using Euclidean distance. We also implemented five Web services (*GenomicWS, DrugWS, ClinicalWS, PharmaWS, and DemogWS*).

Data Sets. For the data sets we created RDF files using concepts from several BioMedical ontologies, including *NCBIGene,PharmKGB, DrugBank, CDT*, and *GeneCDS* available from [1]. We generated synthetic queries to simulate practical cases in which one data type property appears repeatedly in different queries. To generate a set of n queries, we first generated k core queries, each of which has m data type properties. Then, $n - k$ remaining queries were permuted from the k previous ones.

Clustering Evaluation. We used the query sets from the first three iterations to evaluate our clustering model. We first performed clustering based on the first set of queries (training set). For the validation stage we used the second set of queries (validation data) and we evaluated the clustering model based on previous clusters. We compared the robustness of several clustering algorithms using ROC by comparing a partition QS_c of the query set QS obtained by the clustering algorithm to a true partition QS_t labeled based on our knowledge of the queries. The knee of the ROC curve and the area under the curve AUC (Fig. 2) indicate that the single linkage using Euclidean distance gave us the best results. After evaluating the clustering model, we used another set of queries (test data) to perform real clustering based on unseen labels and we used the resulting clusters as labels to train our classification model.

Classifier Evaluation. We investigated two classification algorithms: Naive Bayes and KNN. We used the clustered queries (third labeled query set) to train our classifiers. The ROC curve in Fig. 3 indicate that the NB algorithm gave us better results than the KNN algorithm.

Results. The results of running the classification model on the query sets from the third, fourth, and fifth iterations indicate that the relative sensitivity results

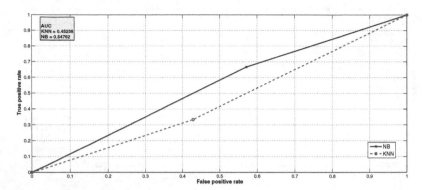

Fig. 3. ROC curves of both classifiers

agree with the classification results 60% of the time for the third iteration, 30% of the time for the fourth iteration, and 40% of the time for the fifth iteration. The classification results indicated 60% of the queries in the third iteration, 60% queries in the fourth iteration, and 80% of the queries in the fifth iteration to be malicious. The relative sensitivity results indicated that 80% of the queries in the third iteration contained a subset of data type properties that is relatively sensitive. Half of those queries contained data that was sensitive relative to D_A while the other half was sensitive relative to some other set D_k in some previously submitted query. In the fourth iteration, only 20% of queries contained relatively sensitive data, half of which where sensitive relative to D_A while the rest were relative to some other set D_k in a previously submitted query. 40% queries in the fifth iteration contained relatively sensitive data, most of which contained data sensitive relative to D_A.

Table 3 provides details of query diversity results for each phase of the third, fourth, and fifth iterations. The table shows both entropy values per phase and diversity changes between phases. We focus on the cases where the entropy values are 0.0 which suggest that all queries had the same value for an attribute and the cases where the entropy is 1.0 which suggest that the queries had equal number of each attribute value. For example, entropy values for the purpose attribute (**HP**) in the second phase of the third iteration match the relative sensitivity results for the last four queries of the third iteration. The diversity results indicate that in the third iteration, the purposes of the submitted queries were equally likely for phase one while in phase two all submitted queries were for the same purpose. The diversity in purposes increased until phase 4 when it remained around 2.0. For the purpose and data combined (**HPD**) the results did not indicate interesting entropy values or significant increase or decrease in diversity. A closer look at the entropy results for the data per purpose (**HPID**) suggest that for most of the purposes in the third iteration the entropy per phase is 0.0 and so is the diversity which indicate similarities among queries in the initial phases which match the results from the other components.

Table 3. Results from our diversity handler for three of the iterations

Iteration3					Iteration4					Iteration5				
Ph	HP	HPD	HDIP		Ph	HP	HPD	HDIP		Ph	HP	HPD	HDIP	
1	0.97	1.92	P1	1.58	3	1.5	2.32	P1	0.0	5	1.37	2.32	P1	1.6
			P2	0.0				P2	1.0				P2	0.0
			P3	0.0				P3	1.0				P3	0.0
2	0.0	0.72	P1	0.0	H	2.0	3.22	P1	1.14	H	2.08	4.1	P1	1.5
			P2	0.0				P2	1.53				P2	1.56
			P3	0.0				P3	0.13				P3	1.0
H	1.48	2.32	P1	0.47	4	1.37	2.32	P1	0.0	6	1.44	3.32	P1	1.6
			P2	0.86				P2	1.6				P2	0.0
			P3	0.0				P3	0.0				P3	0.0
					H	2.15	3.64	P1	1.1	H	2.08	3.28	P1	1.8
								P2	1.6				P2	2.0
								P3	1.0				P3	1.1

5   Related Work

We present some of the proposed approaches for privacy policy management.

Context-Based Privacy Policy Management. Most context-aware systems do not take the dynamicity at the rule level when they make their decisions [6,9,11,3]. Some of these approaches dynamically handle a user request by applying techniques that regulate rather than prevent the data access such as HDB [11,3]. The dynamic trust adjustment model proposed in [6] also dynamically handles context. Their approach focuses on access control, where the focus is mainly on who has access to the information as opposed to what actually have been collected. Also, their approach relies on inferring context using sensed spatial and temporal information and they do not achieve dynamicity at rule level. Among the relatively few researchers who took dynamicity of a context to a higher level by considering dynamicity of a rule is Pallapa et al. [19]. Pallapa et al. proposed a context aware scheme for privacy preservation by maintaining a model of the user's environment, which is characterized by user's *activities* and *situations*. Their solution accounts for fine grained rules and they apply a dynamic rule generator. However, both the rule and the context types are still predetermined based on a set of activities, states, and contexts in which the user could be. Also, these rules are not defined in semantic terms and do not govern what is potentially sensitive data. Our approach transparantly updates policy rules based on dynamically inferring a query's classification, what is considered *relatively* sensitive data, and diversity of queries.

Semantic-Based Privacy Policy Management. To our knowledge, few recent researchers used semantic concepts for defining privacy policies. For instance, Ferrini et al. [9] used XACML obligations to add axioms to an ontology using semantic functions, and they checked for inconsistencies introduced in the ontology due to adding those axioms. We use a similar approach to dynamically add contextual information to an instance to make the rules that govern the data type properties of that instance smarter. However, our approach adds the inferred context to the instance and uses the inferred context to impose more strict rules in the corresponding policy. Thus, both the ontology instance and the policy definition of that instance stay in sync. Among the approaches that proposed solutions for defining policies on top of domain ontologies are [8,20,4,22]. The work by Rahmouni et al. [20] stemmed from issues of diversity, complexity, and dynamicity of the rules governing privacy protection. They proposed a modeling approach to abstract rule complexities and facilitate the automation and enforcement of rules at the process level. The closest approach to ours is the one by Barhamji et al. [4]. While they ensure the dynamicity of a decision by rewriting a query, they still rely on predefined user preferences and they do not incorporate dynamically inferred context.

Privacy Policy Enforcement. Several technologies have been applied to achieve privacy policy enforcement by considering the requester's permission and the owner's consent [17,11,3]. Similar to our approach, those approaches do not rely on a third party for enforcement purposes. Grandison [11] and Agrawal [3] have both leveraged the Active Enforcement module of the Hippocratic Database technology (HDB) by transforming an original query to another query that is policy-compliant. They also track the purpose of a query to determine if a query is suspicious or not, but they detect that only after the fact.

Sensitive Data Detection. Some notable techniques that applied machine learning, data mining, or information theory for sensitive data detection include [2,21,15,13]. These techniques focus on data privacy from a "mining" view. Agrawal et al. [2] have done valuable work on privacy preserving data mining. In their work they defined conditional privacy using conditional entropy and information loss. We leverage similar data mining techniques to define our context and we incorporate the context into our privacy policy rule enforcement. However, their notion of conditional privacy compares the distribution of the original data to that of the perturbed data to test if the original value can be guessed from the perturbed value. In contrast, our approach uses only the original data by comparing the data that appears in a newly submitted query to both the previously determined set of sensitive data and all the previously submitted data sets to dynamically identify potential breach of more sensitive data. Nguyen and Choi [18] used Bayesian classification to classify Web services. Agrawal et al. have developed a classification model which is then sent to the user to rank search results to maintain their privacy. If a user generates a search request, the website returns to him complete search results along with a data mining model for ranking search results based on classification. The user's computer then uses the model to process the original data to return a classification which is then used to rank search results as a convenience for the user. Thus, private user data is not accessed by the website and is used only to produce the model then the model is sent back to the user's machine.

Several approaches have been proposed to achieve privacy by generalizing data to form more abstract information to avoid breaching sensitive data. Sweeny et al. [21] proposed the k-Anonymity approach. Some issues in their approach include utility loss and high dimensionality of anonymized data. To overcome some of those limitations, Wang et al. [23] proposed confidence bounding and Machanavajjhala et al.[15] proposed the l-diversity approach. Mohammad et al. [16] proposed an enhancement over the aforementioned approaches by introducing LKC-privacy, which is a generalization of both approaches with more reasonable constraints on parameters. Fung et al. [10] have recently developed an algorithm that uses LKC-privacy to preserve privacy of data mashup by multiple service providers.

Our approach for sensitive data detection is complementary to other approaches including the notion of cover in association rule mining to determine the set of representative association rules [13] and the notion of perfect privacy using query containment mapping [14]. Query containment have been proposed by Machanavajjhala et al. [14] for ensuring perfect privacy for relational data. Based on that, Barhamji et al. [4] extended the approach to data mashups. They used query containment mapping in their query rewriting approach but they applied it to RDF views.

6   Conclusion

We provided a formal model for privacy policy management that incorporates dynamic context handling into privacy policy rule enforcement. Our future work includes addressing other issues that service-oriented environments entail, incorporating other adversarial scenarios, enhancing our approach by exploring other data mining techniques, and implementing our approach using a privacy policy standard. Future work also includes extensive evaluation of the implementation to assess the performance and the accuracy of the approach. Also, we will incorporate security techniques and support them with soundness proofs. Based on our evaluation, we hope that future enhancements of the proposed model will serve as a foundation for health records infrastructures and inspire productive research in information sharing and management.

References

1. The bio2rdf project, http://s4.semanticscience.org/bio2rdf/
2. Agrawal, D., Aggarwal, C.C.: On the design and quantification of privacy preserving data mining algorithms. In: SIGMOD-SIGACT-SIGART Symposium on Principles of Database Systems (2001)
3. Agrawal, R., Johnson, C., et al.: Securing electronic health records without impeding the flow of information. International Journal of Medical Informatics (2007)
4. Barhamgi, M., Benslimane, D., Ghedira, C., Gancarski, A.L.: Privacy-preserving data mashup. In: International Conference on Advanced Information Networking and Applications (AINA) (2011)
5. Barhamgi, M., Benslimane, D., Medjahed, B.: A query rewriting approach for web service composition. Transactions on Services Computing (2010)
6. Bhatti, R., Bertino, E., Ghafoor, A.: A trust-based context-aware access control model for web-services. In: International Conference on Web Services (2004)

7. Brown, G., Pocock, A., Zhao, M.-J., Luján, M.: Conditional Likelihood Maximisation: A Unifying Framework for Information Theoretic Feature Selection. The Journal of Machine Learning Research (2012)
8. Christopoulou, E., Goumopoulos, C., Zaharakis, I., Kameas, A.: An ontology-based conceptual model for composing context-aware applications. Research Academic Computer Technology Institute (2004)
9. Ferrini, R., Bertino, E.: Supporting rbac with xacml+ owl. In: Symposium on Access Control Models and Technologies (2009)
10. Fung, B., Trojer, T., Hung, P.C., Xiong, L., Al-Hussaeni, K., Dssouli, R.: Service-oriented architecture for high-dimensional private data mashup. IEEE Transactions on Services Computing (2012)
11. Grandison, T., Ganta, S.R., Braun, U., Kaufman, J., et al.: Protecting privacy while sharing medical data between regional healthcare entities. Studies in Health Technology and Informatics (2007)
12. Hall, M., Frank, E., Holmes, G., Pfahringer, B., Reutemann, P., Witten, I.H.: The weka data mining software: An update (2009)
13. Kasthuri, S., Meyyappan, T.: Detection of sensitive items in market basket database using association rule mining for privacy preserving. In: International Conference on Pattern Recognition, Informatics and Medical Engineering (PRIME) (2013)
14. Machanavajjhala, A., Gehrke, J.: On the efficiency of checking perfect privacy. In: SIGMOD-SIGACT-SIGART Symposium on Principles of Database Systems (2006)
15. Machanavajjhala, A., Kifer, D., Gehrke, J., Venkitasubramaniam, M.: l-diversity: Privacy beyond k-anonymity. Transactions on Knowledge Discovery from Data (TKDD) (2007)
16. Mohammed, N., Fung, B., Hung, P.C., Lee, C.-K.: Anonymizing healthcare data: a case study on the blood transfusion service. In: Proceedings of the 15th ACM SIGKDD International Conference on Knowledge Discovery and Data Mining (2009)
17. Mont, M.C., Thyne, R.: Privacy policy enforcement in enterprises with identity management solutions. In: International Conference on Privacy, Security and Trust: Bridge the Gap Between PST Technologies and Business Services (2006)
18. Nguyen, H.-V., Choi, Y.: Proactive detection of ddos attacks utilizing k-nn classifier in an anti-ddos framework. International Journal of Electrical, Computer, and Systems Engineering (2010)
19. Pallapa, G., Di Francescoy, M., Das, S.K.: Adaptive and context-aware privacy preservation schemes exploiting user interactions in pervasive environments. In: International Symposium on a World of Wireless, Mobile and Multimedia Networks (WoWMoM) (2012)
20. Rahmouni, H.B., Solomonides, T., Mont, M.C., Shiu, S.: Privacy compliance in european healthgrid domains: An ontology-based approach. In: International Symposium on Computer-Based Medical Systems, CBMS (2009)
21. Sweeney, L.: Achieving k-anonymity privacy protection using generalization and suppression. International Journal of Uncertainty, Fuzziness and Knowledge-Based Systems (2002)
22. Villata, S., Costabello, L., Delaforge, N., Gandon, F.: A social semantic web access control model. Journal on Data Semantics (2013)
23. Wang, K., Fung, B.C., Philip, S.Y.: Handicapping attacker's confidence: an alternative to k-anonymization. Knowledge and Information Systems (2007)

Ephemeral UUID for Protecting User Privacy in Mobile Advertisements

Keisuke Takemori, Toshiki Matsui, Hideaki Kawabata, and Ayumu Kubota

KDDI R&D Laboratories Inc., Security Department, Japan
(takemori,to-matsui,hi-kawabata,kubota)@kddilabs.jp

Abstract. Recently, a universally unique identifier (UUID) for targeting ad services is supported by a smartphone OS, which can be reset and/or halted by a user. However, when a user resets the UUID, all targeted ad libraries are initialized. It means that the user cannot control ad libraries one at a time. As the user interests managed by the same UUID are easily exchanged between the ad service provider and another ad service provider, the user is anxious about the privacy violation. In addition, as the UUID can be tapped on the unencrypted network, the replay attack using the tapped UUID violates the user's privacy. In this paper, we propose a privacy enhanced UUID that is generated by each ad service provider. As the UUID is encrypted with the time information by each access, the value of *"Enc(Time, UUID)"* is changed frequently. Thus, we call it an ephemeral UUID. The ephemeral UUID is shared through the same ad libraries in any applications, but it cannot be shared through the other ad libraries. Our UUID can be reset by the user and cannot be extracted on the network.

Keywords: Mobile Ad., UUID, Right to be Forgotten, Do Not Track.

1 Introduction

About 60% of applications in a smartphone include ad libraries [1]. In the case of a targeted ad library [2-6], an ID is sent and used for a management key of web-view history, which tracks a user's interests in order to select ad contents. When a user clicks the ad contents displayed in the application, the ad service provider rewards the application developer. Hence, many applications that include ad libraries are provided for free. Figure 1 shows the ecosystem of the mobile ad business [7].

Step 1) An application developer acquires an ad library from an ad service provider and bundles it in his applications.

Step 2) Applications including ad libraries are uploaded to an application market.

Step 3) A user downloads the application to her smartphone.

Step 4) When the application is executed, the ad library retrieves the ID from the smartphone and sends it to the ad service provider.

Step 5) The ad service provider profiles the user's interests by tracking the ad view history and sends back recommended ad contents to the ad library.

Step 6) According to the click counts of the ad contents, the ad service provider rewards the application developer.

H. Decker et al. (Eds.): DEXA 2014, Part II, LNCS 8645, pp. 263–272, 2014.
© Springer International Publishing Switzerland 2014

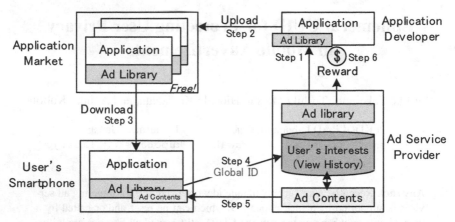

Fig. 1. Ecosystem of Mobile Applications Including Ad Libraries

As the ad libraries are bundled in different applications that are isolated by the executing sandbox, it is difficult for a third party cookie to be exchanged through applications. In the case of Android applications, some ad libraries use device IDs, i.e., international mobile equipment identity (IMEI), media access control (MAC) address, international mobile subscriber identity (IMSI), SIM serial number (ICCID), and telephone number, for user tracking, which can be retrieved by calling the standard API. As the device IDs cannot be written by the user, "Right to be Forgotten" that is regulated by the European Commission [8, 9] will be violated. According to the sandbox, it is difficult to share an opt-out status through the ad libraries in any applications. The "Do Not Track" that is regulated by the Federal Trade Commission (FTC) [10-12] will be violated.

Recently, the advertising.Identifier [13] and the Advertising ID [14] are supported by the iOS and the Android OS for user tracking ID, which can be reset and halted by a user. As they are generated as a common UUID for all ad libraries in a smartphone, both "Right to be Forgotten" and "Do Not Track" can be achieved.

However, when a user resets the OS-supported UUID, all targeted ad libraries are initialized. It means that when the user favors some ad libraries but has concerns about another ad library, she cannot control ad libraries one at a time. As it is easy to exchange the user interests between the ad service provider and another ad service provider using the common UUID, the user is anxious about the unfavorable data exchange [15]. As the common UUID can be tapped on the unencrypted network, the replay attack using the tapped common UUID violates the user's privacy.

In this paper, we propose a privacy enhanced UUID that is issued by each ad server. The UUID is shared through the same ad libraries in any applications via a shared memory, but it cannot be shared through the ad library of another ad service provider as the following mechanism. The ad server encrypts the UUID with the time information *"Enc(Time, UUID)"* and issues it to the ad library in the smartphone. As the ad server decrypts and encrypts the UUID with the time information on every access, the value of *"Enc(Time, UUID)"* is changed frequently. We call it an ephemeral UUID. In the case of the ad library of another ad service provider, as the value of the ephemeral UUID is changeable, the ephemeral UUID in the shared memory cannot be used for the user tracking ID. In addition, the replay attack by the wiretapping is suppressed. When the ephemeral UUID is removed, a new UUID that is not related to the removed UUID is issued. The data of user interests in the ad server, which are managed by the removed UUID, will not be used for the targeted ad service for its user. As the ephemeral UUID is generated by each ad service provider, the user can control the opt-out status one at a time.

The rest of this paper is organized as follows. Section 2 reviews a third party cookie in a web browser. Section 3 shows problems regarding ad libraries in mobile applications. Section 4 proposes the ephemeral UUID. Section 5 evaluates the latencies and Section 6 concludes the paper.

2   Current Ad-Tracking Mechanisms

2.1   Cookie for Web Browser

In the case of ad contents for a web browser, a third party cookie is issued as a user-tracking ID by an ad server to an ad content in the web browser. As the cookie is a randomly generated number, it is not directly related to a user or a device.

The cookie is saved in the browser's folder. When the cookie is removed by the user, a new cookie that is not related to the removed cookie is issued. The data of user interests in the ad server, which are managed by the removed cookie, will not be used for the targeted ad service for its user. The user data managed by the removed cookie are considered to be forgotten. In addition, as the cookie is not designed for a sharing ID between the issuer and the other websites, it is sent in reply only to an issuer website by the web browser.

2.2   OS-Supported UUID

The ad library for mobile applications is implemented with a webview component that are isolated by the application sandbox. In the cases of iOS and Android OS, the UUIDs are provided for ad libraries in the different applications, which can be implemented with the following APIs [13], [14]

-    iOS; ASIdentifierManager.advertisingIdentifier.

-    Android OS; AdvertisingIdClient.Info getAdvertisingIdInfo.

When the ad library calls the API, a common UUID for all targeted ad libraries is issued by the OS. A user can reset the UUID and can set the "Do Not Track" flag via the ad-setting interface.

3   Problems of Ad Library

3.1   Threats of User Data Exchange

Both the iOS and the Android OS generate and support a common UUID for user tracking [13, 14]. It is easy to exchange the user interests between the ad service provider and another ad service provider, when ad service providers use the OS-supported UUID for user tracking. The user is anxious about the unfavorable data exchange.

3.2   Replay Attack by Wiretapping

We investigated how the user data are sent via the 50 ad libraries in February 2014. 78% (39/50) of ad libraries sent user data via an unencrypted network. As the UUID can be extracted from unencrypted packets, some users feel anxious about the replay attack using the tapped UUID.

3.3   Permanent ID

Some ad libraries track user interests using the device IDs. Because the user cannot reset the device IDs, user interests managed by the device IDs in the ad server will not be invalid. Some ad service providers issue the device binding IDs that are generated from the device fingerprints, i.e., CPU name, language, time-zone, and display size [17]. As most fingerprinting parameters are not replaced, the device binding ID is considered as a permanent ID.

3.4   Lack of Privacy Policy

It is important to notify users of the collected data on the ad servers. Japan's Ministry of Internal Affairs and Communications (MIC) suggests that when the collected data of user interests increases, the possibility of personal identification increases [1, 18]. Therefore, a privacy policy for an ad library should explain "what data are collected," "how the data are used," and "who uses the data."

We investigated 50 ad libraries for Android OS in February 2014. The applications were executed to check the data that is sent via the ad libraries. Also, we reviewed the privacy policy for each ad library. The results of our investigation are shown in Tables 1 and 2. According to Table 1, 70% (35/50) of ad libraries sent the device IDs. In contrast, 30% (15/50) of ad libraries sent no IDs, the OS-supported UUID, the ad provider's UUID, or OS fingerprints. 52% (26/50) of ad libraries notify users of "what data are sent" correctly via a privacy policy. It means that 48% (24/50) of ad libraries do not have a privacy policy or notify users incorrectly. According to Table 2, 54% (27/50), 30% (15/50), 10% (5/50), and 2% (1/50) of ad libraries send Android ID, IMEI MAC address, and telephone number for the management keys of user interest. 14% (7/50) of ad libraries provide the location-based ad service.

Table 1. "What Data are Sent" via Ad Libraries and Consistency with Privacy Policy

Investigation date	Feb/2014
Number of investigated ad libraries	50
Sending device ID*	70%(35/50)
Notify via the privacy policy for ad library	86%(43/50)
Correct explanation of sending information	52%(26/50)

Device ID means IMEI, MAC address, IMSI, ICCID, Telephone number, Android ID, and their hash values.

Table 2. Sending Data via Ad Libraries

Android ID	IMEI	MAC address	IMSI	ICCID	Telephone number	Location	Address List
54%	30%	10%	0%	0%	2%	14%	0%
(27/50)	(15/50)	(5/50)	(0/50)	(0/50)	(1/50)	(7/50)	(0/50)

4 Proposal

In this section, we propose a privacy enhanced UUID that is generated by each ad service providers. In Subsection 4.1, we consider the sharing mechanism of the UUID through the ad library in any applications. In Subsections 4.2, we propose safe sharing schemes of the UUID, whereby only the original ad library can use the UUID for user identification.

4.1 How to Share the UUID and Opt-Out Flag

Recently, most smartphones for iOS and Android OS implement internal and/or external SD cards. The SD cards can be used as a shared memory for applications. We consider how to share both the UUID and the opt-out flag through the ad library in any applications. Figure 2 shows the sharing scheme using the shared memory in the smartphone.

Step 1) The ad server issues a UUID when the ad library A accesses without a UUID.
Step 2) The UUID is received by the ad library A.
Step 3) The UUID is written into the shared memory by ad library A.
Step 4) When another application including ad library A is executed, the UUID is read by ad library A from the shared memory.
Step 5) The UUID is sent in reply to the ad server from ad library A. The ad server replies the ad contents of the user's interests to ad library A in another application.
Step a) The user can remove the UUID in the shared memory via ad libraries A. Also, an opt-out flag, which controls the send/stop of the UUID to the ad server, can be set by the user.

In the case of the web browser, a cookie is managed by the access control in the browser's folder. In contrast, as the UUID in the shared memory can be accessed by any ad libraries, the protection mechanisms for the UUID should be implemented against the malicious reuse of another ad library.

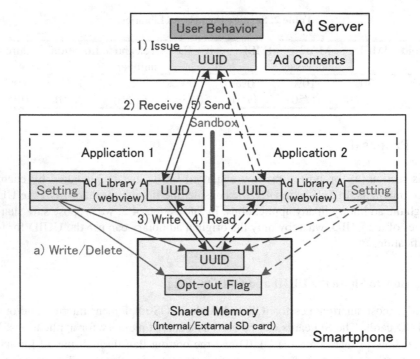

Fig. 2. Sharing Scheme of a UUID and Opt-out Flag Using Shared Memory

4.2   Proposal of an Ephemeral UUID

In this subsection, we propose a privacy enhanced UUID, which encrypts the UUID with current time at every access at the ad server shown in Figure 3. As the value of encrypted UUID changes each access, it is called the "ephemeral UUID". The ephemeral UUID cannot be used for user-tracking ID by another ad library in a smartphone, because the value of the ephemeral UUID is changed frequently. Only the original issuer decrypts the encrypted UUID and extracts the UUID.

Step 1) The ad server issues an UUID when ad library A in an application accesses the ad server without an ephemeral UUID. The ad server adds the current time to the UUID and encrypts it as *"Enc(Time, UUID)."*
Steps 2 - 5) These steps for the ephemeral UUID are the same as in Subsection 4.1.
Step 6) The ad server decrypts the ephemeral UUID and extracts the UUID. The ad server sends both the ad contents of the user's interests and a new ephemeral UUID into which is inserted the new current time to ad library A.
Step a) This step is the same as in Subsection 4.1.

The proposed scheme achieves UUID sharing through the ad libraries in any application via the shared memory. When the user removes the ephemeral UUID in the shared memory, the ad library accesses the server without the ephemeral UUID. Then a new UUID is issued to the ad library. The data of user interests managed by the removed UUID in the ad server are forgotten.

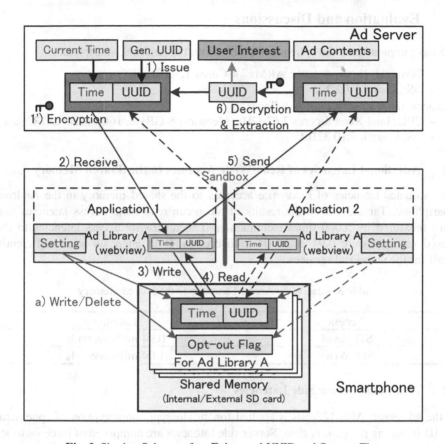

Fig. 3. Sharing Scheme of an Ephemeral UUID and Opt-out Flag

The ad library can set and share an opt-out flag via the shared memory, which controls send/stop of the ephemeral UUID for ad libraries A in any application.

As the ephemeral UUID is saved in the shared memory, any application can read the ephemeral UUID. However, the ephemeral UUID cannot be used for targeting ID for another ad library, because the ephemeral UUID changes at every access to the ad server.

The ad service provider needs to implement the encryption/decryption mechanisms for the ephemeral UUID. As it can be implemented using *"get_current_time"* and standard encryption functions, it is easy to modify the current cookie-based ad server. In addition, the ad library in the application needs to implement management functions for the ephemeral UUID and opt-out flag, which reads/writes the ephemeral UUID/opt-out flag in the shared memory. It is also easy to implement it in the ad library using standard APIs.

5   Evaluation and Discussions

We have implemented the ephemeral UUID using the Android smartphone.

- Terminal: Nexus 4, CPU: ARMv7 4-Cores 1.5GHz, and
- OS: Android 4.2.2

Specs of the ad server are as follows:

- CPU: Intel Xeon 8-cores 2.67-GHz, Memory: 8-GByte, 1600-MHz DDR3, and
- OS: Ubuntu 9.10 32Bit.

5.1   Additional Latencies of Read/Write Accesses to the Shared Memory

We evaluated latencies of read/write accesses to the shared memory in the Android smartphone. Table 3 shows the results of the mean read/write access latencies that were measured by 5,000 times examination. The read/write access latencies to the shared memory were 0.94 and 1.04 milliseconds. Both latencies were sufficiently small for the mobile ad services.

Table 3. Read/Write Access Latencies to the Shared Memory

Access	Latency
SD_Read	0.94 milli-seconds
SD_Write	1.04 milli-seconds

5.2   Additional Server Side Latencies

In the ad server, AES 128 bits is applied for the decryption/encryption of ephemeral UUID including the current time. Server side latencies are composed of three factors:

- Decryption of the ephemeral UUID
- Retrieval of UUID in the SQL-DB, and
- Encryption of the ephemeral UUID with current time.

Here, the latency of the retrieval of UUID in the SQL-DB is the same as the retrieval of the cookie-based ad system. Then, we measured only additional latencies of decryption and encryption processes for the ephemeral UUID. Table 4 shows the result of the mean additional latency that was measured by 5,000 times examination. The total latency of both the decryption and the encryption was sufficiently small.

Table 4. Decryption and Encryption Latencies in the ad server

Processing	Latency
Dec(Eph.UUID) + Enc(Time, UUID)	4.76 micro-seconds

5.3   Comparison with Commercial Ad Libraries

We implemented six test applications. Each application had a different ad library. Only our library had a sharing scheme of the ephemeral UUID. The other five ad libraries use the OS-supported UUID or device IDs for user tracking. We measured the total latency that included from execution of the test application to display of the ad contents. This was just a user experience delay.

Table 5 shows the mean latencies that were measured by 10 times examinations. The latencies of the other libraries including our library are about 3 seconds. As a result, the sharing scheme of the ephemeral UUID via the shared memory achieved reasonable latency for ad libraries in a mobile application.

Table 5. Latencies of Ad Libraries

Ad Library	Latency
AdLantis [3]	2.08 seconds
i-mobile [4]	3.13 seconds
Nend [5]	3.34 seconds
Our ad Library	3.39 seconds
Admob [6]	3.88 seconds

5.4   Opt-Out Mechanism

Recently, ad libraries can use the OS-supported UUID. When a user reset the OS-supported UUID, the user interests on all of the ad service providers are forgotten. Also, a "Do Not Track" flag is effected for all ad libraries on the smartphone.

In contrast, the proposed sharing scheme of the ephemeral UUID can achieve both the reset and the opt-out one at a time. In addition our scheme suppresses the privacy data exchange. It is the same mechanism as the web browser. In addition, if the user would like to reset and/or to opt-out of all of the ad libraries, it is easy to achieve as follows. Main management folders for ephemeral UUIDs and opt-out flags are located as common directories, i.e., /sd_card/uuid/, /sd_card/opt-out/. Then, a sub-folder is implemented by each ad library in the common directories. When the user would like to reset and/or to opt-out of all of the ad libraries, the common directory "/sd_card/uuid/" is removed and/or all opt-out flags in the sub-folders are set.

6   Conclusion

Both "Right to be Forgotten" and "Do Not Track" are very important for mobile ad libraries because the smartphone is considered to be a privacy-sensitive terminal. Recently, a UUID for targeting ad services is supported by the iOS and the Android OS, which can be reset and/or halted. However, when a user resets the UUID, all targeted ad libraries are initialized. In addition, as the user interests managed by the OS-supported UUID are easily exchanged between the ad service provider and another ad service provider, the threat of privacy violation will increase as long as there are no resets of the OS-supported UUID.

In this paper, we proposed a sharing scheme of the ephemeral UUID for ad libraries in any applications. As the ephemeral UUID does not depend on the user and can be reset, our scheme achieves "Right to be Forgotten" for each ad library. In addition, as the opt-out status can be controlled, "Do Not Track" is satisfied for each ad library. As the ephemeral UUID can be decrypted on only the issuer ad server, both the malicious reuse of another ad library and the wiretapping are suppressed. Furthermore, as our schemes are designed simply, they can easily be applied to current ad commercial services. In fact, our scheme has been applied to a commercial ad service for 10 million users (UUIDs) in the world, and reasonable load was achieved for the ad server and the smartphone.

References

1. MIC, Japan: Smartphone Privacy Initiative (2012), `http://www.soumu.go.jp/main_sosiki/joho_tsusin/eng/presentation/pdf/Initiative.pdf`
2. Stevens, R., Gibler, C., Crussell, J., Erickson, J., Chen, H.: Investigating User Privacy in Android Ad Libraries. IEEE Mobile Security Technologies (2012)
3. AdLantis, `http://sp.www.adlantis.jp/`
4. i-mobile, `http://i-mobile.co.jp/en/index.aspx`
5. Nend, `http://nend.net/`
6. Admob by Google, `http://www.google.com/ads/admob/`
7. Farahat, A., Sunnyvale, Bailey, M.C.: How effective is targeted advertising? In: 21st International Conference on World Wide Web, pp. 111–120 (2012)
8. European Commission: A comprehensive approach on personal data protection in the European Union: Right to be Forgotten (2010), `http://ec.europa.eu/justice/news/consulting_public/0006/com_2010_609_en.pdf`
9. European Commission's data protection reform (2013), `http://ec.europa.eu/commission_2010-2014/reding/pdf/news/20131022-libe-data-protection_en.pdf`
10. Federal Trade Commission (FTC): Endorses 'Do Not Track' to facilitate consumer choice about online tracking (2010), `http://www.ftc.gov/opa/2010/12/privacyreport.shtm`
11. FTC Staff Report: Self-Regulatory Principles For Online Behavioral Advertising (2009), `http://www.ftc.gov/os/2009/02/P085400behavadreport.pdf`
12. Do Not Track: Implementations, `http://donottrack.us/implementations`
13. iOS Developer Library, ASIdentifierManager Class Reference, `http://developer.apple.com/library/ios/#documentation/AdSupport/Reference/ASIdentifierManager_Ref/ASIdentifierManagerhtml#//apple_ref/doc/uid/TP40012654`
14. Android advertising ID, `http://developer.android.com/google/play-services/id.html`
15. Grace, M.C., Zhou, W., Jiang, X., Sadeghi, A.-R.: Unsafe exposure analysis of mobile in-app advertisements. In: ACM Conference on Security and Privacy in Wireless and Mobile Network, pp. 101–112 (2012)
16. Wireshark, `http://www.wireshark.org/`
17. AdTruth, Device Fingerprinting, `http://www.adtruth.com/buzz/press/cookie-less-tracking-device-fingerprinting-vs-device-identification-technology`
18. MIC, Japan: Smartphone Privacy Initiative II (2013), `http://www.soumu.go.jp/main_sosiki/joho_tsusin/eng/presentation/pdf/Summary_II.pdf`

Voronoi-Based Spatial Cloaking Algorithm over Road Network

Xiao Pan[1], Lei Wu[1], Zhaojun Hu[1], and Zheng Huo[2]

[1] Shijiazhuang Tiedao University, Shijiazhuang, 050043, China
[2] Hebei University of Economics and Business, Shijiazhuang, 050043, China

Abstract. Privacy preservation has recently received considerable attention in location-based services. A large number of location cloaking algorithms have been proposed for protecting the location privacy of mobile users over road networks. However, most existing cloaking approaches traverse the entire road network repetitively, which has a significant negative impact over the communication and anonymization efficiency. To address this problem, we propose a network Voronoi-based location anonymization algorithm. We employ the network Voronoi diagram to pre-divide the road network into network Voronoi cells. The problem of anonymization over global road networks is reduced to finding cloaking sets in local network Voronoi cells. Experimental results validate the efficiency and effectiveness of the proposed algorithm.

Keywords: Location privacy protection, road network, network Voronoi diagram.

1 Introduction

With advances in wireless communication and mobile positioning technologies, location-based services (LBSs) have seen wide-spread adoption. Academia and industry have pay attention to location privacy preserving in LBSs for about ten years. All research efforts have been put into investigating how to protect the location privacy of mobile users, while still ensure services with high quality. However, most of the existing work only applies to Euclidean space, with the assumption of users' free movement. Obviously, this assumption is not practical. In real life, mobile users are restricted to move on a constrained network (*e.g.*,users walk in city streets, and cars run on a road network). Ignoring the constrained network would lead to user privacy leak.

Road segments [1,5] are the most popular shapes used as the cloaking areas on road networks. With the set of road segments as the spatial area, most existing cloaking algorithms are based on the global graph transversal [5,1,8]. Every time a user issues a query, then the global network data would be retrieved from the disk even if a small part of them is needed. As we known, the whole road network data is very large. The global road network traversal has a significant negative impact over the communication and anonymization efficiency.

H. Decker et al. (Eds.): DEXA 2014, Part II, LNCS 8645, pp. 273–280, 2014.

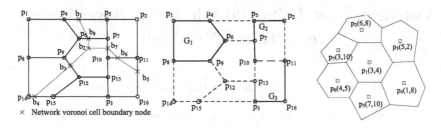

Fig. 1. Network Voronoi diagram   **Fig. 2.** r-path neighbors   **Fig. 3.** A sample of NVD

In order to resolve the problem, we proposed a network Voronoi-based location anonymization algorithm (NVLA) on road networks. As we known, the road network topology is usually static. Thus, we employ the network Voronoi diagram (NVD) to divide the road network into several network Voronoi cells (NVCs). Then, we reduce the problem of anonymization over the global road network into finding cloaking sets in local NVCs. The pre-dividing idea is benefit for improving the cloaking efficiency. For example, Fig. 1 is an example of the NVD for a road network, using points of interests (POIs) $\{p_1, p_2, p_3\}$ as the generators[1]. If the NVC of p_1 NVC(p_1) satisfies mobile users' privacy requirements, NVC(p_1) is used as the cloaking area directly without the global road network traversal.

The remainder of this paper is organized as follows. The problem under study is formally defined in Section 2. Section 3 presents our proposed NVLA algorithm. Section 4 shows the performance evaluation results. Finally, the paper is concluded in Section 5.

2 Preliminary

Like most existing work [1,5], we employ a centralized system architecture, which consists of mobile users, a trusted anonymizing proxy (TAP), and un-trusted LBS servers. The TAP is responsible for storing the NVD, finding cloaking set, and the genuine result refinement. Our work belongs to the anonymizaton algorithm for finding cloaking sets.

2.1 Road Network and Network Voronoi Diagram

Definition 1. *(Road network) We define a road network as a un-directional graph $G(V, E)$, where*

- *V is a vertices set, representing three cases: intersections ($d(v) \geq 3$), terminal points($0 < d(v) \leq 2$), and POIs, where $d(v)$ is the vertex degree.*
- *E is a edges set, representing road segments. Each edge $e \in E$ is a pair $< v_i, v_j >$, where $v_i, v_j \in V$. We denote a point p on edge e as $p \in e$.*

---

[1] p_4 to p_{13} in the Fig. 1 are the intersections of the road network.

A NVD [2] is defined for graphs and is a specialization of Voronoi diagrams where the location of objects is restricted to the edges that connect the vertices of the graph. Before giving the definition of NVD, we give some relevant definitions first. Given a graph $G(V, E)$, nodes set $V = \{p_1, \ldots, p_n, p_{n+1}, \ldots, p_o\}$, where the first n nodes represent the Voronoi generators, the number of nodes in V is o, and the number of edges in E is m. Then the set dominance region and border points are defined as follows [2].

Definition 2. *(Set dominance region) Let I be the set of positive integers. $\forall j \in I$ and $j \neq i$, p_i, $p_j \in V$, the set dominance region of p_i over p_j is*

$$Dom(p_i, p_j) = \{p | p \in \bigcup_{o=1}^{m} e_o, d_n(p, p_i) \leq d_n(p, p_j)\}$$

Definition 3. *(Border points) The border points between p_i and p_j*

$$b(p_i, p_j) = \{p | p \in \bigcup_{o=1}^{m} e_o, d(p, p_i) = d(p, p_j)\}$$

$b(p_i, p_j)$ *is called border points between p_i and p_j. Its physical meaning is that all points on all edges in E that are equally distanced from p_i and p_j.*

Definition 4. *(Voronoi edge set) The Voronoi edge set associated with p_i*

$$V_{edge}(p_i) = \bigcap_{j \in I_n \wedge j \neq i} Dom(p_i, p_j)$$

The Voronoi edge set specifies all the points on all the edges in E that are closer to p_i than any other generator point in E.

Based on Definition 4, the NVD is defined as follows.

Definition 5. *(Network Voronoi diagram, NVD) The network Voronoi diagram are defined as*

$$NVD(P) = \{V_{edge}(p_1), \ldots, V_{edge}(p_n)\}$$

where $P = \{p_1, \ldots, p_n\}$ is the generators set. The elements of NVD are collectively exhaustive and mutually exclusive except for their border points. By properly connecting adjacent border points of a generator p_i to each other without crossing any of the edges on road network, we can generate a bounding polygon cell, called network Voronoi cell, we term $NVC(p_i)$, for that generator.

Obviously, the vertices and edges in NVCs constitute a sub-graph of the road network. We define two subgraph as r-path neighbors as follow.

Definition 6. *(r-path neighbors) Let $G_1(V_1, E_1)$, $G_2(V_2, E_2)$ be two sub-graphs of a graph $G(V, E)$, where $V_1 \subset V$, $V_2 \subset V$, $E_1 \subset E$, and $E_2 \subset E$. A path from v_s to v_d is denoted as $path_{v_s \_ v_d}$, and the path length is denoted as $length(path_{v_s \_ v_d})$. If*

$$\min_{\forall v_s \in V_1, \forall v_d \in V_2, v_s \neq v_d} length(path_{v_s \_ v_d}) = r$$

We call these two sub-graphs r-path neighbors.

For example, in Fig. 1, the set of POIs $\{p_1, p_2, p_3\}$ is the generators set. The road network is divided into three NVCs, NVD(P)=$\{V_{edge}(p_1), V_{edge}(p_2),$ $V_{edge}(p_3)\}$. Fig. 2 shows three sub-graphs of the graph in Fig. 1, that is G_1, G_2, G_3. Edges in the sub-graphs are solid lines, and the dash lines are the original edges in G but not in any sub-graph. G_1 and G_2 are 1-path neighbors through path (p_4, p_5), and G_1 and G_3 are 2-path neighbors through path (p_9, p_{12}, p_{13}).

2.2   Privacy Model

To protect the location privacy in road networks, we adapt location k-anonymity and cloaking granularity as privacy metrics. Thus, we define a (K, L)-location privacy model as follows.

Definition 7. ((K, L)-*location privacy model*) *Let* CS *be the users set and* RS *be the set of road segments, if the following conditions are satisfied*

- $|CS| \geq K$, *where* $|CS|$ *is the number of users in* CS.
- $|RS| \geq L$, *where* $|RS|$ *is the number of road segments in* RS.
- $\forall u \in CS, \exists e_u \in RS, u$ *lies on* e_u.
- *The users in* CS *all report* RS *as the cloaked spatial area.*

we say users in CS *satisfy* (K, L)-*location privacy model.* CS *is termed as the cloaking set, and* RS *is the cloaked segments.*

According to the (K, L)-location privacy model, we classify the NVCs in a NVD into *unsafe cells, safe-medium cells, safe-large cells*.

Definition 8. (*unsafe cells, safe-medium cells, safe-large cells*) *Let* vc *be a NVC and num_edge (num_user) be the number of edges (users) in it. If* $num\_edge < L$ *or* $num\_user < K$, vc *is termed as an* unsafe cell; *if* $L \leq num\_edge < 2L$ *and* $num\_user \geq K$, vc *is a* safe-medium cell; *if* $num\_edge \geq 2L$ *and* $num\_user \geq K$, vc *is a* safe-large cell.

Fig. 3 shows a sample of the NVD. The first number in the parenthesis is the number of users in the cell, and the second one is the number of segments. If K=3 and L=5, NVC(p_7) is a safe-large cell, NVC(p_6) is a safe-medium cell, and NVC(p_3) is an unsafe cell .

3   NVLA:Network Voronoi-Based Location Anonymization Algorithm on Road Network

3.1   Main Idea of NVLA

We apply the parallel Dijkstra algorithm [2] to generate a first order NVD offline. POIs are the generators. An adjacency list is used to store the NVD. If two NVCs share the same border points, the two ones are neighbor cells. In order to locate the cell which the query user is in, we approximate each cell by a minimum bounding rectangle (MBR). Then, the MBRs are indexed by an R-tree. If a query locates at the boundary of two cells, the cell which the query belongs to is selected randomly.

Recall that we divide NVCs are into safe and unsafe cells. For a safe cell, cloaking sets can be found directly (see section 3.2). Now the problem is how to transform an unsafe cell to a safe one. The main idea of NVLA algorithm is as follows. Let num_user and num_edge be the number of users and number of segments in a NVC. Put all NVCs in a min-heap st with num_user as key. Pop the top cell $topc$. If $topc$ is unsafe, merge $topc$ with a neighbor cell in the NVD randomly. Otherwise, cloaking sets are found from $topc$. Then pop the new top cell, and repeat above steps until st becomes empty.

Taking NVCs in Fig. 3 as a running example, $st=\{\text{NVC}(p_4), \text{NVC}(p_1), \text{NVC}(p_7), \text{NVC}(p_6), \text{NVC}(p_3), \text{NVC}(p_2), \text{NVC}(p_5)\}$. Suppose that $K=3$ and $L=5$. Now $topc$ is $\text{NVC}(p_4)$, which violates (3, 5)-location privacy model. The neighbor cells of $\text{NVC}(p_4)$ are $\{\text{NVC}(p_1), \text{NVC}(p_3), \text{NVC}(p_5)\}$. Assume $\text{NVC}(p_1)$ is selected. Merge $\text{NVC}(p_1)$ and $\text{NVC}(p_4)$ to a new cell $\text{NVC}(p_1p_4)$ with 4 users. Now $st=\{\text{NVC}(p_7), \text{NVC}(p_6), \text{NVC}(p_1p_4), \text{NVC}(p_3), \text{NVC}(p_2), \text{NVC}(p_5)\}$. New $topc$ $\text{NVC}(p_7)$ is safe, and the density-aware cloaking algorithm in safe cells is applied to generate cloaking sets.

3.2   Density-Aware Cloaking in Safe Cells

Recall that we divide safe cells into safe-medium cells and safe-large cells. For a safe-medium cell, users in it constitute a cloaking set, and the segments are used as the cloaked segments. Now the main problem is how to find cloaking sets in a safe-large cell. The solution depends on the user density.

On the basis of the user density, safe-large cells are further divided into *sparse* and *dense* ones. We evaluate the user density by $\frac{num\_user}{num\_edge}$.

Definition 9. *(sparse cells, dense cells) For a NVC vc, if $\frac{num\_user}{num\_edge} \leq \delta$, we term vc as a sparse cell; otherwise vc is a dense cell, where δ is a system threshold.*

User-Based Cloaking Algorithm. In sparse cells, there exist many empty edges, that is edges without users lying on. In order to scan empty edges scarcely, we propose a user-based cloaking algorithm (UCA). The main idea is as follows. All non-empty edges in the cell constitute a new graph ug. Find all connected components cc_set of ug. If a connected component cc doesn't satisfies (K, L)-location privacy model, find a minimum r-path neighbor rnc for cc and merge rnc and cc together. Otherwise, the users in the connected component cc constitute a cloaking set.

Continuing the above running example, $topc=\text{NVC}(p_7)$. $\text{NVC}(p_7)$ is a safe-large cell. The local road network in $\text{NVC}(p_7)$ is shown in Fig. 4. Assume $\delta = 1$, $\text{NVC}(p_7)$ is sparse. ug includes two non-empty edges $< o_1, p_7 >$ and $< o_4, b_6 >$. $G_1(V_1, E_1)$ (where $V_1 = \{o_1, p_7\}, E_1 = \{< o_1, p_7 >\}$), and $G_2(V_2, E_2)$ (where

$V_2 = \{o_4, b_6\}$, $E_2 = \{< o_4, b_6 >\}$), are two connected components of G. G_1 and G_2 are 1-path neighbors through path (o_4, p_7). Then merge G_1 with G_2 with path (o_4, p_7). Now the new connected component cc includes 3 edges $\{< o_1, p_7 >, < o_4, b_6 >, < o_4, p_7 >\}$ with 3 users in. Open vertices $V_{open}(cc)=\{o_1, p_7, o_4, b_6\}$. Suppose o_1 is selected. There are three edges, which are adjacent to o_1 but not in cc. That is $\{< b_1, o_1 >, < b_2, o_1 >, < o_1, b_3 >\}$. Assume that $< b_1, o_1 >$, $< b_2, o_1 >$ are inserted into cc. Finally, three users constitute a cloaking set, and $\{< o_1, p_7 >, < p_7, o_4 >, < o_4, b_6 >, < b_1, o_1 >, < b_2, o_1 >\}$ are the cloaked segments.

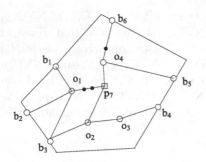

Fig. 4. Road network in NVC(p_7)   **Fig. 5.** Road network in NVC(p_5)

Edge-Based Cloaking Algorithm. For a dense cell, where almost every segment has a user locating on, we propose an edge-based cloaking algorithm (ECA). The basic idea is as follows. We divide the segments into $\lfloor \frac{num\_edge}{L} \rfloor$ groups. Each group contains L segments. The last group contains $2L-1$ segments. Then check the number of users in each group. If the number of users is not less than K, the group is returned as a cloaking set. Otherwise, merge the unsafe group with another one until the group covers more than K users.

Taking NVC(p_5) as an example, the road network is shown in Fig. 5. Suppose that start from o_2, do a breadth-first traversal and label the edges with numbers. The labeled number is shown on each edge. Segments in NVC(p_5) is partitioned into two groups: $\{< o_2, b_4 >, < o_2, p_5 >, < p_5, b_3 >, < p_5, b_5 >, < p_5, o_1 >\}$ and $\{< o_1, b_2 >, < o_1, b_1 >, < o_1, o_3 >, < o_3, b_7 >, < o_3, b_6 >\}$. Both groups meet (3, 5)-location privacy model, they are both returned as cloaking sets.

4   Experiments

We compare our algorithm NVLA with GTCA. GTCA is revised from the DFS-based cloaking algorithm in [5], which incurs fewer query cost than the cloaking algorithms. Both cloaking algorithms are implemented in C++ and run on a desktop PC with a dual AMD 780MHz processor and 2GB main memory.

The evaluation metrics we used include the average success rate and the average cloaking time. The success rate is the portion of users anonymized successfully in the total users issuing queries. And the average cloaking time for successful requests is the time period from when the request is received to when the request is successfully cloaked.

We adapt a real data set, California Road Network and Points of Interest [4], to validate the effectiveness of our cloaking algorithm. The road map contains 21,048 nodes and 21,693 edges; moreover, it is associated with a real dataset of 32,399 POIs. We employ 1% of the real POIs to create the first order NVD, and all the POIs to simulate LBS queries. By default, both the anonymity level (K) and the segment diversity requirement (L) are set to 10. System density threshold δ is 1.

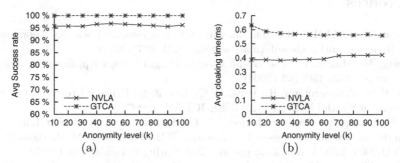

Fig. 6. Impact of anonymity level (a) avg. successs rate (b) avg. cloaking time

We investigate the impact of anonymity level K on the performance of cloaking algorithms. Increasing anonymity level K indicates each cloaking set should cover more users. From Fig. 6(a), we observe that the anonymity level has little influence on the success rates of both cloaking algorithms. The success rates of GTCA remain to be 100% . The success rates of NVLA fluctuate between 95% and 97%. Comparing GTCA and NVLA, the success rates of NVLA are lower than GTCA. That is mainly because unsafe cells in NVLA only check the neighbor cells for merging. When neighbor cells have no user, the anonymization process fails.

Comparing GTCA and NVLA on cloaking time, from Fig. 6(b), NVLA is faster than GTCA, since NVLA finds cloaking set in local NVC, while GTCA is based on global graph traversal. With K increasing, NVLA spends more time on merging cells and finding r-path neighbors . Therefore, the cloaking time of NVLA increases with K increasing.

5   Conclusion

In this paper, we have investigated the network Voronoi-based anonymization algorithms over road networks for location privacy protection. We observed that most existing location cloaking algorithms on road networks need repetitive re-trieve global network information, which has a significant negative impact over the anonymization efficiency. To address this problem, we have employed the NVD to pre-divide the road network into NVCs and proposed a Voronoi-based cloaking algorithm. The average cloaking time is only $0.4ms$ and the cloaking success rate is about 96% for most cases, which validates the efficiency and effectiveness of the proposed algorithms.

Acknowledgement. This research was partially supported by the grant from the Natural Science Foundation of China (No. 61303017), and the Hebei Education Department (No. Q2012131).

References

1. Chow, C., Mokbel, M.F., Bao, J., Liu, X.: Query-aware Location Anonymization for Road Networks. Geoinformatical 15(3), 571–607 (2010)
2. Erwig, M., Hagen, F.: The Graph voronoi Diagram with Applications. Journal of Networks 36(3), 156–163 (2000)
3. Ku, W.S., Zimmermann, R., Peng, W.C., Shroff, S.: Privacy Protected Query Processing on Spatial Networks. In: IEEE ICDEW (2007)
4. Li, F., Cheng, D., Hadjieleftheriou, M., Kollios, G., Teng, S.-H.: On Trip Planning Queries in Spatial Databases. In: Medeiros, C.B., Egenhofer, M., Bertino, E. (eds.) SSTD 2005. LNCS, vol. 3633, pp. 273–290. Springer, Heidelberg (2005)
5. Mouratidis, K., Yiu, M.L.: Anonymous Query Processing in Road Networks. TKDE 22(1), 2–15 (2009)
6. Mouratidis, K., Yiu, M.L.: Shortest Path Computation with No Information Leakage. In: 38th International Conference on Very Large Data Bases (2012)
7. Palanisamy, B., Liu, L., Lee, K., Singh, A., Tang, Y.: Location Privacy with Road network Mix-zones. In: MSN (2012)
8. Yigitoglu, E., Damiani, M.L., Abul, O.: Privacy-preserving Sharing of Sensitive Semantic Locations under Road-network Constraints. In: MDM (2012)

Unique Solutions in Data Exchange

Nhung Ngo and Enrico Franconi

KRDB Research Centre
Free University of Bozen-Bolzano, Italy
lastname@inf.unibz.it
http://www.inf.unibz.it/krdb/

Abstract. Data exchange is the problem of transforming data structured according to a source schema into data structured according to a
target schema, via a mapping specified by rules in the form of source-
to-target tuple generating dependencies. In this context, given a source
instance and a mapping, there might be more than one valid target instance that satisfies the mapping. This issue contradicts the main goal of
exchanging data, namely to have a materialised target instance that can
be used to answer queries over the target schema without reference to the
original source instance. In this paper we introduce and solve the novel
problem of definability abduction, which aims at finding extensions to the
initial schema mappings to guarantee the uniqueness of the materialised
target instance. We consider several semantic criteria to select reasonable extensions and provide provably sound and complete algorithms to
generate these additions. We also do a complexity analysis in different
data exchange settings, also with source and target dependencies.

1 Classical Data Exchange

The problem of data exchange was formally defined in [7] as the problem of
transforming data structured under a source schema into data structured under
a target schema. Given a source instance, the main task in data exchange is
to materialise a valid target instance (called a solution) respecting the schema
mapping, specifications that describe the relationship between data in the two
heterogeneous schemas – the source and the target.

The theory on data exchange has been implemented in the Clio project, a
joint project between the IBM Almaden Research Center and the University
of Toronto [6]. Clio's goal is to radically simplify information integration, by
providing tools that help in automating and managing one challenging piece
of that problem: the conversion of data between representations. Clio pioneered
the use of schema mappings: from this high-level, non-procedural representation,
it can automatically generate either a view, to reformulate queries against one
schema into queries on another for data integration, or code, to transform data
from one representation to the other for data exchange.

In relational data exchange, schema mappings are written in the language of
source-to-target tuple generating dependencies (s-t tgds) to specify that if some
positive facts hold in the source, then some other positive facts must hold in

H. Decker et al. (Eds.): DEXA 2014, Part II, LNCS 8645, pp. 281–294, 2014.

the target. Source-to-target tuple generating dependencies look like rules with positive conjunctive body and positive conjunctive head; the head may contain existentially quantified variables.

As an example, consider the source database with two relations with the schema {*Employee(EMPid), Manager(MANid)*}, and a target database with one relation with the schema *Staff(STid)*. In order to move the data from each of the source relation to the target relation we could state the following mappings (full tgds):

$$Employee(x) \rightarrow Staff(x)$$
$$Manager(x) \rightarrow Staff(x)$$

These mappings guarantee that in the target database all the employees and all the managers will be in the target *Staff* relation. However, the *Staff* relation could contain more tuples, since this case would be still consistent with the specification.

To continue the example, consider in addition to have in the source database a relation with the schema {*Person(PERid)*} and in the target database a relation with the schema {*Phone(PERid,PhNum)*}. In order to populate the target *Phone* relation we need to fill the phone number attribute with unknown values; we could do that with the following mapping (embedded tgd):

$$Person(x) \rightarrow \exists y.Phone(x, y)$$

Intuitively, the schema mapping specifies the fact that all the persons from the source data must have a phone number entry in the target database. However, phone numbers in the target database can be arbitrary, since only their existence is stated.

As we have observed above, under a s-t tgds mapping, there might be more than one solution corresponding to a given source database because the target instances might contain additional facts – like in the first example – or unknown facts – like in the second example. That is, the target database is actually an *incomplete database*, in the sense of classical database theory, namely it is a *set* of possible databases. As a matter of fact, the problem of query answering the target data is inherently complex and non-intuitive for general (non-positive) relational or aggregate queries, since it is basically comparable to entailment with open-world semantics (namely the computation of *certain answers*), and standard relational database technologies can not be used.

Certain answer semantics gives non intuitive answers to non-monotone queries [3,7,1,10], as nicely summarised by [12]: answer to negative queries may be incomplete, and answering queries with aggregation may become trivial and non-informative.

Consider for the above source and target database schemas the following example data:

	EMPid
Employee:	John
	Mary

	MANid
Manager:	Lisa

A negative query on the target database may give an unexpected answer; for example the negative query:

$$\neg Staff(\text{Paul})$$

will have a negative answer, even if we expect it to have a positive answer. This is because the target *Staff* relation may have additional arbitrary tuples, according to the schema mappings.

Similarly, the boolean aggregation query:

$$\#\text{COUNT}\{x : Staff(x)\} = 3$$

will have a false answer, since according to the schema mappings there may be more tuples in the *Staff* relation than the ones coming from the source relations.

In order to recover the expected semantic properties and the good computational behaviour of standard relational database technologies, the data exchange framework restricts the target query language to just monotone queries (i.e., positive queries or union of conjunctive queries). It turns out that the *certain* answer to monotone queries over the incomplete target database are the same as the answers of the same query over a representative specific database (one of the so called *universal solutions* – the *core* being a minimal among them) [7]. With this restriction on the query language query answering over the target database becomes meaningful and efficient (since it is reducible to query answering over the source data).

To give meaning to more expressive queries (with negation or aggregation), various extensions to the data exchange setting have been proposed. They are all based on the idea to restrict the extent and the uncertainty of the target database.

Libkin [11] proposed a notion of Close World Assumption (CWA) solution to overcome some of those issues such as non rewritability and trivial semantics for queries with negation. Intuitively, only facts which are a direct translation of source data are allowed in the target. When this intuition is properly formalised, the CWA-solutions can be characterised as homomorphic images of the canonical solution that have a homomorphism back into the canonical solution. To deal with *aggregate queries*, Afrati and Kolaitis [1] proposed a stricter version of CWA solution, namely the set of all endomorphic images of the canonical universal solution. In order to overcome some anomalies of the previous approaches, Hernich introduced a generalised version of CWA-solution called GCWA*-solution [10] that is basically a union of inclusion-minimal solutions. In comparison with other solutions, query answering over GCWA*-solutions is more intuitive and it is invariant under logically equivalent schema mappings; however, there are data exchange settings and Boolean queries for which query evaluation over GCWA*-solutions is undecidable.

Practitioners did solve this problem – as expected – in a more pragmatic way. It has been reported that users of the Clio data exchange system, based on the classical data exchange theory, fix the uncertainty in the target database by *forcing* the target relations to be closed. In practice, users do store the *core* representing the incomplete target database as an actual relational database, therefore assuming that there can not be more tuples than the ones explicitly coming from the source, and replacing the unknown values (Skolem constants) generated by the existential quantifiers with actual values chosen in an ad-hoc way. Our proposal gives formal grounds to this idea, and it actually generalises in an elegant and abstract way both Libkin's and Afrati and Kolaitis' proposals.

2   Introducing Unique Solutions in Data Exchange

We believe that the main problem of the extensions summarised above is their inherently syntactic nature, namely they fail in defining the problem they aim to solve in a clear and intuitively convincing semantic way. In contrast to the afore-mentioned results, the goal of our work is to enrich the data exchange framework to allow for general relational and aggregate queries, by suggesting "reasonable" amendments to the initial mapping so that the newly obtained schema mapping will then produce a *unique* materialised target instance depending only on the given source database instance and the schema mapping – so we do not have anymore an incomplete database in the target schema. To this end, we intro-duce and solve the novel problem of *definability abduction*, which aims at finding extensions to the initial schema mapping which guarantee the uniqueness of tar-get instance, in a way that the intended semantics of the original mapping is *minimally* changed, for various reasonable semantic notions of minimality.

Let's consider again the examples introduced above to understand informally our proposal. It is intuitive that by adding to the schema mapping the target-to-source dependency

$$Staff(x) \rightarrow (Employee(x) \vee Manager(x))$$

we get that the target relation *Staff* becomes now *exactly* equal to the union of the relation *Employee* and the relation *Manager*. Therefore, the relation *Staff* will contain all and only the tuples we expect in first place: the content of that target relation depends only on the content of the source relations. The schema mapping we have added can be obtained as a solution to the definability abduction problem mentioned above: it is a reasonable addition to the schema mappings making the target relations *definable* from the source relations. Note that the extension to the schema mapping is conservative: the answer to *positive* queries in the system with the original schema mappings will be the same as in the system with the extended schema mappings.

There is also an intuitive *fix* for the target *Phone* relation, which has the problem that the exact phone number of the person is left unspecified. Suppose we actually have among the source relations an additional binary relation stor-ing the contact numbers of people with the schema {*Contact(CONid,PhNum)*}. Then, a reasonable solution forcing the target database to be unique could be:

$$Contact(x, y) \leftrightarrow Phone(x, y).$$

Note that, with this extension to the original schema mapping, it is also implied – as expected – that every source database instance must satisfy the foreign key constraint $Person(x) \rightarrow Contact(x, y)$. Each person should have a known contact number, and this number will populate the target relation *Phone*, eliminating the ambiguity originally introduced by the existential quantifier. Also in this case the schema mapping we have added can be obtained as a solution to the definability abduction problem mentioned above: it is a reasonable addition to the schema mappings making the target relations *definable* from the source relations. Note that also in this case the extension to the schema mapping is conservative: the answer to *positive* queries in the system with the original schema mappings will be the same as in the system with the extended schema mappings.

We say that the target is uniquely defined from a given source instance when it is *definable* from the source, according to a very precise notion introduced by the logician Beth in the 50'ies [4,13,8,16]. Intuitively, a target predicate t is definable from the set of source predicates S under a set of mappings (and possibly constraints) Σ once the extensions of the source predicates from S are fixed in a model of Σ then we are certain that the extension of the target predicate t is fixed as well. This property may not hold for some target predicates in the case of the classical data exchange, since as we have noted with the examples above there is no guarantee of unique solutions. This leads us to the idea of adding some constraints to Σ (in the form of extended mappings) in order to enforce the definability for *every* target predicate. As a matter of fact, this is an abductive reasoning task [15,2] in which an explanation of a given fact (in this case, the definability of target predicates) must be found based on the theory Σ. We call this problem *definability abduction*.

Similar to classical abduction problems, a definability abduction problem may have many possible explanations. Therefore in our framework we propose several semantic criteria of "reasonable" explanations: *minimal explanations* guarantee that the intended meaning of the original mapping is minimally changed, and *general explanations* guarantee that the extended mapping can be applied for any source instance. These "reasonable" explanations can be also used to ensure that the unique solution is actually a "good" one according to aforementioned CWA semantics. It can be seen that our proposed minimal explanation admits a unique CWA-solution, therefore CWA-solutions are a special case of our framework.

We also consider some extensions of the definability abduction problem where specific source instances, source dependencies, and target dependencies may be involved. We do a complexity analysis and show that the problem of checking explanations in these cases are hard in general, and may be undecidable.

This paper vastly extends [9,14], where only a general semantic framework was introduced, by adding source and target dependencies, by considering the explicit presence of a source database instance, by introducing several semantic criteria of "reasonable" explanations, together with sound and complete algorithms, and further complexity and undecidability results.

3   Formal Preliminaries

Firstly, we recall the basic notions from data exchange that we will need.

A *schema* is a finite set of predicate names with associated arities. Let $\mathbf{R}$ be a schema, $\mathbf{R} = \{R_1, ..., R_k\}$, an *instance* I over $\mathbf{R}$ is an union of $R_1^I, R_2^I, ..., R_k^I$ such that each R_i^I is a finite set of tuples having the same arity as R_i. If I is an instance of schema $\mathbf{S}_1$ and J is an instance of schema $\mathbf{S}_2$, we use (I, J) to denote an instance of schema $\mathbf{S}_1 \cup \mathbf{S}_2$.

If I is an instance and ϕ is a logic formula, we write $I \models \phi$ if I satisfies ϕ. If Σ is a set of formulas, we write $I \models \Sigma$ to mean $I \models \phi$ for every $\phi \in \Sigma$.

We denote the source and target schemas by $\mathbf{S}$ and $\mathbf{T}$ respectively. A tuple generating dependency (tgd) is a sentence of the form $\forall \bar{x}(\varphi(\bar{x}) \rightarrow \exists \bar{y} \psi(\bar{x}, \bar{y}))$, where $\bar{x}$ are free variables of both φ and ψ, φ and ψ are conjunctive queries. For the sake of readability, we write $\varphi(\bar{x}) \rightarrow \exists \bar{y} \psi(\bar{x}, \bar{y})$ instead of the full formula.

A tgd is a *source-to-target tgd* (s-t tgd) if φ and ψ are formulas over $\mathbf{S}$ and $\mathbf{T}$ respectively, and vice versa, a tgd is a *target-to-source tgd* (t-s tgd) if φ and ψ are formulas over $\mathbf{T}$ and $\mathbf{S}$ respectively. A tgd is *full* if $\bar{y} = \emptyset$, otherwise, it is *embedded*. A tgd is a local-as-view (LAV) dependency if φ is an atom.

A data exchange setting is a tuple $\mathcal{M} = (\mathbf{S}, \mathbf{T}, \Sigma)$, where Σ is a set of s-t tgds. The set Σ is referred to *schema mapping*. In the following we assume that all the predicates from T appear in the schema mapping, i.e. the mappings tell us some information about each target predicate. By a *solution* to the data exchange setting $\mathcal{M}$ for a source instance I, we mean a target instance J such that $(I, J) \models \Sigma$. In data exchange setting where source and target dependencies are involved, Σ is extended to $\Sigma \cup \Sigma_s \cup \Sigma_t$ where Σ_s (Σ_t) is a set of tgds over source (target) schema.

Given a set of sentences Φ, we use $\sigma(\Phi)$ to denote the signature of Φ, i.e. the set of all non-logical symbols in Φ.

For the rest of this section, we shall review the concept of definability in first order logic (FOL) and then introduce the definability abduction problem.

Definition 1. *Let Σ be a set of sentences in FOL. We say that a predicate p is definable from the set of predicates $\mathbf{P}$ under Σ if for every two interpretations $\mathcal{I} = \langle D^{\mathcal{I}}, \cdot^{\mathcal{I}} \rangle$ and $\mathcal{J} = \langle D^{\mathcal{J}}, \cdot^{\mathcal{J}} \rangle$ such that they are models of Σ, it holds that $\mathbf{P}^{\mathcal{I}} = \mathbf{P}^{\mathcal{J}}$ implies $p^{\mathcal{I}} = p^{\mathcal{J}}$.*

In case Σ contains some functional symbols, we can extend the above definition as follows.

Definition 2. *Let Σ be a set of sentences in FOL. We say that a predicate p is* functional definable *from the set of predicates $\mathbb{P}$ under Σ if for every two interpretations $\mathcal{I} = \langle D^{\mathcal{I}}, \cdot^{\mathcal{I}} \rangle$ and $\mathcal{J} = \langle D^{\mathcal{J}}, \cdot^{\mathcal{J}} \rangle$ such that they are models of Σ, it holds that if $\mathbf{P}^{\mathcal{I}} = \mathbf{P}^{\mathcal{J}}$ and for every function f appearing in Σ, $f^{\mathcal{I}} = f^{\mathcal{J}}$ then $p^{\mathcal{I}} = p^{\mathcal{J}}$.*

From the definition, one can verify definability property by checking whether or not

$$\Sigma \cup \widetilde{\Sigma} \models \forall \bar{x}.p(\bar{x}) \leftrightarrow \tilde{p}(\bar{x})$$

where $\widetilde{\Sigma}$ is obtained from Σ by replacing every target predicate p_t with new predicates with the same arity $\tilde{p}_t$ and $\tilde{p}$ is a new predicate having the same arity as p. That leads to the following definition about an abduction problem that aims to explain the definability of predicates.

Let $\mathbf{P}$ and $\mathbf{P}_1$ be sets of predicates and Σ a set of sentences.

Definition 3. *A triple* $(\mathbf{P}_1, \mathbf{P}, \Sigma)$ *is said to be a* definability abductive problem *(DAP) if there is no* $p \in \mathbf{P}$ *which is definable from the set of predicates* $\mathbf{P}_1$ *under* Σ

A solution of the abduction problem is defined as follow.

Definition 4. *A set of sentences* Δ *is called a* solution *to* $(\mathbf{P}_1, \mathbf{P}, \Sigma)$ *if every predicate in* $\mathbf{P}$ *is definable from* $\mathbf{P}_1$ *under* $\Sigma \cup \Delta$.

In order to distinguish with *solution* of a data exchange setting, for the sake of readability, we use the notion of *d-extension* to refer to solution of definability abduction.

4   Definability Abduction in Data Exchange

First observe that if we are given a schema mapping specified by a set of s-t tgds, then apparently the target predicates are *not* definable from the source schema under this mapping.

Proposition 1. *Let* Σ *be a set of s-t tgds. Then for every* $p \in \mathbf{T}$ *it holds that* p *is not definable from* $\mathbf{S}$ *under* Σ.

Corollary 1. *Every data exchange setting* $(\mathbf{S}, \mathbf{T}, \Sigma)$ *is a* definability abductive problem

In case the data exchange setting contains source dependencies, the above corollary still holds where Σ is replaced by $\Sigma \cup \Sigma_s$. However, the corollary may not hold if target dependencies are involved.

4.1   Solution-Selection Criteria

As in the case of classical abduction, a definability abduction problem may have many d-extensions. Among them, those d-extensions that are less informative with respect to the schema mapping are preferred to others. Put it in other words, those solutions which have more common models with the schema mapping are better than others in the sense of bringing minimal change to the original mapping.

Definition 5. *Given a data exchange setting* $(\mathbf{S}, \mathbf{T}, \Sigma)$, *a d-extension* Δ *is called* minimal *if for every* Δ' *such that* Δ' *is also a d-extension, it holds that*

$$\Sigma \cup \Delta \models \Delta' \;\Rightarrow\; \Sigma \cup \Delta' \models \Delta.$$

Besides, by adding a d-extension Δ to a schema mapping Σ, the space of Σ's models may reduce. That is, given a source instance I, there might be no target instance J such that $(I, J) \models \Sigma \cup \Delta$. This fact motivates us to introduce the following criteria.

Definition 6. *Given a data exchange setting* $(\mathbf{S}, \mathbf{T}, \Sigma)$, *a d-extension* Δ *is called general if for every source instance* I, *there is always* J *such that* $(I, J) \models \Sigma \cup \Delta$

We use the following theorem to characterise general d-extension. Intuitively, general d-extensions are d-extensions that bring nothing except the definability of target predicates.

Theorem 1. Δ *be a general d-extension of* $(\mathbf{S}, \mathbf{T}, \Sigma)$ *iff* $\Sigma \cup \Delta \equiv \bigwedge_{p \in \mathbf{T}} (\forall \bar{x} p(\bar{x}) \leftrightarrow p_s(\bar{x}))$ *where* p_s *does not contain any symbol in* $\mathbf{T}$.

Corollary 2. *Every general d-extension is a minimal one.*

The above corollary shows the connection between the two criteria. Notice that the reverse direction does not hold. As we will show in Example 3 there are minimal d-extensions that are not general.

4.2   Generating Minimal and General d-Extension

We now present some methods to generate minimal and/or general d-extension of a given data exchange setting. Let us first restrict our attention to the case when the schema mapping Σ contains only *full* s-t tgds. Then we can always assume that such a mapping has the following form:

$$\Sigma = \bigcup_{p_i \in \mathbf{T}} \bigcup_{j=1}^{n_i} \{\varphi_j^{p_i}(\bar{x}, \bar{z}_i) \to p_i(\bar{x})\}, \qquad (*)$$

i.e. it consists of Horn clauses.

The following theorem provides an intuitive minimal and general d-extension of this setting.

Theorem 2. *Let* $\mathcal{M} = (\mathbf{S}, \mathbf{T}, \Sigma)$ *be a data exchange setting where* Σ *is a set of full tgds as in* $(*)$. *Then* $\Delta = \bigcup_{p \in \mathbf{T}} \{p(\bar{x}) \to \bigvee_j \exists \bar{z}_j \varphi_j^p(\bar{x}, \bar{z}_j)\}$ *is a minimal and general d-extension of* $\mathcal{M}$.

Example 1. Consider $\mathcal{M} = (\mathbf{S}, \mathbf{T}, \Sigma)$ where $S = \{Employee(\cdot), Manager(\cdot)\}$, $\mathbf{T} = \{Staff(\cdot)\}$, and Σ is the set of following full tgds:

$$Employee(x) \to Staff(x)$$
$$Manager(x) \to Staff(x)$$

Then $\Delta = \{Staff(x) \to (Employee(x) \vee Manager(x))\}$ is a minimal d-extension of $\mathcal{M}$

The situation is dramatically different if the initial schema mapping contains embedded tgds. Next example shows that there might be no intuitive d-extension in this case.

Example 2. Let us consider data exchange setting $\mathcal{M} = (\mathbf{S}, \mathbf{T}, \Sigma)$, in which $\mathbf{S} = \{Person(\cdot)\}$, $\mathbf{T} = \{Phone(\cdot, \cdot)\}$ and $\Sigma = \{Person(x) \rightarrow \exists y.Phone(x, y)\}$. Intuitively, the schema mapping specifies the fact that all the persons from the source data must have a phone number entry in the target database. It turns out that this setting does not have intuitive d-extension. For instance, using the available language at hand, we can propose the d-extension $\Delta = \{Person(x) \wedge Person(y) \leftrightarrow Phone(x, y)\}$ which can hardly be acceptable.

One way to solve the problem with non-intuitive d-extension is to relax a definability abductive problem by allowing additional predicates in the source.

Definition 7. *Let* $\mathcal{M} = (\mathbf{P_1}, \mathbf{P}, \Sigma)$ *be a DAP and* Δ *a set of sentences. Then* Δ *is called a* weak *d-extension of* $\mathcal{M}$ *if it is a d-extension of* $\mathcal{M}' = (\mathbf{P_2}, \mathbf{P}, \Sigma)$ *for some* $\mathbf{P_2}$ *such that* $\mathbf{P_1} \subseteq \mathbf{P_2}$.

That is, weak d-extensions are those that give definability of predicates $\mathbf{P}$ from a set of predicates that is richer than $\mathbf{P_1}$. Clearly, any d-extension of $\mathcal{M}$ is also its weak d-extension. One can see the motivation of introducing a weak notion of d-extension in data exchange. Namely, such additional source predicates can tell us some information about existential values in the target described by embedded schema mappings. Thus, we might be able to find though weak, but minimal d-extensions which could be intuitive from practical point of view. It appears that a d-extension where we ask the user for the existential values in the target by using *fresh* predicates in the source schema, is minimal. Intuitively, a predicate is consider as a *fresh* one if it does not appear in Σ. We extend $\mathbf{S}$ to $\mathbf{S}'$ by adding a set of fresh predicates such that for each $p \in \mathbf{T}$, there is a corresponding fresh $p_s \in \mathbf{S}'$.

Theorem 3. *Let* $\mathcal{M} = (\mathbf{S}, \mathbf{T}, \Sigma)$ *be a data exchange setting where* Σ *is a set of embedded s-t tgds. Then* $\Delta = \bigcup_{p \in \mathbf{T}} \{p_s \leftrightarrow p\}$ *is a minimal weak d-extension of* $\mathcal{M}$ *where* $\{p_s \mid p \in \mathbf{T}\}$ *is a set of different fresh predicates.*

Notice that not every source instance satisfies $\Sigma \cup \Delta$ in this case, which means Δ is not a weak general d-extension. In fact, let Σ' be Σ where p is replaced by p_s, then source instances should satisfy Σ'. Besides, given a source instance, the requirement about *fresh* p_s can be relaxed in the sense that users can pick a source table (which might not be fresh) p_s', copy it to a fresh table p_s and then give its value for an embedded table p if the source instance satisfies $\Sigma' \cup \{p_s \leftrightarrow p_s'\}$.

Example 3. Consider a DAP $\mathcal{M} = (\mathbf{S}, \mathbf{T}, \Sigma)$ where $\mathbf{S} = \{Person(\cdot), Phone(\cdot, \cdot)\}$, $\mathbf{T} = \{Contact(\cdot, \cdot)\}$, and $\Sigma = \{Person(x) \rightarrow Contact(x, y)\}$.
Then $\Delta = \{Contact(x, y) \leftrightarrow Phone(x, y)\}$ is a minimal d-extension of $\mathcal{M}$. It also implies that every source instance must satisfy the source constraint $Person(x) \rightarrow Phone(x, y)$

Algorithm 1. (General d-extension with function symbols for embedded mapping)

Input: Data exchange setting $\mathcal{M}$ = $(\mathbf{S}, \mathbf{T}, \Sigma)$ such that Σ = $\bigcup_{i=1,n}\{\forall \bar{x}_i, \bar{z}_i \phi_{si}(\bar{x}_i, \bar{z}_i) \rightarrow \exists \bar{y}_i \psi_{ti}(\bar{x}_i, \bar{y}_i)\}$.

Output: A minimal general d-extension Δ .

1: Let $\Delta = \emptyset$
2: **for all** $p \in T$ **do**
3:     Let $\delta_p = $ **false**
4: **end for**
5: **for** $i = 1$ to n **do**
6:     Let f_i be the Skolem function corresponding to the mapping i^{th}
7:     **for** Each appearance of target predicate p in the mapping i^{th} **do**
8:         $\delta_p = \delta_p \vee (\exists \bar{z}_i \phi_{si}(\bar{x}_i, \bar{z}_i) \wedge \bar{y}_i = f_i(\bar{x}_i, \bar{z}_i))$
9:     **end for**
10: **end for**
11: **for all** $p \in T$ **do**
12:     $\Delta = \Delta \cup \{p \leftrightarrow \delta_p\}$
13: **end for**

In order to get intuitive general d-extensions of a DAP with embedded tdgs, we suggest another way that uses functional symbols in the source signature and allows this type of equality $y = f(\bar{x})$ where y is a variable and f is a function in the language of schema mapping.

Theorem 4. *Let $\mathcal{M} = (\mathbf{S}, \mathbf{T}, \Sigma)$ be a data exchange setting such that Σ contains only embedded tgds. Then Δ generated by Algorithm 1 is a minimal general d-extension of $\mathcal{M}$*

Example 4. Consider $\mathcal{M} = (\mathbf{S}, \mathbf{T}, \Sigma)$ where $\mathbf{S} = \{A(\cdot, \cdot)\}$, $\mathbf{T} = \{B(\cdot, \cdot)\}$, and $\Sigma = \{A(x, y) \rightarrow \exists z B(x, z) \wedge B(z, y)\}$. Then $\Delta = \{B(x, y) \leftrightarrow \exists y_1 (A(x, y_1) \wedge y = f(x, y_1)) \vee \exists x_1 (A(x_1, y) \wedge y = f(x_1, y))\}$ is a minimal general d-extension of $\mathcal{M}$.

Now we consider a general Σ which may contains both full and embedded tgds by proposing Algorithm 2 and Algorithm 3.

Algorithm 2 generates a (weak) minimal d-extension of data exchange setting $\mathcal{M}$. It works correctly based as the following proposition says.

Proposition 2. *Let $\mathcal{M} = (S, T, \Sigma)$ be a data exchange setting. Then Δ generated by Algorithm 2 on $\mathcal{M}$ is a (weak) minimal d-extension of $\mathcal{M}$.*

Algorithm 3 generates a general d-extension of data exchange setting $\mathcal{M}$ based on the following proposition.

Proposition 3. *Let $\mathcal{M} = (\mathbf{S}, \mathbf{T}, \Sigma)$ be a data exchange setting. Then Δ generated by Algorithm 3 on $\mathcal{M}$ is a general d-extension of $\mathcal{M}$.*

Algorithm 2. (Weak) Minimal d-extension

Input: Data exchange setting $\mathcal{M} = (\mathbf{S}, \mathbf{T}, \Sigma)$.
Output: A weak minimal d-extension Δ to the corresponding definability abductive problem.

1: Let $\Delta = \emptyset$
2: Decompose Σ into the union $\Sigma_f \cup \Sigma_e$ where Σ_f and Σ_e are sets of full (with one predicate in the heads) and pure embedded tgds respectively;
3: For every $p \in \mathbf{T}$ such that $p \in \sigma(\Sigma_f) \setminus \sigma(\Sigma_e)$ do $\Delta := \Delta \cup \{p \to \bigvee_i \exists z_i \varphi_i^p(z_i) \mid \varphi_i^p(z_i) \to p \in \Sigma_f, \text{for every } i\}$
4: For every $p \in \sigma(\Sigma_e) \setminus \sigma(\Sigma_f)$ do $\Delta := \Delta \cup \{p_s \leftrightarrow p \mid p_s \text{ is a fresh source predicate}\}$.

5: For every $p \in \sigma(\Sigma_f) \cap \sigma(\Sigma_e)$ do $\Delta := \Delta \cup \{p \to p_s \vee \bigvee_i \exists z_i \varphi_i^p(z_i) \mid \varphi_i^p(z_i) \to p \in \Sigma_f, \text{for every } i\} \cup \{p_s \to p \mid p_s \text{ is a fresh source predicate}\}$

Algorithm 3. General d-extension with function symbols

Input: Data exchange setting $\mathcal{M} = (\mathbf{S}, \mathbf{T}, \Sigma)$.
Output: A minimal solution Δ to the corresponding definability abductive problem.

1: Let $\Delta = \emptyset$
2: **for all** $p \in \mathbf{T}$ **do**
3:     Decompose Σ_p-set of tgds containing predicate p into the union $\Sigma_{pf} \cup \Sigma_{pe}$ where Σ_{pf} and Σ_{pe} are sets of full (with one predicate in the heads) and embedded tgds respectively;
4:     Let Δ_{pe} be the d-extension of (S, T, Σ_{pf}) as in Theorem 2.
5:     Let δ_{pf} be the the final value of δ_p in Algorithm 1 with the input (S, T, Σ_{pe})
6:     $\delta_p = head(\Delta_{pe}) \vee \delta_{pf}$ where $head(\alpha \to \beta) = \beta$
7: **end for**
8: **for all** $p \in \mathbf{T}$ **do**
9:     $\Delta = \Delta \cup \{p \leftrightarrow \delta_p\}$
10: **end for**

4.3   Involving Source Instances

Now let us look at the definability abduction problem where source instances are part of the input. As we have mentioned before, a d-extension Δ maybe not work for a specific source instance I. On the other hand, some set of sentences Δ, which is not a d-extension but still can play the role of a d-extension w.r.t to I.

Example 5. Consider $\mathcal{M} = (\mathbf{S}, \mathbf{T}, \Sigma)$ where $S = \{A(\cdot), B(\cdot)\}$, $T = \{C(\cdot)\}$, and $\Sigma = \{A(x) \wedge B(x) \to C(x)\}$. Let $\Delta = \{C(x) \to B(x)\}$. Δ is not a d-extension of $\mathcal{M}$. However, if we consider source instance $I = \{A(1), B(1), A(2)\}$, then there is unique $J = \{C(1)\}$ that $(I, J) \models \Sigma \cup \Delta$

The above example leads us to the following definition of d-extension w.r.t a source instance.

Definition 8. *Given a data exchange setting $\mathcal{M} = (\mathbf{S}, \mathbf{T}, \Sigma)$ and a source instance I, a set of sentences Δ is called d-extension with respect to I ,written as d-extension$(\mathcal{M}, I)$, if there is only one target instance J that satisfies $(I, J) \models \Sigma \cup \Delta$*

Obviously, general d-extension are d-extension w.r.t any source instances.

We study the complexity of d-extension checking problem. Theorem 5 shows that the problem is hard even with a simple input.

Theorem 5. *Given a data exchange setting $\mathcal{M} = (\mathbf{S}, \mathbf{T}, \Sigma)$, a source instance I and a set of sentences Δ where Σ is a set of full tgds and Δ is a set of t-s LAV, the problem of checking $\Delta = $ d-extension$(\mathcal{M}, I)$ is Π_2^P-complete.*

4.4 Involving Source and Target Dependencies

Now let us consider an extension of d-extension w.r.t to a source instance, namely, d-extension w.r.t to a set of source instances specified by a set of source dependencies.

Definition 9. *Given a data exchange setting $\mathcal{M} = (\mathbf{S}, \mathbf{T}, \Sigma \cup \Sigma_s)$, a set of t-s tgds Δ is called d-extension with respect to Σ_s, written as d-extension$_s$ if for every source instance I such that $I \models \Sigma_s$ there is only target instance J that satisfies $(I, J) \models \Sigma \cup \Delta$*

It is straightforward to see that the above definition does not coincide with the definition of d-extension of the DAP $\mathcal{M} = (\mathbf{S}, \mathbf{T}, \Sigma \cup \Sigma_s)$.

The below theorem says that without any restriction in syntax of Σ_s (for example: weakly-acyclic set of tgds), the d-extension checking problem is undecidable. Its proof is based on a reduction from the problem of checking conjunctive query containment under constraints which is known to be undecidable [5].

Theorem 6. *Given a data exchange setting $\mathcal{M} = (\mathbf{S}, \mathbf{T}, \Sigma \cup \Sigma_s)$, a set of t-s tgds Δ, the problem of checking whether Δ is a d-extension$_s$ of $\mathcal{M}$ is undecidable.*

Besides, one can also extend the DAP by considering a set of target dependencies in the sense that one need to extend the original mapping so that the final unique solution must satisfy some constraints. Under this intuition, the definition of d-extension w.r.t target dependencies coincides with the definition of d-extension of the DAP in which its theory contains also these target dependencies.

Definition 10. *Given a data exchange setting $\mathcal{M} = (\mathbf{S}, \mathbf{T}, \Sigma \cup \Sigma_t)$, a set of t-s tgds Δ is called d-extension with respect to Σ_t, written as d-extension$_t$ if Δ is a d-extension of the DAP $(\mathbf{S}, \mathbf{T}, \Sigma \cup \Sigma_t)$*

Similar to the case with source constraints, the d-extension checking problem is also undecidable in this case.

Theorem 7. *Given a data exchange setting $\mathcal{M} = (\mathbf{S}, \mathbf{T}, \Sigma \cup \Sigma_t)$, a set of t-s tgds Δ, the problem of checking whether Δ is a d-extension$_t$ of $\mathcal{M}$ is undecidable.*

5   Conclusion

We have considered the problem of gaining definability of target predicates over source predicates in data exchange. We have defined the problem as an abduction task and approached it by stating some criteria to select the "good" d-extensions such as minimality and generation and then provided algorithms to generate such d-extensions. We also have started study the complexity of the d-extension checking problem in different settings. In the general settings we have undecidability results; once we restrict the syntax of input dependencies, the problem becomes decidable.

Beside studying the complexity of the d-extension checking problem by considering different restrictions on input source and target dependencies, in the future, we are also interested in characterising a semantic order among d-extensions to specify the "best" extension. At the end, our ultimate goal is to implement a data exchange tool which provides users with suggestions on how to change their schema mapping in order to achieve the uniqueness of target instances.

We would like to thank Renée Miller for having introduced to us the issues of incomplete target instances in data exchange, and together with Patricia Rodríguez-Gianolli and Evgeny Sherkhonov for the fruitful discussions with us at the early stages of this work.

References

1. Afrati, F.N., Kolaitis, P.G.: Answering aggregate queries in data exchange. In: PODS, pp. 129–138 (2008)
2. Aliseda-Llera, A.: Seeking explanations: abduction in logic, philosophy of science and artificial intelligence. PhD thesis, Stanford, CA, USA, UMI Order No. GAX98-10072 (1998)
3. Arenas, M., Barceló, P., Fagin, R., Libkin, L.: Locally consistent transformations and query answering in data exchange. In: PODS, pp. 229–240 (2004)
4. Beth, E.: On Padoa's method in the theory of definition. Indagationes Mathematicae 15, 330–339 (1953)
5. Calì, A., Gottlob, G., Kifer, M.: Taming the infinite chase: Query answering under expressive relational constraints. J. Artif. Intell. Res. (JAIR) 48, 115–174 (2013)
6. Fagin, R., Haas, L.M., Hernández, M., Miller, R.J., Popa, L., Velegrakis, Y.: Clio: Schema mapping creation and data exchange. In: Borgida, A.T., Chaudhri, V.K., Giorgini, P., Yu, E.S. (eds.) Conceptual Modeling: Foundations and Applications. LNCS, vol. 5600, pp. 198–236. Springer, Heidelberg (2009)
7. Fagin, R., Kolaitis, P.G., Miller, R.J., Popa, L.: Data exchange: semantics and query answering. Theor. Comput. Sci. 336(1), 89–124 (2005)
8. Franconi, E., Kerhet, V., Ngo, N.: Exact query reformulation over databases with first-order and description logics ontologies. Journal of Artificial Intelligence Research (JAIR) 48, 885–922 (2013)
9. Franconi, E., Ngo, N., Sherkhonov, E.: The definability abduction problem for data exchange - (extended abstract). In: Krötzsch, M., Straccia, U. (eds.) RR 2012. LNCS, vol. 7497, pp. 217–220. Springer, Heidelberg (2012)
10. Hernich, A.: Answering non-monotonic queries in relational data exchange. In: ICDT, pp. 143–154 (2010)

11. Libkin, L.: Data exchange and incomplete information. In: Proceedings of the Twenty-Fifth ACM SIGMOD-SIGACT-SIGART Symposium on Principles of Database Systems, PODS 2006, pp. 60–69. ACM, New York (2006)
12. Libkin, L., Sirangelo, C.: Open and closed world assumptions in data exchange. In: Proceedings of the 2009 Description Logics Workshop (2009)
13. Nash, A., Segoufin, L., Vianu, V.: Views and queries: Determinacy and rewriting. ACM Trans. Database Syst. 35, 21:1–21:41 (2010)
14. Ngo, N.: Getting unique solution in data exchange. PVLDB 6(12), 1440–1443 (2013)
15. Paul, G.: Approaches to abductive reasoning: An overview. AI Review 7, 109–152 (1993)
16. ten Cate, B., Franconi, E., Seylan, İ.: Beth definability in expressive description logics. Journal of Artificial Intelligence Research (JAIR) 48, 347–414 (2013)

Translatable Updates of Selection Views under Constant Complement

Enrico Franconi and Paolo Guagliardo

Free University of Bozen-Bolzano, Italy
{franconi,guagliardo}@inf.unibz.it

Abstract Given a lossless view associating a source relation with a set of target relations defined by selection queries over the source, we study how updates of the target relations can be consistently and univocally propagated to the underlying source relation. We consider a setting where some of the attributes in the schema are *interpreted* over some specific domain (e.g., the reals or the integers) whose data values can be compared beyond equality, by means of special predicates (e.g., smaller/greater than) and functions (e.g., addition and subtraction). The source schema is constrained by *conditional domain constraints*, which restrict the values that are admissible for the interpreted attributes whenever a certain condition is satisfied by the values taken by the non-interpreted ones.

We show how to decide whether insertions, deletions and replacements, as well as sequences of insertions and deletions, can be univocally propagated through lossless selection views. In general, a lossy view, which does not preserve the whole informative content of the source, can always be turned into a lossless one by means of a *view complement*, which provides the missing information. For lossy selection views, we show how to find complements that provide the smallest amount of information needed to achieve losslessness, so as to maximise the number of updates that can be propagated under the so-called *constant complement principle*, prescribing that the complement be invariant during update propagation.

1 Introduction

Given a view that maps instances of a source schema (possibly under constraints) to instances of a target schema, the view update problem consists in translating updates issued on the target schema into corresponding updates on the source schema, so that the changes of the source relations reflect exactly the changes introduced into the target ones. The propagation of updates through views is a classical problem in database research [1], which has received renewed attention in recent years.

A view mapping, defining how the the target relations depend on the source one, is *lossless* if the converse is also true, i.e., when the source relations depend on the target ones. Whenever such an inverse relationship from target to source can be expressed constructively (i.e., by defining each source relation in terms of

H. Decker et al. (Eds.): DEXA 2014, Part II, LNCS 8645, pp. 295–309, 2014.

the target ones by means of a query), "good" view updates that satisfy certain conditions can be effectively propagated in an unambiguous way [5]. Obviously, lossless database decompositions have an important role in this context, because they are associated with the existence of a *reconstruction operator* that explicitly prescribes how a source relation can be rebuilt from the fragments into which it has been decomposed.

Decomposing a relation horizontally consists in splitting it into relations with the same arity and over the same attributes, each of which contains a subset of the tuples in the original relation. This kind of decomposition, called *horizontal decomposition*, is lossless precisely when the union of the fragments yields back the original source relation. In [3], lossless horizontal decomposition is studied in a setting where some of the attributes are interpreted over a specific domain with a rich structure (e.g., ordering) that allows its data values to be compared beyond equality, which is of practical importance for real-world applications.

The results in [3] provide the groundwork for univocal propagation of updates in the context of selection views under constraints, mentioned by the authors as future work. The present paper is a follow-up to [3] in which we investigate the propagation of updates through lossless selection views. In particular, given a lossless selection view, as a first contribution we will show how to decide whether insertions, deletions and replacements can be univocally propagated from target to source. Our second contribution is concerned with attaining losslessness from a given lossy selection view, which can be done by means of another selection view that "complements" the original lossy one by providing the missing information.

Outline. After basic preliminaries and notation (Sect. 2), we formally introduce (Sect. 3) the class of views and integrity constraints we consider in this paper; in Sect. 4, we provide algorithms for deciding whether insertions, deletions and replacements can be unambiguously propagated through lossless selection views; in Sect. 5, we present a technique for turning a lossy selection view into a lossless one, so as to be able to apply the results on update propagation also in this case; we conclude (Sect. 6) by discussing future and related work.

2 Preliminaries

We start by introducing some basic notions and necessary notation. We assume the reader to be familiar with formal logic and its application to database theory.

Basics. A *schema* (a.k.a. *relational signature*) is a finite set $\mathbf{R}$ of relation symbols. Each $R \in \mathbf{R}$ is associated with a positive arity $|R|$ indicating the total number of *positions* in R, which are partitioned into *interpreted* and *non-interpreted* ones. We consider an arbitrary, possibly infinite set **dom** of values that can only be compared for equality, and a set **idom** of values from a specific domain (e.g., the integers $\mathbb{Z}$) equipped with a set of predicates (e.g., $\leq$) and functions (e.g., $+$), according to a first-order language $\mathfrak{C}$ closed under negation. An *instance* over a schema $\mathbf{R}$ maps each $R \in \mathbf{R}$ to a relation R^I of appropriate arity on $\mathbf{dom} \cup \mathbf{idom}$, called the *extension* of R under I, such that the values for the interpreted and non-interpreted positions of R come from **idom** and **dom**, respectively. A *fact* is given by the association, written as $R(\bar{t})$, between a relation symbol R and a tuple $\bar{t}$ of values of appropriate arity; instances can be conveniently represented as sets of facts. All instances in this paper are finite. Given an instance I over a schema **R** consisting only of relation symbols with the same arity, we denote by $\bigcup(\mathbf{R}, I)$ the set $\{\bar{t} \mid \bar{t} \in R^I,\, R \in \mathbf{R}\}$, that is, the union of the extensions under I of the symbols in **R**.

Constraints. The sets of relation symbols occurring in a formula φ are denoted by $\mathsf{sig}(\varphi)$; $\mathsf{sig}(\cdot)$ is extended to sets of formulae in the natural way. A *constraint* is a closed formula (i.e., one without free variables) in some fragment of first-order logic (FOL) with constants $\mathbf{dom} \cup \mathbf{idom}$ under the standard name assumption, that is, the interpretation of each constant is the name of the constant. Given a set Γ of constraints and an instance I over $\mathsf{sig}(\Gamma)$, we say that I is a *model* of (or *satisfies*) Γ, and write $I \models \Gamma$, if the relational structure $\langle D, I \rangle$, where D is the set of constants appearing in I and Γ, makes every formula in Γ true under the standard FOL semantics. We write $I \models \varphi$ as short for $I \models \{\varphi\}$, and we say that I satisfies φ. A set of constraints Γ *entails* (or *logically implies*) a constraint φ, written $\Gamma \models \varphi$, if every finite model of Γ also satisfies φ. All sets of constraints considered in this paper are finite.

Specific Languages. The results presented in this paper are independent of the language $\mathfrak{C}$ used for specifying constraints on the interpreted positions, as long as $\mathfrak{C}$ is closed under negation. In Section 5, we will require that the satisfiability in $\mathfrak{C}$ be decidable. A specific language that enjoys these properties, which we will use in examples throughout the paper, is the fragment of linear arithmetic over the integers consisting of *Unit Two Variable Per Inequality* constraints (UTVPIs). Formally, a UTVPI formula has the form $ax + by \leq d$, where x and y are integer variables, $a, b \in \{-1, 0, 1\}$ and $d \in \mathbb{Z}$. UTVPIs can express comparisons between two variables and between a variable and an integer, but also compare the sum or difference of two variables with an integer. The satisfiability of a set of UTVPIs can be checked in polynomial time [11,8,7]. A Boolean combination of UTVPIs is referred to as BUTVPI; the problem of deciding whether a set of BUTVPIs is satisfiable is NP-complete [12].

3   Selection Views under Constraints

We work under the universal relation assumption, that is, we consider a *source schema* **S** that consists of a single relation symbol S. For the sake of simplicity, w.l.o.g. we assume that the first $\|S\|$ positions of S are non-interpreted, while the remaining ones are interpreted. Under this assumption, for $n = |S|$ and $k = \|S\|$, instances over **S** map each relation symbol to a subset of $\mathbf{dom}^k \times \mathbf{idom}^{n-k}$. W.l.o.g. we further assume that a variable associated with the i-th position of S is named x_i if $i \leq k$, and y_{i-k} otherwise. By default, $\bar{x}$ and $\bar{y}$ denote the tuples $(x_1, \ldots, x_k)$ and $(y_1, \ldots, y_{n-k})$, respectively.

The source schema $\mathbf{S}$ is constrained by formulae of the form:

$$\forall \overline{x}, \overline{y} . S(\overline{x}, \overline{y}) \wedge \lambda(\overline{x}) \rightarrow \delta(\overline{y}) \ , \tag{1}$$

where $\lambda(\overline{x})$ is a Boolean combination of equalities $x = a$, with x from $\overline{x}$ and a from $\mathbf{dom}$, and $\delta(\overline{y}) \in \mathfrak{C}$. We use $x \neq a$ as short for $\neg(x = a)$ and we write (1) simply as $S \wedge \lambda \rightarrow \delta$. Constraints of this form, which by means of a formula in $\mathfrak{C}$ restrict the values that are admissible at interpreted positions whenever the values at non-interpreted ones satisfy a certain condition, are called *conditional domain constraints* (CDCs) [3].[1] We refer to an instance over $\mathbf{S}$ that satisfies the integrity constraints as *source instance*.

A *selection view* is a mapping f that associates each source instance with an instance over a target schema $\mathbf{T}$ consisting of relation symbols of the same arity of S, each of which is defined as follows:

$$\forall \overline{x}, \overline{y} . T(\overline{x}, \overline{y}) \leftrightarrow \big(S(\overline{x}, \overline{y}) \wedge \lambda(\overline{x}) \wedge \sigma(\overline{y}) \big) \ , \tag{2}$$

where $\lambda(\overline{x})$ is as in (1) and $\sigma(\overline{y}) \in \mathfrak{C}$. We write (2) simply as $T \leftrightarrow S \wedge \lambda \wedge \sigma$.[2] The set $\Sigma_{\mathbf{ST}}$ consisting of formulae of the form (2), one for each $T \in \mathbf{T}$, is called the *specification* of f.

A selection view f induces a horizontal decomposition on S; so, for a set $\Sigma_{\mathbf{S}}$ of source constraints, we say that f is *lossless* under $\Sigma_{\mathbf{S}}$ if the target symbols in $\mathbf{T}$ form a lossless horizontal decomposition of S under $\Sigma_{\mathbf{S}}$. That is, f is lossless if and only if $S^I = \bigcup_{T \in \mathbf{T}} T^I$ for every source instance I, which in turn means that the query $S(\overline{x}, \overline{y})$ is equivalent to $\bigvee_{T \in \mathbf{T}} T(\overline{x}, \overline{y})$ under $\Sigma_{\mathbf{S}} \cup \Sigma_{\mathbf{ST}}$, where $\Sigma_{\mathbf{ST}}$ is the specification of f.

Example 1. Consider the source schema $\mathbf{S} = \{S\}$, with S of arity 4 and its last two positions interpreted over the integers. Let $\mathfrak{C}$ be the language of UTVPIs and consider the following set $\Sigma_{\mathbf{S}}$ of CDCs over $\mathbf{S}$:

$$S \wedge (x_1 = a) \rightarrow (y_1 + y_2 \leq 5) \ ; \qquad S \wedge (x_1 \neq a \vee x_2 = b) \rightarrow (y_2 \geq 2) \ .$$

Let f be the selection view specified by the following set $\Sigma_{\mathbf{ST}}$ of definitions:

$$T_1 \leftrightarrow S \wedge (x_1 \neq a \vee x_2 \neq b) \ ; \qquad\qquad T_2 \leftrightarrow S \wedge y_1 \leq 4 \ .$$

Using the technique devised in [3], it can be checked that f is lossless under $\Sigma_{\mathbf{S}}$.

An *update* on a view f with target schema $\mathbf{T}$ is a mapping from the set of instances over $\mathbf{T}$ into itself. An update u is *translatable on* an instance $J \in \mathsf{img}(f)$ if $u(J) \in \mathsf{img}(f)$, that is, there exists a source instance I such that $f(I) = u(J)$; if the underlying view is lossless, the instance I is univocally determined (i.e., it is unique). An update is *globally translatable* if it is translatable on every source instance. From a practical viewpoint, when the target is materialised, the notion of translatability on a single instance allows to determine "on-the-fly" whether to accept or reject incoming update requests. On the other hand, the notion of

---

[1] In [3], the condition $\lambda(\overline{x})$ is restricted to a conjunction of possibly negated equalities, but the results carry over to the variant used in this paper.

[2] See note 1.

global translatability is useful to determine offline which updates can be safely made available online, but it could also be employed on-the-fly when the target is/can not be materialised. Clearly, the feasibility of this latter application relies on the complexity of checking for global translatability.

Observe that for translatable updates w.r.t. lossy views there exists in general more than one source instance that reflects exactly the changes introduced by the update on the target instance. This means that such updates can be propagated to the source in more than one way, thus generating ambiguity and raising the problem of choosing a suitable alternative, which in general calls for user intervention. To overcome this problem, we proceed as follows: in the upcoming Section 4, we first investigate translatability w.r.t. lossless selection views; then, in Section 5, we will show how to construct a lossless selection view from a lossy one, in order to be able to apply the translatability results obtained previously.

4 Update Propagation through Lossless Selection Views

In this section, we study the translatability of insertions, deletions and replacements w.r.t. lossless selection views under CDCs. In general, a view that defines each target symbol by means of a (safe) query over the source is *updatable* if each source symbol can be similarly defined by a (safe) query over the target. Thus, for selection views, the notions of losslessness and updatability coincide. Here, we will apply the general criterion for the translatability of updates introduced in [4] for updatable views to the specific case of lossless selection views.

Denote by $\mathsf{repl}(\Sigma)$ the set of constraints obtained from Σ by replacing each occurrence of $S(\overline{x}, \overline{y})$ by $\bigvee_{T_i \in \mathbf{T}} T_i(\overline{x}, \overline{y})$. Then, we have that $\mathsf{repl}(\Sigma_\mathbf{S}) = \{T_i \wedge \lambda \to \delta \mid (S \wedge \lambda \to \delta) \in \Sigma_\mathbf{S}\}$ and $\mathsf{repl}(\Sigma_\mathbf{ST}) = \{\phi, \psi \mid T_i, T_j \in \mathbf{T}\}$, where:

$$\phi = \left(\quad \forall \overline{x}, \overline{y} \, . \quad T_i(\overline{x}, \overline{y}) \qquad\qquad\qquad \to \lambda_i(\overline{x}) \wedge \sigma_i(\overline{y}) \quad \right) ; \qquad (3)$$

$$\psi = \left(\quad \forall \overline{x}, \overline{y} \, . \quad T_i(\overline{x}, \overline{y}) \wedge \lambda_j(\overline{x}) \wedge \sigma_j(\overline{y}) \to T_j(\overline{x}, \overline{y}) \qquad\qquad \right) . \qquad (4)$$

Note that (3) is in fact equivalent to two CDCs: $T_i \to \sigma_i$ and $T_i \wedge \neg\lambda_i \to \neg\sigma_i$. For short we write (4) as $T_i \wedge \lambda_j \wedge \sigma_j \to T_j$ and we call constraints of this form full *conditional inclusion dependencies* (CINDs). We refer to $\mathsf{repl}(\Sigma_\mathbf{S}) \cup \mathsf{repl}(\Sigma_\mathbf{ST})$ as the set of *derived* target constraints, which intuitively require that every tuple in T_i must satisfy the source CDCs, the selection conditions for T_i and, if it also satisfies the selection conditions for T_j, must be in T_j as well. The derived target constraints can be partitioned into a set of CDCs, consisting of $\mathsf{repl}(\Sigma_\mathbf{S})$ and the CDCs corresponding to each formula of the form (3) in $\mathsf{repl}(\Sigma_\mathbf{ST})$, and a set of CINDs, consisting of the formulae in $\mathsf{repl}(\Sigma_\mathbf{ST})$ of the form (4).

Example 2. Consider the selection view f of Example 1. As f is lossless, $S(\overline{x}, \overline{y})$ is equivalent to $T_1(\overline{x}, \overline{y}) \vee T_2(\overline{x}, \overline{y})$ under $\Sigma_\mathbf{S} \cup \Sigma_\mathbf{ST}$. Replacing every occurrence of the former in $\Sigma_\mathbf{S} \cup \Sigma_\mathbf{ST}$ by the latter, we obtain the following set $\Sigma_\mathbf{T}$ of derived target constraints:

$$T_1 \to (x_1 \neq a \vee x_2 \neq b) \; ; \quad T_2 \wedge (x_1 \neq a \vee x_2 \neq b) \to T_1 \; ;$$
$$T_2 \to y_1 \leq 4 \; ; \qquad\qquad\qquad T_1 \wedge y_1 \leq 4 \to T_2 \; ;$$

$$T_1 \wedge (x_1 = a) \rightarrow (y_1 + y_2 \leq 5) \; ; \qquad T_1 \wedge (x_1 \neq a \vee x_2 = b) \rightarrow (y_2 \geq 2) \; ;$$
$$T_2 \wedge (x_1 = a) \rightarrow (y_1 + y_2 \leq 5) \; ; \qquad T_2 \wedge (x_1 \neq a \vee x_2 = b) \rightarrow (y_2 \geq 2) \; .$$

Let f be a lossless selection view under CDCs, and let $\Sigma_\mathbf{T}$ the corresponding set of derived constraints; we denote by $\mathsf{cdc}(\Sigma_\mathbf{T})$ the set of CDCs in $\Sigma_\mathbf{T}$. In the following, by "translatable" we mean "translatable w.r.t. f", and when we say "translatable on J" we implicitly assume $J \in \mathsf{img}(f)$. We have that an instance J over the target schema is in $\mathsf{img}(f)$ if and only if it is a model of $\Sigma_\mathbf{T}$ [4], and that an update u is translatable on J if and only if $u(J)$ satisfies $\Sigma_\mathbf{T}$ [4].

4.1 Insertions and Deletions

A target update consisting in either the insertion or the deletion of n tuples can be specified by means of the target instance containing precisely the n facts to be inserted or deleted. More precisely, we call *insertion* (*deletion*) the unconditional addition (removal) of a fixed set of facts J into (from) any instance to which the update is applied, and we indicate whether J is to be interpreted as an insertion or a deletion by preceding it with a "+" or a "−", respectively. Thus, $+J$ means that the facts in J are to be inserted, so $+J$ represents the update u such that $u(I) = I \cup J$ for every target instance I; on the other hand, $-J$ means that the facts in J are to be deleted, so $-J$ represents the update u such that $u(I) = I \setminus J$ for every target instance I.

An insertion is translatable on an instance precisely if the set of *new facts* that are actually inserted into the instance satisfies the derived target constraints.

Proposition 1. $+J'$ *is translatable on* J *if and only if* $J' \setminus J \models \Sigma_\mathbf{T}$.

The global translatability of an insertion depends only on whether the set of facts the insertion consists of satisfies the derived target constraints.

Theorem 1. $+J'$ *is globally translatable if and only if* $J' \models \Sigma_\mathbf{T}$.

Example 3. Consider the selection view f from Example 1, let $J = \{T_1(c, b, 3, 3), T_2(c, b, 3, 3)\} \in \mathsf{img}(f)$, and let u be the insertion $+J'$, where $J' = \{T_1(c, b, 3, 3), T_2(a, b, 0, 5)\}$. Then, u is translatable on J, because $J \setminus J' = \{T_2(a, b, 0, 5)\}$ satisfies $\Sigma_\mathbf{T}$ of Example 2; indeed, $u(J) = J \cup J' = \{T_1(c, b, 3, 3), T_2(c, b, 3, 3), T_2(a, b, 0, 5)\}$ is a model of $\Sigma_\mathbf{T}$. However, u is not globally translatable, as $J' \not\models \Sigma_\mathbf{T}$.

A deletion is translatable on an instance iff the set of *existing facts* that are actually deleted from the instance satisfies the derived target constraints.

Proposition 2. $-J'$ *is translatable on* J *if and only if* $J \cap J' \models \Sigma_\mathbf{T}$.

Example 4. Consider the selection view f from Example 1, let $J = \{T_1(c, b, 3, 3), T_2(c, b, 3, 3)\} \in \mathsf{img}(f)$, and let u be the deletion $-J'$, where $J' = \{T_1(c, b, 3, 3), T_2(a, b, 0, 5)\}$. Then, u is not translatable on J, as $J \cap J' = \{T_1(c, b, 3, 3)\}$ does not satisfy $\Sigma_\mathbf{T}$; indeed, $u(J) = J \setminus J' = \{T_2(c, b, 3, 3)\}$ is not a model of $\Sigma_\mathbf{T}$.

The global translatability of a deletion depends only on a subset of the facts the deletion consists of, namely those that satisfy the derived CDCs.

Theorem 2. $-J'$ *is globally translatable if and only if* $J' \setminus J'' \models \Sigma_\mathbf{T}$, *where* J'' *is the maximal subset of* J *that does not satisfy* $\mathsf{cdc}(\Sigma_\mathbf{T})$.

Example 5. Consider the selection view f from Example 1, and let u be the deletion $-\{T_2(a, b, 3, 3), T_2(a, b, 0, 5)\}$. The fact $T_2(a, b, 3, 3)$ violates one of the CDCs in $\Sigma_\mathbf{T}$, and so it cannot belong to any instance in $\mathsf{img}(f)$. Then, as $\{T_2(a, b, 0, 5)\}$ is a model of $\Sigma_\mathbf{T}$, we have that u is globally translatable.

4.2 Sequences of Insertions and Deletions

We now extend our investigation to arbitrary (finite) sequences of deletions and insertions (in any order), that is, composite updates of the form $u_n \cdots u_1$ where each u_i is either $+J_i$ or $-J_i$. We will show that any such sequence is equivalent to a single update that inserts some tuples and, at the same time, deletes some others, which are different from the inserted ones. We refer to this kind of update as an *unconditional update*. Formally, an unconditional update u is denoted by an unordered pair $\langle +J_1, -J_2 \rangle$ consisting of an insertion $+J_1$ and a deletion $-J_2$ such that $J_1 \cap J_2 = \varnothing$; then, we have that $u(J) = (J \cup J_1) \setminus J_2 = (J \setminus J_2) \cup J_1$ for every target instance J. Obviously, insertions and deletions are special cases of unconditional updates.

Note that an unconditional update $\langle +J_1, -J_2 \rangle$ is equivalent to $+J_1$ followed by $-J_2$, as well as $-J_2$ followed by $+J_1$. Intuitively, since the insertion and the deletion are disjoint, it does not matter which one we apply first. The following is not surprising:

Proposition 3. $\langle +J_1, -J_2 \rangle$ *is globally translatable iff* $+J_1$ *and* $-J_2$ *are.*

Clearly, a sequence of insertions and deletions is translatable (globally or on a specific instance) whenever each update in the sequence is; however, the converse is not true in general, because translatability requires the combined effect of the update to yield an instance that satisfies the derived target constraints, and this can happen even when some step in the sequence results in an illegal state.[3]

Example 6. Let f be the selection view of Example 1, and let u be the sequence $+J_1 - J_2$, where $J_1 = \{T_1(a, b, 4, 1)\}$ and $J_2 = J_1 \cup \{T_2(a, b, 4, 1)\}$. For every J in $\mathsf{img}(f)$, we have that $u(J) = (J \cup J_1) \setminus J_2 = J \setminus J_2 = J$ because $J_2 \supseteq J_1$ and $J_2 \cap J = \varnothing$. So u is globally translatable, despite the fact that $+J_1$ is not.

A sequence of insertions is equivalent to a single insertion that, in one shot, inserts all of the facts inserted in successive chunks along the sequence; the same holds for sequences of deletions. So, in turn, an arbitrary sequence of insertions and deletions is equivalent to an alternating sequence of one insertion and one deletion, as every subsequence consisting of insertions (deletions) can be replaced by a single insertion (deletion). The following shows that such an alternation of insertions and deletions is equivalent to a single unconditional update.

---

[3] This is the standard semantics of database transactions.

Theorem 3. $+J_1-J_2\cdots+J_{n-1}-J_n$ *is equivalent to* $\langle +A, -D\rangle$ *where*

$$A = \bigcup_{i=1}^{n/2} Z_{2i-1} \quad and \quad D = \bigcup_{i=1}^{n/2} Z_{2i} \ , \quad with \ Z_k = J_k \setminus \bigcup_{j=1}^{\lceil \frac{n-k}{2} \rceil} J_{k+(2j-1)} \ .$$

Intuitively, a fact is in A (resp., D) iff it is inserted (resp., deleted) at some point of the update sequence and it is not deleted (resp., inserted) afterwards.

The above theorem, in combination with Proposition 3, allows us to check for the (global) translatability of an arbitrary sequence of insertions and deletions by checking for the (global) translatability of an insertion and a deletion.

Example 7. Let f be the selection view of Example 1, and let u be the sequence $-J_1+J_2+J_3-J_4-J_5+J_6-J_7$ with $J_1 = \{T_2(a, b, 1, 4)\}$, $J_2 = \{T_1(b, c, 3, 1), T_2(a, b, 0, 5)\}$, $J_3 = \{T_1(c, b, 2, 4), T_2(a, b, 1, 4)\}$, $J_4 = \{T_2(a, b, 0, 5)\}$, $J_5 = \{T_1(c, b, 2, 5)\}$, $J_6 = \{T_2(c, b, 2, 4)\}$ and $J_7 = \{T_1(b, c, 3, 1), T_2(c, b, 2, 5)\}$. Then, u is equivalent to $u' = +J_1'-J_2'+J_3'-J_4'+J_5'-J_6'$, where $J_1' = \varnothing$, $J_2' = J_1$, $J_3' = J_2 \cup J_3$, $J_4' = J_4 \cup J_5$, $J_5' = J_6$ and $J_6' = J_7$. By Theorem 3, u' is equivalent to $\langle +A, -D\rangle$ where:

$$A = \left(J_1' \setminus (J_2' \cup J_4' \cup J_6')\right) \cup \left(J_3' \setminus (J_4' \cup J_6')\right) \cup \left(J_5' \setminus J_6'\right) =$$
$$= \{T_1(c, b, 2, 4), T_2(a, b, 1, 4), T_2(c, b, 2, 4)\} \ ;$$
$$D = \left(J_2' \setminus (J_3' \cup J_5')\right) \cup \left(J_4' \setminus J_5'\right) \cup J_6' =$$
$$= \{T_2(a, b, 0, 5), T_1(c, b, 2, 5), T_1(b, c, 3, 1), T_2(c, b, 2, 5)\} \ .$$

It can be verified that both $+A$ and $-D$ are globally translatable, hence so is u' by Proposition 3 and, in turn, u.

4.3   Replacements

Differently from all the updates we have considered so far, replacements have a semantics where insertion is inherently conditional on deletion: replacing a fact $T_i(\bar{t})$ with $T_j(\bar{t}')$ in an instance J means inserting $T_j(\bar{t}')$ into J *only when* $T_i(\bar{t})$ is effectively deleted from it, that is, only if $T_i(\bar{t})$ belongs to J. A replacement u is specified by means of a surjective mapping $r \colon D \to A$ where D (for "deleted") and A (for "added") are disjoint sets of facts; then, $u(J) = (J \setminus D) \cup r(J \cap D)$ for every target instance J. Since replacing D by A in J amounts to first deleting D from J and then inserting $r(J \cap D)$ into J, the following is immediate.

Proposition 4. *A replacement specified by $r \colon D \to A$ is translatable on J if and only if $-D+r(J \cap D)$ is.*   □

Obviously, a replacement is not globally equivalent to a deletion followed by an insertion, which in turn would make it equivalent to an unconditional update by Theorem 3, because the set of replaced facts depends on the specific instance to which the update is applied. A replacement is globally translatable precisely if it replaces consistent sets of facts with consistent sets of facts, as shown below.

Theorem 4. *Let u be the replacement specified by $r: D \rightarrow A$, and let $\tilde{D}$ be the maximal subset of D that satisfies $\mathsf{cdc}(\Sigma_{\mathbf{T}})$. Then, u is globally translatable iff $\tilde{D} \models \Sigma_{\mathbf{T}}$ and, for each subset D' of $\tilde{D}$, $r(D')$ satisfies $\Sigma_{\mathbf{T}}$ whenever D' does.*

Example 8. Consider the selection view f of Example 1, and let u be the replacement specified by $r = \{T_1(a, c, 2, 0) \mapsto T_2(a, c, 2, 1), T_2(a, c, 2, 0) \mapsto T_1(a, c, 2, 1), T_2(a, b, 3, 2) \mapsto T_2(a, b, 4, 2)\}$. Then, u is not globally translatable because, even though $D = \tilde{D} = \{T_1(a, c, 2, 0), T_2(a, c, 2, 0), T_2(a, b, 3, 2)\}$ satisfies $\Sigma_{\mathbf{T}}$, we have that $D' = \{T_2(a, b, 3, 2)\} \subseteq \tilde{D}$ is a model of $\Sigma_{\mathbf{T}}$ whereas $r(D') = \{T_2(a, b, 4, 2)\}$ is not. Indeed, u is not translatable on any instance that contains D'.

4.4   Complexity Results

In this section, we study the complexity of the decision problem associated with global translatability, measured w.r.t. the size of the update. Given a fixed source schema under CDCs and a fixed lossless selection view, the *translatability problem* takes as input an update u and answers the question: "Is u globally translatable?" Notably, we will show that for insertions, deletions, unconditional updates and sequences of insertions and deletions the problem is in AC^0, that is, it can be solved by (a uniform family of) Boolean circuits of constant depth and polynomial size consisting of gates with unbounded fan-in [10].

By Theorem 1, the problem of deciding whether an insertion is globally translatable reduces to model checking in FOL, which is AC^0 in data complexity [2].

Theorem 5. *The translatability problem for insertions is in AC^0.*

By Theorem 2, whether a deletion $-J$ is globally translatable can be decided by checking whether the maximal subset $\tilde{J}$ of J that satisfies $\mathsf{cdc}(\Sigma_{\mathbf{T}})$ is a model of $\Sigma_{\mathbf{T}}$.[4] This set can be computed by an AC^0 circuit C1 by independently going through each fact in J and checking whether it satisfies $\Sigma_{\mathbf{T}}$. Whether $\tilde{J}$ satisfies $\Sigma_{\mathbf{T}}$ can be solved by an AC^0 circuit C2, and by connecting the output of C1 to the input of C2 we still get an AC^0 circuit. Thus, we have the following:

Theorem 6. *The translatability problem for deletions is in AC^0.*

Given an unconditional update $\langle +J_1, -J_2 \rangle$, by Theorem 5 and Theorem 6 there exist two AC^0 circuits for deciding the translatability of $+J_1$ and $-J_2$. Since by connecting the outputs of these circuits to an AND gate we get a circuit with constant depth and polynomial size in $|J_1| + |J_2|$, the following holds:

Theorem 7. *The translatability problem for unconditional updates is in AC^0.*

The global translatability of an arbitrary sequence of insertions and deletions can be checked by constructing the sets A and D of Theorem 3, and checking whether $\langle +A, -D \rangle$ is globally translatable. This check can be done in AC^0 by Theorem 6 and, as A and D can be constructed in AC^0, we get the following:

---

[4] $\tilde{J} = J \setminus J'$, where J' is the maximal subset of J that does *not* satisfy $\mathsf{cdc}(\Sigma_{\mathbf{T}})$.

Theorem 8. *The translatability problem for sequences consisting of insertions and deletions is in* AC^0.

Theorem 4 yields a naive algorithm for checking the global translatability of a replacement specified by $r: D \to A$ that runs in exponential time in the size of D (by going through each subset of $\tilde{D}$). Actually, things are much easier than that, as it turns out that we do not need to explore the whole powerset of $\tilde{D}$.

Theorem 9. *The global translatability of a replacement specified by* $r: D \to A$ *can be decided in linear time in the size of* D.

5   Attaining Losslessness from Lossy Selection Views

In general, given a lossy view, losslessness can be achieved by means of a so-called *view complement* [1], which is another view that provides at least the amount of information that is missing from the original one.

Definition 1 (View complement [1]). *Let f and g be views. We say that g is a* complement *of f if, for every pair of distinct source instances I, I', it holds that $g(I) \neq g(I')$ whenever $f(I) = f(I')$.*

Observe that the notion of complement is symmetric: if g is a complement of f, then f is a complement of g; for this reason, we say that f and g are *complementary*. We assume w.l.o.g. that complementary views have disjoint target schemas; it is easy to see that two complementary views f and g can be combined into an injective[5] view $f \cup g$ associating each source instance I with $f(I) \cup g(I)$.

Update propagation in the presence of view complements is regulated by the *constant complement principle* [1], prescribing that updates must not affect, directly or indirectly, the information provided by the complement. The motivation for this requirement is that complementary information is provided only for the sake of updating the original lossy view. The connection between translatability w.r.t. lossy views under constant complement and translatability w.r.t. lossless views is as follows:

Theorem 10 ([4]). *Let f and g be complementary views with target schemas* **T** *and* **C**, *respectively. Let u be an update over* **T**, *and let v be the update over* **C** *such that $v(J) = J$ for every instance J over* **C**. *Then, u is translatable w.r.t. f under constant g if and only if $u \cup v$ is translatable w.r.t. $f \cup g$.*

In light of the above theorem, we are interested in finding complements for which the resulting lossless view $f \cup g$ can be expressed in the same language as f. In the following, given a lossy selection view f (under CDCs), we will show how to find a selection view that is a complement of f, so that the translatability results established in the previous section carry over to translatability under constant complement, since the union of two selection views is also a selection view.

---

[5] In general, for arbitrary views, losslessness is equivalent to injectivity; in particular, for selection views, losslessness also coincides with updatability.

5.1   Minimal Complements

A complement assigns univocal semantics to updates, issued on lossy views, that would otherwise be ambiguous. Thus, under the constant complement principle, the choice of a complement determines which updates are translatable and which are not, so an update may be translatable under one complement but not under some other. As the complement must be invariant during the propagation process, if a complement provides more information than another, less updates are translatable under the former than under the latter. The following allows us to compare how much information a view transfers w.r.t. an other.

Definition 2 (Information-transfer order [1,5]). *Let f and g be views. We say that f is less informative than g (written $f \leq g$) if, for every pair of source instances I, I', we have that $g(I) = g(I')$ implies $f(I) = f(I')$.*

We write $f \equiv g$ if and only if $f \leq g$ and $g \leq f$; in such a case, we say that f and g are equally informative, are equivalent under $\leq$, or transfer the same amount of information. A view is lossless if it provides all of the informative content of the source, i.e., it transfers the same amount of information as the identity view.

Since the less information view complements provide the more likely updates are to be translatable, the amount of information transferred by a view complement should be kept as small as possible, which leads us to the following notion of minimality:

Definition 3 (Minimal complement [1]). *Let f and g be complementary views. We say that g is a minimal complement of f if, for every complement h of f, it is the case that $g \leq h$ whenever $h \leq g$.*

Unfortunately, the existence of a *unique* (up to information transfer) minimal complement is limited to trivial cases with no practical relevance, namely when the original view is either injective or constant [1].

When considering a specific class of views that satisfy certain properties, such as that of being expressible in a concrete language, we are interested in finding a complement that is minimal among all complements within that class. We refer to minimal complements within a class $\mathbb{C}$ of views as *minimal in* $\mathbb{C}$. If for a view f in $\mathbb{C}$ there exists a unique minimal complement in $\mathbb{C}$, this is called the *perfect* complement for f (in $\mathbb{C}$).

5.2   Complements of Selection Views

We now turn back our attention to the class of selection views. All of the results presented in the following hold for selection views over a source schema $\mathbf{S} = \{S\}$ constrained by CDCs; for such a view f and a source instance I, U_f^I denotes the set $\bigcup(\mathbf{T}, f(I)) = \{\bar{t} \mid \bar{t} \in T^{f(I)}, T \in \mathbf{T}\}$, where $\mathbf{T}$ is the target schema of f.

The information-transfer order of Definition 2 can be characterised in terms of containment between the sets of selected tuples: intuitively, f transfers less information than g if f selects a subset of the source tuples selected by g.

Proposition 5. $f \leq g$ if and only if $U_f^I \subseteq U_g^I$ for every source instance I.

Complementarity can also be characterised in terms of the selected tuples: f and g are complementary if each source tuple is selected by at least one of them.

Proposition 6. f and g are complementary if and only if $S^I = U_f^I \cup U_g^I$ for every source instance I.

We now show that, when it does exist, the perfect complement of a selection view f under CDCs is the view that selects all and only the source tuples not captured by f, that is, not appearing in any of the target relations defined by f.

Proposition 7. g is the perfect complement of f if and only if $U_g^I = S^I \setminus U_f^I$ for every source instance I.

Note that, for selection views, the notion of perfect complement is symmetric because, when f and g are complementary, for every source instance I we have that $U_f^I = S^I \setminus U_g^I$ if and only if $U_g^I = S^I \setminus U_f^I$. So, if g is a perfect complement of f, then f is a perfect complement of g, and we say that f and g are *perfectly complementary*.

A selection view f is lossless if and only if it selects all of the tuples in each source instance I, that is, $S^I = U_f^I$. In general, for every selection view f and every source instance I, it always holds that $U_f^I \subseteq S^I$, hence f is lossy if and only if there is a source instance $\hat{I}$ containing some fact $S(\bar{t})$ where $\bar{t}$ does not belong to the extension of any target symbol under $f(\hat{I})$, that is, for which $\bar{t} \notin U_f^{\hat{I}}$. So, for each $T \in \mathbf{T}$ defined by $T \leftrightarrow \lambda \wedge \sigma$, the values in $\bar{t}$ at non-interpreted positions do not satisfy λ, or the values in $\bar{t}$ at interpreted positions do not satisfy σ.

Given a model I of a set $\Sigma_\mathbf{S}$ of CDCs, every subset of I is a model of $\Sigma_\mathbf{S}$ as well, thus we can restrict our attention to source instances consisting of a single tuple. With each equality or inequality between a variable x_i and a constant a we associate a propositional variable p_i^a, whose truth-value indicates whether the value in the i-th position is a. To each valuation of such variables corresponds a (possibly infinite) set of tuples; e.g., a valuation of $\{p_1^a, p_2^b\}$ that assigns true to p_1^a and false to p_2^b identifies all the tuples in which the value of the first element is a and the value of the second is different from b.

We say that a CDC $S \wedge \lambda \to \delta$ is *applicable* under a valuation α if the propositional representation of λ is true under α; similarly, a definition $T \leftrightarrow \lambda \wedge \sigma$ in $\Sigma_\mathbf{ST}$ is applicable under a valuation α if the propositional representation of λ is true under α. A valuation α is *consistent* if

(1) no two variables referring to the same position (i.e., with the same subscript) but to different constants (i.e., with different superscripts) are both true under α, and (2) no definition in $\Sigma_\mathbf{ST}$ of the form $T \leftrightarrow \lambda$ (that is, without condition on the interpreted attributes) is applicable under α. Given a valuation α, we define the set Π^α consisting of

– each of the $\mathfrak{C}$-formulae δ appearing in the consequent of a CDC in $\Sigma_\mathbf{S}$ that is applicable under α, and

- the *negation* of each of the $\mathfrak{C}$-formulae σ (if any) appearing in a definition in $\Sigma_{\mathbf{ST}}$ that is applicable under α.

Then, a selection view f specified by $\Sigma_{\mathbf{ST}}$ is lossless under $\Sigma_{\mathbf{S}}$ if and only if Π^α is unsatisfiable for every consistent valuation α [3]. We denote by $\mathsf{CVSAT}(f)$ the set of all consistent valuations α for which Π^α is satisfiable.

Theorem 11. *The perfect complement g of f exists if and only if, for each α in $\mathsf{CVSAT}(f)$, there is a formula $\pi^\alpha \in \mathfrak{C}$ equivalent to $\bigwedge_{\phi \in \Pi^\alpha} \phi$. In such a case, g is the view that, for each $\alpha \in \mathsf{CVSAT}(f)$, defines $C(\overline{x}, \overline{y}) \leftrightarrow \lambda(\overline{x}) \wedge \sigma(\overline{y})$ where*

$$\lambda(\overline{x}) = \left[\bigwedge_{\alpha(p_i^a) = \mathtt{true}} (x_i = a) \right] \wedge \left[\bigwedge_{\alpha(p_i^a) = \mathtt{false}} (x_i \neq a) \right] \quad ; \qquad \sigma(\overline{y}) = \pi^\alpha \ .$$

When satisfiability in $\mathfrak{C}$ is decidable, as is the case for UTVPIs and BUTVPIs, Theorem 11 directly yields an algorithm for computing the perfect complement of any selection view. Observe that, when $\mathfrak{C}$ is the language of BUTVPIs, every selection view has a perfect complement (which can be constructed as in Theorem 11) because BUTVPIs are closed under conjunction; however, when $\mathfrak{C}$ is the language of UTVPIs, there exist selection views that do not have a perfect complement, as shown below.

Example 9. Consider the source schema $\mathbf{S} = \{S\}$ without integrity constraints, where S is binary and its second position is interpreted over the integers. Let f be the selection view specified as follows:

$$T_1(x_1, y_1) \leftrightarrow x_1 = a \wedge y_1 > 1 \ ; \qquad T_2(x_1, y_1) \leftrightarrow x_1 = a \wedge y_1 < 0 \ .$$

The view f is lossy, because there exist two consistent valuations $\alpha_1 = \{p_i^a \mapsto \mathtt{true}\}$ and $\alpha_2 = \{p_i^a \mapsto \mathtt{false}\}$ for which $\Pi^{\alpha_1} = \{y_1 \leq 1, y_1 \geq 0\}$ and $\Pi^{\alpha_2} = \varnothing$ are satisfiable. Let g be the view constructed as in Theorem 11, specified as:

$$C_1(x_1, y_1) \leftrightarrow x_1 = a \wedge y_1 \leq 1 \wedge y_1 \geq 0 \ ; \qquad C_2(x_1, y_1) \leftrightarrow x_1 \neq a \ .$$

While f can be expressed when $\mathfrak{C}$ is the language of UTVPIs, g cannot. However, both f and g can be expressed when $\mathfrak{C}$ is the language of BUTVPIs.

If the perfect complement does not exist, there is not a unique minimal complement, but multiple incomparable ones. Below, we show how to find a minimal complement (or the perfect one, should it exist) when $\mathfrak{C}$ consists of UTVPIs.

Theorem 12. *Let $\mathfrak{C}$ be the language of UTVPIs, let f be a selection view, and let g be a view that, for each $\alpha \in \mathsf{CVSAT}(f)$, defines $C(\overline{x}, \overline{y}) \leftrightarrow \lambda(\overline{x}) \wedge \sigma(\overline{y})$, where $\lambda(\overline{x})$ is as in Theorem 11 and $\sigma(\overline{y})$ is a formula in Π^α that does not entail any formula in $\Pi^\alpha \setminus \{\sigma(\overline{y})\}$. Then, g is a minimal complement of f.*

When $\mathfrak{C}$ is the language of UTVPIs, the selection view f of Example 9 has the following minimal complements: g_1 specified by $C_{11}(x_1, y_1) \leftrightarrow x_1 = a \wedge y_1 \leq 1$ and $C_{12}(x_1, y_1) \leftrightarrow x_1 \neq a$, and g_2 specified by $C_{21}(x_1, y_1) \leftrightarrow x_1 = a \wedge y_1 \geq 0$ and $C_{22}(x_1, y_1) \leftrightarrow x_1 \neq a$.

6 Discussion and Outlook

In this paper, we investigated the problem of propagating updates through selection views defined over a source schema constrained by CDCs, where some of the attributes are interpreted on a domain consisting of data values that can be compared beyond equality, according to a FO language $\mathfrak{C}$ closed under negation.

In the first place, focusing on lossless selection views, we have provided necessary and sufficient criteria for determining whether insertions, deletions, replacements and arbitrary sequences of insertions and deletions can be unambiguously propagated on a specific target instance (within the image of the underlying selection view) as well as globally (i.e., on every target instance). We have studied the complexity of the decision problem associated with global translatability of such updates and we have shown that it is in AC^0 for insertions, deletions, sequences of insertions and deletions, and for combined updates consisting in the simultaneous insertion and deletion of two disjoint sets of facts. With regard to replacements, we have shown that their global translatability can be checked in linear time in the size of the update. These results provide strong evidence of the practical feasibility of a fully-automatic system for the consistent and univocal propagation of updates through selection views.

In the second place, we considered lossy selection views and studied the problem of attaining losslessness by means of a view complement that provides the information missing from the original view. We have shown how to find complements that are selection views as well, so as to be able to apply the translatability results obtained for lossless views also under the constant complement principle, requiring that the complement be invariant during update propagation. In particular, we gave a general technique for finding, if any, the perfect complement of a selection view, that is, the unique selection view that provides the smallest amount of information needed to attain losslessness. We have shown that, when $\mathfrak{C}$ is the language of BUTVPIs, there exists a perfect complement of any given selection view, while this is not the case when $\mathfrak{C}$ is the language of UTVPIs.

To the best of our knowledge, [9] is the only work where the problem of update propagation through selection views has been previously studied in some depth. The setting of [9] differs from ours mainly in that their language allows to express equalities between variables associated with non-interpreted attributes (i.e., the x_i's). How the insertion/deletion (replacements and sequences are not considered) of a tuple into/from a relation must be propagated to the others is determined by constructing the so-called *complete transfer matrix*, which essentially plays the role of the CINDs in our derived target constraints. In [9], only translatability on a specific instance is addressed, global translatability is not.

Most of the work about view complements is concerned with views defined by projections over a source schema constrained by functional dependencies (FDs). In [5] it is shown that there are projective views under FDs for which the perfect complement does not exist, while in [6] sufficient conditions for its existence are given.

We conclude by briefly discussing two main research directions that we deem worthy of further investigation.

Update Independence. Updating a target relation independently of the others is important in a distributed setting, where it might not be possible to simultaneously update all of the peers that must be modified in order to maintain global consistency after a change happens in one of them, and so one wants to update a peer without affecting the others. We conjecture that T_i and T_j are update independent iff either (the propositional representation of) $\lambda_i \wedge \lambda_j$ is unsatisfiable, or $\{\delta_i, \delta_j\} \cup \{\delta \mid (S \wedge \lambda \rightarrow \delta) \in \Sigma_{\mathbf{S}}, \ \lambda_i \wedge \lambda_j \rightarrow \lambda\}$ is unsatisfiable (in $\mathfrak{C}$).

Additional Source Constraints. In [3], losslessness of selection views is investigated also in the presence of more traditional integrity constraints on the source schema, namely FDs and unary inclusion dependencies (UINDs), in addition to CDCs. Suitable restrictions on the CDCs are proposed to "separate" them from FDs and UINDs, and thus exploit the losslessness results obtained under CDCs in isolation. It would be of interest to study the translatability of updates (under constant complement) in the case when the source schema is constrained by a separable combination of CDCs with FDs and UINDs. Note that the techniques for finding complements of selection views described in Section 5 would not be affected by the presence of such additional source constraints. However, for what concerns translatability, we would obtain a set of derived target constraints that includes also FDs and UINDs, so the results of Section 4 do not automatically carry over to this case.

References

1. Bancilhon, F., Spyratos, N.: Update semantics of relational views. ACM Trans. Database Syst. 6(4), 557–575 (1981)
2. Barrington, D.A.M., Immerman, N., Straubing, H.: On uniformity within NC^1. Journal of Computer and System Sciences 41(3), 274–306 (1990)
3. Feinerer, I., Franconi, E., Guagliardo, P.: Lossless horizontal decomposition with domain constraints on interpreted attributes. In: Gottlob, G., Grasso, G., Olteanu, D., Schallhart, C. (eds.) BNCOD 2013. LNCS, vol. 7968, pp. 77–91. Springer, Heidelberg (2013)
4. Franconi, E., Guagliardo, P.: On the translatability of view updates. In: AMW 2012. CEUR Workshop Proceedings, vol. 866, pp. 154–167. CEUR-WS.org (2012)
5. Guagliardo, P., Pichler, R., Sallinger, E.: Enhancing the updatability of projective views. In: AMW 2013. CEUR Workshop Proceedings, vol. 1087 (2013)
6. Hegner, S.J.: FD covers and universal complements of simple projections. In: Lukasiewicz, T., Sali, A. (eds.) FoIKS 2012. LNCS, vol. 7153, pp. 184–202. Springer, Heidelberg (2012)
7. Jaffar, J., Maher, M.J., Stuckey, P.J., Yap, R.H.C.: Beyond finite domains. In: Borning, A. (ed.) PPCP 1994. LNCS, vol. 874, pp. 86–94. Springer, Heidelberg (1994)
8. Lahiri, S.K., Musuvathi, M.: An efficient decision procedure for UTVPI constraints. In: Gramlich, B. (ed.) FroCoS 2005. LNCS (LNAI), vol. 3717, pp. 168–183. Springer, Heidelberg (2005)
9. Maier, D., Ullman, J.D.: Fragments of relations. SIGMOD Rec. 13(4), 15–22 (1983)
10. Papadimitriou, C.H.: Computational Complexity. Addison Wesley (1994)
11. Schutt, A., Stuckey, P.J.: Incremental satisfiability and implication for UTVPI constraints. INFORMS Journal on Computing 22(4), 514–527 (2010)
12. Seshia, S.A., Subramani, K., Bryant, R.E.: On solving boolean combinations of UTVPI constraints. JSAT 3(1-2), 67–90 (2007)

Data Integration by Conceptual Diagrams

Loredana Caruccio, Vincenzo Deufemia,
Mara Moscariello, and Giuseppe Polese

Department of Management and Information Technology
Università di Salerno
via Giovanni Paolo II, 132, Fisciano(SA), Italy
{lcaruccio,deufemia,gpolese}@unisa.it

Abstract. We present a visual language based approach and tool to perform data integration at conceptual level, aiming to reduce the complexity of such task when integrating numerous and complex data sources. The visual language provides iconic operators to manipulate the constructs of conceptual data schemas of database sources, in order to specify how to merge and map them to a reconciled schema. The proposed tool allows not only to generate the relational reconciled schema, but also to automatically generate metadata and inference mechanisms to guarantee the loading and periodical refresh of data from source databases. Finally, we evaluated CoDIL through a usability study.

Keywords: Data Integration, Data Source, Data Reconciliation, Conceptual Data Schema.

1   Introduction

The reconciliation of multiple heterogeneous data sources entails detecting correspondences among concepts represented in source schemas, and resolving conflicts in order to create a unified global schema whose constructs are related (mapped) to constructs of local source schemas [5,6,10]. This is a complex operation originally carried out by means of manually coded ETL procedures, yielding a time consuming activity, which required the involvement of people with programming skills. In the literature, we can find several software tools providing automated support to the data integration process [7,13,16]. Since this type of tools suffers from scale-up problems when integrating complex data sources, researchers have started investigating higher level approaches [9,15,18], aiming to derive conceptual-level, platform-independent design approaches, including techniques for their automatic implementation [14,17]. With respect to these approaches, in this paper we propose an integrated approach using a visual language based front-end to simplify the specification of the data integration process [8], and back-end logic inference mechanisms to support implementation [3]. We also present a user study in which we have evaluated front-end usability and the overall contribution to complexity reduction and productivity improvement. The underlying visual language has resulted easy to learn for accomplishing sufficiently complex data integration tasks.

H. Decker et al. (Eds.): DEXA 2014, Part II, LNCS 8645, pp. 310–317, 2014.

The paper is organized as follows. Section 2 describes the visual front-end, whereas Section 3 describes the implemented tool. Language and tool evaluation is presented in Section 4. Finally, summary and concluding remarks are included in Section 5.

2   The Visual Front-End

The visual front-end has its backbone in the visual CoDIL [8], which provides icon operators specifying mappings between pairs of conceptual data subschemas. In addition, the visual front-end provides layering mechanisms for help focusing on the current subschemas to be mapped, and a simulator showing the evolution of the reconciled schema as the data integration process progresses, as shown in Figure 1. Notice that the central part of the window contains the palette of icon operators that have been applied. A more detailed description of main icon operators of CoDIL is shown in Table 1.

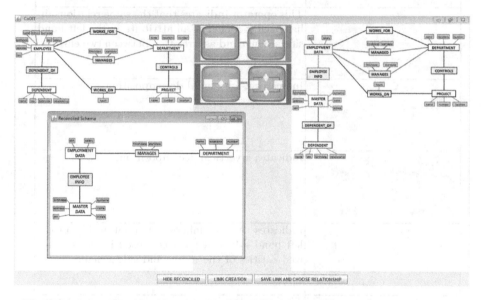

Fig. 1. Application of a horizontal relationship partitioning equivalence operator

In particular, Figure 1 shows a scenario in which the DBA has compared two pairs of subschemas from the compared source subschemas. In the first case, s/he has selected the vertical partitioning operator to indicate that the relationship between entities Employment and Master Data in the right window must be adopted in the reconciled schema, and that the entity Employee from the left window must be mapped to it. In the second case, s/he has selected the *horizontal-partitioning* operator to indicate that the relationship Manages from the left window must be adopted in the reconciled schema, and the pair of

Table 1. A portion of the icon operators dictionary of CoDIL

derivable	Operator indicating that the same concept has been modeled as a simple attribute in the left hand side schema and as a derivable attribute in the right hand side schema.
	This operator indicates that the same concept has been modeled as a simple attribute in the left hand side schema and as a composite attribute in the right hand side schema.
	Indicates that the same concept has been modeled as an attribute in the left hand side schema and as a relation between two entities.
	This operator indicates that the same relationship has been modeled as an attribute in the left hand side schema and as a relation between two entities.
	Indicates a horizontal decomposition.
	Indicates a vertical decomposition.
	Indicates that an inheritance relationship in the left hand side schema is collapsed in the specialized entities of the right hand side schema.
	This operator indicates an inheritance relationship in the left hand side schema that is collapsed in the generalized entity of the right hand side schema.
	This operator indicates that an inheritance relationship in the left hand side schema is implemented using relationships in the right hand side schema.

relationships **Manages** and **Managed** from the right window must be mapped to it. Notice that color patterns are used both to highlight the currently compared sub-schemas, and the window containing the solution to be adopted for the reconciled schema (e.g., highlighting through the green color the border of the icon operator half that is closest to the window containing the adopted solution).

In order to avoid cluttering the screen, the DBA can disable the coloring of a previously selected operator, similarly to the technique of layers used in image editing tools [1].

The semantics of CoDIL has been modeled by means of Description Logic [3], which provided us the basis to construct inference mechanisms to support the implementation of the visually specified data integration processes.

3 System Prototype

The visual front-end and back-end modules have been implemented within the Conceptual Data Integration Tool (CoDIT), whose architecture is shown in Figure 2.

The *Visual Editor* supports the data integration specification process described in the previous section. The *Reconciled Schema Generator* provides automated ER to relational mapping like most conceptual modeling CASE tools [4]. Its output is the relational schema of the reconciled database, and it is stored in the system catalog. The *Data Loader* is a module responsible for extracting the data from sources and for loading them in the reconciled database, after transformations accomplished by means of metadata and scripts generated from the visual editor. Finally, the system relies on third-party tools to perform relational to ER reverse engineering when needed [2].

4 Evaluation

The experiments we present in this section aim to evaluate the usability of the visual front-end, and its contribution to the effectiveness of the data integration process in terms of simplicity of use and time requested.

4.1 Methodology

The tool has been evaluated by means of several experiments involving fifteen graduate students in Computer Science attending an advanced database course. They were between 25 and 29 years old and had attended an undergraduate course on database, hence they were all familiar with ER modeling. Three of them had also an industrial experience in information system development, but none had attended a course or had a professional experience in data integration topics.

The study has been organized in steps by using the think-aloud technique, an approach originating in cognitive psychology [11]. Before starting the experiments, we spent two hours to introduce participants with some notions of

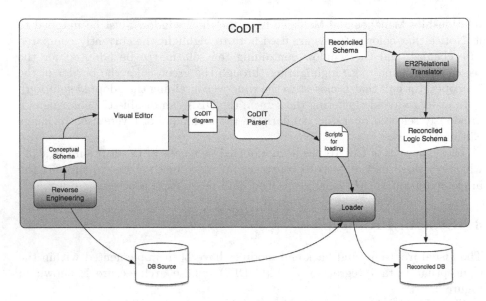

Fig. 2. System architecture and the process for generating the reconciled schema

database schema level integration, the visual language CoDIL, and CoDIT functionalities, followed by two hours of exercises on some sample data integration tasks with CoDIT. Successively, participants were asked to fill a pre-task questionnaire, to perform a complete data integration task involving two Clinical data sources, and finally to fill a post-task questionnaire. During the task we did not provide support to participants in order to avoid biasing the experiment.

The goal of the pre-task questionnaire was to assess the acquired familiarity with the tool CoDIT, and contained three questions. In the first one, participants were requested to observe a CoDIT screenshot, with a pair of highlighted subschemas from the two windows, and to select a proper icon operator to express the conceptual link between them. In the second question, participants were given an icon operator, and were requested to highlight at least a pair of subschemas that could be related based on it. Finally, in the third question, participants were shown twelve pairs of conceptually equivalent subschemas, and were requested to specify for each pair the icon operator most suitable to express their conceptual relationships.

Regarding the data integration task, participants were given two source conceptual schemas of clinical records, and a possible reconciled version of them. Then, participants were given two hours to derive the reconciled schema from the two source schemas by using CoDIT. Successively, they were asked to fill the post-task questionnaire in Figure 3 as a feedback of their experience with CoDIL and CoDIT. Answers to questions from 1 to 9 were given according to a Likert scale [12]: from 1 (very low) to 5 (very high); answers to questions 10 and 11 required specifying one of the icon operators; finally the remaining were open answer questions.

Usability Questionnaire
1. How easy has been solving the data integration task?
2. Was your background knowledge (including training on CoDIT) adequate to face the assigned task?
3. Did you feel you could complete the task without external help?
4. How easy is using CoDIT?
5. Is the user interface of CoDIT pleasant and easy to understand?
6. How effective are the mechanisms (coloring, layers) for highlighting the current pair of subschemas on which to apply an operator?
7. Were messages and warnings easy to understand?
8. Is the semantics of icon operators intuitive?
9. How easy is the task of applying an icon operator?
10. Which is the icon operator you felt more complex among those you applied during the data integration task?
11. Which is the icon operator you felt less complex among those you applied during the data integration task?
12. What did you like least about CoDIT?
13. What did you like most about CoDIT?

Fig. 3. Usability questionnaire

4.2   Results

Pre-task Questionnaire. All participants provided a correct answer to question one. Regarding question two, eleven participants provided a correct solution, whereas four provided a wrong solution. For question three, eight participants correctly associated all the twelve pairs of sub-schemas through the right icon operators, whereas a participant made one error, three made two errors, two made three errors, and one made five errors. The errors revealed that main difficulties were with the application of operators concerning generalization/specialization hierarchies and with the transitive closure of relationships.

Task Accomplishment. Eight participants successfully completed the task, whereas five of them completed the task with some errors. Approximately, their solutions were at least eighty percent correct. Finally, two participants completed about half of the whole task with some errors.

The process followed by the thirteen participants completing the task had some differences in the selected icon operators and/or in their application order. It is worth to note that although two solutions might be semantically equivalent, they might yield different complexity for the back-end phases to be successively handled by CoDIT, e.g., ER to relational schema mapping and consequently, periodical data refresh from sources. Thus, although equivalent, different solutions might be of different quality. In order to compare the quality of the eight correct solutions, for each of them we have automatically generated the integrated database, and have run a data refresh session using two source databases.

Regarding the five complete solutions with errors, they were wrong in the specification of constraints for some icon operators, especially for transitive closure and generalization/specialization hierarchies.

Post-task Questionnaire. Figure 4 shows for the average values of user answers to the post-task questionnaire. In particular, the minimum and maximum peaks show that most participants considered the task demanding (question 1), and the semantics of icon operators intuitive (question 8). Moreover, the peak of question 7 reveals that the tool still needs some engineering work to improve

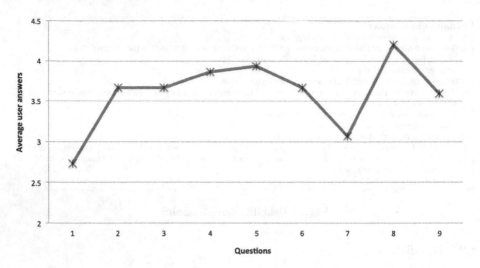

Fig. 4. The average values of user answers to the post-task questionnaire

messages and warning dialogues. Regarding question 10, most of them indicated the transitive closure operator and its application as the most complex one, except two who indicated generalization/specialization hierarchy operators. For question 11, five of them indicated an attribute mapping operator as the simplest one, six indicated a conflict operator, and two the entity partitioning operators. Regarding the responses to questions 12 and 13, the participants considered the intuitiveness of icon operators and the layering mechanisms as the most appealing feature of CoDIT, whereas the graphical layout management and the specification of constraints for icon operators as the least liked one.

Based on these results, we can conclude that CoDIL, and the associated tool CoDIT, appear to be sufficiently intuitive to use, and have the potential of making the data integration process quite effective.

5 Conclusion and Future Work

We have proposed an approach and a system to simplify the specification of the data integration process. It provides a visual front-end based on the visual language CoDIL and several lower level modules supporting implementation tasks. A usability study involving undergraduate students in computer science confirmed language intuitiveness and potential for reducing the inherent complexity of the data integration process.

In the future we would like to investigate the possibility of using the principles of design patterns to define composed operators representing good integration practices to be used in future data integration projects. Moreover, we would like to investigate the possibility of introducing gestures in CoDIL, and to verify the possibility of a Sketch-based implementation of the whole approach.

References

1. Adobe: Photoshop creative cloud,
   http://www.adobe.com/products/photoshop.html (last accessed May 20, 2014)
2. Alalfi, M.H., Cordy, J.R., Dean, T.R.: SQL2XMI: Reverse engineering of uml-
   er diagrams from relational database schemas. In: Proceedings of 20th Working
   Conference on Reverse Engineering (WCRE), pp. 187–191 (2008)
3. Baader, F., Calvanese, D., McGuinness, D.L., Nardi, D., Patel-Schneider, P.F.:
   Description Logic Handbook. Cambridge University Press (2003)
4. Batini, C., Ceri, S., Navathe, S.B.: Conceptual Database Design: An Entity-
   relationship Approach. Benjamin-Cummings Publ., Redwood (1992)
5. Calvanese, D., De Giacomo, G., Lenzerini, M., Nardi, D., Rosati, R.: Data integration in data warehousing. International Journal of Cooperative Information
   Systems 10(3), 237–271 (2001)
6. Castano, S., De Antonellis, V.: Global viewing of heterogeneous data sources. IEEE
   Transactions on Knowledge and Data Engineering 13(2), 277–297 (2001)
7. Deufemia, V., Giordano, M., Polese, G., Tortora, G.: A visual language-based system for extraction-transformation-loading development. Software: Practice and Experience (2013), http://dx.doi.org/10.1002/spe.2201
8. Deufemia, V., Moscariello, M., Polese, G.: Visually integrating databases at conceptual level. In: Proceedings of the International Working Conference on Advanced
   Visual Interfaces (AVI 2014), pp. 359–360 (2014)
9. El Akkaoui, Z., Zimanyi, E.: Defining ETL worfklows using BPMN and BPEL. In:
   Proceedings of the ACM 12th International Workshop on Data Warehousing and
   OLAP (DOLAP), pp. 41–48 (2009)
10. Halevy, A., Rajaraman, A., Ordille, J.: Data integration: The teenage years. In:
    Proceedings of the 32nd International Conference on Very Large Data Bases
    (VLDB), pp. 9–16 (2006)
11. Norman, K.L., Panizzi, E.: Levels of automation and user participation in usability
    testing. Interacting with Computers 18(2), 246–264 (2006)
12. Oppenheim, A.N.: Questionnaire design, interviewing, and attitude measurement.
    Martin's Press, London (1992)
13. Pentaho: Kettle, http://community.pentaho.com/projects/data-integration/
    (last accessed May 20, 2014)
14. Rizzolo, F., Kiringa, I., Pottinger, R., Wong, K.: The conceptual integra-
    tion modeling framework: Abstracting from the multidimensional model. CoRR
    abs/1009.0255 (2010)
15. Simitsis, A., Skoutas, D., Castellanos, M.: Representation of conceptual etl designs
    in natural language using semantic web technology. Data Knowl. Eng. 69(1), 96–
    115 (2010)
16. Talend: Open studio, http://www.talend.com (last accessed May 20, 2014)
17. Vassiliadis, P., Simitsis, A., Skiadopoulos, S.: Conceptual modeling for ETL pro-
    cesses. In: Proceedings of the 5th ACM International Workshop on Data Ware-
    housing and OLAP (DOLAP), pp. 14–21 (2002)
18. Wilkinson, K., Simitsis, A.: Designing integration flows using hypercubes. In: Pro-
    ceedings of the 14th International Conference on Extending Database Technology
    (EDBT/ICDT), pp. 503–508 (2011)

DatalogBlocks: Relational Logic Integration Patterns

Daniel Ritter[1] and Jan Bross[2]

[1] HANA Platform, SAP AG, Dietmar-Hopp-Allee 16, 69190 Walldorf, Germany
[2] DHBW Karlsruhe, Erzbergerstrasse 121, 76133 Karlsruhe, Germany
{daniel.ritter,jan.bross}@sap.com

Abstract. Although most of the business application data is stored in relational databases, the programming languages in integration middleware systems - connecting applications - are not relational data-centric. Due to unnecessary data-shipments and faster computation, some middleware system vendors consider to "push-down" integration operations closer to the database systems.

We address the opposite case, which is "moving-up" relational logic programming for implementing the integration semantics within a standard middleware system. These semantics can be described by the well-known *Enterprise Integration Patterns*. For declarative and more efficient middleware pipeline processing, we combine these patterns with Datalog$^+$ and discuss their expressiveness and practical realization by example.

Keywords: Apache Camel, Datalog, Integration Patterns, Middleware.

1 Introduction

Integration middleware systems address the fundamental need for application integration by acting as the messaging hub between applications. As such, they have become ubiquitous in service-oriented enterprise computing environments in the last years. These middleware systems mediate messages between the applications mostly in formats like XML and JSON. However, the business application data is stored in relational databases. In addition middleware systems usually use relational databases for reliable, asynchronous messaging. For more efficient message processing, efforts have been made to combine (parts of) middleware and database systems on the database level [10,7]. The resulting "push-down" of integration semantics and business logic led to the development of new, mostly proprietary database programming languages (e. g., [4,3]).

On the other hand, former work in the area of logic programming and relational databases resulted in standardized, sufficiently expressive languages for application logic such as Prolog and Datalog [17]. The latter seemed long "forgotten" until academia [1] and industry [12] started to see additional value in its efficient computability for various applications (e. g., declarative networking, business analytics). However, in the context of integration systems, the feasibility

H. Decker et al. (Eds.): DEXA 2014, Part II, LNCS 8645, pp. 318–325, 2014.
© Springer International Publishing Switzerland 2014

of representing integration semantics by relational logic remains open, despite various advantages: (1) a declarative description of integration semantics (instead of, e.g., Java, C#), (2) relational logic data processing closer to its actual storage representation, and (3) efficient query processing, e.g., through data partitioning.

The work on *Enterprise Integration Patterns* (EIP) [14] provides the building blocks for implementing integration logic by defining a comprehensive set of standard patterns like routing and mapping. In this work, we combine these building blocks with Datalog$^+$ (non-stratified Datalog), which we call *DatalogBlocks* (DL$^+$B), and briefly discuss their expressiveness. We decided to use Datalog$^+$ due to its efficient processing capabilities and argue that it is sufficient for representing most of the integration semantics. An open-source system that implements most of the EIPs is *Apache Camel* [2]. It is thus used as an example for introducing and describing integration semantics and embedding Datalog in some selected EIPs throughout the paper. The main contributions of this work are (a) the application of Datalog to the integration domain forming relational logic integration patterns, (b) the definition of some selected core patterns by example, and (c) a brief discussion of the expressiveness of Datalog$^+$ in the integration domain.

Listing 1. "Twitter daily trends using Datalog with `choice`"

```
1  from (" twitter :// trends / daily "). unmarshal (JSON > Datalog )
2     . choice (). when (). expression ( DatalogPred ( trends ,
          =c ( name , "DEXA" ) ) )
3        . marshal ( Datalog > recv format ). to (<recv:a>)
4     . otherwise ()
5        . marshal ( Datalog > recv format ). to (<recv:b>);
```

Listing 2. "Twitter Trends JSON"

```
1  {
2  " name ": " Finally Friday ",
3  " events ": null ,
4  " promoted ": null ,
5  " query ": " Finally Friday "
6  }
```

Listing 3. "Corresponding Datalog Facts"

```
trends4 (name, events,
   promoted, query).
meta (" trends4 ", " name ",
      "1").
meta (" trends4 ", " events ",
      "2"). ...
```

For instance, Listing 1 shows the composition of DL$^+$B statements (e.g., *Content-based Router, (Un)marshal*) to a Camel-Datalog pipeline for the processing of a Twitter feed[1], which subscribes to daily trends. The trends are returned in JSON format (cf. Listing 2), which is unmarshalled to Datalog facts using the DL$^+$B (un-) marshalling capabilities (cf. Listing 3). From now on, the canonical data model [14] is Datalog. That allows for Datalog with `choice` (line 2)). The Apache Camel `choice condition` expression is a Camel–Datalog predicate. Here the condition expression makes use of a "String contains" built-in Datalog predicate =c(X,Y). The routing logic is based on the evaluation of the

---

[1] Top Twitter topics: https://dev.twitter.com/docs/api/1/get/trends/daily

Camel–Datalog predicate `DatalogPred` and passes the facts to one of the two receivers (e. g., other patterns or real message endpoints). The example illustrates the implementation of a *Content-based Router* pattern [14] with relational logic programs and the extension of Datalog$^+$ with `choice` [15], without changing the actual Datalog evaluation.

Section 2 sets the DL$^+$B approach into context with related work. In Section 3 selected core DL$^+$B language EIP constructs are introduced by example, before discussing their expressiveness in Section 4. Section 5 concludes the paper.

2   Related Work

The application of Datalog to integration programming for current middleware systems has not been considered before, up to our knowledge. Related work in the broadest sense can be found in the area of declarative XML message processing (e. g., [5,6]). Using an XQuery data store for defining persistent message queues, the work targets a subset of DL$^+$B (i. e., message queuing).

The data integration domain uses integration systems for querying remote data that is treated as local or "virtual" relations. Starting with SQL-based approaches, e. g., using the `Garlic` integration system [13], the data integration research reached relational logic programming, summarised by [11]. In contrast to remote queries in data integration, DL$^+$B extends integration programming with declarative, relational logic programming for application integration as well as the expressiveness of logic programs through integration semantics.

The expressiveness of the DL$^+$B evaluation, e. g., with respect to parallel processing, can be compared to fundamental work on parallel Datalog [8,9], which requires extensions of the Datalog evaluation instead of parallelisation on an integration pipeline level consisting of composed DL$^+$B statements.

Further related work like Datalog with `choice` for data exchange [15], map/reduce [16], or stream processing [18] requires an extension of Datalog. We argue that these concepts can be represented by composed DL$^+$B constructs, without extending the actual Datalog evaluation.

3   Relational Logic Integration Patterns

The open source integration system Apache Camel is used for the explanation of integration semantics represented by the EIPs and for concrete examples of the DL$^+$B integration language constructs. As most integration systems, Apache Camel offers a set of transport protocol adapters (e. g., HTTP, TCP/UDP) and format handlers (e. g., SOAP, CSV) for message reception, as well as a configurable pipeline that executes integration logic on those messages, while providing extension mechanisms for embedding new language constructs. Subsequently, the relevant Camel–Datalog language embeddings are defined by example. These constructs constitute the building blocks of what we call relational logic integration patterns or DL$^+$B.

Camel–Datalog Expressions. For embedding arbitrary language constructs, Camel allows to implement specific *Expression* and *Predicate* extensions. For DL$^+$B, we define a Camel–Datalog expression by a list of Datalog rules and an optional Datalog query: `expression(list(<rule>)[,<query>])`. The resulting facts from the embedded Datalog evaluation of the rules and the query have the correct format to be used in subsequent DL$^+$B statements. However, some constructs like `choice().when(<expression>)` or `filter(<expression>)` require a boolean output. For that, an empty evaluation result is mapped to `false`, or `true` otherwise. Expressions evaluating to a boolean value are called predicates.

Marshal/Unmarshal. The Camel–Datalog language constructs can only work on messages, whose payload contains Datalog facts. Hence, a format conversion between arbitrary formats (e. g., JSON, XML, CSV) and Datalog facts is required. For `marshal/unmarshal` operations, Camel offers the implementation of a `DataFormatDefinition`, which is used to convert a specific format to Datalog (using a Datalog parser) and vice versa.

3.1   Datalog Programs as Canonical Data Model for Integration

When connecting applications, various integration patterns might be executed on the transferred messages in a uniform way. The arriving messages are transformed into an internal format understood by the pattern implementation, called *Canonical Data Model* [14], before the messages are transformed to the target format. This model is informally defined as *Datalog Program*, which consists of a set of facts, with an optional set of rules and a set of meta-facts that describes the actual data (cf. Listing 3). The meta-facts encode the name of the fact's predicate and all parameter names within the relation as well as the position of each parameter. With that information, parameters can be accessed by name instead of position by Datalog rules, e. g., for selections, projections.

3.2   Common Integration Patterns Using Datalog$^+$

The DL$^+$B patterns represent a synthesis of well-defined EIPs and Datalog evaluation. While messages are routed through integration pipelines, the DL$^+$B patterns process them. From the large amount of integration patterns (mainly from [14]), we selected core patterns, for which we discuss the mapping to DL$^+$B subsequently. The DL$^+$B patterns are based on standard Datalog evaluation on the defined canonical data model.

Message and Content Filters. The *Message Filter* is a special routing pattern with a single message input and output (1:1), which removes messages from the pipeline according to a Camel–Datalog predicate evaluated on the data within the message's payload or routes the original message. Listing 4 shows the usage of the Camel–Datalog predicate in a Camel (message) filter. The predicate is evaluated on the set of facts within the message, while the message is only routed to `<recv>`, if it evaluates to `true`.

Listing 4. "Message Filter Pattern with Datalog"

```
1  from( direct:processing )
2      . filter () . expression (<predicate>) . to (<recv>);
```

Another filter pattern is the *Content Filter*. Comparable to projections in Datalog, the filter allows to remove parts from the original message's payload passing a transformed message to the pipeline. Hence it can be seen as a special *Message Translator* pattern [14], which translates one message format to another. Listing 5 shows the usage of the Datalog construct (`datalog(list( <rule>)[,<query>])`) that allows to evaluate arbitrary filter-/mapping rules.

Listing 5. "Content Filter Pattern with Datalog"

```
1  from( direct:processing )
2      . datalog ( list (<rule>) [ ,<query>]) . to (<recv>);
```

Content-Based Router. The Datalog with `choice` example in Listing 1 introduces the *Content-based Router* pattern (similar to [15]). The router features a choice on the content of the message to determine the route to choose. The routing is a 1:n fork on a pipeline, while the original message is passed 1:1 to the selected receiving route. The integration logic program for the router is shown in Listing 6. The Camel construct `choice().when()...otherwise()` is configured by multiple Camel–Datalog predicates. The router consists of multiple message filters and sends the incoming message to the route of the first positively evaluated Camel–Datalog predicate. If all case-expressions are evaluated to false, the message is routed to the `otherwise` route.

Listing 6. "Content-based Router Pattern with Datalog"

```
1  from( direct:processing )
2      . choice () . when () . expression (<predicate>) . to (<recv:a>)
3      . otherwise () . to (<recv:b>);
```

Datalog Message Splitter and Aggregator. In addition to these basic routing patterns, the two antipodes message *Splitter* and *Aggregator* are widely used. A *Splitter* consumes a message from the pipeline and splits it into multiple new messages according to a split-expression (1:n message fork), while the pipeline branching is 1:1. The simplest case is splitting the set of facts based on the Datalog predicate name into multiple messages. For more sophisticated cases a list of Camel–Datalog expressions can be used. Each expression is evaluated on the incoming message and each evaluation result is sent as a separate message. The *Splitter* shown in Listing 7 splits the set of incoming facts within the original message to several outgoing messages according to a list of Camel–Datalog expressions.

Listing 7. "Splitter Pattern with Datalog"

```
1  from( direct:processing )
2      . split ( list (<expression>)) . to (<recv>);
```

A pattern with a n:1 message cardinality is the *Aggregator*. The *Aggregator*, as shown in Listing 8, receives a stream of messages and correlates messages according to a correlation condition `<correlation condition>`. In case of DL$^+$B, the `<correlation condition>` consists of a list of Camel–Datalog predicates. The messages for which the same predicate evaluates to `true` are correlated. Alternatively a predicate name and a name of a parameter can be specified. Messages with the same parameter value for the given parameter name are then correlated. When a complete set of correlated messages has been received, the aggregator applies an aggregation function `<aggregation function>` and publishes a single, new message containing the aggregated result. The aggregator can be persistent, when storing a set of aggregates. Since the canonical data model consists of Datalog relations, no format conversion for the storage on the database is required. The aggregation function is defined as Camel–Datalog expression, which is evaluated on the combined facts and meta-facts of the messages to be aggregated. The Camel completion conditions `completionSize(<completion condition>` and `completionTimeout(<completion condition>)` can be parametrized by an integer value or a special Camel–Datalog expression.

Listing 8. "Aggregator Pattern with Datalog"

```
1   from( direct:processing )
2      .aggregate(<correlation  condition>, <aggregation
            function>)
3      [. completionSize(<completion  condition>)|
            completionTimeout(<completion  condition>)]
4      .to(<recv>);
```

4   About the Expressiveness of DL$^+$B Patterns

The capabilities and language constructs of integration systems differ. However, the open source system Apache Camel implements most of the "de-facto" standard EIPs, whose definition is the same across these systems. The expressiveness of the defined DL$^+$B patterns is limited to the capabilities of the Datalog system and the mapping to the integration language constructs. This work builds on recursive Datalog$^+$ (unstratified Datalog) with extensions for numeric expressions and String comparison. With the concept of `Datalog Programs` as canonical data model, all DL$^+$B operations can work on the message's Datalog facts payload. Through the created and transferred meta-facts, which describe the actual facts, the facts' parameters can be accessed by name, when specifying rules for Camel–Datalog expressions or predicates. This is especially useful when dealing with long (automatically generated) facts during the (un-) marshalling. Furthermore meta-facts resolve the problems with Data formats like JSON and XML, which do not guarantee the order of parameters and whose relations can vary in the number of their parameters for the same predicate. The meta-facts are used to allow for consistent rule definition in DL$^+$B. Camel–Datalog expressions are the basic constructs for the configuration of the DL$^+$B EIPs. As these expressions and predicates consist of a list of Datalog rules and an optional query, the full capabilities of the underlying Datalog implementation can be leveraged. Making the query optional provides the flexibility to execute multiple rules on a message. The whole intentional database is part of the expression's result as opposed to a single query's result.

These expressions are used in the DL$^+$B *Content Filter* pattern to transform the structure of the original to an output message, e. g., using projections, selections. Thus this pattern can also be used as a *Message Translator* [14]. Compared to standard integration programming, only the lack of calls to filter/mapping programs, available in higher-order programming languages, can be seen as limitation.

The *Message Filter* pattern allows to use arbitrary Camel–Datalog predicates to decide whether a message is left in the pipeline or filtered out. Using predicates it is for example possible to verify the data structure and to check whether a message contains certain facts.

This constitutes the prerequisite of a *Content-based Router*, too, which requires a boolean output for the decision on the route selection. However, one major issue with that are "overlapping" rule evaluation results. That leads to situations in which more than one condition on routes evaluates to *true*. In those cases, either a "take first/last/best" strategy can be used, while "best" means most facts, or, if known beforehand, the cases could be limited to a sequence of boolean route-branchings.

Similarly the `<correlation condition>` of the *Aggregator* pattern consists of a list of Datalog predicates, from which multiple rules can evaluate to true for a single incoming messages, here a "take first/last/best" strategy can also be used. Before evaluating an aggregation DL$^+$B expression the facts and meta-facts of multiple Datalog are combined. This leads to possible merge conflicts, when several messages contain facts with the same predicate name, but different variables. This can be mitigated by renaming the facts prior to aggregation using a message filter.

The *Splitter* pattern uses a list of DL$^+$B expressions to generate multiple messages by evaluating each expression on the message and forwarding each evaluation result as a separate message. This could lead to a non-disjunctive splitting. Subsequently problems can occur when merging the messages as there are duplicate facts. This issue has to be handled within the *Aggregator*.

5   Concluding Remarks

In this work we combined "de-facto" standard integration programming in form of the EIPs with Datalog evaluation and discussed the expressiveness of the resulting DL$^+$B language constructs. The language embedding and integration semantics were defined by example of the open source integration system Apache Camel, which has a broad coverage of EIPs and allows for extensible integration programming language constructs. With Camel–Datalog expressions and DL$^+$B, integration logic can be expressed declaratively (e. g., no Java, C# code required) and the message processing can be executed closer to its storage representation, while leveraging the full processing capabilities of the integration systems.

The composition of the constructs/patterns leads to more complex integration scenarios that will not only allow for declarative integration programming, but for efficient relational data processing within standard middleware systems. For instance, the *Scatter/Gather* pattern [14] is a composition of a splitter and an aggregator that can be used to implement map/reduce-style processing.

References

1. Alvaro, P., Marczak, W.R., Conway, N., Hellerstein, J.M., Maier, D., Sears, R.: Dedalus: Datalog in time and space. In: de Moor, O., Gottlob, G., Furche, T., Sellers, A. (eds.) Datalog 2010. LNCS, vol. 6702, pp. 262–281. Springer, Heidelberg (2011)
2. Anstey, J., Zbarcea, H.: Camel in Action. Manning (2011)
3. Binnig, C., May, N., Mindnich, T.: SQLScript: Efficiently Analyzing Big Enterprise Data in SAP HANA. In: BTW, pp. 363–382 (2013)
4. Binnig, C., Rehrmann, R., Faerber, F., Riewe, R.: Funsql: it is time to make sql functional. In: EDBT/ICDT Workshops, pp. 41–46 (2012)
5. Böhm, A., Kanne, C.-C., Moerkotte, G.: Demaq: A foundation for declarative xml message processing. In: CIDR, pp. 33–43 (2007)
6. Bonifati, A., Ceri, S., Paraboschi, S.: Pushing reactive services to xml repositories using active rules. Computer Networks 39(5), 645–660 (2002)
7. Doraiswamy, S., Altinel, M., Shrinivas, L., Palmer, S., Parr, F., Reinwald, B., Mohan, C.: Reweaving the tapestry: Integrating database and messaging systems in the wake of new middleware technologies. In: Härder, T., Lehner, W. (eds.) Data Management in a Connected World. LNCS, vol. 3551, pp. 91–110. Springer, Heidelberg (2005)
8. Ganguly, S., Silberschatz, A., Tsur, S.: A framework for the parallel processing of datalog queries. In: SIGMOD Conference, pp. 143–152 (1990)
9. Ganguly, S., Silberschatz, A., Tsur, S.: Mapping datalog program execution to networks of procesors. IEEE Trans. Knowl. Data Eng. 7(3), 351–361 (1995)
10. Gawlick, D., Mishra, S.: Information sharing with the oracle database. In: DEBS (2003)
11. Genesereth, M.R.: Data Integration: The Relational Logic Approach. Synthesis Lectures on Artificial Intelligence and Machine Learning. Morgan & Claypool Publishers (2010)
12. Green, T.J., Aref, M., Karvounarakis, G.: LogicBlox, platform and language: A tutorial. In: Barceló, P., Pichler, R. (eds.) Datalog 2.0 2012. LNCS, vol. 7494, pp. 1–8. Springer, Heidelberg (2012)
13. Haas, L.M., Kossmann, D., Wimmers, E.L., Yang, J.: Optimizing queries across diverse data sources. In: VLDB, pp. 276–285 (1997)
14. Hohpe, G., Woolf, B.: Enterprise Integration Patterns: Designing, Building, and Deploying Messaging Solutions. Addison-Wesley Longman Publishing Co., Inc., Boston (2003)
15. Saccà, D., Serra, E.: Data exchange in datalog is mainly a matter of choice. In: Barceló, P., Pichler, R. (eds.) Datalog 2.0 2012. LNCS, vol. 7494, pp. 153–164. Springer, Heidelberg (2012)
16. Shaw, M., Koutris, P., Howe, B., Suciu, D.: Optimizing large-scale semi-naïve datalog evaluation in hadoop. In: Barceló, P., Pichler, R. (eds.) Datalog 2.0 2012. LNCS, vol. 7494, pp. 165–176. Springer, Heidelberg (2012)
17. Ullman, J.D.: Principles of Database and Knowledge-Base Systems, vol. I. Computer Science Press (1988)
18. Zaniolo, C.: A logic-based language for data streams. In: SEBD, pp. 59–66 (2012)

Secure Data Integration: A Formal Concept Analysis Based Approach

Mokhtar Sellami, Mohamed Mohsen Gammoudi, and Mohand Said Hacid

ISETK Kef Tunisie, ISAMM Manouba Tunisie, UCBL Lyon 1 France
sellamimokhtar@yahoo.com, momogammoudi@gmail.com,
mshacid@bat710.univ-lyon1.fr

Abstract. Integrating and sharing information, across disparate data sources, entail several challenges: autonomous data objects are split across multiple sources. They are often controlled by different security paradigms and owned by different organizations. To offer a secure unique access point to these sources, we propose two-step approach based on Formal Concept Analysis. First, it derives a global vision of local access control policies. Second, it generates a mediated schema and the GAV/LAV mapping relations, while preserving the local source properties such as security.

Keywords: Access Control, Data Integration, Formal Concept Analysis.

1   Introduction

Data Integration [7] aims at providing a unique access interface to distributed data sources. This involves several issues: heterogeneous data objects, owned by different organizations, are often controlled through different access control paradigms. Heterogeneity needs the definition of global schema and the mapping. The mapping between the global schema and each local schema can be delineated through one of the prominent approaches [7] (i.e. GAV (Global As View), LAV (Local As View) or GLAV (Global As View)). These data-centric approaches aim at solving the data heterogeneity, query processing, and optimization problems. These approaches don't focus on the security aspects (i.e. availability, confidentiality, and integrity) which are major issues. Hence, access control aims at preventing unauthorized users from accessing sensitive data [4]. Data integration security is an 'open' issue since each source defines its own access control policies. Thus, the integration of the various security policies derives a representative policy to manage access to the whole data sources. To tackle these issues, we propose a two-step approach based on FCA (Formal Concept Analysis) theory [8]. It starts by combining the local policies to generate a synthesis policy at the mediator level. Then, it generates a mediated schema from the global policy. Finally, a mapping between the global schema and the local schemas is performed either by GAV or LAV. The use of FCA is justified by their sound mathematical foundations. FCA is a renowned formalism in data analysis and knowledge discovery because of its usefulness in important domains of knowledge discovery in databases (KDD) [11].

H. Decker et al. (Eds.): DEXA 2014, Part II, LNCS 8645, pp. 326–333, 2014.

Thus, we focus on three issues; i) a policy-centric approach: by investigating the global schema derivation according to the global policy (i. e. taking into account access control policies as the key to define visible parts of the sources to be integrated). ii) The preservation of the local source policies: an access control, enforced at the mediator level, has to preserve the local access control policies. iii) a mapping language-independent approach: by deriving mapping relations based on the different mapping languages (GAV, LAV).

This paper is organized as follows. Section 2 discusses the state of the art on information security and policy integration. Section 3 presents our FCA-based solution for secure data integration. The last section is devoted to the conclusion and future work.

2   Related Work

Information integration security is a challenging process, especially in enforcing access control to data in distributed environment [4, 5]. The authors in [6] present an approach which enforces rules and conditions expressed by privacy policies in the case of Hippocratic databases. Enforcing privacy policies does not require any modification of existing database applications. It is fulfilled by rewriting queries. For instance, a query Q is transformed to Q' in such a way that its result complies with the cell-level disclosure policy P. There is no query modification needed in our approach. This is due to the generation global schema according to the global policy. The authors in [9] propose an approach to integrate data using GAV while taking into account the authorization policies. This approach identifies the combination of virtual relations that could lead to the no preservation of the local authorization while ensuring no conflicts arise at mediator level. Its drawback is that it relies on the GAV as assumption to be applied. Moreover, this work lacks flexibility as it doesn't cover other access controls models and other integration approaches (i.e. LAV, GLAV).

Policy integration approaches aim at specifying policies by more than one policy authors and integrating them to check their compliance with the global requirements. In [1, 10], the authors describe an algebra for composing access control or privacy policies when different enterprises cooperate. Hence, the policy algebra is modeled as a composition language. Rao [12] propose algebra for fine grained integration of XACML policies. The former supports complex integration requirements using the defined algebraic operations. Nevertheless, these approaches were not designed to take into account data integration property. They aim at providing users of one system with access to data of another system, but do not consider how access to combined data, provided by different systems, should be enforced. The integration is still related to the link between the policy elements.

3   Formal Concept Analysis (FCA)

FCA is a branch of mathematical order theory [8], or more precisely a branch of lattice theory that has emerged during the 1980s.

Definition 1. Formal Context. is a triple $K = (G, M, I)$ where G called objects and M are called attributes and $I = G \times M$ is a binary relation. We say that an object g has attribute m if g and m are in relation I (denoted by gIm).

Definition 2. Derivation Operators. Let $K = (G, M, I)$ be a formal context and $A \subseteq G$ be a set of objects. We define $A' = \{ m \in M \mid \forall g \in A : gIm \}$ i. e. A' is the set of all attributes that all objects in G share. Analogously, let $B \subseteq M$ be a set of attributes. We define $B' = \{ g \in G \mid \forall m \in B : gIm \}$; i. e. B' is the set of those objects that have all attributes from B.

Definition 3. The concept lattice of a context (G; M; I) is a complete lattice in which infimum and supremum are given by:

$$\bigwedge_{t \in T}(A_t, B_t) = (\ \bigcap_{t \in T} A_t, (\ \bigcup_{t \in T} B_t))$$
$$\bigvee_{t \in T}(A_t, B_t) = ((\ \bigcup_{t \in T} A_t), \ \bigcap_{t \in T} B_t)$$

Definition 4. Partial order relation between concepts: Let (A1, B1) and (A2, B2) two formal concepts. (A1, B1) << (A2, B2) if and if $A1 \subseteq A2 \Leftrightarrow B2 \subseteq B1$ (A1, B1) is said a sub concept and (A2, B2) is said a super concept.

Definition 5. Galois lattice: the set of formal concepts Ordered by a partial order relation ≪ is said a Galois lattice.

4   A FCA-Based Secure Data Integration Approach

The proposed approach takes as input a set of source schema with its policies. This will be incrementally integrated based on the following 2 steps:

1. Step 1: Global Policy Generation: it starts by translating the schemas and policies to formal contexts. Then, first, it identifies the preserved-rule set of each attribute individually; second, it detects the possible attribute combination to identify the adequate rules that must be added to control this kind of combinations.
2. Step 2: Global Schema and Mapping Generation: it extracts the mediated schema according to the inferred policy. Then, it derives the GAV or LAV mapping relation between the global schema and the local schemas. Finally, it translates the global policy using specific access control model.

Our approach is illustrated through relational data integration as a reference framework. We conventionally assume that:

Data Model: 3 Relational Data Sources use the same attribute definition: S1: Admission: (SSN, AdmissionDate, Department), S2: Disease (SSN, DoctorID, Diagnosis) and DepartmentDoctor(DoctorID, Department) and S3: "Patient"(SSN, Admission-Date, Department Sex).

Access Control Models: ABAC (Figure 1-a), VBAC (Figure 1-b) and Flat RBAC (Figure 1-c) using the same profile name to define the access control rules.

Mediator Level. Global Schema and Global policy aren't defined yet.

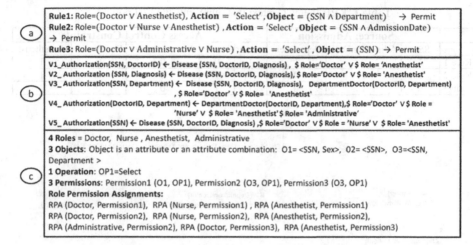

Fig. 1. A snapshot of the local access control policies

4.1 Step 1: Global Policy Generation

The generation of the global policy involves three stages: i) extracting formal contexts; ii) deriving a preliminary access rule set of each individual attributes; iii) identifying the possible attribute combinations and deriving the associated rules that complete the global policy to avoid answering queries of illegal users playing on this combination at mediator level.

Policy and Data Context Extraction. This step begins by identifying the similarities between attributes using existing matching techniques [2]. Then, it respectively generates the formal contexts: Data Context and Policy Contexts (table 1) based on the following definitions.

Definition 6 (Flat-RBAC Policy as a Formal Context (A, B, I) Given a Flat-RBAC Policy P and a transformation function σ_{RBAC}, a formal context is obtained as follows:

$$\sigma_{RBAC}(P) \begin{cases} if\ R_i\ :\ Role\ and\ Att_j\ \in O_j|\ O_j: Object\ then\ A = R_i\ \cup Att_j \\ if\ P_k: Permission\ |\ (k \leq t)\ \ then\ B = Rule_k \\ if\ \exists\ RPA\ (R_i, P_k)\ and\ O_j\ \in P_k\ then\ I = 1\ else\ I = 0 \end{cases}$$

Definition 7 VBAC Policy as a Formal Context (A, B, I). Given a VBAC Policy P and a transformation function σ_{VBAC}, a formal context is obtained as follows:

$$\sigma_{VBAC}(P) \begin{cases} if\ C_i\ :\ Constraints\ and\ Att_j: Attributes\ then\ A = C_i\ \cup Att_j \\ if\ VH_i\ :\ Virtual\ Authorization\ then\ B = Rule_k \\ if\ \ C_i\ \in VH_k\ and\ Att_j\ \in VH_k\ then\ I = 1\ \ else\ \ I = 0 \end{cases}$$

Definition 8 (ABAC Policy as a Formal Context (A, B, I). Given an ABAC Policy P and a transformation function σ_{ABAC}, a formal context is obtained as follows:

$$\sigma_{ABAC}(P) \begin{cases} if\ R_i\ :\ Role\ and\ Att_j: Attribute\ then\ A = R_i\ \cup Att_j \\ if\ Rule_k: Rule\ \ Then\ B = Rule_k \\ if\ \ R_i\ \in Rule_k\ and\ Att_j\ \in Rule_k\ then\ \ I = 1\ else\ I = 0 \end{cases}$$

Table 1. KP1: Policy Context that represents the first policy[1]

	Source: Admission			Access Control Constraints				
	SSN	AdmissionDate	Department	Doctor	Nurse	Administrative	Pharmacist	Anesthetist
Rule 1	1	0	1	1	0	0	0	1
Rule 2	1	1	0	1	1	0	0	1
Rule 3	0	1	0	1	1	0	0	1

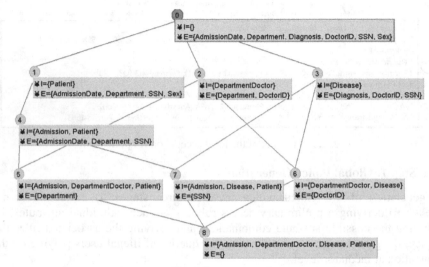

Fig. 2. Concept Data Lattice L^D of local sources and their attributes

The Figure 2 displays the Data Lattice L^D. It describes the local sources and theirs attributes. We can detect that the two attributes 'SSN' and 'AdmissionDate' come from two sources 'Patient' and 'Admission'. Indeed, the attribute 'Department' similarly exists at both sources 'Patient' and 'DepartmentDoctor'.

Preserved Rule Extraction: It considers each attribute individually in the Data Context K^D and extracts the corresponding access control rule set. Then, it identifies the shared profiles by all sources which have access to this attribute. Hence, the algorithm 1 takes as input the Data Context K^D and the Policy Contexts K^{Pi} (see table 1). For each attribute from K^D, it regroups the access control rules (line 2) of the attribute from K^{Pi}, and it respectively splits the obtained Attribute Policy Context on two matrices Attribute Matrix AM and Access Constraints Matrix CM (line 3,4). Then, the different rules obtained from the CM are combined, according to a **supremum** definition, to extract the profiles which must be derived at the global level (line 5-6). For instance, the preserved rule like '**Doctor, Nurse** → **SSN**' is made up of the profiles 'Doctor' and **'Nurse'**, and the attribute "SSN".

The algorithm 1 doesn't focus on the attribute combinations that can appear at the mediator level. So, we apply the following step to detect them and to retrieve rules that control this kind of combinations.

---

[1] A unique formal context is presented due to paper length requirements.

```
Input: Data Context K^D, Policies Context K^Pi
Output: a Preliminary Set Preserved Rules PR
1: foreach Attribute Att_i in K^D
2:    K^P_Atti =Extract rules associated to Att_i from K^Pi
3:    if (|K^P_Atti | > 2 )
4:        K^C_Atti=BinaryDecompose(K^P_Atti)
5:        if (supremum(K^C_Atti )# Ø)
6:            R^SA_G = supremum( K^C_Atti) → Atti (*)
7:        endif
8:    else R^SA_G=ExtractRule(K^P_Atti)
9:    endif
10:   P^R ← P^R U R^SA_G
11:endfor
```
Algorithm 1. Preserved Rule Extraction Algorithm

Attribute Combination Detection: It starts by detecting the intersection areas (Fig 3). Thereafter, it retrieves the shared attributes between sources that can be used for possible combinations. Thus, a combination between Department Doctor and Disease is ensured through DoctorID. A possible combination is {Department, Diagnosis}. By reapplying the algorithm 1, we obtain the rule "Doctor → SSN, Diagnosis" that control this kind of combination.

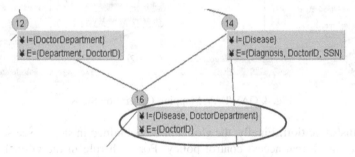

Fig. 3. Example of Attribute Combination

4.2 Step 2: Global Schema and Mapping Generation

In this step, we use the global policy to extract the attributes that belongs to the global schema. The step consists of global lattice generation, mapping derivation, and policy translation.

Global Lattice Generation: We use the work in [3] which presents a method of imposing constraints while extracting formal concepts. Virtual relations must contain the attributes used in the global policy at mediator level. Accordingly, we consider these visible attributes as constraints to generate the global schema.

Definition 9 Visible Attribute: Suppose that 'A' is an attribute. 'A' is a visible attribute, if 'A' has a global authorization rule that governs access to this attribute.

The Global Lattice is made up of a list of interesting concepts used in the global schema and mapping generation. It is composed of these concepts.

—C1:<Intent={SSN}, Extent={Admission, Disease, Patient}>
—C2:<Intent={SSN,Department,AdmissionDate},Extent ={Admission,Patient}>
—C3:<Intent={Diagnosis, DoctorID,SSN} ;Extent={Disease}>
—C4:<Intent={Department,DoctorID} ;Extent={ DoctorDepartment}>

Mapping Generation: It is performed using the GAV or LAV assumptions. The Figure 4-a (-b) describes the steps of the virtual relation generation and the GAV (LAV) mapping derivation using views as conjunctive rules [7]. First, it starts by building the virtual relation **G(X)** from the Concept Intent. Second, a conjunctive query **Q(X)** is built over the local sources (Concept Extent). Finally, a GAV mapping **M=G(X) ⊆ Q(X)** is generated (Fig.4-a-(3)). Whilst, the LAV mapping (Fig.4-b-(3)). **Mi=Si(X) ⊆ G(X)** is generated for each local source **S(X)** where Q(X) is conjunctive query over the global schema. For each lattice concept, the steps are performed while ensuring minimality [2] (i.e. no redundant relations appear in the global schema).

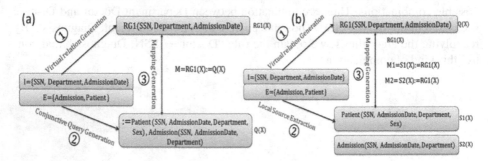

Fig. 4. GAV/LAV Mapping Generation Steps

Policy Transformation: Finally, the global policy obtained in step 1 (see section 4.1) is translated into a real access control policy. For each rule in the Global Policy, it generates a Global Authorization View. The global virtual relation which contains the attribute is a query part, and the profiles of the rule are the constraint part.

```
Input: P^G : Global Policy rules, G:Global Schema
Output:  Policy^G: VBAC Policy
1: for each rule R_i ∈ P^G do  //R has the form R=Cri-> Atti
2:   for each m_j∈ M do //m_i has the form G_i(Att):-S_1,..S_n
3:     if(R_i.att=m_j.G_i.Att) then
4:         R^P_k =GV_i_Authorization(R_i.att):=G_i R_i.C_i
5:           if R^P_k∉ Policy^G then    add  R^P_k to  Policy^G
6:     endif
7:   endfor
8: endfor
```

Algorithm 6. CBAC Policy Translation Algorithm

In this paper, we propose an algorithm that translates a global policy into VBAC policy at the mediator level. This model offers very fine grained access constraints. It is the most suitable model at relational data integration and conjunctive query as mapping. This is a global authorization view example: **GV1_Authorization(SSN, Department):=RG1(SSN, Department, AdmissionDate), $Role= "Doctor"**.

5   Conclusion

Based on the major advantages of our FCA-based approach (an access control policy-centric and a mapping language-independent solution), we intend to deal with semantic data by considering other issues, such as heterogeneities, data dependencies, and semantic constraints. We will also address the problem of consistency and compliance between global and local policies. Although, we will tackle the policy reconfiguration and query revocation to defeat inference problem that may appear at mediator level playing on the data dependencies.

References

1. Backes, M., Dürmuth, M., Steinwandt, R.: An algebra for composing enterprise privacy policies. In: Samarati, P., Ryan, P.Y.A., Gollmann, D., Molva, R. (eds.) ESORICS 2004. LNCS, vol. 3193, pp. 33–52. Springer, Heidelberg (2004)
2. Bellahsene, Z., Bonifati, A., Rahm, E. (eds.): Schema Matching and Mapping. Data-Centric Systems and Applications. Springer, Heidelberg (2011)
3. Belohlavek, R., Vychodil, V.: Closure-based constraints in formal concept analysis. Discrete Appl. Math. 161(13-14), 1894–1911 (2013)
4. Bertino, E., Sandhu, R.: Database security-concepts, approaches, and challenges. IEEE Trans. Dependable Secur. Comput. 2(1), 2–19 (2005)
5. Bertino, E., Jajodia, S., Samarati, P.: Supporting multiple access control policies in database systems. In: Proceedings IEEE Symposium on Security and Privacy, pp. 94–107 (May 1996)
6. Chen, B.C., LeFevre, K., Ramakrishnan, R.: Privacy skyline: Privacy with multidimensional adversarial knowledge. In: Proceedings of the 33rd International Conference on Very Large Data Bases, VLDB 2007, pp. 770–781 (2007)
7. Doan, A., Halevy, A.Y., Ives, Z.G.: Principles of Data Integration. M. Kaufmann (2012)
8. Ganter, B., Wille, R.: Formal Concept Analysis: Mathematical Foundations, 1st edn. Springer- Verlag New York, Inc., Secaucus (1997)
9. Haddad, M., Hacid, M.S., Laurini, R.: Data integration in presence of authorization policies. In: Min, G., Wu, Y., Liu, L.C., Jin, X., Jarvis, S.A., Al-Dubai, A.Y. (eds.) TrustCom, pp. 92–99. IEEE Computer Society (2012)
10. Pincus, J., Wing, J.M.: Towards an algebra for security policies. In: Ciardo, G., Darondeau, P. (eds.) ICATPN 2005. LNCS, vol. 3536, pp. 17–25. Springer, Heidelberg (2005)
11. Poelmans, J., Kuznetsov, S.O., Ignatov, D.I., Dedene, G.: Formal concept analysis in knowledge processing: A survey on models and techniques. Expert Systems with Applications 40(16), 6601–6623 (2013)
12. Rao, P., Lin, D., Bertino, E., Li, N., Lobo, J.: Fine-grained integration of access control policies. Computers & Security 30(2-3), 91–107 (2011)

A Lateral Thinking Framework for Semantic Modelling of Emergencies in Smart Cities

Antonio De Nicola[1], Michele Melchiori[2], and Maria Luisa Villani[1]

[1] Computing and Technological Infrastructure Lab, ENEA
Rome, Italy
{antonio.denicola,marialuisa.villani}@enea.it
[2] Università degli Studi di Brescia,
Dip. di Ingegneria per l'Informazione, via Branze, 38
25123 Brescia - Italy
michele.melchiori@unibs.it

Abstract. Manual definition of models for emergency management scenarios is a demanding activity due to the huge number of different situations to be considered. It requires knowledge related to the crisis and emergency domains, to the context (e.g., a specific city and its current regulations) and to modelling techniques. In this paper, we propose to tackle this problem according to a lateral thinking perspective and, following this line, we present a framework supporting automatic creation of conceptual models concerning emergency management scenarios by means of semantic techniques. In particular, this framework relies on an ontology and on a set of semantic rules to manage, respectively, the domain and contextual knowledge, and on the design patterns approach to support the modelling activity. A software experimentation of the framework based on SPARQL and applied to emergency scenarios in smart cities is proposed to demonstrate the viability of the approach.

Keywords: emergency management, lateral thinking, domain ontology, contextual rules, design pattern, mini-story, software architecture.

1 Introduction

Smart cities are characterized by interconnected physical and virtual services aiming at simplification of citizens activities, consumption of sustainable primary resources, as water and energy, and involvement of people in decisions that could have an impact on their life. Smart cities ecosystems are threatened by several hazards spanning from natural disasters, as earthquakes, to anthropic events, as terrorist attacks.

Promptness and reactiveness of service providers and institutional operators to face and manage emergency situations are becoming an important issue. A promising approach to reach this goal is computer-based simulation of such crisis events and the related management phase. Existing technologies to such purpose are agents-based simulation [4] and workflow systems [13].

H. Decker et al. (Eds.): DEXA 2014, Part II, LNCS 8645, pp. 334–348, 2014.

Precondition to simulation is the development of simulation models. Such conceptual models represent also a valuable support for sharing information about emergency plans in a user-friendly way among different players (e.g., experts, institutional operators, politicians, and engineers) who play a role in an emergency management scenario.

However, crisis and emergencies are usually situations not obvious to be conceived in advance since it is not easy to predict availability of resources, happening of events and behavior of people. Currently, operators spend a lot of time in defining one single scenario to be simulated. They try to imagine the services and people impacted by an event, what its intensity is and how long it lasts. The result is that they come out with partial solutions and, consequently, they feel they are "tilting at windmills". We can say that manual definition of Emergency Management (EM) scenarios is a loosing strategy since the issue here is to foresee as many EM situations as possible.

Lateral thinking [6] is a creative problem-solving method based on not obvious reasoning with the aim of conceiving ideas different from those generated by vertical thinking that uses incremental and straightforward logical processes. According to [6], a lateral thinking approach generates as many alternative solutions as possible, which are obtained for example by analogical reasoning, in order to provide original solutions that may not be otherwise taken into account.

We believe that an approach of this type best fits our needs. For this reason, here we propose a framework towards automatic creation of (parts of) EM models, by relying on formalized knowledge of the problem domain. In this way, we may reduce the number of unforeseen situations in EM models due to human limits. Our proposal is to automatically generate semantically coherent fragments of emergency management scenario models, called mini-stories, which might not be obvious, to be supplied as input for scenarios creation by composition. These EM scenario models will be then simulated by simulation tools or used as a basis to discuss about EM plans.

For the sake of space, in this paper we focus on mini-stories generation whereas the composition problem will be treated in future works. In particular, we have chosen CEML [8,5] as a scenario modelling language derived from SysML [9], we follow the design patterns-based modelling methodology described in [8] and we have adopted the notion of mini-stories from [22]. Design patterns provide the syntactic support in building the models. A description of the idea was presented in the paper [7] together with a case study concerning business ecosystems. In this paper we additionally present a software environment including an ontology focusing on EM in smart cities and we demonstrate the validity of the framework with some experiments.

Our approach integrates three types of knowledge: *structural knowledge*, supporting models construction and provided by design patterns; *domain knowledge*, related to smart cities and emergency management and gathered in an ontology, and *contextual knowledge*, related to a specific location or specific rules and regulations to be applied in the specific temporal period and contained in a rule repository. Furthermore, the proposed approach requires the cooperative work of domain and application experts in order to provide the last two types of knowledge aforementioned.

The paper is structured as follows. Section 2 discusses the related work in the area and Section 3 presents a smart city case study to be used in the rest of the paper as running example. The modelling methodology is presented in Section 4 whereas Section 5 and 6 focus respectively on domain and contextual knowledge. A software application developed to demonstrate the viability of the approach is presented in Section 7. Finally, Section 8 provides conclusions and future research directions.

2   Related Work

Simulation of EM scenarios is considered a relevant topic as demonstrated, for example, by the works [11] and [12]. Ontologies in EM concern mainly three aspects: (i) supporting interoperability among systems [17,10]; (ii) providing semantics to data and models [1,16]; (iii) semantic enrichment of simulation models [2,21]. Supporting interoperability is for example discussed in [3,17]. In particular, the paper [17] presents an intelligent emergency management system (EMS) for an airport emergency management. The EMS is based on interpreting complex events defined as meaningful correlations of simple events collected and communicated by networks of data acquisition systems operating in the airport. The domain specific ontology is used to integrate and provide meaning to the messages generated by the acquisition systems. The process of building an ontology for EM from glossaries and vocabularies, to support communication among software systems for critical infrastructures, is described in [10].

A semantics-based emergency response system is proposed [1] with the aim of supporting emergency decision makers by retrieving and possibly adapting solutions provided for previously events. The system uses real time information provided as short message service (SMS) by on-the-field users. Ontologies provide semantics to data and models. In particular, they are used both to interpret and classify SMS messages, and for solution retrieval from the solution data base.

Ontologies have also been recognized to be potentially useful for simulation modelling [2]. In particular, ontologies are used for semantically enriched descriptions of model components in discrete-event simulation modelling [21]. These descriptions are used for performing components discovery and for determining compatibility of components. Compatibility is established based on the data exchanged between them and the fact that their behaviors are aligned.

Our approach is related to these works but we consider the problems of generation of mini-stories constrained and guided by a domain ontology. At the best of our knowledge, the problem of generating emergency management scenarios by means of semantic techniques has not yet been tackled before by the research community.

3   Smart City Case Study

A smart city aims at providing high quality of life to people living and visiting it by means of a sustainable consumption of primary resources, such as water and energy, and by leveraging on modern ICT infrastructures. There is currently a large agreement on the fact that impact on the environment, social implications and respect for diversity do not hinder innovation but boost it. In this context, needs of patients, tourists, commuters and, more in general, citizens are considered in advance to better organize the activities and infrastructures that can fulfill them. To this aim, a smart city is characterized by a set of services aiming at simplifying citizens' lives through interconnected physical and virtual services. A paradigmatic example is the car sharing service: it implies the existence of cars and the availability of a wired or wireless connection and a web service to offer the service. Hence, problems affecting the operational status of a service can easily propagate to the others and, finally, impact on citizens' life. Examples of possible hazards are either natural catastrophic events, as earthquakes, floods, tsunami, and snowstorms or anthropic events, as terrorist attacks, traffic collisions, and clashes. In case of an emergency, other services are in charge of the recovery operations. Examples are police, civil protection, medical services, army and ICT recovery services.

4   The Emergency Scenario Modelling Framework

The EM modelling framework aims at supporting the modelling activity. It is based on a conceptual model, a methodology leveraging on the use of design patterns and on a reference architecture, a domain ontology and some contextual rules.

4.1   Conceptual Model for Emergency Management Scenarios

The conceptual model, illustrated in Fig. 1, is especially useful to locate the mini-stories generation approach, which is the focus of this paper, inside the wider EM scenarios modelling problem.

Starting from a textual description of the object of study, such as that described in Section 3, one might be interested in the identification of one or more *(EM) Scenarios* in order to define an *EM Plan*, which collects all the foreseen scenarios. An EM Scenario is a still narrative situation of an emergency and of the actions taken to solve it. Such emergency is usually caused by some unpredictable event, occurring in a certain place and impacting one or more specified real worlds objects (such as people, infrastructures, institutions, companies, and so on). In order to evaluate the impact of these events and the efficacy of various emergency resolution measures that could be taken, each EM scenario may be represented through *models*, expressed in machine readable form in order to enable computer supported analysis such as simulation. To facilitate the modelling activity, in our approach these models have a *structure* built with one

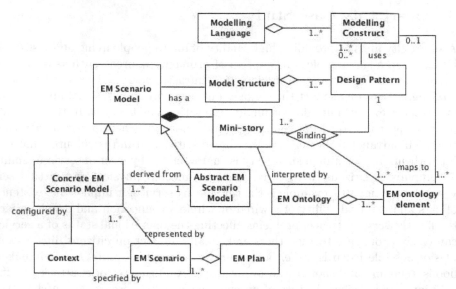

Fig. 1. Conceptual model of the scenarios modelling approach presented as UML diagram

or more *Design Patterns*, that are realized by using *Modelling Constructs* of a certain *Modelling Language*. An *EM ontology* can provide semantics to design patterns by means of a map from a set of *EM ontology elements* to the modelling constructs. Indeed, this map allows to realize a *binding* from a given design pattern to some elements of the EM ontology (one element for each modelling construct used in the pattern), to obtain what we call a *Mini-story*, that is, a fragment of a scenario model meaningful from a semantic perspective. One or more mini-stories can originate from the same design pattern through different bindings. This allows a bottom-up creation of *Abstract* and *Concrete EM Scenario Models*, through patterns instantiations, obtained by semantic composition of mini-stories. In the EM scenario example above, an abstract EM scenario model would have components such as: *earthquake*, *transportation service* and *electricity infrastructure*, and *public institution*. An abstract scenario model can originate one or more *Concrete EM Scenario Models*, which can differ for *context* data, such as the identification of the real objects (e.g., name and location) and their characteristics; for the magnitude of the emergency; and/or for the response measures (e.g., response time).

4.2   Patterns-Based Methodology and Reference Architecture

In this paper, we face the problem of providing automatic support to the construction of as many as possible EM scenario models to the aim of defining an EM plan for a given smart city. For this reason, here we propose a methodology for semantics-based generation of mini-stories since they are the building blocks of EM scenario models, and a suite of tools that can be used to implement it, shown in Fig. 3.

To better illustrate the approach, we have chosen CEML and the pattern-based methodology presented in [8]. As already stated in the introduction, CEML is a domain-specific modelling language derived from SysML, an UML's profile, allowing experts to build formally grounded models in a user-friendly way.

The method towards automatic construction of EM scenarios models starts from pre-defined design patterns and, by means of mini-stories semantic binding and composition and data assignment, produces concrete EM scenario models.

Fig. 2 presents the core EM design pattern edited in the CEML language [8] and used as example in the following of the paper. This design pattern represents the following generic situation. An external event impacts on a service that provides resources to a user. Then a human service sends human resources to recovery the damaged service. Furthermore, the figure presents two examples of mini-stories generated according to this pattern. The mini-story on the left represents an obvious configuration in an emergency scenario, since firefighters are expected and are appropriate to intervene on the building fire. On the contrary, the generated mini-story depicted on the right is not obvious since it describes a not expected configuration where the recovery activity is performed by policemen. Nevertheless, this unexpected mini-story can be considered by experts as possible and relevant in the emergency scenario. The underlying idea that emerges is that in case of large scale emergencies the availability of the most appropriate human resources cannot be granted since they could be occupied elsewhere. Consequently, other human resources can intervene in the crisis scenario. The lateral thinking approach therefore supports in identifying such not obvious situations.

The methodology supporting our approach can be outlined according to the following phases:

1. *Ontology engineering.* The ontology covers knowledge about smart cities and emergencies caused by natural or anthropic events. Such knowledge concerns several domains spanning from descriptions of hazards and events, to critical infrastructures, to services provided to companies and citizens, to recovery and rescue services, and to users. Preliminary examples of emergency ontologies already exist in the literature (see, for instance, [10] and [16]) and can be extended and specialized to be used in a given domain. The ontology is built by domain experts by means of an ontology management system (OMS) (e.g., Protégé [15]).

2. *Contextual rules definition.* Rules concerning the specific context considered such as the location, the temporal period and the current laws and regulations are defined. These rules are specified by application experts through a rule editor and have to be satisfied by the scenario models and, consequently, by the mini-stories.

3. *Model structure definition.* The model structure is defined by means of a design patterns approach. These patterns are defined through a modelling tool by domain and application experts.

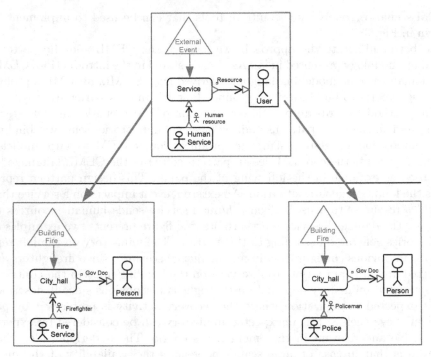

Fig. 2. Core emergency management scenario design pattern

4. *Semantics-based generation of mini-stories.* Mini-stories are semantically coherent fragments of scenario models. These are generated by the binding
   engine starting from design patterns and considering the domain and contextual knowledge. Such process is described in details in Section 7.
5. *Validation of mini-stories.* Mini-stories are collected in a repository once
   they have been validated by domain and application experts. They can use
   a validator module conceived to support the voting activity aimed at validation. In case a generated mini-story describes a configuration considered
   as not valid, the experts can update the knowledge base in order to remove
   the cause of the non acceptance. This can be done either by reviewing the
   ontology or the contextual rules or even the desired patterns.

A key role in the reference architecture is played by the knowledge base containing the smart city and crisis management ontology, the contextual rules, the
design patterns and the validated mini-stories.

5   Domain Knowledge Modelling

5.1   Emergency Management Upper Level Ontology

In the following we describe the task of extending the domain ontology with
CEML concepts. This activity has the purpose to represent explicitly CEML

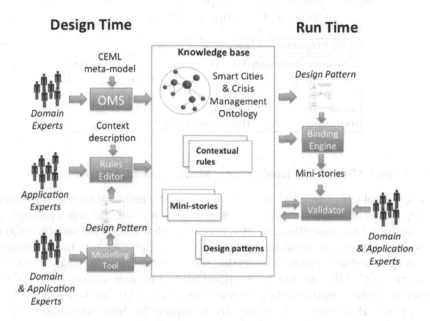

Fig. 3. Reference Architecture

concepts in the ontology and specify their relationship with the ontology concepts. In fact, as we said, the domain ontology extended with CEML concepts, CEML patterns and contextual rules are the bases of the mini-stories generation process.

To this purpose, we implemented an upper level ontology, called *CEML ontology*, in order to provide an OWL representation of the CEML metamodel. Selected CEML concepts of the *CEML ontology* are related to concepts of the domain ontology by means of *RDF:SubClass* relationships. The selection criteria consists of choosing those CEML concepts that either (i) correspond to an ontology concept, or (ii) generalize one ore more ontology concepts.

For example, the CEML element External Event is related to the ontology in order to specify that CEML External Event is more general than some domain concepts, for example Natural Event and Anthropic Event. Additionally, two *RDF:SubClass* relationships are added, i.e., between Natural Event and External Event, and between Anthropic Event and External Event. Moreover a CEML concept is featured by properties corresponding to the ones it owns in the CEML model, for example the External Event concept has a property *impacts on* whose range is the CEML concept Service. The set of properties described in the CEML model are listed in Table 1. The domain ontology extended with CEML concepts is reported in Fig. 3 as Smart Cities & Crisis Management Ontology.

Table 1. Properties of CEML concepts

CEML Property	Domain	Range
hasHumanResource	Human Service	Human Resource
hasResource	Service	Resource
hasUser	Service	User
impactingOn	External event	Service
recovering	Human service	Service

5.2 Smart Cities and Emergency Management Ontology

Usually, ontologies described in emergency management literature, for example in [20], are specialized on different subdomains relevant for our approach and specifically for our case study: events, services, infrastructures, people, organizations. However, in our approach we need to cover these aspects in an integrated way. A quite natural way for integrating these ontologies is to exploit the CEML properties of CEML concepts, as described in the previous section. In fact, CEML properties define relationships between general concepts, as we can note in Table 1, where these concepts are usually belonging to different subdomains and therefore related to different ontologies.

Generally speaking, an ontology describes static and shared knowledge about a scenario in term of concepts, properties and relationships.

The ontology is represented using OWL that allows managing and reasoning (e.g., subsumption and coherence checks) on it with widely available tools. The OWL representation we adopt follows a standard approach where, in particular, a concept is represented as OWL:class; a property is represented as OWL:DatatypeProperty; an is-a relationship is represented as RDF:SubClass; a domain specific relationship is represented as OWL:ObjectProperty.

In particular, for the experimentation on our case study detailed in Section 7, we created a domain ontology for the smart city scenario extended with CEML concepts and properties as explained in the previous Section 5.1. For example, the domain ontology concepts that specialize directly or indirectly the CEML External Event are 49. The extended ontology is made of 166 concepts and 99 properties, and an excerpt of it[1] is presented in Figure 4.

6   Contextual Knowledge Modelling

Contextual rules provide knowledge to be applied on top of domain knowledge for some modelling objective. They depend on the specific context as, for instance, the geographical location (e.g., Rome city does not have a wind power plant), the time period (e.g., the metro service does not operate on holidays) or the existence of laws and regulations (e.g., the army is not authorized to intervene in case of clashes).

---

[1] The full OWL version of the ontology can be visualized by the WebProtégé tool at the address http://tinyurl.com/crisismng

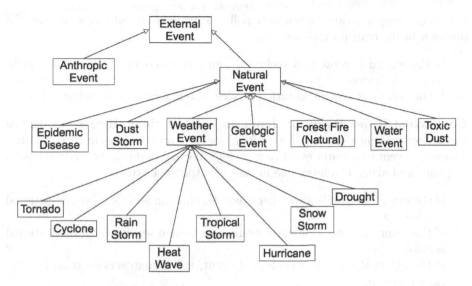

Fig. 4. A graphical representation of an excerpt from the emergency management ontology (represented as oriented arrows)

Contextual rules are defined by application and domain experts starting from the ontology concepts. As a definition strategy we suggest to partition them into some categories presented below. Furthermore, we suggest to define them by focusing on top level concepts or, at least, to consider first concepts that are at the upper levels of the subsumption hierarchies characterizing the domain ontology. In this way the semantic binding can exploit such relationship and build mini-stories more effectively since what is valid for an upper level concept remains valid for its specialized concepts (e.g., what is valid for the *external event* concept, is also valid for both the *earthquake* and the *tsunami* concepts).

In the following the contextual rules which have been identified and used for the smart city test case are presented in natural language. These rules refer to the core EM scenario design pattern presented in Figure 2.

Contextual Rules Based on a Geographical Location. In such example, we model the case of a city without any airport, embassy, museum, wind farm, and incinerator. Furthermore the considered city is not in the nearby of the sea and of any volcanos.

1. The service is not any of: airport, embassy, museum, wind farm, and incinerator.
2. The external event is not any of: flood, sea level rise, and tsunami.
3. The external event is not a subclass of volcanic event.

Contextual Rules Based on Time Period. Here first we model a particular temporal range when only private technical services can intervene to restore a technical fault since institutional operators are not available. Then we model the particular temporal range when only police is available to intervene in case of a problem in the transportation service.

1. If the external event is a technical fault, the human service is a private technical service.
2. If the service is a transportation service, the human service is the police.

Contextual Rules Based on Laws and Regulations. Here we model the regulations stating that only institutional service can intervene in case of either a weather event or a water event or a geological event and only security services, as police and army, can intervene in case of a terrorist attack.

1. If the external event is a weather event, the human service is an institutional service.
2. If the external event is a water event, the human service is an institutional service.
3. If the external event is a geological event, the human service is an institutional service.
4. If the external event is a terrorist attack, the human service is a security service.

7   Software Application and Experimental Results

7.1   Binding Engine

The proposed methodology has been tested on the smart cities case study with the support of prototype tools following the architecture of Fig. 3. The ontology covering concepts from the smart city domain and the emergency management field was edited in OWL/RDF with the Protégé tool [15], and integrated with the CEML metamodel, which is also available in OWL/RDF.

The semantic binding is realized through a SPARQL [18] select query by using three types of knowledge: from the design pattern, from the domain and from the context. Specifically, a CEML pattern-base query is constructed by accounting for the CEML concepts and relationships that are used in a given design pattern, to retrieve all of their specializations from the domain ontology. Fig. 5 shows an excerpt of such a query related to the CEML pattern presented in Fig. 2, where we have highlighted three examples of code blocks. Namely, block 1 is to retrieve all the specializations of the object property *impactingOn*, connecting an **External Event** subclass with a **Service** subclass. The code in block 2 is finalized to retrieving all the **Service** subclasses of the ontology that are leaf concepts, and, through the code of block 3, just those of such subclasses that are involved in the *impactingOn*, *hasResource* and *hasUser* objectProperties specializations are selected for the final result. The effect of a CEML pattern-base query is a set of mini-stories resulting by linking each CEML relationship of the pattern to one of the object properties specializing it, and each CEML construct involved in that relationship to a leaf subclass of the corresponding

```
PREFIX : .....................
SELECT DISTINCT  ?ext_ev ?service ?resource ?human_resource ?human_service ?user
WHERE {
```

```
?ipros a owl:ObjectProperty;
    rdfs:domain ?ext_ev1 ;
    rdfs:range ?service1 ;                                                    1
    rdfs:subPropertyOf CrisisAndEmergency:impactingOnService .
FILTER (! (?service1 = CrisisAndEmergency:Service))
```

```
?pro a owl:ObjectProperty;
    rdfs:domain ?service2 ;
    rdfs:range ?resource1 ;
    rdfs:subPropertyOf CrisisAndEmergency:hasResource
FILTER (! (?service2 = CrisisAndEmergency:Service))
?prou a owl:ObjectProperty;
    rdfs:domain ?service3 ;
    rdfs:range ?user1 ;
    rdfs:subPropertyOf CrisisAndEmergency:hasUser .
FILTER (! (?service3 = CrisisAndEmergency:Service))
```

```
?service rdfs:subClassOf CrisisAndEmergency:Service .
FILTER NOT EXISTS {?sub_ev1 rdfs:subClassOf ?service  .
 ?sub_ev2 rdfs:subClassOf ?service .                                          2
 FILTER (! (?sub_ev1 = ?sub_ev2))  }
```

```
FILTER EXISTS {
?service rdfs:subClassOf ?service1 ;
         rdfs:subClassOf ?service2 ;                                          3
         rdfs:subClassOf ?service3 .
}
```

```
?ext_ev rdfs:subClassOf CrisisAndEmergency:External_event .
FILTER NOT EXISTS {?sub_ev1 rdfs:subClassOf ?ext_ev  .
 ?sub_ev2 rdfs:subClassOf ?ext_ev .
 FILTER (! (?sub_ev1 = ?sub_ev2))  }
FILTER EXISTS { ?ext_ev rdfs:subClassOf ?ext_ev1 . }
.....................

}
```

Fig. 5. The CEML pattern base query

CEML concept. This pattern-base query can be refined with contextual rules, each providing a filter SPARQL statement.

These queries (one for each pattern) are executed by the Binding Engine. This has been implemented in Java, based on the Apache Jena framework including the ARQ library [14], which implements a SPARQL 1.1 engine. The Binding Engine is a configurable component through a XML file specifying the list of CEML patterns to be used to model a given EM scenario, and, for each pattern, the set of contextual rules to be applied for mini-stories generation. All of these rules are then combined and integrated with the base query originated from the CEML pattern, to form a unique SPARQL query. This is run on the ontology to retrieve the mini-stories corresponding to that pattern. We used a PostgreSQL [19] database to persistently save the mini-stories, and to enable the user validation activity. At this stage of the work, this activity simply consists of querying the database to perform a manual check of the results and/or filter the mini-stories with respect to some user defined constraints. As a future work, we intend to provide more automatic support to the minis-stories validation phase.

Table 2. Likely mini-stories referring to the core EM design pattern

External Event	Service	Resource	User	Human Service	Human Resource
Building fire	Stadium	Entertainment	Unspecified person	Medical service	Medical doctor
Tornado	Waste collection service	Waste management	Enterprise	Fire service	Technical operator
Forest fire	Solar power plant	Solar energy	Enterprise	Fire service	Firefighter

7.2 Experimentation

We prepared some experiments with the objective to evaluate the effectiveness and usefulness of our semantics-based approach. To this aim, we used the ontology described in Subsection 5.2, consisting of 166 concepts and 99 relationships, all of them being of *is-a* type and CEML relations specializations. The Binding Engine was run on the CEML pattern shown in Fig. 2, using the following queries:

1. a CEML pattern base query that only considers the part of the ontology including the *is-a* relationships with the CEML concepts;
2. the query in Fig. 5 accounting also of the CEML relations specializations that are of interest for the given pattern;
3. two queries also incorporating the contextual rules.

Furthermore, these queries were formulated to just retrieve the leaf concepts of the ontology to avoid the creation of redundant mini-stories.

For each run, we recorded the number of generated mini-stories and obtained the following results. In case 1, the number of mini-stories, such as 32781000, was prohibitive both for recording and analysis. However, a quite significant number of such mini-stories was semantically wrong according to query 2, which resulted in 155264 mini-stories. Finally, we ran again the query by first adding the law-depending contextual rules and those based on a time period, obtaining 75536 mini-stories, and then added also the location-specific queries, getting the final number of 9190 mini-stories.[2] We think that this is a manageable number for mini-stories browsing by the interested user, and remark that this number can be further reduced by adding other contextual rules during the validation activity. Furthermore, such result even more justifies the need for automatic support to the EM scenario models generation.

Table 2 and Table 3 show, respectively, some examples of likely and unlikely (not obvious) mini-stories we obtained.

---

[2] These mini-stories are available at: http://tinyurl.com/minist3

Table 3. Unlikely mini-stories referring to the core EM design pattern

External Event	Service	Resource	User	Human Service	Human Resource
Building fire	Stadium	Entertainment	Unspecified person	Army	Soldier
Tornado	Waste collection service	Waste management	Enterprise	Army	Soldier
Forest fire	Solar power plant	Solar energy	Enterprise	Private ICT recovery service	Technical operator

8   Conclusions and Future Work

Emergency management models can be used for sharing EM information among experts and private and institutional operators and as an intermediate step towards a computable format for simulation tools. In this paper, we presented an approach for generation of EM scenarios models to increase preparedness and promptness in managing such situations.

From an application-oriented perspective, we face the issue of automatic engineering of not obvious conceptual models of crisis and emergency scenarios. At the best of our knowledge, this is the first proposal addressing this problem.

From a scientific perspective, automatic generation of models requires to first develop three different sources of knowledge: syntactic knowledge, provided by design patterns; domain knowledge, provided by an ontology; and contextual knowledge, provided by means of rules. Although design patterns are widely used to support modelling, our work is the first that considers integration of these types of knowledge to automatize the modelling engineering process.

The viability of the approach has been demonstrated through an empirical evaluation based on an OWL ontology, developed with the Protégé ontology management system, concerning smart cities and EM, and on the SPARQL query engine. As a future work, we plan to add other semantics-based techniques, e.g., similarity reasoning, to support the definition of the contextual rules process. Then we intend to study the adoption of methods originally conceived for Web services composition, and adapt them in order to support EM scenario models definition by means of mini-stories composition.

References

1. Amailef, K., Lu, J.: Ontology-supported case-based reasoning approach for intelligent m-Government emergency response services. Decision Support Systems 55(1), 79–97 (2013)
2. Araujo, R.B., Rocha, R.V., Campos, M.R., Boukerche, A.: Creating Emergency Management Training Simulations through Ontologies Integration. In: Proc. of CSE 2008 - Workshops, pp. 373–378. IEEE (2008)

3. Atluri, V., Shafiq, B., Soon Ae, C., Nabi, G., Vaidya, J.: Uicds-based information sharing among emergency response application systems. In: 12th Annual International Digital Government Research Conference: Digital Government Innovation in Challenging Times, pp. 331–332. ACM (2011)
4. Collier, N.: Repast: An extensible framework for agent simulation (2001)
5. D'Agostino, G., De Nicola, A., Di Pietro, A., Vicoli, G., Villani, M.L., Rosato, V.: A Domain Specific Language for the Description and the Simulation of Systems of Interacting Systems. Advances in Complex Systems 15, 1250072–1250088 (2012)
6. De Bono, E.: Lateral thinking. Harper and Row (1970)
7. De Nicola, A., Melchiori, M., Villani, M.L.: A semantics-based approach to generation of emergency management scenario models. In: Proc. of I-ESA 2014, vol. 7, pp. 163–173. Springer (2014)
8. De Nicola, A., Tofani, A., Vicoli, G., Villani, M.L.: An MDA-based Approach to Crisis and Emergency Management Modeling. International Journal On Advances in Intelligent Systems 5(1&2), 89–100 (2012)
9. Friedenthal, S., Moore, A., Steiner, R.: A practical guide to SysML: The systems modeling language. Elsevier (2011)
10. Grolinger, K., Brown, K.P., Capretz, M.A.M.: From Glossaries to Ontologies: Disaster Management Domain(S). In: Proc. of SEKE 2011, pp. 402–407 (2011)
11. Guo, D., Ren, B., Wang, C.: Integrated agent-based modeling with GIS for large scale emergency simulation. In: Kang, L., Cai, Z., Yan, X., Liu, Y. (eds.) ISICA 2008. LNCS, vol. 5370, pp. 618–625. Springer, Heidelberg (2008)
12. Hawe, G.I., Coates, G., Wilson, D.T., Crouch, R.S.: Agent-based simulation for large-scale emergency response: A survey of usage and implementation. ACM Comput. Surv. 45(1), 8:1–8:51 (2012)
13. Hofmann, M., Sackmann, S., Betke, H.: A novel architecture for disaster response workflow management systems. In: Proc. of ISCRAM 2013 (2013)
14. Apache Jena, version 2.11.1 (2013), http://jena.apache.org
15. Knublauch, H., Fergerson, R.W., Noy, N.F., Musen, M.A.: The Protégé OWL plugin: An open development environment for semantic web applications. In: McIlraith, S.A., Plexousakis, D., van Harmelen, F. (eds.) ISWC 2004. LNCS, vol. 3298, pp. 229–243. Springer, Heidelberg (2004)
16. Li, X., Liu, G., Ling, A., Zhan, J., An, N., Li, L., Sha, Y.: Building a Practical Ontology for Emergency Response Systems. In: Proc. of CSSE 2008, vol. 4, pp. 222–225. IEEE (2008)
17. Mijović, V., Tomašević, N., Janev, V., Stanojević, M., Vraneš, S.: Ontology enabled decision support system for emergency management at airports. In: Proc. of I-Semantics 2011, pp. 163–166. ACM (2011)
18. Pérez, J., Arenas, M., Gutierrez, C.: Semantics and complexity of SPARQL. In: Cruz, I., Decker, S., Allemang, D., Preist, C., Schwabe, D., Mika, P., Uschold, M., Aroyo, L.M. (eds.) ISWC 2006. LNCS, vol. 4273, pp. 30–43. Springer, Heidelberg (2006)
19. PostgreSQL, version 1.14.2 (2012), http://www.postgresql.org
20. Rocha, R.V., Araujo, R.B., Campos, M.R., Boukerche, A.: HLA compliant training simulations creation tool. In: Proc. of DS-RT 2009, pp. 192–198 (2009)
21. Teo, Y.M., Szabo, C.: CODES: An Integrated Approach to Composable Modeling and Simulation. In: Proc. of ANSS 2008, pp. 103–110. IEEE (2008)
22. Thalheim, B., Tropmann-Frick, M.: Mini Story Composition for Generic Workflows in Support of Disaster Management. In: Proc. of ISSASiM 2013, pp. 36–40. IEEE (2013)

An Approach for Automatic Query Expansion Based on NLP and Semantics

María G. Buey, Ángel Luis Garrido, and Sergio Ilarri

IIS Department, University of Zaragoza, Spain
{mgbuey,garrido,silarri}@unizar.es

Abstract. Nowadays, there is a huge amount of digital data stored in repositories that are queried by search systems that rely on keyword-based interfaces. Therefore, the retrieval of information from repositories has become an important issue. Organizations usually implement architectures based on relational databases that do not consider the syntax and semantics of the data. To solve this problem, they perform complex Extract, Transform and Load (ETL) processes from relational repositories to triple stores. However, most organizations do not carry out this migration due to lack of time, money and knowledge.

In this paper we present a methodology that performs an automatic query expansion based on natural language processing and semantics to improve information retrieval from relational databases repositories. We have integrated it into an existing system in a real Media Group organization and we have tested it to analyze its effectiveness. Results obtained are promising and show the interest of the proposal.

Keywords: information retrieval, query expansion, semantic search.

1 Introduction

Search systems in different contexts are adopting keyword-based interfaces due to their success in traditional web search engines, such as Google or Bing. These systems are usually based on the use of inverted indexes and different ranking policies [1]. In the context of keyword-based search on classical relational databases, most approaches retrieve data that exactly match the user keywords [2]. However, they do not consider the semantic contents of the keywords and their relationships. This can lead to losing information (low recall). Moreover, search engines return needless data, that are not interesting (low precision). This problem has received a great deal of attention and several approaches have been proposed to solve it, that are applicable in *Automatic Query Expansion (AQE)* solutions. One of the most natural and successful techniques is to expand the original query with other words that best capture the actual user intent, or that simply produce a more useful query that is more likely to retrieve relevant documents. AQE is currently considered a promising technique to improve the retrieval effectiveness of document ranking and it is being adopted in commercial applications, especially for desktop and intranet searches. However, very little work has been done to review such studies.

H. Decker et al. (Eds.): DEXA 2014, Part II, LNCS 8645, pp. 349–356, 2014.

Under these circumstances, we have developed a methodology called SQX-Lib (Semantic Query eXpansion Library). This paper is an extended version of a demo presented in [3]; the extensions include a detailed description of the methodology and a study of an example of its application in a real environment. SQX-Lib has been implemented as a library to encapsulate its functionality so that it can be fit in any environment. It is focused on automatically and semantically expanding the scope of the searches, and fine-tunes them by taking advantage of the Named Entities (NEs) present in the query string. In more detail, SQX-Lib performs three main tasks: *1) Obtain words with common lemmas*, to extract those words that belong to the same family as the keywords entered as a query. *2) Obtain words with similar meanings*, to return records which contain words that are synonymous to the keywords introduced. *3) Refine the queries with named entities*. NEs are considered as a whole. For example, if we search "Real Madrid" the search engine does not return data that contain the words "Madrid" and "real", but those related to that Spanish football team. To carry out these tasks, we have used a parser and a POS (Part of Speech) tagger as natural language processing (NLP) tools, a NE repository, a lexical database and a set of dictionaries for obtaining the meanings and the synonyms of the keywords, and a disambiguation engine designed to eliminate ambiguities.

Therefore, this paper provides two main contributions: Firstly, we present a new multilingual semantic query expansion methodology for information systems based on relational databases; and secondly, we have experimentally tested the use and the impact of our system in a real company environment and we have surveyed the opinion of the users.

This paper is structured as follow. Section 2 describes the state of the art. Section 3 explains the general architecture of our solution and presents the proposed algorithm. Section 4 discusses the results of our study with real data. Finally, Section 5 provides our conclusions and some lines of future work.

2   Related Work

The relative ineffectiveness of information retrieval (IR) systems is largely caused by the inaccuracy of a query formed by a few keywords, and AQE is a well-known method to overcome this limitation by expanding the original query of the user by adding new elements with a similar meaning. An IR system is likely to return good matches when an user query contains multiple specific keywords that accurately describe the information needed. However, the user queries are usually short and the natural language is inherently ambiguous. This is known as the vocabulary problem [4], compounded by synonymy (different words with similar meaning) and polysemy (a word with different meanings). To deal with this problem, several approaches have been proposed, including: interactive query refinement, relevance feedback, word sense disambiguation, and search results clustering. The survey of Carpineto and Romano [5] presents a wide study of AQE and a large number of approaches that handle various data resources and employ very different principles and techniques.

Most IR systems are based on the computation of the importance of terms that occur in the query and in the documents, and on several widely-used ranking models. Linguistic analysis techniques handle language properties such as morphological, lexical, syntactic, and semantic word relationships, to expand or reformulate query terms. Methodologies such as *corpus-specific global* techniques analyze the contents of a database to identify features used in similar ways [6], and *Query-specific local* techniques take advantage of the local context provided by the query using a subset of documents that is returned with the given query [7]. Other current proposal is [8], that uses relational databases combined with machine learning techniques and semantics. Also, substantial effort has been applied to the task of named entity recognition. A named entity is a word or a set of words that correspond with names of people, places, organizations, etc. Named Entity Recognition (NER) is an important task of information extraction systems and it consists of locating proper nouns in a text [9]. Novel approaches, such as [10], try to find more complex names (e.g., film or book titles) in web text. Knowledge-based proposals rely on the combination of a range of knowledge sources and higher-level techniques (e.g., co-reference resolution). Dictionaries and extensive gazetteer lists of first names, company names, and corporate suffixes are often claimed to be a useful resource. Moreover, one of the problems when working with the semantics of words is that there may be ambiguities when there is more than one possible meaning for each word. Word Sense Disambiguation (WSD) is the ability to identify automatically the meaning of words in a context and it is a natural and well-known approach to the vocabulary problem in IR [11]. Depending on how the *similarity relation* between the words and their meanings [12] is computed, we can distinguish different techniques: *measures based on glosses*, where a gloss is a definition or explanation of a word in a dictionary [13]); *measures based on conceptual trees*, which are based on a tree or is-a hierarchy, such as WordNet [14]; and *measures based on the content of the information*. Several experiments suggested that using WordNet may not be effective for IR [15], at least as long as the selection of the correct sense definition (or synset) is imperfect. However, research based on the work using WordNet for AQE has continued using more sophisticated methods [16].

3 Architecture

This section describes the process that performs the semantic expansion of the input keyword-based query and its associated named entity recognition method and disambiguation process. The semantic expansion process carries out three main tasks that we summarize below:

1. *Analysis of the keywords of the query*: first, SQX-Lib performs an analysis of the keywords introduced as a query. This task consists of several steps:
   - *Obtain the logical query structure*: it performs an appropriate construction of the query with the introduced terms without losing its structure. This task takes into account and processes the logical operators, parentheses and quotes that could appear in the search string.

- *Morphological analysis of the query:* SQX-Lib analyzes syntactically the query string to obtain a preprocessing of the terms that constitute the introduced query, as well as a possible first recognition of the NEs present in it (in case have been introduced capitalized). Specifically, for this purpose we have used the functionalities of the Freeling [17] NLP tool.
- *Obtain named entities:* next, it analyzes again the query string in order to get all the NE that are present. The method is described in Section 3.1.
- *Obtain exact words:* it performs another analysis of the query to search words that have been introduced quoted. These words will be searched as they have been written in the query string and they are not processed to find their semantics.
- *Remove stop words:* then, SQX-Lib eliminates those words from the query string that are considered as noise (articles, prepositions, etc.), as they do not provide any relevant information. The information about the type of each word is provided by the NLP tool.
- *Select terms to process:* once it has obtained the structure of the query, the NE and the exact words present in the query, and eliminated the stop words, SQX-Lib selects the words that are processed in order to find their semantics to enrich the input query. In this case, we consider only those words that are common names or verbs.

2. *Processing of terms:* At this stage, SQX-Lib analyzes the selected terms on the previous step. This task consists of four steps:

   - *Obtain lemmas:* it extracts the lemmas of the terms using the Freeling NLP tool described before.
   - *Obtain synonyms:* it searches and obtains the sets of synonyms for each term, by using the extracted lemmas and the semantic resources that a set of dictionaries and a thesaurus offer.
   - *Disambiguation:* if the terms have more than one set of synonyms, SQX-Lib performs a disambiguation process to select the most appropriate ones. This process is described in Section 3.1.
   - *Select the closest synonyms:* then, from the set selected in the previous step, it selects the synonyms that are the closest to each term.

3. *Expanding the query:* finally, SQX-Lib reconstructs and expands the query through the relevant data extracted in the previous tasks while keeping the initial logical structure of the query.

3.1   Named Entity Recognition Method and Disambiguation Process

We have performed a new knowledge-based method for NER that relies on a lexical thesaurus and on a set of resources that contain information extracted before from the text repository. Our method involves three main tasks:

1. *Collection of the NEs identified by NLP (Natural Language Processing):* First, it collects the named entities identified previously by the NLP tool.

2. *Identification of hidden NEs:* Secondly, it identifies the named entities that are hidden in the text, i.e. those NEs which have not been identified by the NLP tool because they have been written in lowercase. This happen because the NLP tool only identifies NE written in uppercase. So, for this purpose, for each word in the text, the method tests if it is possible the word is an NE by consulting an NE repository constructed before. Then, it also tests if it is a common name in EuroWordNet. In case the word could be both a named entity and a common name, the process has to disambiguate its meaning.

3. *Identification of embedded NEs:* Next, the method analyzes the named entities found to identify if some of them are embedded within another, i.e., whether multiple named entities recognized constitute a greater one. For this purpose, it identifies groups of named entities (two or more named entities that appear together without another word involved). Then, it calculates the joint and separate occurrences of each named entity of a group, and for each set of occurrences it disambiguates between them considering the other query keywords. Finally, it selects the most likely ocurrence of a named entity from each group.

We have also implemented a disambiguation engine that includes a set of methods and techniques that compute the semantic relation between words. It uses a set of dictionaries with semantic information of words and a lexical thesaurus to find the possible sets of synonyms for a word. The lexical thesaurus used has been EuroWordNet [18], which is a multilingual database for several European languages. This process combines the techniques of measures based on conceptual trees and measures based on the content of the information. It uses a given word as input and a set of lists of meanings of that word provided from the aforementioned lexical resources. Specifically, it calculates the degree of semantic similarity between a word and each possible list of meanings by adding up the weight obtained for each term that appears in a list of meanings. This weight is calculated by taking into account the TF-IDF of the term, the length of the shortest path in the EuroWordNet hierarchy between the term with the word that is being analyzed, the measure of specifity that the term has in the hierarchy, and an adaptation of the Normalized Google Distance (NGD) [19]. That is, it takes into account the relation that it has with the word and with the set of documents. Once the weight is computed for each possible list of meanings, the disambiguation engine returns the set whose weight is the smallest.

4   Case Study

The present case is the search engine that is working over the news repository used everyday by all the departments of the Heraldo Group[1], a leading Spanish media. We have carried out this case study due to the difficulty we have found when comparing our proposal with other existing ones, since there are not similar approaches using the Spanish language.

---

[1] http://www.grupoheraldo.com/

The data repository used by the documentation department of the Heraldo Group is a relational database with about 10 million records. It contains news, interviews, articles, photos, infographics, videos and entire pages corresponding to publications of the group. The contents and its metadata are in Spanish plain text format. Text fields are fully indexed, so the search engine can quickly locate records using a query based on keywords. Every day the system receives an average of almost one thousand queries.

Although the system provides advanced search options, users mainly use the basic search interface, which is very similar to that of many search engines that can be found in the Internet, such as Google. This search interface is indeed very fast and practical, but of course, it also has many shortcomings regarding the quality of the results, as explained in Section 1. We find three major groups of queries using this interface: general text-based queries (61%), documentation department professional queries (8%), and queries based on thesauri and specific database fields (31%). The really expandable queries are the text-based ones. Therefore, we decided to link SQX-Lib to this type of search, which would ensure a greater coverage and quality of results for the input queries.

For experimental evaluation, we have carried out a semi-automatic process considering the set of the last queries performed on the system. First, we have discarded the queries made by the documentation department because this kind of queries are based on thesaurus tags. Likewise, we have also discarded all the queries that are using database fields different from the text content, like the date, the author, etc. So, finally, we have select a sampling of 1,000 real queries made by users to perform our analysis. We have applied our methodology and we have obtained the following results: 64% of these queries include NEs and could be optimized using SQX-Lib; 61% of these queries could take advantage of the lemmatization step; 58% of these queries include one or more common names from which the system is able to obtain synonyms and related words; and 13% of the queries could lead to build an incorrect query. We have reached this conclusion by noting that the results of certain expanded queries were incorrect and did not contain the desired information due to language idioms.

In general, we have found that 95% of these queries are expandable, that is, they can take advantage of SQX-Lib and be optimized by our system. So, we obtained the next general conclusions by observing the results obtained:

- The lemmatization expansion over the query leads to an average improvement of 50% in the recall.
- The semantic expansion step translates into obtaining about double results on average.
- The use of an automatic filter when we found a NE implies a reduction of an average of 20% of documentary noise.
- If we use all the options together, on average we obtain almost four times more results than using the original query, so SQX-Lib successfully minimizes the documentary silence problem. Besides, this is not done at the expense of introducing more noise.

Moreover, we have carried out an opinion survey among the workers of the Heraldo Group about their use of the system improved with SQX-Lib in their daily work. They were asked about what features they would like to have in the search system by default. We have seen that 70% of the users would include the lemmatization expansion despite the rest of users prefer this feature to be optional, 58% of the users like the semantic expansion and they would also include it by default, and 88% of the users agree about the need of a default mechanism to filter the results using the name entities embedded in the query. In general, all the users are satisfied with the new functions, although not all of them agree about whether it must be optional or applied by default. This is due to the different search style of each department, which is consequence of the different types of information that they want to find.

5   Conclusions and Future Work

In this paper, we have presented a semantic query expansion methodology called SQX-Lib, that combines different techniques, such as lemmatization, NER and semantics, for information extraction from a relational repository. The process includes a disambiguation engine that calculates the semantic relation between words and selects the best meaning for those words. We have integrated SQX-Lib in a real major Media Group in Spain and we have carried out a case study in this real environment. In our evaluation we have found that about 95% of the user keyword-based queries are optimized by our system, and obtaining up to four times more results than in a normal search. Moreover, we also have carried out an opinion survey in the company about the user experience using SQX-Lib in their daily work. It is a prototype designed to give a solution to the particular environment of Heraldo Group, as future work remains generalizing the solution in other languages, such as English, to enable its evaluation with standard test packages like TREC[2]. This should also facilitate the comparison of our solution with others approximations. Moreover, there is some other applications where it can be useful. The disambiguation techniques implemented in SQX-Lib have been used to disambiguate terms included in ontologies in order to compare between these ontologies, and be able to construct topic maps [20] or recommenders [21,22].

Acknowledgment. This research work has been supported by the CICYT project TIN2010-21387-C02-02 and DGA-FSE. We thank Raquel Trillo and Jorge Gracia for their help.

References

1. Baeza-Yates, R., Ribeiro-Neto, B., et al.: Modern information retrieval, vol. 463. ACM Press, New York (1999)

---

[2] http://trec.nist.gov/

2. Yu, J.X., Qin, L., Chang, L.: Keyword search in databases. Synthesis Lectures on Data Management 1(1) (2009)
3. Granados Buey, M., Luis Garrido, Á., Escudero, S., Trillo, R., Ilarri, S., Mena, E.: SQX-Lib: Developing a semantic query expansion system in a media group. In: de Rijke, M., Kenter, T., de Vries, A.P., Zhai, C., de Jong, F., Radinsky, K., Hofmann, K. (eds.) ECIR 2014. LNCS, vol. 8416, pp. 780–783. Springer, Heidelberg (2014)
4. Furnas, G.W., Landauer, T.K., Gomez, L.M., Dumais, S.T.: The vocabulary problem in human-system communication. Communications of the ACM 30(11), 964–971 (1987)
5. Carpineto, C., Romano, G.: A survey of automatic query expansion in information retrieval. ACM Computing Surveys 44(1) (2012)
6. Qiu, Y., Frei, H.-P.: Concept based query expansion. In: Annual International ACM SIGIR Conference on Research and Development in Information Retrieval, pp. 160–169 (1993)
7. Buckley, C., Salton, G., Allan, J.: Automatic retrieval with locality information using SMART, pp. 59–72 (1993)
8. Bergamaschi, S., Guerra, F., Interlandi, M., Trillo-Lado, R., Velegrakis, Y.: QUEST: A keyword search system for relational data based on semantic and machine learning techniques. Proceedings of the VLDB Endowment 6(12), 1222–1225 (2013)
9. Sekine, S., Ranchhod, E.: Named Entities: Recognition, Classification and Use. John Benjamins (2009)
10. Downey, D., Broadhead, M., Etzioni, O.: Locating complex named entities in web text. In: IJCAI, vol. 7, pp. 2733–2739 (2007)
11. Sanderson, M.: Retrieving with good sense. Information Retrieval 2(1), 49–69 (2000)
12. Navigli, R.: Word sense disambiguation: A survey. ACM Computing Surveys 41(2), 10 (2009)
13. Vasilescu, F., Langlais, P., Lapalme, G.: Evaluating variants of the lesk approach for disambiguating words. In: LREC (2004)
14. Miller, G.A.: WordNet: A lexical database for English. Communications of ACM 38(11), 39–41 (1995)
15. Voorhees, E.M.: Using WordNet to disambiguate word senses for text retrieval, pp. 171–180 (1993)
16. Schütze, H., Pedersen, J.O.: Information retrieval based on word senses (1995)
17. Carreras, X., Chao, I., Padró, L., Padró, M.: FreeLing: An open-source suite of language analyzers. In: Fourth International Conference on Language Resources and Evaluation, pp. 239–242. European Language Resources Association (2004)
18. Vossen, P.: EuroWordNet: A multilingual database with lexical semantic networks. Kluwer Academic Boston (1998)
19. Cilibrasi, R.L., Vitanyi, P.M.: The Google similarity distance. IEEE Transactions on Knowledge and Data Engineering 19(3), 370–383 (2007)
20. Garrido, A.L., Buey, M.G., Escudero, S., Ilarri, S., Mena, E., Silveira, S.B.: TMgen: A topic map generator from text documents, pp. 735–740 (November 2013)
21. Garrido, A.L., Pera, M.S., Ilarri, S.: SOLER-R, a semantic and linguistic approach for Book Recommendations. In: 14th IEEE International Conference on Advanced Learning Technologies, ICALT (July 2014)
22. Garrido, A.L., Ilarri, S.: TMR: A Semantic recommender system using topic maps on the items descriptions. In: 11th European Conference of Web Semantic (May 2014)

Mapping Construction Compliant with Schema Semantics

Verena Kantere

University of Geneva
verena.kantere@unige.ch

Abstract. A dominant characteristic of autonomous information sources is their heterogeneity, in terms of data formats and schemas. We need schema and data mappings between such sources in order to query them in a uniform and systematic manner. Guiding the discovery of mappings employing automatic tools is one of the fundamental unsolved challenges of data interoperability. In this work we consider the problem of discovering mappings for schemas of autonomous sources that can gradually revealed. Using as an example setting an overlay of peer databases, we present a mapping solution that discovers mappings, which can be adapted to new and gradually revealed schema information. Mapping discovery is schema-centric and incorporates new semantics as they are unveiled.

1 Introduction

Automating the discovery of schema (and data) mappings is one of the fundamental challenges of data interoperability. The research literature on schema mapping includes significant works [1], [2], [3], [4], [5] etc. All of these are complementary and tackle different aspects of the problem, making various assumptions about the setting. Clio [2], [6] is a tool that provides possible mappings for a pair of schemas based on user-defined data value correspondences. The aim of Clio is to produce all or most of the possible mappings, and not to choose or rank the alternative solutions. The Clio mappings can be put in ToMAS [3], [7] which preserves their semantics as the schemas may evolve. In addition, Muse [8], [4] offers a technique for the disambiguation of mappings based on data examples, whereas the technique in [5] explores the benefits of possibly available additional semantic information. The focus in [9] is the inference of mapping rules for pairs of peer DTDs, based on node correspondences. The aim is to spot common groups of nodes that can be bound to single tree expressions. In [10] an algorithmic approach to automating the discovery of mappings is presented. The prerequisite is that schema and meta-schema information is available in specific formats and that schema matching implies exact matching of included concepts.

As discussed in [1], a challenge is to maintain mappings in a dynamic autonomous environment, where any kind of information (data/meta-data and schemas) change independently. Towards this direction, we consider an environment where communication ends would like to establish schema mappings, but remote schemas, and, thus, their implied semantics, are revealed gradually; therefore, a matching with the local schema is piecemeal and the discovery of correct mappings is progressive. Assuming autonomous

H. Decker et al. (Eds.): DEXA 2014, Part II, LNCS 8645, pp. 357–365, 2014.

communicating sources, it is required that such a process can take place in shortage of data and their semantics.

Our motivation derives from the needs of the applications of distributed autonomous information sources such as federated databases in Peer-to-Peer, Cloud Computing or Big Data environments:

Dynamic data sharing. Autonomous data sources are apt to retaining their autonomy as long as their provisional requirements for data integration and exchange are fulfilled. The needs for data sharing as well as the stored data and their semantics at the sources may change dynamically. For example, a source may choose to reveal a different part of its data or change the structure of the revealed data in time. Therefore, the schema mappings between pairs of sources have to be flexible and adaptable to new data semantics.

Gradual data sharing. Even if sources keep the semantics and the part of the shared data static, they may choose to gradually match their schema with the schema of other sources. Schema matching is a costly procedure; if a pair of sources decides to perform the entire procedure at once and then discovers that they cannot fulfil each other's data sharing needs, both of them will have spent time and human resources on the establishement of non-useful schema mappings. The sources would benefit from an incremental schema integration procedure that goes on as long as the discovered schema mappings promise data sharing of high quality.

This work provides a schema mapping solution fine-tuned to the described settings of distributed autonomous data sources and contributes the following:

- A technique that discovers schema mappings in a semi-automatic manner for pairs of revealing relational schemas. The technique is schema-centric instead of mapping-centric, meaning that the mapping is adapted to the incrementally disclosed semantics, as the schemas are revealed. The disclosed semantics are directly determined by schema matching or indirectly inferred based on the schema structure.
- The discovered mapping includes attribute correspondences and join paths between such attributes that can be changed individually and give a new mapping, always compliant with the schema semantics. This allows the employment of the technique for semi-revealed schemas in order to create incomplete mappings, and, reemploy the technique as more schema elements and semantics are revealed, in order to increment the mappings *while* the remain compliant with the new semantics.

Moreover, the technique can produce mappings with value conditions, exploiting normal query traffic between the matched remote schemas or available data. The tool that implements the mapping mechanism is demonstrated in [11] and is integrated and presented in [12][1].

In the rest of this paper, Section 2 formalizes the problem and overviews the mapping mechanism. Section 3 introduces the mapping discovery procedure, which is analysed in Section 4. Section 5 concludes the paper.

---

[1] More about GroupPeer and the mapping tool can be found in
http://www.dblab.ece.ntua.gr/~vkante/groupeer/

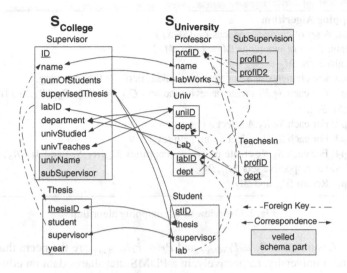

Fig. 1. A pair of revealing schemas

2   The Mapping Mechanism

For ease of presentation and without loss of generality we present the mapping mechanism applied in a peer data management system (PDMS). Peer databases store and manage their data locally and establish acquaintances with other peers so that they can exchange data. An acquaintance between two peers involves the agreement on a set of mappings that match information between the two peer schemas. A peer query posed in the system is expressed on the local schema of the posing peer and is rewritten through the respective mappings on the local schema of its acquaintees, in order for the latter to provide answers with local information. The query is propagated along overlay paths and answered by remote peers, which do not hold direct mappings with the query-posing peer. Pairs of remote peers may exchange queries and data, which gradually reveal local schema and data semantics to each other. Such peers may want to construct mappings between their schemas, in order to become directly acquainted in the overlay.

Peer databases hold relational data, which comply with relational schemas. A mapping between two relational schemas should encapsulate the schema semantics while adhering to the schemas' structure. Each peer holds with its acquainted peers schema mappings of the forms GAV (Global-As-View) or LAV (Local-As-View), inherited from the data integration domain [13]. The mappings are of the well-known conjunctive form (equivalent to SPJ queries). In GAV/LAV definitions for the P2P setting, the global schema is the schema of the peer on which the query is initially posed and the local schema is the schema of the peer on which it is rewritten. We assume a pair of remote peers that exchange queries and intent to establish mappings with each other. Without loss of generality, we take the view of one of the peers. P_S is the query-posing peer and P_T is the answering one. The schema of P_S, S_S, is gradually revealed to P_T and is progressively mapped to the local schema S_T.

Mapping Algorithm

Input: A set of correspondences $C_U(S_S, S_T)$

Output: A set of mappings $\mathcal{M} = \{\mathcal{M}_G, \mathcal{M}_L\}$

Initialization: $\mathcal{M}_G = \mathcal{M}_L = \emptyset$

When new elements of S_S or S_T are revealed, perform the following:

Step1: For each $S_S.R_S$ create or revise the sets $C_{\mathcal{D}_G}(R_S, S_T)$, $C_{\mathcal{D}_L}(R_S, S_T)$ from $C_U(S_S, S_T)$.

Step2: For each $S_S.R_S.A$ select a correspondence C

Step3: For each $S_S.R_S$ select a join path set $\mathcal{JP}$

Step4: For each $S_S.R_S$ create or evolve mappings $M_G(R_S, S_T)$ and $M_L(R_S, S_T)$ and add to respective sets $\mathcal{M}_G, \mathcal{M}_L$.

Step5: Return $\mathcal{M}_G$ and $\mathcal{M}_L$.

Fig. 2. Overview of the mapping algorithm

Example 1. Assume that $P_S \equiv P_{College}$ and $P_T \equiv P_{University}$ are the peers that belong to a college and a university, respectively, in a PDMS that shares data on education, (depicted in Figure 1). The schema $S_{College}$ of the college is gradually revealed to $P_{University}$. The mapping mechanism of $P_{University}$ maps $S_{College}$ and the local schema, $S_{University}$, as the first is revealed and matched with the latter. □

P_T can use an automatic schema matching tool [14], [15] in order to find conceptual simple matchings between the revealed part of S_S, and S_T. Such a tool produces correspondences of schema concepts.

Definition 1. *A schema element E of a relational schema S is a relation R, denoted else as S.R or an attribute A of a relation R denoted as R.A or S.R.A. The concept of a schema element E is denoted as concept(E).* □

Definition 2. *An undirected conceptual correspondence C_U is an undirected matching between the concepts that correspond to two schema elements E_1, E_2 and is denoted as $C_U(concept(E_1), concept(E_2))$. The undirected correspondence holds under a confidence value, which is denoted as $|C_U(concept(E_1), concept(E_2))|$.* □

Whenever not confusing, hereafter, we omit prefixes and the function *concept*(.). An automatic schema matching tool produces sets of undirected correspondences C_U).

The classical definitions of GAV and LAV mappings denote information subsumption between the local and the global schema [13]. Concerning a PDMS, a GAV mapping is defined as $M_{GAV} : R_S \supseteq view(S_T)$ and a LAV mapping is defined as $M_{LAV} : view(S_S) \supseteq R_T$; R_S, R_T are relations of S_S, S_T, respectively, and *view*(.) denotes a schema view. In this spirit, we require that each conceptual correspondence employed in a mapping is compliant with the respective information subsumption. Intuitively, the structural matching of schema elements in a mapping encompasses the matching of their semantics. The latter should be coherent with the information flow represented by the mapping. For example, if a correspondence $C_U(R_S.A, R_T.A)$ is included in a GAV mapping, $R_S \supseteq view(S_T)$, this correspondence should imply that $R_T.A$ is conceptually subsumed by $R_S.A$; yet, in case of a LAV mapping $R_S \subseteq view(S_T)$ (in this case S_T is revealed to S_S), the correspondence should imply the opposite. To support conceptual matching that encapsulates information subsumption, we define the *directed correspondence*:

Definition 3. *A directed correspondence* C_D *is a matching between the concepts* $concept(E_1)$ *and* $concept(E_2)$ *of schema elements* E_1 *and* E_2 *that denotes subsumption of* $concept(E_1)$ *by* $concept(E_2)$, *and is denoted as* $C_D(concept(E_1), concept(E_2))$. *The directed correspondence holds under a confidence value, denoted as* $|C_D(concept(E_1), conceptl(E_2))|$. □

The mapping discovery mechanism is represented by a quatruple $< S_S,$ $S_T, C, \mathcal{M}, C_\mathcal{P} >$, where S_S and S_T are the source and target revealing schemas. C comprises the correspondences that conceptually match S_S and S_T and are used in current mappings. $\mathcal{M}$ comprises the current mappings. More specifically, C includes two sets of directed correspondences, $C_{\mathcal{D}_G}$ and $C_{\mathcal{D}_L}$ that are employed in GAV and LAV mappings, respectively. $\mathcal{M}$ includes two sets of mappings $\mathcal{M}_G$ and $\mathcal{M}_L$. $\mathcal{M}_G$ includes the GAV mappings, M_G w.r.t. S_S, i.e. $M_G : R_S \supseteq view(S_T)$, where $R_S \in S_S$; and $\mathcal{M}_L$ includes the LAV mappings, M_L w.r.t. S_T, i.e. $M_L : R_S \subseteq view(S_T)$, where $R_S \in S_S$. The mappings may be incomplete, meaning that not all involved attributes of R_S may be matched in S_T. Furthermore, $C_\mathcal{P}$ is the set of possible alternative correspondences. Intuitively, the join path JP for a directed correspondence C_D is the combination of joined attributes of relations of the target schema S_T that is used so that the correspondence can be included in the specific mapping. The mapping algorithm (Figure 2) is iterative and constructs the mappings taking as input C.

3   Mapping Discovery

As explained in Section 2, the mapping mechanism produces sets of GAV and LAV mappings, $\mathcal{M}_G(S_S, S_T)$ and $\mathcal{M}_L(S_T, S_S)$, respectively. A mapping $M_G \in \mathcal{M}_G$ or a mapping $M_L \in \mathcal{M}_L$ concerns a relation $S_S.R_S$ that is mapped partially or thoroughly to relations of S_T, employing directed correspondences from $C_{\mathcal{D}_G}$ or $C_{\mathcal{D}_L}$, respectively. We continue without distinguishing between GAV and LAV mappings, since they are produced in the same way.

The novelty of the proposed mapping discovery procedure is two-fold. First, the procedure is centered on semantics and assisted by schema constraints. It starts with a matching of semantics and builds mappings that encapsulate the matching but also adhere to the schema structure. Second, the mappings are able to evolve with time, as more schema parts are revealed, and adapt to the new semantics. Overall, mapping discovery follows these guidelines:

- Mappings should include the relations that encapsulate the necessary semantics and as few as possible others.
- Most of the semantics encapsulated in a mapping are usually concentrated in one or a few relations.

Section 4 discusses these guidelines as the mapping discovery procedure is presented.

The construction of the mapping of R_S can be easily monitored using an associative structure, M_{atrix}. The latter saves the current form of the mapping and allows for the user to change the composition of the mapping if the schema semantics evolve.

Definition 4. *The* $M_{atrix}(R_S, S_T)$ *structure consists of the following set of elements. For each attribute* $R_S.A$:

Table 1. $M_{atrix}(Supervisor, S_{University})$

Attribute	Correspondence	Alias	C_{D_p}	JP
ID	Professor.ID	-	Student.supervisor	-
name	Professor.name	-	-	-
numofstudents	-	-	-	-
supervisedThesis	Student.supervisor	-	Student.thesis	Professor,Student
labID	Lab.ID	-	Student.lab	Professor, Lab
department	Lab.department	-	TeachesIn.dept, Univ.dept,	TeachesIn, Professor
univStudied	Univ.unID	-	-	Professor, Univ
univTeaches	Univ.unID	Univ\$1	-	Professor, TeachesIn, Univ\$1

- *the selected correspondence* $C_D(R_S.A, S_T.R_T.A)$,
- *an alias name for* R_S *w.r.t.* C_D, *alias*(C_D, R_S),
- *a set of possible correspondences* $C_{D_p}(\mathcal{R}_S, \mathcal{S}_T)$,
- *the join path for the chosen correspondence* $JP(R_S, S_T)$,

Keeping a value for the above elements is optional.   □

Briefly, for each attribute of R_S, $M_{atrix}(R_S, S_T)$ keeps track of the currently chosen correspondence as well as of the rest of the correspondences among which the first is selected. For the chosen correspondence of each attribute, $M_{atrix}(R_S, S_T)$ also indicates the respective join path in S_T that should be incorporated in the mapping. Intuitively, the join path for a directed correspondence is the combination of joined attributes of relations of the target schema that is used so that the correspondence can be included in the specific mapping. The $M_{atrix}(R_S, S_T)$ is visualized in the form of a matrix *rows* × *columns*, where the *rows* correspond to the attributes of R_S and the *columns* to the items (or sets of items) that are logged. The information contained in the column that refers to the selected correspondences is stored in the C set of the mapping mechanism and the information in *JP* is stored in the set of produced mappings $\mathcal{M}$. Table 1 shows the initialized $M_{atrix}(Supervisor, S_{University})$.

4   Construction of Mappings

The construction of the mapping $M(R_S, S_T)$ is performed through the construction of the respective matrix, $M_{atrix}(R_S, S_T)$, described in the following.

The selected correspondences for all attributes of R_S have to be incorporated in the mapping, in a way that the structure of the latter encapsulates their correlations. Essentially, we have to determine a *join path*, i.e. one or more chains of relation joins in S_T such that all relations that are involved in the mapping have at least one join with at least one other relation that is also involved in the same mapping. Formally, a join path is defined as follows:

Definition 5. *Assume   that $R_S.A$ is   corresponded   to $S_T.R_T.A$,   i.e.* $\exists C_D(R_S.A, S_T.R_T.A) \in C$ *or* $\exists C_D(S_T.R_T.A, R_S.A) \in C$. *A join path JP for the attribute $R_S.A$ and a schema S_T is a chain of joins on attributes of S_T declared as an* *ordered set:* $JP(R_S.A, S_T) = (R_{T_1}, ..., R_{T_n})$, *where* $R_{T_i}, R_{T_{i+1}}$ *are relations of* S_T, *such that they share a foreign key constraint. It holds that* $R_T \equiv R_{T_i}, i = 1, ..., n$.   □

Visually, if the relations of S_T involved in the initializing mapping are nodes of a graph, and the joins between pairs of these relations are edges between the respective pair of nodes, we would like this graph to be connected (i.e. no natural joins). A join path $JP(R_S.A, S_T)$ connects the corresponded attribute of $R_S.A$ in S_T with at least one more corresponded attribute of R_S in S_T, through joins on relations of S_T. Since the available schema constraints are foreign keys, paths are limited on foreign key joins.

The goal is to find the most *appropriate* join paths for each one of the selected correspondences for the attributes of R_S, in terms of semantics encapsulation in the mapping. Let us visualize again the relations of S_T to be involved in the mapping as nodes of a connected graph; we make two controversial requirements, concerning the graph connectivity:

– The graph is as connected as possible
– The graph is connected with the fewer possible nodes

Intuitively, a high graph connectivity, results in short paths between pairs of the graph nodes, which can be translated as tight conceptual correlation among the relations involved in the mapping. However, optimizing the length of each path separately, can lead to a graph with a high number of nodes (remember that nodes correspond to relations). This means that many relations that do not hold correspondences are involved in the mapping, introducing in it many concepts of S_T that are irrelevant to the concepts of R_S. Thus, many associative relations may obscure the mapping semantics.

The two requirements can be compromised in the *minimum sum* of all the join paths that are necessary in order to join all the relations involved in the mapping. Thus, we define the appropriateness of a *set* of join paths, w.r.t. R_S as follows:

Definition 6. *The appropriateness of a set of join paths* $\bigcup JP(R_S.A_i, S_T)$ *is the sum of the length of them:* $\biguplus |JP(R_S.A_i, S_T)|$, *in which common parts of join paths are considered only once; the length of a join path* $JP(A, S_T) = (R_{T_1}, ..., R_{T_n})$ *is its arity, i.e.* $|JP(A, S_T)| = n$ *and denotes the number of consecutive joins between* R_{T_1}, R_{T_n}.   □

The selection of join paths has to comply with the principle that no two schema elements refer to the exact same concept. This necessitates that two attributes $R_S.A_1$ and $R_S.A_2$ that are corresponded to the same target attribute $S_T.R_T.A$, should probably not use the same join path, if they use any. Intuitively, a join path from $R_T.A$ to a relation R_O indicates a specific semantic correlation of $R_T.A$ and R_O. The fact that the two attributes A_1 and A_2 belong to the same source relation, but also, they are corresponded to (i.e. subsumed by) the same target attribute $R_T.A$, implies that the respective concepts are very close. In effort to distinguish between the concepts of A_1 and A_2 the mapping algorithm prefers different join paths between R_T and R_O, if this is possible.

The algorithm proceeds as follows:

1. Include all correspondences C in the target set of join paths search.
2. Select a relation R_O in the target schema as the origin of the join path search and remove the correspondences $C(R_S.A, R_O.A) \in C$ from the target search set.

Selection of Join Paths
Input: A set of directed correspondences C for relation R_S and schema S_T
Output: The selected join paths for the attributes of R_S, $\mathcal{JP}$
Initialization: $\mathcal{JP}(R_S, S_T) = \emptyset$
Step1: Initialize the target set $C^t = C$.
Step2: Select the relation that is the origin of the join paths R_O
Step3: For each attribute A of R_S do:
- Pick $C(A, R_T.A) \in C$;
- if $R_T \equiv R_O$ do: $C^t = C^t - C(A, R_T.A)$
Step4: Check for join paths of length $i = 1, .., |S_T|$ as follows:
For each relation R_T in S_T:
- If possible, create a temporary join path $JP_i^t = (R_O, R_T)$.
Step5: While a join path of i length is not found, do:
- i++
- Create the ordered set of i-1 length temporary join paths $\mathcal{JP}_i^t$ as
i-permutations of the set of relations in S_T, s.t. each $JP^t \in \mathcal{JP}_i^t$ is feasible.
Step6: For each $C(A, R_T.A) \in C^t$ do:
- For each $JP \in \mathcal{JP}$ do:
- if $JP.R_n \equiv R_T$ then $C^t = C^t - C(A, R_T.A)$; continue from Step6.
- while not found, pick a $JP^t \in \mathcal{JP}_i^t$; if $JP^t.R_n \equiv R_T$ do: $\mathcal{JP} = \mathcal{JP} \cup JP^t$ and
$\mathcal{JP}^t = \mathcal{JP}^t - JP^t$.
Step7: Create the ordered set of i-1 length temporary join paths $\mathcal{JP}_i^t$ as i-permutations of the relations in S_T, s.t. each $JP^t \in \mathcal{JP}_i^t$ is feasible.
Step8: If $C^t \neq \emptyset$, (it holds that i++$\leq |S_T|$) continue from Step5.
Step9: Return $\mathcal{JP}$.

Fig. 3. Selection of Join Paths

3. Iteratively search join paths of increasing length for the rest of the correspondences in C; create the possible join paths by considering all relations (including R_O) in the target schema.
4. Add the discovered join paths to the result set $\mathcal{JP}$. Check this set before searching for a new join path w.r.t. another correspondence of C that is still in the target set.

Figure 3 presents the pseudocode of the algorithm.[2] The following example gives a flavor of the algorithm's execution.

Example 2. Assume that *Supervisor* is to be matched on $S_{University}$; we start with a relation that serves as origin for the join paths, e.g. $R_O = Professor$. R_O is used to seek the most appropriate join paths. All three relations *Student*, *Lab* and *Univ* can be directly joined with *Professor*. Yet, if the join path $JP = (Professor, Univ)$ is selected for $C(univStudied, Univ.unID)$, it cannot be re-used for $C(univTeaches, Univ.unID)$. For the latter it is necessary to find another join path, if there is such one. From S_T we have to choose between: $JP_1 = (Professor, TeachesIn, Univ\$1)$ and $JP_2 = (Professor, Project, Univ\$1)$. Since *TeachesIn* is already in selected join paths, JP_1 is chosen. □

---

[2] For simplicity the algorithm does not show implementation improvements such as keeping track of join paths and storing parts of join paths in order to avoid creation of join sub-paths from scratch.

5   Conclusions

We present a novel schema mapping solution targeted to autonomous environments where relational data sources gradually reveal their schemas. The mechanism discovers mappings in a schema-centric manner, adapting their semantics to the progressively unveiled schema semantics. The proposed solution can create incomplete mappings, and, increment these mappings as more schema elements and semantics are revealed, while keeping them compliant with the new semantics[3].

Acknowledgement. The research leading to these results has received funding from the European Union 7th Framework Programme (FP7/2007-2013) under Grant Agreement n° 619706 ASAP.

References

1. Andritsos, P., Fuxman, A., Kementsietsidis, A., Miller, R.J., Velegrakis, Y.: Kanata: Adaptation and evolution in data sharing systems. SIGMOD Record 33(4), 32–37 (2004)
2. Hernández, M.A., Miller, R.J., Haas, L.M.: Clio: A semi-automatic tool for schema mapping. In: SIGMOD Conference, p. 607 (2001)
3. Velegrakis, Y., Miller, R.J., Popa, L., Mylopoulos, J.: Tomas: A system for adapting mappings while schemas evolve. In: ICDE 2004, p. 862 (2004)
4. Alexe, B., Chiticariu, L., Miller, R.J., Pepper, D., Tan, W.C.: Muse: A system for understanding and designing mappings. In: SIGMOD Conference, pp. 1281–1284 (2008)
5. An, Y., Borgida, A., Miller, R.J., Mylopoulos, J.: A semantic approach to discovering schema mapping expressions. In: ICDE, pp. 206–215 (2007)
6. Miller, R.J., Haas, L.M., Hernández, M.A.: Schema mapping as query discovery. In: VLDB, pp. 77–88 (2000)
7. Velegrakis, Y., Miller, R.J., Popa, L.: Preserving mapping consistency under schema changes. VLDB J. 13(3), 274–293 (2004)
8. Alexe, B., Chiticariu, L., Miller, R.J., Tan, W.C.: Muse: Mapping understanding and design by example. In: ICDE, pp. 10–19 (2008)
9. Bonifati, A., Chang, E.Q., Ho, T., Lakshmanan, L.V.S., Pottinger, R.: Heptox: Marrying xml and heterogeneity in your p2p databases. In: VLDB, pp. 1267–1270 (2005)
10. Fletcher, G.H.L., Wyss, C.M.: Data mapping as search. In: Ioannidis, Y., et al. (eds.) EDBT 2006. LNCS, vol. 3896, pp. 95–111. Springer, Heidelberg (2006)
11. Kantere, V., Bousounis, D., Sellis, T.: A Tool for Mapping Discovery over Revealing Schemas, http://www.dbnet.ece.ntua.gr/~vkante/mapping (submitted for publication)
12. Kantere, V., Bousounis, D., Orfanoudakis, G., Tsoumakos, D., Sellis, T.: GrouPeer: A system for clustering PDMSs, http://www.dbnet.ece.ntua.gr/~vkante/approx (submitted for publication)
13. Lenzerini, M.: Data Integration: A Theoretical Perspective. In: 21th ACM PODS (2002)
14. Do, H., Rahm, E.: Coma - A System for Flexible Combination of Schema Matching Approaches. In: VLDB (2002)
15. Madhavan, J., Bernstein, P., Rahm, E.: Generic Schema Matching with Cupid. In: VLDB (2001)

---

[3] Due to lack of space, this paper gives only some highlights of the proposed mapping mechanism. For more information please refer to
http://www.dblab.ece.ntua.gr/~vkante/groupeer/

Web Materialization Formulation: Modelling Feasible Solutions

Srđan Zagorac and Russel Pears

Auckland University of Technology, Auckland, New Zealand
{srdan.zagorac,russel.pears}@aut.ac.nz

Abstract. Performance of multi domain web search applications is typically hindered by the availability and accessibility of the web data sources. In this paper we consider web data materialization as a solution. The web data services are modelled via binding schema patterns – access patterns – thereby defining input and output dependencies between the participating data sources. Web materialization is formulated as a set of interdependent blocks, each being a deciding factor in formulating an obtainable materialization. In this work consideration is given to the feasibility of the proposed set of web sources for the given materialization task. The model for analysing the feasible materialization solution in terms of reachability and bound is established. To demonstrate the effectiveness of such a feasibility analysis model, an empirical study is performed on a set of materialization tasks ranging in their schema dependency complexity.

1 Introduction

Structured data on the Web exists in several forms; hidden in back-end Deep Web databases (such as Amazon.com or IMDB.com) and fronted by HTML forms and various web APIs. Web data can help improve Web search in a number of ways; for example, Deep Web databases are not generally available to search engines, and, by surfacing this data, a heterogeneous multi-domain Web search can expand the scope and quality of the Web-search index [1][6].

Search computing (SeCo) [2] aims to answer multi domain queries by accessing heterogeneous web based data sources typically found in the Deep Web. A major difficulty facing multi-domain query search is our inability to control query execution upon its submission to a remote data source. Often, the search process is affected by high latency due to Internet network congestion, routing problems or simply slow responsiveness of the 'other' side. The services involved may restrict the number of daily accesses or impose time delays between search requests. Further the data obtained might suffer from low coverage and a high level of duplicates, thus requiring invocations of multiple sources, if available, in order to harvest the required data corpus. Evidently, such constraints act detrimentally to the quality of the search process in terms of result delivery time and the result quality itself as the decreased availability of web sources might cause just a subset of the required result to be obtained. In order to alleviate such problems this research elaborates on a system for persistent

H. Decker et al. (Eds.): DEXA 2014, Part II, LNCS 8645, pp. 366–374, 2014.

materialization [7] of remote Web Data Sources. The research presented in this paper builds on web data materialization characterization presented in [5].

In this research we make the following novel contributions: 1) Formulation and definition of the model for the analysis of the materialization feasibility; 2) A detailed empirical study conducted on a set of materialization tasks ranging in their schema dependencies and complexity.

The rest of the paper is organized as follows: In Section 2 we define the fundaments of the challenge and introduce the proposed model. Section 3 presents a service description model devoted to the problem of Web data materialization. The materialization feasibility context is also defined and presented. Section 4 brings the detailed analysis of the materialization feasibility problem together with the model for feasibility analysis definition. In section 5 we evaluate the effectiveness of the feasibility analysis in a series of experiments conducted over a set of materialization tasks ranging in the access pattern (AP) interdependency complexities. Finally, section 6 draws some conclusions and paves the way for future research activities.

2   Problem Statement

Our approach to SeCo materialization is to solve the problem of finding a sequence of access patterns, which, upon execution of a given set of input values, produces a particular materialization.

Materialization formulation deals with building factors of the materialization solution. First, it considers all access patterns (AP) in the service description framework – described in section 3 - that bind the wanted materialization output in their output domains. Second, the access pattern selection is analyzed in terms of their input and output domain dependences - feasibility. Third, the analysis produces the feasible combination of AP – a reachability graph i.e., a combination of access patterns for which the full materialization is possible. Fourth, all the services mapped to the selected AP are taken into consideration and used during the materialization process, according to the run-time optimization procedure. Focus of this work is on defining the model for the feasibility analysis by assessing the factors that contributes to the feasibility of the solution.

The actual feasibility analysis is performed with two distinct objectives lay out.

Firstly, the objective of the analysis is to establish which of the AP combinations in the potential selection is capable of producing desired materialization output for a given set of input dictionaries. In order to perform the analysis the concept of reachability is explored [8]. A combination of AP is observed as a network in which input and outputs of the AP represent the nodes and the query execution instigates propagation of the values - network tokens through the network. By determining the furthest access pattern that can be reached by a single query execution the reachability of the solution is established.

Secondly, all reachable combinations of AP are further analyzed to determine the network nodes whose position shapes the network coverage – Boundedness - in terms of number of queries this AP combination executes [8].

3   Preliminaries

The notion of web service materialization is observed through the Search Computing (SeCo) multi domain multi query framework. SeCo defines multi domain services within its own service description framework (SDF) [3] which is based on a multi-layered model. The service materialization model is defined and implemented as an SDF extension. The model is a dependent of SDF and operates in a seamless and complementary manner to it. The service materialization formulation and its feasibility analysis are expressed in terms of those models.

3.1   SeCo Service Model

In SeCo, data services are described in a three-tier Service Description Framework, as represented in Figure 2. The topmost level is the conceptual level, at which services are represented as an abstraction of the real world objects, and referred to as a Domain Diagram (DD). At the logical level, services are described in terms of data retrieval patterns, or Access Patterns (AP). Each AP holds references to the related domains and denotes a number of input-output attributes and a set of ranking attributes that specify the ordering imposed on the service results. Access patterns are linked at a logical level through a connection pattern (CP) which describes the pair-wise attribute relations that enable the logical connections between data sources and execute a join operation between their invocations in search time. Each AP is related to one or more Service Interfaces (SI), a representation defined at the physical level of the concrete data sources. Service interfaces also describe profile information about data sources, such as their invocation signature, response time, reliability or service level agreement.

3.2   Feasibility Analysis SDF Transformation

As the logical level of the SeCo Service description framework may contain many access patterns with overlapping input and output attributes domains, analysis of reachability within the selected set of access pattern is necessary in order to elect the combination of access patterns with the full reach.

Figure 4 presents a multi pattern multi service scenario MPMS [5] - the materialization is performed on a set of sources connected by a schema where each AP may be mapped to multiple services. The access patterns share input and output attribute domains thereby facilitating the data surfacing [1] with reseeding [4] i.e., discovered output attribute values of one access pattern materialization result feed input attributes of another access pattern materialization via the underlying connection patterns.

The above access pattern interaction are further represented as a bipartite graph in Figure 5 where the input attribute domains form a set of vertices of the graph and the output attribute domains form another set of vertices of the graph.

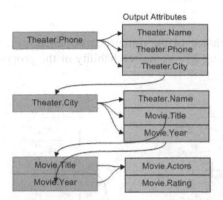

Fig. 1. Bipartite graph representation of the access patterns

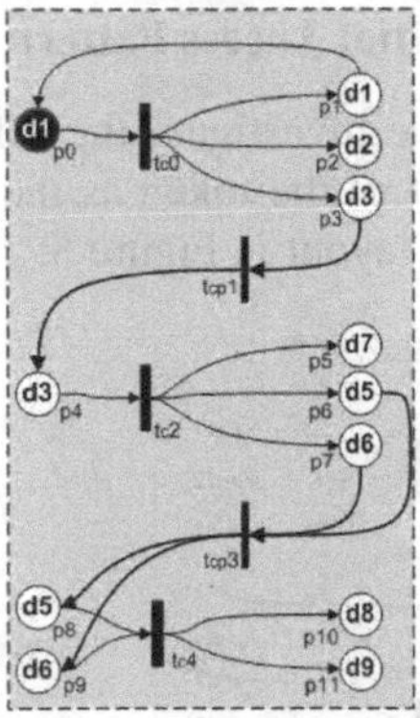

Fig. 2. Petri Net representation of the MPMS scenario with **token** in **p0**

In the context of the MPMS scenario, we model the directed bipartite graph as a Petri Net, shown in Figure 5 in which:

- The input attributes and their domains are a set of input places as well as the output attribute and their domains are a set of output places.
- The input and output attribute domain values are represented as **tokens** that propagate through the network.
- The query is a transformation from input to output domain (tc) as well as the process of supplying output domain values to the input domains via CP (tcp).
- The places and transitions constitute the nodes of the graph connected via directed arcs. Places may contain zero or more tokens, labelled with input and output attribute domain values and their types.
- The state referred to as **marking** M, is the distribution of tokens over places.

By analysing the token propagation through the MPMS net we deduce the *reachability* of the proposed access pattern combination. In the presented scenario the tokens will propagate through all places of the net if and only if there is an input dictionary with at least one value for places P0, and if the top and middle AP query execution resulted in values at P3, P6 and P7. Further, by analysing the size of the input dictionaries and the size of the discovered output attribute domain we define the number of materialization calls the proposed access pattern combination can execute - *Boundedness*. The number of materialization calls for the presented scenario depends on number of tokens supplied to P0, P4, P8 and P9, as produced by P1, P3, P6 and P7.

4   Feasibility Analysis Model

Service materialization feasibility model defines concepts novel to the SDF and web sources materialization in general.

4.1   Initial Access Pattern

Firstly, we investigate the effect of the choice of the initial access pattern i.e., the placement of the token as the initial marking M0 to the reachability of the proposed Multi AP layout in Figure 8.

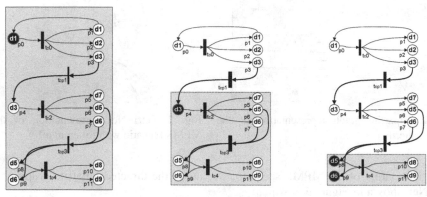

Fig. 3. TheaterByPhone –
initial AP

Fig. 4. TheaterByCity –
initial AP

Fig. 5. MovieByTitle – initial AP

The choice of the *initial access pattern* place in the given topology affects the total materialization product. In case of TheaterByPhone – token is placed to **P0** as state M_0 being the initial pattern, the self-driven materialization of all three access patterns is limited by reseeding volume of the place **P1** - **d1** output domain attribute - as illustrated in Figure 8. A partial materialization of dependant access patterns is limited by the size of their input attribute domain dictionaries – **d3**, **d5** and **d6** – i.e., number of tokens placed to **P4**, **P8** and **P9**, and produced non-duplicate output attribute values - tokens by **P6 and P7**, as illustrated in Figures 9 and 10.

4.2   Transformation Direction

A further factor affecting the reach of the materialization is observed through the directions in the connection pattern transformations in the elected MPMS topology. In the output attribute domain of the MovieByTitle we added Theater.City(d3) attribute also present in the input attribute domain of TheatreByCity, thus forming new connection pattern between these two access parterns – transition tcp5. Newly added connection pattern – transition tcp5 – between MovieByTitle and TheatreByCity Fig. 10, removes dependencies – transition tcp1 – between the top – TheatreByPhone – access pattern and the rest of the network. As an addition to the example in Figure 11, it becomes evident that the materialization of this layout may be observed as two separate materializations, where the top one is used to reinforce the d3 input attribute value supply via connection pattern transition tcp1 in case the supply from connection pattern tcp5 is unavailable or degraded by duplicate values or any other constraint.

Figure 11, shows how the change of the arc direction between TheaterByCity and TheaterByPhone – transition *t1* – disconnects MovieByTitle and TheaterByPhone sub-topology from the rest of the network, making it impossible to reach each other.

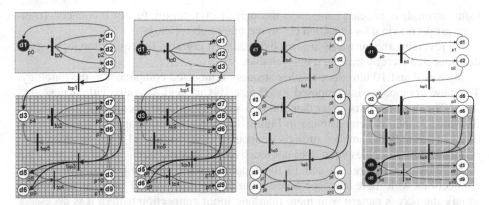

Fig. 6. Two sub-topologies derived on the base of the new connection pattern

Fig. 7. Reversal of the connection pattern flow between sub-topologies

Fig. 8. More than one connection pattern per access pattern

Fig. 9. Two initial access patterns as a seed

4.3   Connection Patterns between Access Patterns

Figure 12 shows the topology where an access pattern shares connections with more than one other access pattern – connection pattern transitions tcp1 and tcp5. Here it is necessary to choose more than one access pattern as a starting seed of the multi AP topology. As demonstrated in Figure 13, the only way to reach all parts of the graph is to start both the top and the bottom access patterns – places p0 and p7, p8 as state M0.

As demonstrated, the choice of the initial access patterns, the direction of the connection pattern arcs and number of connection patterns per AP all play a crucial role in reachability of the whole network.

4.4   Proposed Feasibility Analysis Model

As demonstrated in the previous section the reachability of MPMS topology (MPMSNet) is affected by three factors: Initial Access Pattern Choice, Arcs Direction of the connection patterns and Number of connection patterns per AP in the examined MPMS topology.

Strongly Connected [8] and Boundedness [8] properties of the Petri Net are used to model reachability analysis of the MPMS materialization scenario. By using these properties two model propositions for the analysis of the MPMS reachability are derived and proved.

Proposition 1. Initial access patterns are to be chosen from the transient strongly connected components of the given MPMS topology, which outwards paths lead from the output domain to the input domain.

As depicted in Figures 8 & 9 strongly connected transient component is formed by TheaterByPhone access pattern driven by the reseeding data surfacing process via input and output attribute Theater.Phone (d1) – places p0 and p1. TheaterByPhone forms strongly connected component through d1 connection, further it connects (transient property) via d3 – transition t1 - to the other sub-topology.

By reversing the outwards arc i.e., by making it ergodic, the rest of the topology is disconnected thereby making it unreachable.

Figures 9 and 10 introduce another strongly connected component to the topology. The new component is composed by TheaterByCity and MovieByTitle access patterns. The component is larger than the first strongly connected component - TheaterByPhone – and, as demonstrated in Figure 11, once this component outwards arc is directed towards the smaller strongly connected component – t1 - the topology connects the sub-topologies yet it does not reach all the places.

As shown in Figures 12 and 13 the topology where access pattern shares input connections with more than one access pattern the Proposition1 holds as in order to satisfy the access pattern with more than one input connection pattern it is necessary to choose initial access patterns in both sub-topologies of the graph. Hence, as demonstrated in Figure 13 the only way to reach all parts of the graph is to start both the top and the bottom access patterns, that is to define places p0, p7 an dp0 as state M0.

Proposition 2. Boundedness of each sub graph (strongly connected component - SCP) is determined by the highest bound of the attribute (place) in the output domain making *transient relation* with any sub graph's input domain.

To illustrate, let us consider the topology in Figure 13. The bound of the top sub graph – TheaterByPhone – is defined by the bound of d1 output domain attribute – place p1 (transient place **tp1**) - since this attribute holds transient relation with the input domain. In other words as many non-duplicate values it delivers to the input domain that many queries will be executed.

Bottom sub graph's bound is determined with highest of the bounds of d3, d5 or d6 - places p6, p7, p10 (transient places tp6, tp7, tp10) - as they all equally contribute to the input domain of the sub graph.

Once the reachability graph determined via proposition 1 is established, proposition 2 is used to determine how many times this layout can be reached i.e., the maximum number of materialization calls that can be executed in the given MPMS net.

5   Empirical Study

In this section we demonstrate the performance of our approach in the context of the feasible solutions delivered by the proposed model. The performance of the model was evaluated by the number of executed queries (firing sequences) against MPMS layouts with different number of bound places and strongly connected components.

5.1   Experimental Settings

To evaluate the proposed solution we apply the model to a set of materialization tasks designed to support a 'reallocation search' multi domain search scenario illustrated by real life queries such as 'I am looking for a Java developer position in area where housing is in a certain price range and there are good Italian and Thai restaurants nearby', against a service repositories all which deal with a variety of domains across jobs, real estate, entertainment and general shopping domain. Each task is constrained by a predefined set of input dictionaries and wanted materialization coverage. The feasibility analysis is performed against a set of potential access patterns |AP|=120, derived from the aforementioned DD. The empirical study considered the access patterns 1) *jobByLocation* (AP1) which takes job position location and job type as input attributes ;2) *realEstateByLocation* (AP2) which takes a real estate type, price range and the location as the input attributes 3) *restaurantByLocation* (AP3) which takes a cuisine type and location as inputs. For practicality reasons all the solution recommended by the feasibility analysis were restricted to MPMSNet with strongly connected component size of 2<=|SCP|<=5, and the number of transient places 1<|tp|<=10. For each AP at least one feasible solution was chosen for each combination of feasibility properties; for each feasible solution 10 runs were performed; the materialization goal was set to 50 executed queries / retrieved result sets. The input dictionaries for attributes that were not populated by a reseeding or a connection pattern were supplied by a static dictionary of 50 values belonging to their respective domains; to avoid biases, the input dictionaries were initialized at each run by a randomly selected subset. To perform the evaluation, we created a master databases composed of ~100K listed items from several existing online real estate, restaurant rating and job sites.

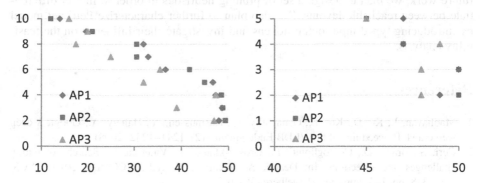

Fig. 10. No Of **tp** vs No Of Executed Queries   **Fig. 11.** No Of **scp** vs No Of Queries

Influence of Transient Places to the Materialization. Figure 14 examines the effect of the number of transient (bound) places to the number of executed queries (sequences). The result reveals an interesting trend as the number of queries decreases with increase of bound places. The layouts with less than 5 transient places seem to perform adequately, on the whole reaching the given goal. The obtained materialization goal decreases with increase of transient places. This is most likely imposed by the unpredictable nature of the seeding output attributes 'capacity' as it is hard to predict the number of valid input values produced by the transient place due to the unknown data distribution of the remote source.

Influence of Strongly Connected Places to the Materialization. Figure 15 examines the effect of the number of strongly connected components of the given feasible layout to the obtained materialization. The results are consistent in all 3 AP, as there is no significant shift in the obtained materialization volume as the number of *scp* increases. This further validates the importance of the positioning of the input values, as there was no deviation in the input dictionary sizes and the executions performed close to 100% in each case.

6   Conclusion

In this paper we have designed a model for analysing a potential section of access patterns in the given service description framework in order to determine a feasible combination of AP for which the given materialization task succeeds. Furthermore, we have introduced and characterized feasible solution properties – reachability and boundedness – building blocks of the proposed analysis.

We have performed a range of live materialization tasks using the presented algorithms and confirmed validity of the analytic step in the materialization formulation. We have also come to an interesting observation in regards to the bound AP attribute in respect to the final materialization outcome. The experiment results have shown that our approach provides better performance in terms of final materialization coverage where the number of bound places is smaller in equally reachable AP layouts. As future work, we plan to design a set of pruning heuristics in order to further differentiate between reachable layouts. We also plan to further characterize Petri net model by introducing typed input output tokens and investigate their influence on the feasibility analysis.

References

1. Madhavan, J., Ko, D., Kot, Ł., Ganapathy, V., Rasmussen, A., Halevy, A.: Google's deep web crawl. Proceedings of the VLDB Endowment 1(2), 1241–1252 (2008)
2. Ceri, S., Braga, D., Corcoglioniti, F., Grossniklaus, M., Vadacca, S.: Search computing challenges and directions. In: Dearle, A., Zicari, R.V. (eds.) ICOODB 2010. LNCS, vol. 6348, pp. 1–5. Springer, Heidelberg (2010)
3. Brambilla, M., Campi, A., Ceri, S., Quarteroni, S.: Semantic resource framework. In: Ceri, S., Brambilla, M. (eds.) Search Computing II. LNCS, vol. 6585, pp. 73–84. Springer, Heidelberg (2011)
4. Wu, P., Wen, J.R., Liu, H., Ma, W.Y.: Query selection techniques for efficient crawling of structured web sources. In: Proceedings of the 22nd International Conference on Data Engineering, ICDE 2006, p. 47. IEEE (April 2006)
5. Bozzon, A., Ceri, S., Zagorac, S.: Materialization of web data sources. In: Ceri, S., Brambilla, M. (eds.) Search Computing III. LNCS, vol. 7538, pp. 68–81. Springer, Heidelberg (2012)
6. Cafarella, M.J., Madhavan, J., Halevy, A.: Web-scale extraction of structured data. ACM SIGMOD Record 37(4), 55–61 (2009)
7. Gupta, A., Mumick, I.S. (eds.): Materialized views: Techniques, implementations, and applications. MIT Press, Cambridge (1999)
8. Murata, T.: Petri nets: Properties, analysis and applications. Proceedings of the IEEE 77(4), 541–580 (1989)

Efficient R-Tree Based Indexing for Cloud Storage System with Dual-Port Servers*

Fan Li, Wanchao Liang, Xiaofeng Gao**, Bin Yao, and Guihai Chen

Department of Computer Science and Engineering,
Shanghai Jiao Tong University, P.R. China

Abstract. Cloud storage system such as Amazon's Dynamo and Google's GFS poses new challenges to the community to support efficient query processing for various applications. In this paper we propose RT-HCN, a distributed indexing scheme for multi-dimensional query processing in *data centers*, the infrastructure to build cloud systems. RT-HCN is a two-layer indexing scheme, which integrates HCN-based routing protocol and the R-Tree based indexing technology, and is portionably distributed on every server. Based on the characteristics of HCN, we design a special index publishing rule and query processing algorithms to guarantee efficient data management for the whole network. We prove theoretically that RT-HCN is both query-efficient and space-efficient, by which each server will only maintain a tolerable number of indices while a large number of users can concurrently process queries with low routing cost. We compare our design with RT-CAN, a similar design in traditional P2P network. Experiments validate the efficiency of our proposed scheme and depict its potential implementation in data centers.

Keywords: Distributed Index, R-Tree, Data Center Network.

1 Introduction

Cloud storage systems have been continuously drawing attentions from both academia and industry in recent years. From classical systems for general data services, such as Google's GFS [1], Amazon's Dynamo [2], Facebook's Cassandra [3], to newly designed systems with specialities, such as Haystack [4], Megastore [5], Spanner [6], various distributed storage systems were constructed to satisfy the increasing demand of online data-intensive applications that require massive scalability, efficient manageability, reliable availability, and low latency in the storage layer. Correspondingly, many works are proposed for designing

* This work has been supported in part by the National Natural Science Foundation of China (Grant number 61202024, 61202025, 61133006), China 973 project (2014CB340303, 2012CB316200), Shanghai Educational Development Foundation (Chenguang Grant No.12CG09), Shanghai Pujiang Program 13PJ1403900, and the Natural Science Foundation of Shanghai (Grant No.12ZR1445000).
** Corresponding author.

H. Decker et al. (Eds.): DEXA 2014, Part II, LNCS 8645, pp. 375–391, 2014.

new indexing scheme and data management system to support large-scale data analytical jobs and high concurrent OLTP queries [7–10].

In a cloud DB system, datasets are partitioned among distributed servers, while users may hop among multiple servers to process their query requests. To provide an efficient indexing scheme, a common feature of the above literature is that they split indices into two categories: global index and local index, and then portioned their global indices to each server according to an overlay network architecture. Following the direction of indices, users route queries among servers based on the underlying routing protocols of the network connecting servers together. However, all these designs are constructed on P2P networks [10–12], while nowadays cloud systems are usually built on an infrastructure called *data center*, which consists of large number of servers interconnected by a specific *Data Center Network* (DCN) [13–20]. For instance, Cisco applies Fat-Tree topology as its DCN architecture for provably efficient communication [13]. Different from P2P network, DCN is more structured with low equipment cost, high network capacity, and support of incremental expansion. It is natural that such infrastructures bring new challenges for researchers to design efficient indexing scheme to support query processing for various applications.

In this paper, we propose our RT-HCN, a distributed indexing scheme for multi-dimensional query processing in *Hierarchical Irregular Compound Networks* (HCN) [21, 22], which is the latest designed DCN structure using dual-port servers. HCN has many attractive features including low diameter, high bisection width, large number of node-disjoint paths for pairwise traffic, and supports low overhead and robust routing mechanisms. Additionally, in many online data-intensive services users tend to query data with more than one key, e.g., in Youtube video system users may want to find videos via both video ids and size ranges. Therefore, designing an indexing scheme for multi-dimensional query processing in HCN is useful and meaningful for real-world cloud applications, which has both theoretical and practical significance in this area. To search the data efficiently, the R-tree [23] based multi-dimensional index is used in our system. RT-HCN integrates HCN-based routing protocol and the R-Tree based indexing technology.

Similar to previous works, RT-HCN is a two-layer indexing scheme with a global index layer and a local index layer. Since datasets are distributed among different servers, we can use an R-Tree like indexing structure to index locally stored data for each server. Next, RT-HCN portionably distributes these local indices across servers as their global indices. To avoid single master server bottleneck, each server only maintains partial global index for its potential index range. Based on the characteristics of HCN, we design an index publishing rule to guarantee an "onto" mapping from global index to local stored data. We also propose the corresponding query processing algorithms to achieve query efficiency and load balancing for each node in the network. Finally, we prove theoretically that RT-HCN is both query-efficient and space-efficient, by which servers will not maintain redundant indices while a large number of users can concurrently process queries with low routing cost. We compare our design with

RT-CAN [7], a similar design for traditional P2P network. Experiments validate the efficiency of our proposed scheme and depict its potential implementation in data centers.

Our contribution of this paper is threefold: (1) to the best of our knowledge, we are the first to propose a distributed multi-dimensional indexing scheme for a specific DCN structure to improve query efficiency and system QoS; (2) noticing and taking advantage of the topology of HCN, we present a specialized mapping technique to improve global index allocation in the network, resulting query-efficiency and load-balancing for the cloud system; and (3) we theoretically prove the efficiency of RT-CAN, and compare our model numerically with RT-CAN [7], an indexing scheme for P2P network. Simulation results show that our scheme costs less index space for each node while provides faster query processing speed with higher bandwidth.

The rest of this paper is organized as follows. Section 2 summarizes the related work in this research area. Section 3 introduces the overview of our system, including coding of HCN and meta-servers. In Sec. 4 we illustrate the two-layer index construction with global index publishing rules. In Sec. 5 we depict the query processing algorithms with corresponding performance analysis theoretically. Section 6 discusses the maintenance and improvement of RT-HCN. Section 7 compares our design with the latest work RT-CAN in [7] and proves the efficiency of our construction. Finally, Section 8 provides a conclusion.

2   Related Works

Nowadays, there are lots distributed storage systems which assist to manage big data for cloud applications. Among them, we have excellent commercial implementations like key-value based system Amazon's Dynamo [2], Google's Google File System [1] (GFS), and BigTable [24], which aim at dealing with large scales of data. Meanwhile, some open source systems such as HDFS, HBase and HyperTable also provided a good platform for research use. Cassandra [3] is one non-rational database that combines features of BigTable and Dynamo. Some other systems such as Ceph [25], are designed to provide high performance in objects retrieval.

We want to build a second level overview index and our work follows the framework proposed in [8]. It offers an idea to build a two level index in cloud system for data retrieval on top of a physical layer. Moreover, an efficient and extensible framework for index in P2P based cloud system was put forward in [10]. However, more specialized topology has been designed to meet the requirement of today's cloud system and that is why we want to apply the two-level index design to specific data center network and discuss its improvement. As the topology of DCN is known, we can guarantee the processing time by calculating out the physical hops needed for a given query. While in P2P network only logical hops of the overlay network can be estimated.

Data center network (DCN) is the network infrastructure inside a data center, which connects a large number of servers via high-speed links and switches.

Compared to traditional cloud system which is usually based on P2P network, specially and carefully designed DCN topologies fulfill the requirements with low-cost, high scalability, low configuration overhead, robustness and energy-saving. DCN structures can be roughly divided into two categories, one is switch-centric such as VL2 [14] and Fat-Tree [13]. The other is server-centric like BCube [16], DCell [15], FiConn [17, 18], MDCube [19] and uFix [20]. They usually have more advantages than the former designs. HCN [22], the topology chosen in our system falls into the server centric topology. It is a well-designed network for data center and offers a high degree of regularity, scalability, and symmetry. Different from traditional P2P network, we have to be aware of the physical topology when we are discussing DCN and that is why we need to improve the mapping technique for distributing global index to fix a given network.

3   System Overview

In this section, we will introduce the topology of HCN and illustrate some basic features of it. Then, we explain some preliminary definitions for further illustration of index construction strategy.

3.1   Hierarchical Irregular Compound Network

HCN (Hierarchical irregular Compound Network) is a well-designed network for data center and offers a high degree of regularity, scalability, and symmetry. A level-h HCN with n servers in every single unit is denoted as $HCN(n, h)$. HCN is a recursively defined structure. A high-level $HCN(n, h)$ employs a low level $HCN(n, h - 1)$ as a unit cluster and connects many such clusters by means of a complete graph. $HCN(n, 0)$ is the smallest module (basic construction unit) that consists of n dual-port servers and an n-port miniswitch. For each server, its first port is used to connect with the miniswitch while the second port is employed to interconnect with another server in different smallest modules for constituting larger networks. Figure 1 illustrates an example of HCN with $n = 4$ and $h = 2$, which consists of 64 servers. Each server is labeled according to a coding process, which will be introduced in Sec. 3.2.

It is easy to see that there is always multiple routes between any two servers in HCN and this is called multi-path routing which provides good features like high bandwidth, good balancing and error tolerance. This is the main aspect we want to concern during index construction and also becomes a main reason why special designed index scheme for specific network is well worth being discussed.

For clarity, we summarize the symbols with their meanings in Table 1. Some of them will be described in the following sections.

3.2   Meta-Server and Coding Strategy

Given an $HCN(4, h)$, there are 4^{h+1} servers in total and are coded by n ranging from 0 to $4^{h+1} - 1$. Thus we use S_n to denote the nth server in the HCN.

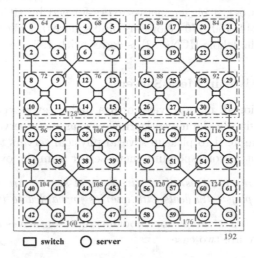

Fig. 1. HCN(4,2) with Coding for Meta-Server

Table 1. Symbol Description

Sym	Description	Sym	Description
n	Code for server and meta-server	h	The highest level of HCN
S_n	The server with code n	$\mathbf{B}$	Data boundary
M_n	The meta-server with code n	$\mathbf{B}_n$	Potential index range for M_n
$\mathbf{R}_n$	Representatives for M_n	R_n^i	The ith representative for M_n
$\mathbf{N}_n$	S_n's node set for publishing.	N_n^i	The ith publishing node for S_n

Since HCN itself is a recursively defined structure, there are also 4^{h-l} different HCN(4, l)'s ($0 \leq l \leq h$) in the same HCN(4, h). We consider each S_n and HCN(4, l) ($0 \leq l \leq h$) as a meta-server in our system.

Definition 1. *A Meta-server is a single server or an HCN(4, l) ($0 \leq l \leq h$) considered entirety.*

Meta-servers together constitute the overlay network and facilitate global queries, which is going to be explained in detail later. Meta-servers are also coded by n and denoted as M_n. The coding strategy of M_n is given by:

$$M_n = \begin{cases} S_n, & 0 \leq n < 4^{h+1} \\ \text{HCN}(4, q-1) \text{ consists of } S_r, S_{r+1}, \cdots, S_{r+4^q-1}, & n \geq 4^{h+1} \end{cases}, \quad (1)$$

where q and r from the above equation represent the quotient and the reminder when the given n is divided by 4^{h+1}. From the function, we know that M_n is equivalent to S_n when $n < 4^{h+1}$, and when $n \geq 4^{h+1}$ M_n represents a specific HCN(4, l) ($0 \leq l \leq h$). For example, Fig. 1 shows that in an HCN(4, 2), 64 meta-servers consist of only one server are coded with the number ranging from 0 to 63, 16 meta-servers formed by HCN(4, 0) are coded with 64, 68, $\cdots$, 124 (shown as red squares), 4 bigger meta-servers formed by HCN(4, 1) are coded with number 128, 144, 160, 176 (shown as blue squares), while the biggest green square formed by the whole HCN(4, 2) is coded with 192.

3.3 Representatives in Meta-Server

As is already mentioned, meta-servers form a higher-level overlay network and assist query processing in the network. However, as meta-servers are merely an abstract concept, we need to pick up several physical servers to be in charge of queries that are sent to corresponding meta-server from the overlay network.

Definition 2. *Representatives of a given meta-server are several carefully picked servers that physically deals with queries forwarded to that meta-server.*

We now provide our strategy for choosing representatives of a given meta-server M_n and there are two cases:

1. If $n < 4^{h+1}$, we chose S_n as the representative of M_n since $M_n = S_n$.
2. If $n \geq 4^{h+1}$, M_n, now, is actually an HCN(4, l) $(0 \leq l \leq h)$, and we provide a special bit manipulation to find out its representatives. First we calculate the quaternary form of n, denoted as $q_0 q_1 \cdots q_m$. Here $m > h$ stands since $n \geq 4^{h+1}$. Secondly, we pick up the first $m - h$ bits as shown in Eqn. (2) and calculate the decimal number b. Then we replace each of the last b bits with q_*, where $q_* \neq q_{m-b}$ and will get several newly formed number. Finally we calculate the decimal form of the last $h + 1$ bits of the new numbers and servers coded with those numbers are the representatives for this M_n.

$$\underbrace{\overbrace{q_0 q_1 \cdots q_{m-h-1}}^{m-h \text{ bits}}}_{=b \text{ (decimal)}} \Big| \overbrace{q_{m-h} \cdots \underbrace{q_{m-b} q_{m-b+1} \cdots q_m}_{\text{replace part}}}^{h+1 \text{ bits}} \tag{2}$$

For example, the grey nodes in Fig. 2 (a), (b) illustrates the representatives for corresponding meta-server HCN(4, l) when l equals to 0 and 1. It is obvious that these representatives actually takes the advantages of HCN topology and offer good connectivity which will facilitate query process.

We denote the representatives of M_n as set $\mathbf{R}_n$ and $\mathbf{R}_n$ has different number of entities in different situation. In case 1, $\mathbf{R}_n = \{R_n^1\}$ while in case 2, $\mathbf{R}_n = \{R_n^1, R_n^2, R_n^3\}$. Here R_n^i stands for the server which is the ith representative of meta-server M_n, and R_n^i is called an lth-level representative if and only if M_n is a meta-server formed by an HCN(4, l) $(0 \leq l \leq h)$.

Now we give some explanation on choosing representatives. It is obvious to choose S_n as representatives for M_n when $n < 4^{h+1}$ because they are exactly the same. So we focus our discussion on case 2 here. According to the topology of HCN, it is not hard to find that all of the representatives we choose for M_n $(n \geq 4^{h+1})$ are servers that are more closely connected to other HCN(4, l)'s in the same level. Thus, our strategy cut the cost of queries forwarding and since there are three representatives for a single M_n, it also offers flexibility to choose the closet representative or the dullest one. This strategy also fits quite well with

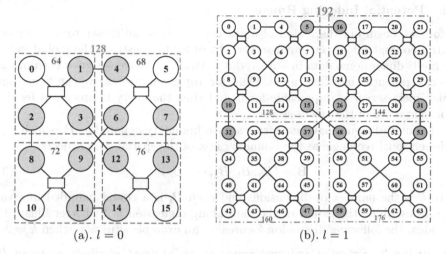

Fig. 2. The Representatives for Meta-servers in HCN(4,2)

the multi-path routing of HCN. What's more, the following lemma and theorem show that our strategy also offers good scalability and balancing property.

Lemma 1. *Each meta-server M_n ($n \geq 4^{h+1}$) has exactly three representatives.*

Proof: There are four different bits in quaternary number and our strategy replace the bits with a same bit different with q_{m-b} in Eqn. (2), so the result is three. The biggest $M_{(h+1) \cdot 4^{h+1}}$ is the only special one that has four representatives since the bit q_{m-b} does not exist in this case. $\square$

Theorem 1. *Each server S_n will be the representative for exactly two different meta-servers.*

Proof: Firstly, it is obvious that each server should be chosen as a representative for M_n where $n < 4^{h+1}$. Secondly, according to Lemma 1, the representative-time in the whole system is $2 \cdot 4^{h+1}$ in total. Thirdly, no server can be selected for three times. Since the quaternary form of the coding for a given server is fixed, there exist only one way for it to be split according to the form given by Eqn. (2). Combining the above discussion, we can draw a conclusion to our proof according to the Pigeonhole Principle. $\square$

4   Indexing Construction

In this section, we first introduce the vital component of the higher level overlay network, which is essential for constructing the global index. Then we introduce the construction our two-layer index in detail with index publishing rules.

4.1   Potential Indexing Range

Before we begin our discussion of index construction, we illustrate another essential concept about our meta-servers. In order to construct the global index for multi-dimensional data in our overlay network, we have assigned each meta-server a potential indexing range. Thus, for any given queries, we can figure out which meta-server is responsible for it and then the query is processed by the representatives of that meta-server.

The multi-dimensional data forms a data boundary denoted as $\mathbf{B}$, which is a k-dimensional rectangle as the bounding box of the spatial data objects:

$$\mathbf{B} = (B_0, B_1, B_2, \cdots, B_k) \ . \tag{3}$$

Here k is the number of dimensions and each B_i is a closed bounded interval $[l_i, u_i]$ describing the extent of the data along dimension i. To better illustrate our idea, the following discussion focuses on an example situation when $k = 2$.

Definition 3. *Potential indexing range is an abstract attribute assigned to meta-server and indicates which meta-server (indeed the subordinate representatives) is (are) responsible for processing a coming query.*

Suppose the data is bounded by $\mathbf{B} = (B_0, B_1)$, where B_0 is $[l_0, u_0]$ and B_1 is $[l_1, u_1]$. We calculate a quaternary number Q_n for each meta-server M_n and use it to help figure out the potential index range $\mathbf{B}_n$ for M_n. Suppose $q_0 q_1 \cdots q_m$ is the corresponding quaternary form for n, then Q_n is calculated according to the following two cases:

1. If $m \leq h$, we add $m - h$ consecutive 0's to the front of $q_0 q_1 \cdots q_m$ and construct an $h + 1$ bit quaternary number $Q_n = 00 \cdots 0 q_0 q_1 \cdots q_m$.
2. If $m > h$, $q_0 q_1 \cdots q_m$ can be split as the form explained in Eqn. (2). In this situation, we pick out the last $h + 1$ bits and delete the replace part to get $Q_n = q_{m-h} q_{m-h+1} \cdots q_{m-b}$.

Now, we use the following iteratively defined function to calculate $\mathbf{B}_n$.

$$\mathbf{B}_n = pir(\mathbf{B}, Q_n) = \begin{cases} pir(([l_0, \frac{l_0+u_0}{2}], [l_1, \frac{l_1+u_1}{2}]), Q'_n), & q_0 = 0 \\ pir(([\frac{l_0+u_0}{2}, u_0], [l_1, \frac{l_1+u_1}{2}]), Q'_n), & q_0 = 1 \\ pir(([l_0, \frac{l_0+u_0}{2}], [\frac{l_1+u_1}{2}, u_1]), Q'_n), & q_0 = 2 \\ pir(([\frac{l_0+u_0}{2}, u_0], [\frac{l_1+u_1}{2}, u_1]), Q'_n), & q_0 = 3 \\ ([l_0, u_0], [l_1, u_1]) & Q_n = \varnothing \end{cases}, \tag{4}$$

where Q_n is a quaternary number denoted as $q_0 q_1 \cdots q_i$ and Q'_n is a newly constructed quaternary number $q_1 q_2 \cdots q_i$. For higher dimensional data, the data range can be seen as a hyper-rectangle and the potential index ranges for different meta-servers are sequenced according to their coding in a similar way.

For example in Fig. 3 (a), two axes B_0 and B_1 denote the range of data in two dimensional space (w.l.o.g., we assume it generates a square area, rather than a rectangle). Then the potential indexing range for M_{80} (denoted as red square) should be $([\frac{l_0+u_0}{2}, \frac{l_0+3u_0}{4}], [l_1, \frac{3l_1+u_1}{4}])$; the *pir* for M_{144} (denoted as blue square) is $([\frac{l_0+u_0}{2}, u_0], [l_1, \frac{l_1+u_1}{2}])$, while the *pir* for M_{192} is $([l_0, u_0], [l_1, u_1])$.

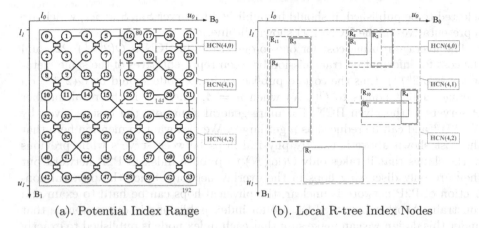

(a). Potential Index Range   (b). Local R-tree Index Nodes

Fig. 3. Illustration of Range Mapping and Index Distribution

4.2 Global Index in Overlay Network

As is discussed above, each data server S_n has built an R-tree index for its local data to facilitate multi-dimensional search. Then, S_n adaptively selects a set of index nodes $\mathbf{N}_n = \{N_n^1, N_n^2, \cdots, N_n^{d_n}\}$ from its local R-tree and publishes each N_n^i to the representatives of a specific meta-server whose potential indexing range just covers the minimum bounding range of N_n^i. We initially choose the last but one level index nodes from a given R-tree to be published since these nodes are usually not frequently updated and do not introduce too many false positives either. The format of the published R-tree nodes is (n, mbr), where n indicates the origin server for storing the data and mbr is the minimal bounding range of the published R-tree node. After receiving the published nodes, representatives buffers the index in memory. In this way, the global index composes of several R-tree nodes from the local indexes and is distributed over the data center.

Algorithm 1. Index Publishing (For S_n)

---

1    $\mathbf{N}_n$ =getSelectedRTreeNode(S_n)
2   **for each** $N_n^i \in \mathbf{N}_n$ **do**
3   Find the least n' s.t. $\mathbf{B}_{n'}$ fully covers $N_n^i.mbr$
4   Get the representatives $\mathbf{R}_{n'}$ for $M_{n'}$
5   **for each** $S_k \in \mathbf{R}_{n'}$ **do**
6   S_k inserts $(n, N_n^i.mbr)$ into its global index set
7   **end**
8   **end**

Fig. 3 (b) provides a simple example of a local R-tree. If the node R_1 is selected to be published, it should be published to server S_{17}, S_{18}, S_{19}, which are representatives of the HCN(4, 0) shown in red. Similarly, if the node R_3 is selected to be published, it should be published to server S_{16}, S_{26}, S_{31}, which are representatives of the HCN(4, 1) shown in blue.

The average routing cost for one-to-one traffic in HCN is $O(2^{h+1})$ [22], and the cost for information transmission between representatives of the same meta-server is $o(2^{h+1})$, thus the cost to publish an index node should be $O(2^{h+1})$ on average, and it equals to $O(\sqrt{N})$ when $n = 4$, where N is the total number of servers in the given HCN. For more general situation, the cost is given by $N^{-\log_2 n}$ and can be reduced as n get larger. We also want to mention here that the cost shown above is exactly physical hops between servers while previous works claims that it takes only $O(\log N)$ to publish index in P2P network but they are only discussing hops in the overlay network. Since the physical connection of P2P network is unclear, the physical hops can be hard to exam and constrain. Another improved feature for index publishing in our system is that under this design we can make sure that each index node is published to exactly only three servers in the system. However, the strategy used in [7] publish the nodes to other servers as long as the range of the index node overlaps with the potential index range of the given servers. Then, the amount of index nodes published in the system is hard to control.

5   Query Processing

In this section we show how our global index can be applied to process queries and reduce the cost of physical hop counts in an HCN.

5.1   Point Query

A query is denoted as $Q(value)$ where $value = (v_0, v_1)$, indicating a multi-dimensional data point. Given an indexed R-tree node N_i^n, N_i^n should be searched to retrieve possible results if and only if $N_i^n.mbr$ covers the query point (v_0, v_1). Such node N_i^n, however, can be published to several different representatives in our system. The following theorem shows that we have to search a set of servers' global index to get the full results. Our design splits the processing of a query into two phases. In the first phase, the query is done by the response server S_{n_1} of the meta-server that is responsible for the query and the representatives look up the global index, search the buffered R-tree nodes and return the entries that satisfy the query. In the second phase, based on the received index entries, the query is forwarded to the corresponding physical server S_{n_2} and retrieve the results via the local R-tree.

Theorem 2. *The published index node that may contain results for a given value (v_0, v_1) are in representatives of $h + 2$ meta-servers. Moreover, exactly $h + 1$ servers need to be searched in total to get the full results.*

Proof: Suppose an indexed R-tree node N_i^n contains the search value (v_0, v_1), it is certain that the destination for publishing this index node must be the representative of some meta-servers whose potential index range covers point (v_0, v_1), other wise it will contradict with our publishing rule. Therefore, we can easily figure out there are $h + 2$ different meta-servers that meet the above condition. They respectively consist of a single server, an HCN(4, 0), an HCN(4, 1), $\cdots$, and the biggest HCN(4, h). Since the meta-server formed by a single server takes itself as a representative and is also one of the representatives for the meta-server formed by HCN(4, l) that meet the above condition for some l, the total number of servers to search is then $h + 1$. It is also good to notice that the $h + 1$ servers are actually representatives of different levels that is from 0th level to h^{th} level, this will facilitate our explanation for query forwarding. □

Although it seems that $h + 1$ is a quite big number while the total number of servers are 4^{h+1}, we provide our strategy for forwarding the query based on a greedy algorithm and show that the cost of forwarding is really small even when there are $h + 1$ servers to be searched. As we have discussed above, there are $h + 1$ representatives to search and they range from level 0 to level h. To process a query $Q(v_0, v_1)$, we first forward the query to the nearest h^{th} level representative S_{n_h} which is responsible for the given query. S_{n_h} searches its buffered global index and returns matched results to the user. Then, it forwards the query to the nearest $h - 1^{th}$ level representative $S_{n_{h-1}}$, respectively. Until a 0th level representative has been searched, we can finally get the full results. The head of the routing message follows the format of $\{l, value\}$, where l is the level for representative and $value$ is the querying point. Alg. 2 describes such process in detail.

Algorithm 2. Point Query Processing

   Input : $\{l, value\}$
   Output: N (the result set of indexed R-tree nodes)
1 $\mathbf{N} = \varnothing$
2 for $\forall N \in S.globalIndex$ do
3   if $N.mbr$ covers $value$ then
4   Add N into $\mathbf{N}$
5   end
6 end
7 $S_{next} = $ the nearest $l - 1^{th}$ level representative for $value$
8 Forward query message $\{l - 1, value\}$ to S_{next}
9 return $\mathbf{N}$

We can see from the above algorithm that the shortest query path for a given query is $O(\log N)$ and if we want to maintain the multi-path routing of HCN, our index publishing scheme does bring out some extra cost for specific queries. But since the cost of forwarding queries from any $l + 1^{th}$ level representative to the nearest l^{th} level representative will never exceed the maximum cost of point-to-point routing in an HCN(4, l) which is $O(2^{l+1})$, and noticing that under this design only with a probability of $\frac{1}{4}$ will a query pass result in redundant routing, thus, the average cost of query in our system will only be $\frac{5}{4}$ times of the cost of point-to-point routing in an HCN(4, h) and then is still constrained by a logarithmic scale. What's more, with a similar analysis, it is easy to figure out that the parameter $\frac{5}{4}$ will decrease to $\frac{n+1}{n}$ when considering an HCN(n, h). We mention again that the cost we were discussing above are all physical connections rather than hops in the overlay network.

5.2   Range Query

A range query is denoted as $Q(range)$, where $range = ([l_0, u_0], [l_1, u_1])$ is a multi-dimensional hypercube. If an R-tree node's boundary $N_i^n.mbr$ overlaps with $range$, it should be searched to retrieve the possible query results. Let us think about an easy case first, if the $range$ is small enough and can be fully covered by some $\mathbf{B}_n$, where n is smaller than 4^{h+1}, then $Q(range)$ actually goes in the same way we do for point queries in the last subsection. Inspired by this special case, we can figure out that range query is an variation of point query, with only a small modification.

We can first find out the smallest HCN(4, l) that fully covers the queried range, then just as what we did in point query, we first forward the query to the nearest h^{th} level representative S_{n_h}, which is responsible for the given range. Then, S_{n_h} forwards the query to the nearest $h - 1^{th}$ level representative $S_{n_{h-1}}$, respectively, until an lth level representative has been searched. Then, different from what we did in point query, from the lth level on, the representatives will forward the query to all the lower level representatives, whose potential index range overlaps with $range$, to guarantee the correctness and completeness of results. Since routing cost in HCN(4, l) where $l < h$ is $o(\log N)$, with similar analysis for point query, it is easy to figure out that the cost for a range query is still $O(\log N)$, and the detail of the range query algorithm is omitted here as it is similar to Alg. 2.

6   Further Discussion and Optimization

We only used HCN(4, 2) and 2-dimensional data for discussion in the above sections, this does not mean these two parameters should be fixed. When considering the implementation of an HCN(n, h), the function to map each meta-server a different potential index range should be changed since n is different. Noticing that h is a parameter that indicates the scalability of the whole system, it does not affect much the design of the mapping function because the system is a recursively designed architecture. In a situation where $n \neq 4$, we can also build a function similar to Equation (4), the only difference is that we are now dealing with an n-dimensional number rather than a quaternary number. For higher dimensional data, we can take advantage of the space-filling curve [27] to reduce dimension. Taking advantage of these space filling curves, it is possible to generate one-to-one mappings between spaces of different dimensions, that is, we can apply our strategy similarly to higher-dimension data.

We also offer an optimization for data distribution. It is sure that the system will be more balanced if the data with close features are allocated closer. It is also good to keep the locality coherence of data for retrieval, especially for multi-dimensional data. Locality sensitive functions offered by [26] can facilitate data distribution and offer good performance of query processing. Another significant improvement can be done is about queries with specific frequency. It is a common scene in real world that some data are more hot and queried more frequently than others. In this case, it is sure that some servers which are responsible for the hot data are more busy and are in the risk of being bottleneck for the system. A heuristic strategy for this problem is that we may want to assign a dynamic potential index range for each server and discuss the cost of maintenance and the improvement of performance.

7 Performance Evaluation and Numerical Experiments

In this section, we evaluate the performance of RT-HCN. We simulate an $HCN(4, 2)$, generate the multi-dimensional data and randomly allocate them in different servers. We consider randomly generated data together with data following Zipf distribution to prove the scalability of RT-HCN. The queries including both point queries and range queries are randomly generated from any server. Since the cost of query processing is proved to takes only $N^{-\log_2 n}$ physical hops for any given query, performance of our system discussed in this section focus mainly on two metrics: Heat of each server and Index nodes in each server, both of which reflects the high scalability and balance of the system. Table 2 is the parameters used in our experiments.

Table 2. Parameters Used in Our Experiments

Parameter	Default	Range	Meaning
D	320000	[320000, 1280000]	Number of data items in total
h	2		Highest level of HCN
N	64		Total number of servers

First, we estimate the number of published index nodes in different servers in the network. We compare the results of our publishing strategy with the one used in RT-CAN to show that we provide a more scalable and balanced publishing rule. Our evaluation is first carried out with randomly generated data and then with data that obey centralized distribution. Fig. 4 and Fig. 5 show the data distribution in the experiments.

We then vary the number of data items distributed to each server and investigate the changes in number of published index nodes. Fig. 6-8 show the number of global indices on each server according to RT-HCN and RT-CAN publishing rules, where number of data item varies from $320,000$ to $1,280,000$. From the results we can see that our system maintains less index entries at each server compared to RT-CAN under uniform data distribution. Similarly, results in Fig. 9-11 show that when data are not randomly distributed, our design still provides a better result with tolerate number of indices.

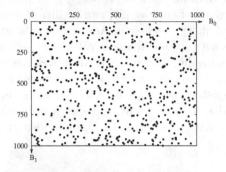

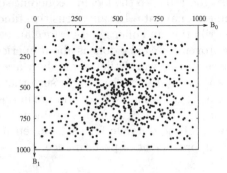

Fig. 4. Uniform Data Distribution   **Fig. 5.** Zipf Data Distribution

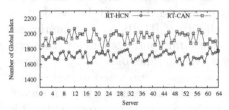

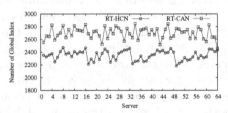

Fig. 6. D=320,000   **Fig. 7.** D=640,000

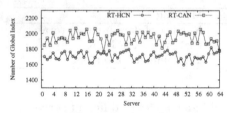

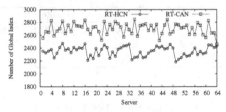

Fig. 8. D=320,000   **Fig. 9.** D=640,000

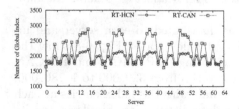

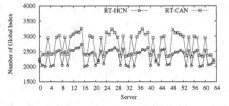

Fig. 10. D=640,000   **Fig. 11.** D=1,280,000

Finally, we estimate the visiting frequency of different servers in the network (denoted as a heatmap) when a given number of randomly generated queries are processed. The heatmap is drawn according to the access frequency of each server and directly reflects the processing condition of servers in the system. When a multiple-paths routing is applied, the variance of access frequency of the whole system is reduced, indicating a more balancing performance. Moreover, when a larger HCN is built, good balancing performance of smaller HCNs reduces the risk of any server being the bottleneck. Thus, in other words, good results from heatmap also reflect good scalability of our system design. The results are shown in Fig. 12 and Fig. 13. We can see that shortest path routing offers a not bad performance since most servers are almost as busy as each other and when we apply multiple paths for one-to-one traffic we get a even more balanced performance since our publishing strategy is not locality based.

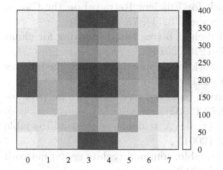 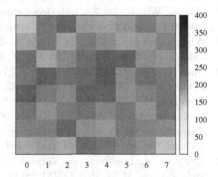

Fig. 12. Heatmap Shortest-Paths   **Fig. 13.** Heatmap Multiple-Paths

8   Conclusion

In this paper, we proposed an indexing scheme named RT-HCN for multi-dimensional query processing in data centers, which are the infrastructures for building cloud storage systems and are interconnected using a specific data center network (DCN). RT-HCN is a two-layer indexing scheme, which integrates HCN-based routing protocol [22] and the R-Tree based indexing technology, and is portionably distributed on every server. Based on the characteristics of HCN, we design a special index publishing rule and query processing algorithms to guarantee efficient data management for the whole network. We prove theoretically that RT-HCN is both query-efficient and space-efficient, by which each server will only maintain a constrained number of indices while a large number of users can concurrently process queries with low routing cost. We compare our design with RT-CAN [7], a similar design for traditional P2P network. Experiments validate the efficiency of our proposed scheme and depict its potential implementation in data centers.

References

1. Ghemawat, S., Gobioff, H., Leung, S.-T.: The Google file system. ACM SIGOPS 37(5), 29–43 (2003)
2. DeCandia, G., Hastorun, D., Jampani, M., et al.: Dynamo: amazon's highly available key-value store. ACM SIGOPS 41(6), 205–220 (2007)
3. Lakshman, A., Malik, P.: Cassandra: a decentralized structured storage system. ACM SIGOPS 44(2), 35–40 (2010)
4. Beaver, D., Kumar, S., Li, H.C., et al.: Finding a Needle in Haystack: Facebook's Photo Storage. In: USENIX OSDI, pp. 47–60 (2010)
5. Baker, J., Bond, C., Corbett, J.C., et al.: Megastore: Providing Scalable, Highly Available Storage for Interactive Services. ACM CIDR 11, 223–234 (2011)
6. Corbett, J.C., Dean, J., Epstein, M., et al.: Spanner: Google's globally-distributed database. ACM TOCS 31(3), 8 (2013)
7. Wang, J., Wu, S., Gao, H., et al.: Indexing multi-dimensional data in a cloud system. In: ACM SIGMOD, pp. 591–602 (2010)
8. Wu, S., Wu, K.-L.: An Indexing Framework for Efficient Retrieval on the Cloud. Bulletin of TCDE of the IEEE Computer Society 32(1), 75–82 (2009)
9. Wu, S., Jiang, D., Ooi, B.C., Wu, K.-L.: Efficient b-tree based indexing for cloud data processing. ACM VLDB 3(1-2), 1207–1218 (2010)
10. Chen, G., Vo, H.T., Wu, S., et al.: A Framework for Supporting DBMS-like Indexes in the Cloud. ACM VLDB 4(11), 702–713 (2011)
11. Jagadish, H.V., Ooi, B.C., Vu, Q.H.: Baton: A balanced tree structure for peer-to-peer networks. In: ACM VLDB, pp. 661–672 (2005)
12. Ratnasamy, S., Francis, P., Handley, M., et al.: A scalable content-addressable network. ACM SIGCOMM 31(4), 161–172 (2001)
13. Al-Fares, M., Loukissas, A., Vahdat, A.: A scalable, commodity data center network architecture. ACM SIGCOMM 38(4), 63–74 (2008)
14. Greenberg, A., Hamilton, J.R., Jain, N., et al.: VL2: a scalable and flexible data center network. ACM SIGCOMM 39(4), 51–62 (2009)
15. Guo, C., Wu, H., Tan, K., et al.: Dcell: a scalable and fault-tolerant network structure for data centers. ACM SIGCOMM 38(4), 75–86 (2008)
16. Guo, C., Lu, G., Li, D., et al.: BCube: a high performance, server-centric network architecture for modular data centers. ACM SIGCOMM 39(4), 63–74 (2009)
17. Li, D., Guo, C., Wu, H., et al.: FiConn: Using backup port for server interconnection in data centers. In: IEEE INFOCOM, pp. 2276–2285 (2009)
18. Li, D., Guo, C., Wu, H., et al.: Scalable and cost-effective interconnection of data-center servers using dual server ports. IEEE/ACM TON 19(1), 102–114 (2011)
19. Wu, H., Lu, G., Li, D., et al.: MDCube: a high performance network structure for modular data center interconnection. In: ACM CoNEXT, pp. 25–36 (2009)
20. Li, D., Xu, M., Zhao, H., Fu, X.: Building mega data center from heterogeneous containers. In: IEEE ICNP, pp. 256–265 (2011)
21. Guo, D., Chen, T., Li, D., et al.: BCN: expansible network structures for data centers using hierarchical compound graphs. In: IEEE INFOCOM, pp. 61–65 (2011)
22. Guo, D., Chen, T., Li, D., et al.: Expandable and cost-effective network structures for data centers using dual-port servers. IEEE Trans. Comp. 62(7), 1303–1317 (2013)
23. Antonin, G.: R-trees: A dynamic index structure for spatial searching. ACM SIGMOD 14(2), 47–57 (1984)

24. Chang, F., Dean, J., Ghemawat, S., et al.: Bigtable: A distributed storage system for structured data. ACM TOCS 26(2), 4:1–4:26 (2008)
25. Weil, S.A., Brandt, S.A., Miller, E.L., et al.: Ceph: A scalable, high-performance distributed file system. In: USENIX OSDI, pp. 307–320 (2006)
26. Charikar, M.S.: Similarity estimation techniques from rounding algorithms. In: ACM STOC, pp. 380–388 (2002)
27. Sagan, H.: Space-filling curves. Springer, New York (1994)

Design and Evaluation of a Storage Framework for Supply Chain Management

Tatsuo Tsuji, Ryota Shimono, and Ken Higuchi

Information Science Department, Faculty of Engineering, University of Fukui
Bunkyo 3-9-1, Fukui City, 910-8507, Japan
{tsuji,shimono,higuchi}@pear.fuis.u-fukui.ac.jp

Abstract. Radio-frequency identification (RFID) technology is widely available and actively used in business areas such as supply chain management. We propose a storage framework for a supply chain management system to administer the movement and time history of a product with an RFID tag. The storage structure is adapted to handle movement and time history data based on an encoding scheme for multidimensional datasets. Three types of encoded value compactly keep all the routing path and temporal information of a product from factory shipment, and the information can be quickly traced by using these encoded values. This paper describes our storage framework for supply chain management and evaluates the database constructed based on the encoding scheme.

1 Introduction

Radio-frequency identification (RFID) technology has been applied to a wide range of areas. Recently, it has been adopted in various business areas such as supply chain management. Since companies or even users can easily obtain movement information about products using RFID, significant improve of the supply chain management schemes can be expected. In supply chain management, RFID tags are attached to products, and RFID readers in the detection region generate RFID data (e.g., *tag identifier, location, time*) when the products move or stay near a detection region. RFID data generated in various locations are sent to a central server via a network and stored in a database. By analyzing product movement paths or stay times, enterprises can obtain valuable data, which can be used to improve logistics strategy.

RFID data is generated in the form of streaming data and is then stored in a database for data analyses. One approach is on-line processing for RFID data stream (e.g. [4][6][7]) and another approach is off-line processing for analysis of stored RFID data (e.g. [5][8][9]). For better design and implementation of RFID database, in both approaches efficient scheme is required for handling the moving location paths and temporal information of the products with RFID tags since their factory shipment.

[1] proposes a path encoding scheme to encode flow information for products based on the prime number labeling scheme[2] and range labeling scheme[16] used in XML technology. [3] presents a path encoding scheme which is based on Euler's formula

H. Decker et al. (Eds.): DEXA 2014, Part II, LNCS 8645, pp. 392–408, 2014.

for primes and integer tuple compression to a real number. However, the important drawbacks of these encoding schemes include that the encoded results might often overflow the machine word size (e.g. 64 bits) for long moving paths. This leads to the performance deterioration of RFID systems.

In this study we design a storage scheme to administer all routing and temporal information related to the delivery of products after shipment from the production factory. The scheme employs *history-pattern* encoding[14] for dynamically increasing multidimensional datasets. The proposed storage scheme maintains compactly the *moving location path, arrival time,* and *departure time* histories at known locations on the routing path. The above performance deterioration is significantly alleviated in our storage scheme even if the encoding results exceed the machine word size. This paper describes our new design of a storage framework for supply chain management. It also describes and evaluates the database constructed using our storage scheme compares with an existing scheme based on RDB[1].

2   History-Pattern Encoding

In this section, we describe the *history-pattern* encoding scheme presented in [14] for dynamic multidimensional datasets. Fig. 1 illustrates the required core data structures and the encoding scheme. When an extendible array A extends, a fixed-size sub-array equal to the size of the current A in every dimension is attached to the extended dimension.

2.1   Data Structures for History-Pattern Encoding

The core data structures of an n-dimensional extendible array A for history-pattern encoding consist of *history tables* and a *boundary vector table* that maintain extension history of A. Fig.1 shows an example of a two dimensional extendible array.

(1) History Table
For each dimension i ($i = 1, \ldots , n$), the history table H_i is maintained. Each history value in H_i represents the extension order of A along the i-th dimension. A counter called *history counter h* is maintained. It is initialized to 0 and incremented by one each time A is extended. H_i is a one-dimensional array, and each subscript k ($k > 0$) of H_i covers the subscript range from $2^{k-1} - 1$ to $2^k - 1$ of the i-th dimension of A, i.e., the bit pattern size of each subscript value in the range is k. If A is extended along dimension i, h is incremented and the value is stored in the next element of H_i.

For example, in Fig. 1, the bit pattern size of each subscript value in the range 4~7 of dimension 1 is 3. Therefore $H_1[3]$ stores the incremented history counter value when the sub-array including the element (4,3) is attached at the extension along the dimension 1.

(2) Boundary Vector Table
The *boundary vector table* B is a single one-dimensional array whose subscript is a history value. Each element of B maintains the past *shape* of A when the history counter was a given value. The past shape is represented by the *boundary vector*

included in the element of B. Together with the boundary vector, B also maintains the dimension of A extended at the given history counter value.

At initialization A includes only the element $(0, 0, \ldots, 0)$, and the history counter is initialized to 0. B[0] includes 0 as its extended dimension and $<0,0, \ldots,0>$ as its boundary vector. Assume that the current history counter value is h, and B[h] includes $<b_1, b_2, \ldots, b_i, \ldots, b_n>$ as its boundary vector. When the current A extends along the dimension i, B[h+1] includes i as its extended dimension and $<b_1, b_2, \ldots, b_i+1, \ldots, b_n>$ as its boundary vector.

In the history-pattern encoding, an extendible array A has two size types: *real size* and *logical size*. Assume that the tuples in an n-dimensional dataset M have been converted to the set of coordinates in A. Let s be the largest subscript of dimension k, and $b(s)$ be the bit size of s. Then, the real size of dimension k is $s + 1$, and the logical size is $2^{b(s)}$. The real size is the cardinality of the k-th attribute. In Fig. 1, the real size and the logical size are [5, 4] and [8, 4] respectively.

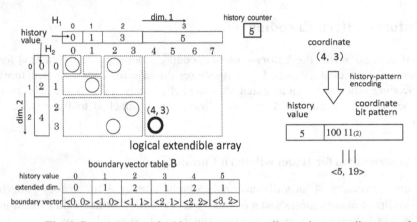

Fig. 1. Data structures for history-pattern encoding and an encoding example

2.2   Array Extension

The history-pattern encoding scheme aims encoding for a multi-dimensional dataset M. Each attribute of a tuple in M has various data type as in a relational table.

Suppose that a tuple is inserted in M, whose k-th attribute value is new,. This insertion increases the real size of A in dimension k by one. If the increased *real size* + 1 of dimension k does not exceed the current logical size $2^{b(s)}$, then A is not physically extended, and neither history table H_k nor the boundary vector table B is updated. However, if the *real size* of dimension k exceeds the current logical size, then A is logically extended. That is, the current history counter value h is incremented to $h + 1$, and this value is set to $H_k[b(s + 1)]$. Moreover, the boundary vector in B[h] is copied to B[$h + 1$], and the dimension k of the boundary vector is incremented by one. k is set to the *extended dimension* slot in B[$h + 1$]. Observe the change of the boundary vectors in Fig. 1 when the history value increases .

Note that h has one-to-one correspondence with its boundary vector in B[h] and uniquely identifies the past (logical) shape of A when the history counter value was h.

To be more precise, for the history value h, if the boundary vector in B[h] is $<b_1, b_2, ..., b_n>$, the shape of A at h was $[2^{b_1}, 2^{b_2}, ..., 2^{b_n}]$. For example, as shown in Fig. 1, because the boundary vector for the history value 3 is $<2,1>$, the shape of A when the history counter value was 3 is $[2^2, 2^1] = [4,2]$. Note that h also uniquely identifies the sub-array that is attached to A at extension when the history counter value was $h - 1$, and vice versa.

2.3 Encoding/Decoding

Using the data structures described in Section 2.1, an n-dimensional coordinate $I = (i_1, i_2, ..., i_n)$ can be encoded to the pair $<h,p>$ of *history value h* and *bit pattern p*. The history tables H_i ($i = 1, ..., n$) and the boundary vector table B are used for the encoding. The history value h is determined as the maximum value in $\{H_k[b(i_k)] \mid 1 \leq k \leq n\}$, where $b(i_k)$ is the bit size of the subscript i_k in I. For each history value h, the boundary vector in B[h] gives the required bit pattern size of each subscript in I. According to this boundary vector, the coordinate bit pattern p can be obtained by concatenating the subscript bit pattern of each dimension in descending order of the dimensions (from the lower to the higher bits of p). The storage unit for p can be one-word length, e.g., 64 bits.

In Fig. 1, as an example, the history-pattern encoding $<h,p>$ of the array element (4,3) is shown. $H_1[b(4)] = H_1[b(100_{(2)})] = H_1[3] = 5$ and $H_2[b(3)] = H_2[b(11_{(2)})] = H_2[2] = 4$. Since $H_1[b(4)] > H_2[b(3)]$, h is $H_1[3] = 5$. So element (4,3) is known to be included in the *sub-array* on dimension 1 at history value 5. Therefore, the boundary vector to be used is $<3,2>$ in B[5]. In (4,3) to be encoded, the subscript 4 of the first dimension and the subscript 3 of the second dimension form the upper 3 bits and lower 2 bits of p, respectively. Therefore p becomes $10011_{(2)} = 19$. Eventually, the element (4,3) is encoded to $<5,19>$. Generally, the bit size of history value h is rather small compared to that of pattern p; if the storage size for the pair is assumed to be 16 bits, typically the upper 4 bits are for h, and the lower 12 bits are for p.

Conversely, to decode the encoded pair $<h,p>$ to the original n-dimensional coordinate $I = (i_1, i_2, ..., i_n)$, the boundary vector in B[h] is known. Then, the subscript value of each dimension is sliced out from p according to the boundary vector. Note that the procedure for extending an extendible array described in Section 2.2 assures the following important property on $<h, p>$.

[**Property 1**] History value h denotes the bit size of its coordinate bit pattern p.

2.4 Implementation of a Multidimensional Dataset

As well as the core data structures (the history tables and the boundary vector table) presented in Section 2.1, two additional types of data structures are required to implement a multidimensional dataset M using history-pattern encoding.

CVT_i (attribute subscript ConVersion Tree) converts the attribute values of each dimension $i (1 \leq i \leq n)$ to their subscript values of the corresponding extendible array. CVT_i is implemented using a B^+ tree. It has a subscript counter sc_i initialized to be 0. If a new attribute value v emerges by the insertion of a tuple, sc_i is incremented and $<v. sc_i>$ is inserted to CVT_i.

ETF (Encoded Tuple File) is a sequential file for storing <history, pattern> pairs encoded for the tuples in M.

It should be noted that from Property 1 presented in Section 2.3, *h* can be used as the header of the encoded tuple <*h*, *p*>. This enables the ETF to be a sequential file of packed variable-length records. We call the above implementation scheme of *M* as HPMD (History Pattern implementation of Multidimensional Datasets).

3   Storage Scheme for Supply Chain Management

In our supply chain management storage scheme, moving path and time information of a product are represented by tree graphs. In this section, we describe the mapping scheme of a dynamic tree to HPMD data structures. Note that an RFID tag is attached to each product when it is shipped from the production factory, and a product is identified by its tag ID (tid).

3.1   Mapping Dynamic Tree

Each depth level of a tree is mapped to a dimension of an extendible array. The number that represents each node's relative position among its siblings is mapped to the subscript value of the corresponding array dimension. Therefore, each node can be uniquely specified by its corresponding *n*-dimensional coordinate in the array (Fig. 2(a)). The coordinate can be encoded to its corresponding <*history value, pattern*>, which was explained in Section 2.3. Note that each level of a tree is mapped to the dimension number of the array in ascending order. When the tree height increases by one, the HPMD dimensionality also increases by one, as is shown in Fig. 2(a)(b).

The dimensional extension of HPMD can be efficiently handled. Fig. 2(b) shows the addition of a new dimension against a two-dimensional HPMD in Fig. 2(a). When a new dimension is added to an *n*-dimensional HPMD, history table H_{n+1} (Section 2.1) of the new dimension would be created and initialized. All the existing elements in the original HPMD become *n*+1 dimensional, and the subscripts of the newly added dimension of them can be treated as 0. Note that the re-encoding of the elements in the HPMD before extension is not necessary, because the modification of the <history value, pattern> of nodes is not necessary.

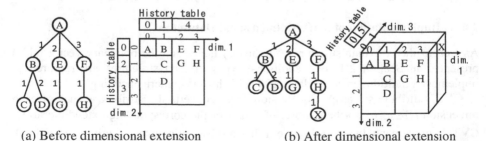

(a) Before dimensional extension   (b) After dimensional extension

Fig. 2. Dimensional extension in HPMD

3.2   Moving Path Graph and Time Graph

Assume that the products with RFID tags whose *tid*s are tid_1, tid_2, tid_3, and tid_4 are shipped from a factory located at A and each product is then moved along the following location path. The arrival date and the departure date are indicated in the bracket. These dates are normalized as the days elapsed from the base date, e.g. "3/14/2014". For example, product tid_1 arrived at location B on date 4 and departed from B on date 5.

tid_1: A[- ,1] -> B[4, 5] -> C[7, -] tid_2: A[- ,1] -> B[4,4] -> D[6, -]
tid_3: A[- ,1] -> E[2, 2] -> C[3,-] tid_4: A[- ,1] -> D[4, -]

According to the above location paths, for example, the locations that are directly visited from location A are {B, E, D} and they are visited in the date order E -> B -> D or E -> D -> B. Fig. 3 shows the arrival date sequences of the four products. These sequences can be represented by the moving graph *MG* also shown in Fig. 3. Each node of the graph is labeled with the corresponding location name, and each edge is labeled with the arrival date order among its siblings; in Fig.3, E -> D -> B is adopted.

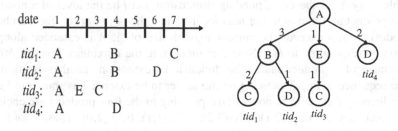

Fig. 3. Arrival date sequence of the four products and its moving graph (*MG*)

The number of products shipped from A continues to increase with time, and the number of locations where these products arrive, stay, and leave from also continues to increase with time. In other words, *MG* is a dynamically growing tree graph in terms of both the numbers of nodes and edges.

On the other hand, time information, such as arrival time, staying time, and departure time, can be represented or computed using the two types of time graphs shown in Fig. 4. Fig. 4(a) is the departure time graph called *D-TG* and Fig. 4(b) is the arrival time graph called *A-TG*. Edges of *D-TG* and *A-TG* are labeled with the normalized departure date and arrival date from/to the related location respectively. The staying time of a product at some location can be computed as "*departure time - arrival time*". Both *D-TG* and *A-TG* are dynamically growing tree graphs as *MG*.

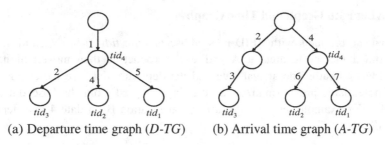

(a) Departure time graph (*D-TG*)   (b) Arrival time graph (*A-TG*)

Fig. 4. Two types of time graph for the four products

3.3   Implementation Scheme

Here the tree graph *MG* and the two types of the time graphs *D-TG* and *A-TG*, will be implemented in computer storage using HPMD to take advantage of HPMD implementation.

(1) Implementation of *MG*

A moving path graph *MG* can be mapped to the HPMD data structures. The number on each edge of graph *MG* represents the arrival date order among its destination sibling nodes. Therefore, these numbers represent the subscripts of the underlying extendible array A on the corresponding dimension. Let l be the level of a node n in *MG*. If a product at n moves to the next location and the *fan out* (i.e., the number of child nodes) from n exceeds the current logical size of A, A is extended along the dimension corresponding to $l+1$. Note that increasing the maximum level of *MG* by the movement of a product causes the dimensional extension stated in Section 3.1. The time sequence in Fig. 3 causes A on the server to be extended as shown in Fig. 5. The coordinates of the current nodes corresponding to the four products are encircled in Fig. 5 and they are tid_1: (3,1), tid_2: (3,2), tid_3: (1,1), tid_4: (2,0). These coordinates can be encoded to <history value, pattern> pairs as follows based on the extensions of A in Fig. 5: tid_1: <3,7>, tid_2: <4,14>, tid_3: <2,3>, tid_4: <3,4>.   See the encoding procedure described in Section 2.3.

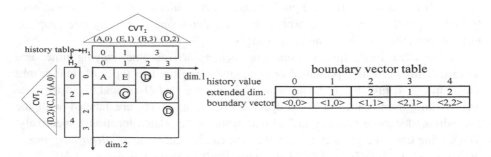

Fig. 5. Extensions of the extendible array for *MG* in Fig. 3

(2)   Implementation of *D-TG* and *A-TG*

Using the normalized date stated in Section 3.2, we can directly represent both *D-TG* and *A-TG* using a single extendible array, i.e., each date labeled on a level *l* edge of *D-TG* or *A-TG* in Fig. 4 is used as the corresponding subscript of the dimension *l* of the extendible array. Thus the CVTs can be avoided unlike the implementation of *MG*. Note that the extendible arrays for both time graphs are rather sparse using the above subscripting method with the normalized date.

Both *D-TG* and *A-TG* are also dynamically growing tree graphs, they are mapped to the underlying extendible arrays as in *MG*. The encoding of the nodes of the two graphs can be performed as discussed in (1) of this section.

(3) Storing Encoded Results

For each product, the encoded results of the location path in *MG* and the time histories in *D-TG* and *A-TG* are combined and stored sequentially in the ETF, as was described in Section 2.4. See Fig. 6. Such combined encoded results are referred to as a *tracking record* in the following. The corresponding tag ID is prefixed to this tracking record as its header. To quickly obtain the tracking record of a product, a B^+ tree with tag ID key called *tag index* is maintained. Each time a product moves to the next location, the corresponding *tracking record* for location and time information is re-encoded and updated. Therefore, all moving history and information of the product since the factory shipment up to the present is packed into the only one tracking record

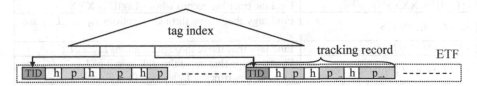

Fig. 6. Sequentially stored tracking records in ETF and its index

4   Retrieval Query

There are three types of retrieval for tracking records stored in the ETF; tag *ID retrieval* using the tag index described in (3) of Section 3.3, and *path-oriented retrieval* traversing *MG*, *D-TG*, or *A-TG* tree graphs. The latter retrieval is specified by a path expression query and it may include an aggregation function, in which case it is called *path-oriented aggregated retrieval*.

4.1   Templates for Retrieval Query

Based on [1] and XPath [15], which is a query language for selecting nodes from an XML document, we defined templates for the three query types, tag ID query, path-oriented query, and path-oriented aggregated query. Fig.7 shows the syntax for these query types. PathCondition consists of location steps, and each step has a parent axis (/) or an ancestor axis (*). In addition, each step may have TimeCondition, which is a predicate for arrival time and departure time at the specified location such as "L_1[arrive >= 20 and leave <=22]". TimeSelection is the argument of AggregateFunction. It consists of location steps and specifies arrival time (arrive), departure time (leave), and stay time (stay). Loc denotes a location name. Table 1 shows some examples of the three query types.

Tag ID Retrieval Query = <TID = id>
Path-Oriented Retrieval Query = <PathCondition>
Path-Oriented Aggregate Query = < AggregateFunction, PathCondition>

PathCondition → /LocStep ‖ /LocStep PathCondition
LocStep → Loc[TimeCondition] ‖ *
AggregateFunction → count(TimeSelection)‖sum(TimeSelection) ‖ avg(TimeSelection) ‖
                    max(TimeSelection) ‖ min(TimeSelection)
TimeSelection → Loc[TimeCondition]/LA ‖   Loc[TimeCondition]/stay ‖
                Loc[TimeCondition]/LA – Loc[TimeCondition]/LA
    LA → leave ‖ arrive

Fig. 7. Templates for three types of query

Table 1. Query examples

Query	Semantics
<TagID = XXX>	Find the tracking record whose TagID is XXX
</*/ L_1/*/.../*/ L_n>	Find tags that move through locations $L_1, ..., L_{n-1}$ and stop at L_n
</*/ L_1/*/.../*/ L_n/*>	Find tags that move through locations $L_1, ..., L_n$
</*/ L_1[stay<T]/*/.../*/ L_n/*>	Find the tags that move through locations $L_1, ..., L_n$, whose stay times at L_1 are less than T
<avg(L_2 /stay), /*/ L_1/ L_2>	For the tags that arrives at L_2 directly from L_1, get their average stay time at L_2
<min(L_2 /arrive), /*/ L_1/*/ L_2/*>	For the tags that move to L_2 via L_1, get the earliest arrival time at L_2
<max(L_2/leave-L_1/arrive),/*/L_1/*/ L_2[stay<3]/*>	For the tags that move to L_2 via L_1, and the stay time at L_2 is less than 3, get the elapsed maximum time from the arrival at L_1 to the leave from L_2

4.2   Query Processing Using HPMDs for *MG, D-TG* and *A-TG*

For a retrieval query, each tracking record stored in the ETF in Fig.6 is sequentially fetched and checked to determine if it satisfies the query condition. Let n be the current dimensionality of the HPMDs for the three types of the tree graph. In the following, we will describe the retrieval process using </*/A/*/B[stay<10]/C/*> as a typical example query, which searches all the tag IDs of the products that went through locations A, B, C and stayed at location B within 10 days.

[**Syntax Analysis Phase**] The query is decomposed into the six location steps separated by "/". These steps are stored in variables S_1~S_6. If S_i ($i=1,\cdots,6$) includes a location name, it is converted to the subscript of dimension i by using CVT_i (see Section 2.4) in HPMD data structures for *MG*.

[Retrieving Phase] For each of the tracking records in the ETF file, the retrieving phase of the query is executed sequentially. This phase consists of the two parts, namely, the decoding part and the checking part.

(Decoding Part) In this part, the tracking record is decoded into subscripts according to the procedure shown in Section 2.3 by using the HPMD data structures for MG, D-TG and A-TG. The decoded subscripts for the location names, departure times, and arrival times are set in variables $L_1 \sim L_n$, $D_1 \sim D_n$ and $A_1 \sim A_n$ respectively.

(Checking Part) This part checks that the tracking record decoded in the decoding part satisfies the query condition converted in the above syntax analysis phase.

Since $S_1("*")$ represents the repetition of the location steps, the comparison of the location subscript for "A" in $S_2("A")$ is performed for the dimensions $1 \sim n$ one by one. Therefore, the subscripts in $L_1 \sim L_n$ are compared with the subscript in S_2. If it matches at L_k, the next step $S_3("*")$ is checked. Since S_3 is "*", the comparison of the subscripts in $L_k \sim L_n$ is performed one by one with the location subscript for "B" in S_4 ("B[stay<10]"). If this matches at L_j, then if $(D_j - A_j) < 10$, the process goes to the next step $S_5("C")$. In this step, if the subscript of C matches at L_{j+1}, the final step $S_6("*")$ is checked. Since S_6 is "*", the tag ID corresponding to this tracking record can be matched and included in the result set of the query.

5   Implementation Using Relational Tables

In this section, we briefly describe the implementation of the moving paths and time information for RFID tags using the relational tables presented in [1] in order to compare our storage scheme.

5.1   Encoding Moving Paths and Time Information

In [1], the *prime number labeling scheme* [11] for XML trees is employed to encode the moving location paths. This scheme is based on the unique factorization property of a natural number and the *Chinese remainder theorem* [17]. It encodes a path from the root node to an arbitrary node into the two values ELEN and OEN. For each location L_i ($i = 1, \cdots, k$) on the tag moving path $L_1 \rightarrow L_2 \rightarrow \ldots \rightarrow L_k$, let Prime($L_i$) be the prime number corresponding to L_i. ELEN is the product of the prime numbers Prime(L_1)$\times$Prime(L_2)$\times \ldots \times$Prime(L_k), and OEN provides the order of the location L_i on the path as OEN *mod* Prime(L_i). [1] and [11] can be referred to for more details. In our study, OEN is computed faster using the extended Euclidean Algorithm [18].

In order to handle the history of time information (*arrival* and *departure* times) for a moving tag, these times are encoded to a pair of values <Start, End> by employing *range labeling* [16], which is used for labeling static XML trees. Start and End are consecutively assigned during the depth-first search. The region labeling has the following property. Node A is an ancestor of node B if and only if A.Start<B.Start and B.End<A.End. This property enables to track the time history of the moving tag.

5.2   Using Relational Tables and Query Retrieval

In [1], the movement of tags described in Sections 5.1 are represented by the three relational tables shown in Fig. 8. PATH_TABLE stores ELEN and OEN for moving path information using the encoding scheme discussed in Section 5.1. Path ID (PATH_ID) is attached to each record in PATH_TABLE. In TIME_TABLE, the time information START and END encoded by the scheme discussed in Section 5.1 is stored. LOCATION is a location name, and ARRIVAL_TIME and DEPARTURE_TIME represent arrival and departure times. TAG_TABLE has identifiers for moving path (PATH_ID) and time information(<START, END>).

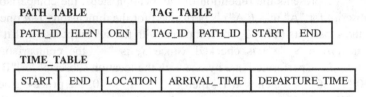

Fig. 8. Relational tables for supply chain management

The relational tables specified above can be retrieved using an SQL query. From Section 5.1, a path contains locations L_1, L_2, ..., L_k if and only if "ELEN mod $Prime(L_1) \times Prime(L_2) \times ... \times Prime(L_k)$ = 0". Therefore, we insert this expression in the *where* clause of the SQL *select* query. To determine the ancestor-descendant or the parent–child relationship, we use OEN. For example, if L_a and L_b are in the ancestor–descendant relationship, "OEN mod $Prime(L_a)$ < OEN mod $Prime(L_b)$" is inserted in the *where* clause, and if L_a and L_b are in the parent–child relationship, from the Chinese Remainder Theorem, "OEN mod $Prime(L_a)$+1= OEN mod $Prime(L_b)$" is inserted in the *where* clause. After finding the paths that satisfy the path condition, we join PATH_TABLE and TAG_TABLE on PATH_ID to obtain tags. If the queries have time conditions, we join TAG_TABLE and TIME_TABLE. We can efficiently retrieve the time information using the property of the region labeling scheme stated in Section 5.1. A query < /*/A/*/B[stay<10]/C/* >, whose retrieval process in HPMD was described in Section 4.2, can be represented by the SQL query shown in Fig. 9. In the figure, p_A, p_B, and p_C represent Prime(A), Prime(B), and Prime(C), respectively.

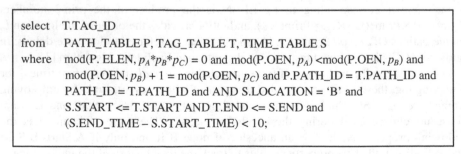

Fig. 9. SQL Query for </*/A/*/B[stay <10]/C/* >

6 Related Work

RFID data is generated in the form of streaming data and is then stored in a database for data analyses. One approach is on-line processing for RFID for data stream processing [4][6][7]. [4] presents the design of a scalable and distributed stream processing system for RFID tracking and monitoring. [6] discusses a stream query language that provides temporal event detection, and [7] takes an event-oriented approach to process RFID data. Another approach is off-line processing for stored data processing [5][8][9]. [5] proposes an RFID warehousing model that preserves object transitions while providing significant compression. [8] proposes a method to construct a warehouse of commodity flows, and [9] addresses the problem of efficiently modeling identifier collections occurring in RFID-based item-tracking applications and databases.

Some studies of efficient path encoding schemes can also be found. [1] proposes a very enlightening scheme described in Section 5, but it involves some problems. Firstly a tree structure for the *range labeling* described in Section 5.1 should be traversed to obtain the time history of a moving tag. This causes time-consuming table searching using SQL commands. Secondly, the range labeling scheme used for encoding time information is only for a static tree, thus re-encoding <START, END> and reorganization of TAG_TABLE and TIME_TABLE would be required for dynamic situation. This is inappropriate for on-line and real time processing. Thirdly, a cycle is not allowed in a moving location path because more than one prime number cannot be assigned to the same location. In a practical situation, this limitation implies that *send back* of products is not allowed. [3] presented a scheme to resolve this problem. The scheme partitions the whole set of locations into clusters and enables reuse of prime numbers in the different clusters to allow a cycle. However, considerable overhead would arise by the partition, and the encode/decode cost would increase.

In contrast, our history-pattern encoding does not share the above three problems. For the first problem above, in our encoding scheme all the moving and time histories of a product are kept in a single tracking record, hence the traversing is unnecessary. For the second problem, our scheme in HPMD is completely dynamic; re-encoding of any tracking records in ETF is unnecessary, thus enables on-line and real time processing. For the third problem, since the same location can appear on different level of a moving path tree *MG*, our encoding scheme admits a cycle.

The encoded size of a location path is much less than that of [1] as will be demonstrated by the experimental results that will be shown in (1) of Section 7.2. Moreover, since our encoding scheme requires only *shift* and *mask* instructions and does not require *multiplication* and *division* machine instructions, it can encode/decode with no multi-precision library functions, even for a long path that exceeds a machine word size.

The *history-offset* encoding scheme presented in [12][13] is similar to the history-pattern encoding presented in [14]. However, the size limitation of the encoded result also exists as it does in [1] and [11]. Thus, it requires multi-precision library functions for a long path.

7   Experimental Evaluations

We constructed two systems for supply chain management. One is a prototype system based on HPMD(denoted by hpmd in the following). hpmd includes the HPMD data structures for *MG, D-TG* and *A-TG*. The other system is based on the relational tables described in Section 5.2(denoted by rmd), which were implemented using the PostgreSQL RDBMS[19]. These systems are quite different in both their encoding schemes and implementation schemes. The employed computing environment is as follows.

CPU: Xeon X5690 (3.47 GHz),   Memory: 48 GB,
OS: Cent OS Release 6.4   PostgreSQL: Version   8.4.13

(Index Assignment) In hpmd, *tag index* for accessing a tracking record is assigned to ETF (See Fig.6). In rmd, an index for TAG_ID is assigned to TAG table and an index for PATH_ID is assigned to PATH table.

7.1   Test Datasets and Retrieval Queries

Typically, products do not move singularly, but move together in groups. Table 2 shows the two kinds of grouping, i.e., grpouing A and grouping B. *Grouping factor* $(g_1, g_2, ..., g_k)$ means that in the i-th $(1 \le k \le i)$ movement, the number of products in a group is g_i. Since g_i in *grouping A* is much larger than g_i in *grouping B*, products in grouping A move together in much larger groups than those of grouping B. The generation for the next location is stopped or the next location is generated according to the "probability to move next location". Table 3 shows the three kinds of parameter used for generating the test datasets to be used. G is the kinds of grouping (2 kinds) shown in Table 2, P is the total number of delivered products shipped from the factory (3 values) and L is the total number of locations to be visited in these product deliveries (2 values). Total 12(=2x3x2) datasets are generated by the combination of these 3 kinds of parameter. In the following, we specify a dataset as xyz, where x is a or b (grouping A or grouping B) for G, y is 4, 5, or 6 (10^4, 10^5, 10^6) for P, and z is 2 or 3 (10^2, 10^3) for L. For example 'b53' means a dataset of grouping B whose total number of products and locations are 10^5 and 10^3 respectively.

Table 4 shows eight queries used to measure retrieval times. Query 1 is a tag ID query, and queries 2~8 are path-oriented queries. Queries 5~8 specify time conditions, and Queries 6~8 include aggregate functions.

Table 2. Parameters for groupings A and B

grouping	grouping factor	minimum path length	maximum path length	probability to move next location
grouping A	(1000, 500, 200, 100, 30, 10, 3, 1)	4	8	0.5
grouping B	(50, 20, 10, 5, 3, 1, 1, 1)	4	8	0.5

Table 3. Parameters for test datasets

G: grouping factor	P: total number of products	L: total number of locations
grouping A, grouping B	$10000(10^4)$, $10000(10^5)$, $1000000(10^6)$	$100(10^2)$, $1000(10^3)$

Table 4. Measured retrieval queries

query no.	query
q1	<TagID = 543>
q2	</*/F/*>
q3	</A/B/C/D/*>
q4	</*/B/*/D/*/E/*>
q5	</*/B/*/D[arrive<200]/*/E>/*>
q6	<COUNT(),/*/B/*/D[arrive<200]/*/E/*>
q7	<AVG(B/stay), /*/B/*/D/*/E/*>
q8	<MIN(B/stay), /*/B/*/D/*/E/*>

7.2   Experimental Results

We constructed eight databases with the twelve test datasets specified in Table 3 using our prototype system (denoted by hpmd) and PostgreSQL RDBMS (denoted by rmd). hpmd includes the HPMD data structures for *MG*, *D-TG* and *A-TG*.

(1) Database Size and Construction Time

Fig. 10 shows the sizes and construction times of the databases for the test datasets. Both the size and time in hpmd and rmd increase if P value(the number of products) increases for the same G value (grouping) and L value (the number of locations). Construction time for rmd increases if L value increases for the same G and P values.

We can also observe that the database size of hpmd is less than that of rmd for the same test dataset. The reason is that in hpmd the moving history of a product is compactly stored in a single tracking record in the ETF.   While in rmd, the information in the moving history is separately stored in the related records of the three tables in Fig. 8.

As for database construction time, the database size in hpmd is much smaller than that of rmd and this provides the advantage of hpmd in the construction time, But, the disadvantage of hpmd is that in order to guarantee the real time processing, update of the tracking record in the ETF file is necessary each time a product moves. While in rmd, in fact the maximum lengths of ELEN and OEN in the test data sets b62 and b63 are both 117 bits, which is exceeding one-word length. Thus the disadvantage of rmd is that much time is spent in computing ELEN and OEN by using multiple precision arithmetic library GMP[20]. We can observe in Fig. 10 the tradeoff between these disadvantage and advantage of hpmd and rmd.

Here we consider the effect of the grouping factor. The database size for grouping B is larger than that of grouping A in both hpmd and rmd for the same P and L values. This dues to that the products in grouping B move together in a smaller group than those of grouping A. Therefore the total number of records for moving path and the time information becomes larger for grouping B than that for grouping A.

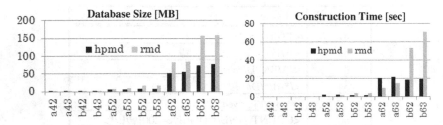

Fig. 10. Constructed database size and construction time

(2) Query Retrieval

For the eight queries in Table 4, retrieval times were measured using the twelve test datasets. Fig. 11 shows the results in msec. The retrieval time were measured 10 times and averaged for each query in both hpmd and rmd.

Query q1 is a tag ID query for getting the tracking record of the specified tag ID product. The record keeps all the moving history of the product. In rmd, join operations among the three tables in Fig. 8 are necessary to collect the moving history of the specified tag ID. In contrast, in hpmd such collection is not necessary since the newest moving history is always kept in the tracking record of the tag ID. We can see in Fig.11 that the retrieval time of q1 is negligibly small in hpmd.

Query q2~q8 are path-oriented queries. Firstly, all the graphs for q2~q8 in Fig. 11 show that the retrieval time of hpmd is dependent only on the total number of products (P value); in fact, the retrieval times are nearly the same irrespective of the query. This dues to that in hpmd all the tracking records in ETF are sequentially fetched and inspected whether they meet the specified query condition as was stated in Section 4.2. Since a product is one to one correspondent with its tracking record, the retrieval time is directly dependent on the P value.

Secondly, we can see that the retrieval times for hpmd are less than those of rmd in all test datasets for all queries. This advantage of hpmd over rmd owes to the history pattern encoding which provides compact storage for moving information and its decoding method. Thirdly we consider the effect of the grouping factor. The retrieval time difference between hpmd and rmd is larger in grouping B than in grouping A for the same P and L values. In rmd, join operations among the three tables are necessary. In particular, PATH_TABLE and TIME_TABLE in grouping B have considerably more records than those in grouping A, which makes the join operations more time-consuming.

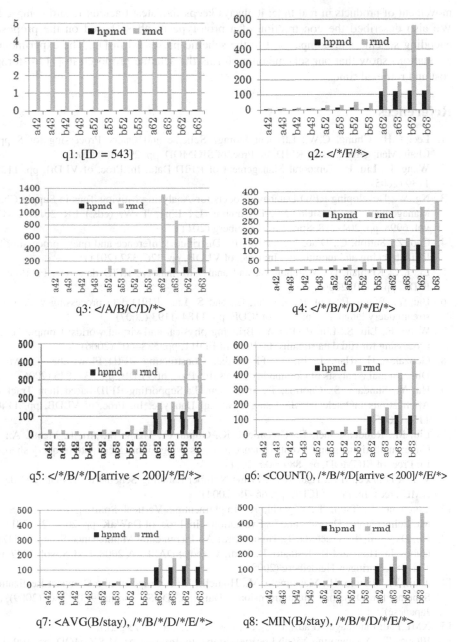

Fig. 11. Retrieval time [msec]

8 Conclusion

We presented a new storage framework for supply chain management based on history-pattern encoding. The storage scheme was designed to treat the dynamic movement of products in real time; it always keeps the latest tracking record compactly. We also described the construction of a prototype database based on the proposed encoding scheme and compared it to the scheme presented in [1]. The experimental evaluations show that our scheme outperforms the scheme proposed in [1] in storage cost and retrieval time.

References

1. Lee, C.H., Chung, C.W.: Efficient Storage Scheme and Query Processing for Supply Chain Management using RFID. In: Proc. of SIGMOD, pp. 291–302 (2008)
2. Wang, F., Liu, P.: Temporal Management of RFID Data. In: Proc. of VLDB, pp. 1128–1139 (2005)
3. Ng, W.: Developing RFID Database Models for Analysing Moving Tags in Supply Chain Management. In: Jeusfeld, M., Delcambre, L., Ling, T.-W. (eds.) ER 2011. LNCS, vol. 6998, pp. 204–218. Springer, Heidelberg (2011)
4. Cao, Z., Sutton, C., Diao, Y., Shenoy, P.: Distributed inference and query processing for RFID tracking and monitoring. In: Proc. of VLDB, pp. 326–337 (2011)
5. Li, J.X., Klabjan, D.: Warehousing and analyzing massive RFID data sets. In: Proc. of ICDE, pp. 1–10 (2006)
6. Bai, Y., Wang, F.: Liu. P., Zaniolo, C., and S. Liu.: RFID data processing with a data stream query language. In: Proc. of ICDE, pp. 1184–1193 (2007)
7. Wang, F., Liu, S., Liu, P., Bai, Y.: Bridging physical and virtual worlds: Complex event processing for rfid data streams. In: Proc. of EDBT, pp. 588–607 (2006)
8. Gonzalez, H., Han, J., Li, X.: FlowCube: Constructing RFID FlowCubes for Multi-Dimensional Analysis of Commodity Flows. In: Proc. of VLDB, pp. 834–845 (2006)
9. Hu, Y., Sundara, S., Chorma, T., Srinivasan, J.: Supporting RFID-based Item Tracking Applications inOracle DBMS Using a Bitmap Datatype. In: Proc. of VLDB, pp. 1140–1151 (2005)
10. Liu, M., Rundensteiner, E., Greenfield, K., Gupta, C., Wang, S., Ari, I., Mehta, A.: E-Cube: multi-dimensional event sequence analysis using hierarchical pattern query sharing. In: Proc. of SIGMOD, pp. 889–900 (2011)
11. Wu, X.D., Lee, M.L., Hsu, W.: A Prime Number Labeling Scheme for Dynamic Ordered XML Trees. In: Pro of ICDE, pp. 66–78 (2004)
12. Tsuchida, T., Tsuji, T., Higuchi, K.: Implementing Vertical Splitting for Large Scale Multidimensional Datasets and Its Evaluations. In: Proc. of DaWaK, pp. 208–223 (2011)
13. Jin, D., Tsuji, T., Tsuchida, T., Higuchi, K.: An Incremental Maintenance Scheme of Data Cubes. In: Haritsa, J.R., Kotagiri, R., Pudi, V. (eds.) DASFAA 2008. LNCS, vol. 4947, pp. 172–187. Springer, Heidelberg (2008)
14. Tsuji, T., Mizuno, H., Matsumoto, M., Higuchi, K.: A Proposal of a Compact Realization Scheme for Dynamic Multidimensional Datasets. DBSJ Journal 9(3), 1–6 (2009) (in Japanese)
15. XML Path Language(XPath), http://www.w3.org/TR/xpath/
16. IllGust, T.: Accelerating XPath Location Steps. In: Proc. of ACM SIGMOD, pp. 109–120 (2002)
17. http://en.wikipedia.org/wiki/Chinese_remainder_theorem
18. http://en.wikipedia.org/wiki/Extended_Euclidean_algorithm
19. PosgreSQL, http://www.postgresql.org/docs/8.4/static/index.html
20. GMP, https://gmplib.org

On Using Requirements Throughout the Life Cycle of Data Repository

Stéphane Jean[1], Idir Ait-Sadoune[2], Ladjel Bellatreche[1], and Ilyès Boukhari[1]

[1] LIAS/ISAE-ENSMA - University of Poitiers
1, Avenue Clement Ader, 86960 Futuroscope Cedex, France
{jean,bellatreche,ilyes.boukhari}@ensma.fr
[2] SUPELEC - Computer Science Departement
Gif-sur-Yvette, France
idir.aitsadoune@supelec.fr

Abstract. Requirements engineering aims at providing a requirement specification with some nice properties such as completeness or accuracy. In the lifecycle of a Data Repository ($\mathcal{DR}$), user requirements are usually assumed to be homogenous and used mainly to define the conceptual model of a $\mathcal{DR}$. In this paper, we study the interest of the requirements in the other phases of the life cycle of a $\mathcal{DR}$. We propose a generic model based on ontologies to unify the used vocabularies and requirements languages. Then we extend this model using the formal method B to check the consistency of the requirements w.r.t. the integrity constraints defined on the logical schema. Finally we propose to select optimization structures of a $\mathcal{DR}$ using the user requirements instead of SQL queries. Several experiments on the Star Schema Benchmark (SSB) confirm the interest of our proposition.

1 Introduction

Data Repositories ($\mathcal{DR}$s) which include databases and data warehouses are used to store in an efficient way a large amount of data. The development of a $\mathcal{DR}$ is based on a life cycle based on seven main steps [1,2]: (a) user requirement analysis, (b) conceptual design, (c) logical design, (d) Extraction, Transformation, Loading (ETL) procedures, (e) deployment phase, (f) physical phase and (g) exploitation phase. By exploring the literature, we figure out that the database community mainly focused on the ETL, logical, deployment and physical phases. The user requirement phase is largely studied but outside of the database community and at the same time it is considered a pre-condition for the success of the different phases of the $\mathcal{DR}$ life cycle [1]. Our goal is to study the interest of the user requirements for the other phases of the $\mathcal{DR}$ life cycle.

In this paper, we show the strong connection between user requirements and the ETL, logical and physical phases. With the globalization, the number of autonomous participants which design a $\mathcal{DR}$ has increased. And, as a consequence, each designer is free to use its own vocabulary and formalism to express his/her requirements. In order to reduce the heterogeneity of data and thus alleviate

H. Decker et al. (Eds.): DEXA 2014, Part II, LNCS 8645, pp. 409–416, 2014.
© Springer International Publishing Switzerland 2014

the task of the ETL developers, the requirements need first to be integrated. To satisfy this recommendation, we propose a generic user requirement model based on ontologies. In the logical phase, we consider the problem of $\mathcal{DR}$ consistency. By enhancing our generic user requirement model with the B formal method, our approach is used to check the consistency of the requirements and if they satisfy the constraints defined on the logical model. Finally, in the physical phase, we consider the problem of the selection of optimization structures directly from the user requirements. We show the interest of our approach through an experiment on the Star Schema Benchmark (SSB).

This paper is structured as follows. Section 2 presents our approach to integrate heterogeneous requirements which is extended in Section 3 with the B formal method. In Section 4 we show how the requirements can be used for the physical phase. Finally Section 5 describes related work and Section 6 concludes.

2   Integration of Heterogeneous Requirements

A domain ontology is a formal and shared representation of the concepts of a domain. Our approach relies on this notion by making the following assumptions: (1) a *Global (or shared) Ontology (GO)* exists on the domain targeted by the application to be designed. This assumption may seem strong. However, it is realistic in domains such as engineering where standard ontologies already exist (e.g., the Electronic Components ontology: IEC 61360-4). (2) Each designer extracts a *Local Ontology (LO)* from the GO. Our approach allows the designer to define concepts in his/her LO that do not exist in the GO as long as they inherit from concepts in the GO.

For unifying the vocabularies used to define requirements, our approach consists in describing the requirements of each designer by the concepts of his/her LO. However, each designer may have used different formalisms for defining his/her requirements. As a lot of such formalisms exist, we have chosen to define a generic requirement formalism based on three frequently used formalisms: goal-oriented languages, use cases of UML and the MCT model of the MERISE method [3]. We have studied these three formalisms to identify their core notions. Among them, we find the *actors*, the *requirements* defined by a set of *actions*, *results* and *criteria* and the *relationships between requirements*. Based on these core notions we have defined our generic requirements model by merging the metamodels of the studied requirements formalisms. The result is a generic metamodel and each requirement defined by a designer in one of the three studied formalism, is an instance of this metamodel. Figure 1(b) presents a part of our generic metamodel. Each requirement *(Requirement)* is described by an identifier *(IdReq)*, a name *(NameReq)*, a textual description *(DescriptionReq)*, an object *(PurposeReq)*, a context *(ContextReq)* and a priority *(PriorityReq* which takes one of the following values: *Mandatory, High* or *Medium)*.

As our approach consists in describing the requirements with the concepts of the ontologies, we have defined the links between the ontology metamodel and our generic requirements metamodel. Figure 1 illustrates these links. The

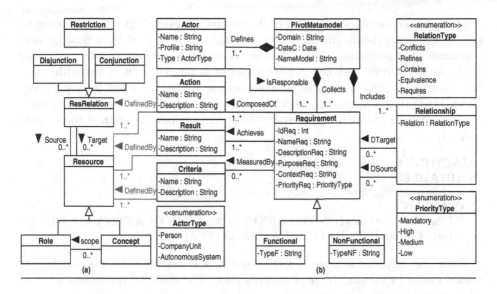

Fig. 1. The link between the ontology and the generic requirements metamodels

ontology metamodel is presented in the left part (part (a)) while our generic requirements metamodel is shown in the righ part (part (b)). These links are done by associating the characteristics of the requirements *(Action, Result and Criteria)* with the metaclass *rdfs:Resource* of the OWL language. The result of these links is a new metamodel called *OntoGoal*. By defining instances of *OntoGoal*, the requirements of the designers are described by the LO by linking the characteristics *(Action, Result and Criteria)* of each requirement to the corresponding concepts and properties in the LO. An example of an instance of OntoGoal is presented in Figure 2 and will be described later.

3 Inconsistencies Detection Using the B Formal Method

The integrated requirements may be contradictory and/or inconsistent with the integrity constraints defined on the logical schema. To deal with this problem, we propose an approach based on the B method to formalize the user requirements. We have chosen the B method because it proposes an incremental approach which is well adapted to the definition of requirements. Let us consider the following requirement extracted from the SSB benchmark: *this requirement quantifies the amount of revenue increase (extended price*discount) that would have resulted from eliminating company-wide discounts in the 2-3 % range for products sold in a quantity lower than 25 and shipped in 1993*[1]. As our approach

---

[1] The extended price is represented as the *lo_extendedPrice* property defined on the *Lineorder* class in the ontology. The quantity and discount are represented by the *lo_discount* and *lo_quantity* properties of the *Lineorder* class.

requires a domain ontology, we have first defined an ontology corresponding to the SSB logical schema (in particular, a table is represented by a concept and a column by a property). The following B model formalizes the two ontology concepts *LineOrder* and *Date*. Each concept is modelled by a **variable** with the same name. Each variable is a set and it is characterized by a property (**invariant**) that defines the structure of its elements with the keyword **struct**. Each element contains a list of attributes. An attribute is defined by a name and a type.

> **MACHINE** StarSchemaBenchmark
> **VARIABLES**
>   LineOrder, ..., Date
> **INVARIANT**
>   LineOrder ⊂ **struct**(LO_ORDERKEY ∈ NAT, LO_QUANTITY ∈ 1..50, ...)∧
>   Date ⊂ **struct**(D_DATEKEY ∈ NAT, D_YEAR ∈ NAT, ...)

Once the database schema is formalized in a B machine, we extend the obtained model by adding the requirements part. The chosen requirement is modelled by the following B operation:

> sum ⟵ requirement_q1(yy,dd,qq) =
> **PRE**
>   yy ∈ NAT ∧ dd ∈ NAT ∧ qq ∈ NAT
> **THEN**
>   sum := ∑ (lo,date).(lo ∈ LineOrder ∧ lo'LO_DISCOUNT ∈ dd-1 .. dd+1 ∧ ...
> | lo'LO_EXTENDEDPRICE * lo'LO_DISCOUNT)
> **END**

This operation takes as input three parameters: *yy* representing the year, *dd* representing the reduction, and *qq* representing the quantity. It computes as output the *sum* parameter. The **PRE** clause defines the operation precondition and the **THEN** clause describes the action to be performed by the operation. In the *requirement_q1* operation, the precondition defines the parameters type and the action defines the method for computing the *sum* parameter.

The proposed approach formally defines the requirements. It can also be used to check if the expressed requirements satisfy or not the integrity constraints. Indeed, using the *Action* class of the ontology, we can check if two requirements are contradictory or not by reasoning on the requirement actions.

4 The Interest of Requirements for $\mathcal{DR}$ Physical Design

4.1 From Traditional to Requirements-Based $\mathcal{DR}$ Physical Design

During the physical design of a $\mathcal{DR}$, the DBA selects a set of optimization structures to improve the performance of a set of queries while respecting a set of constraints. This problem is currently based on a set of frequent queries. One difficulty is that these queries can only be identified once the $\mathcal{DR}$ has been used for a while. Yet these queries are often the result of the requirements which are available from the early stage of the design life cycle. Moreover, the constraints (storage space, maintenance cost) are often expressed as non-functional requirements. As a result, our approach proposes to use requirements for an early physical design of a $\mathcal{DR}$. If $\bar{SO} = \{\mathcal{I}, \mathcal{MV}, \mathcal{HP}\}$ represents a set of optimization techniques such as indexes, materialized views or horizontal partitioning and $contr(SO)$ the constraints associated to these optimization techniques, the physical design problem can be defined as follows.

Inputs:

_ a database schema $R = \{R_1, R_2, \cdots, R_r\}$;

_ a set of requirements $Rq = \{Rq_1, Rq_2, \cdots, Rq_k\}$ which imply the execution of a set Q of queries on the $\mathcal{DR}$;

_ a set of considered optimization techniques $\bar{SO}$;

_ a set of constraints $contr(SO)$.

Output: a set of optimization structures, instance of $\bar{SO}$, which improves the performance of Q and satisfies the constraints $contr(SO)$.

Thus, the interest of our approach depends on the possibility to identify the queries corresponding to the defined requirements.

4.2   SQL Queries Generation from the User Requirements

We use the *measurable requirements*, i.e., the requirements that have a quantifiable result and possibly a set of criteria. The transformation of such a requirement into a query corresponds to a Model-to-Text transformation in the Model Driven Engineering vocabulary. It is roughly defined as follows.

```
SELECT (Property Expr List) [Result.Measure(Result.DefinedBy)]
   FROM (Join Distinct Classes) [Result.DefinedBy, Criterion.DefinedBy]
  WHERE (Conjunctive Criteria)[Criterion.DefinedBy Criterion.Opr
                                Criterion.Literal]
```

The `SELECT` clause of the query is defined as a set of expressions on the properties corresponding to the requirement *Result*. Each expression is a function given by the attribute *measure* of the *Result* class applied on the properties of the expression. The `FROM` clause is defined as join operations between the different classes corresponding to the requirement *Result* and to its *Criteria*. Finally the `WHERE` clause is defined as a conjunction of the criteria. Each criterion is composed of a left operand, of a comparison operator and of a right operand.

For example, let us consider the SSB requirement used in section 3. We have represented this requirement as an instance of OntoGoal in Figure 2. For clarity, the links between the requirement and the corresponding ontology concepts (*LineOrder* for the result and *Date* for the criterion *shipped in 1993*) are omitted. The *Result* is an expression with the property *lo_extendedPrice* as the left operand, the *lo_discount* property as right operand and the operator ***. This requirement is transformed into the following query:

```
SELECT sum (lo_extendedprice*lo_discount)
   FROM lineorder INNER JOIN date on lo_orderdate = d_datekey
  WHERE d_year=1993 AND lo_discount<3 AND lo_discount>1 AND lo_quantity<25
```

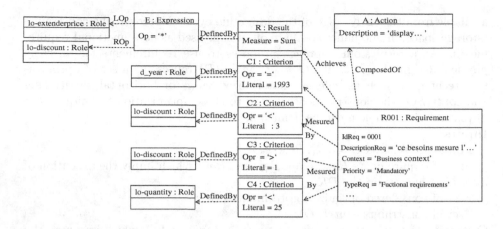

Fig. 2. A requirement of SSB represented as an instance of OntoGoal

Clearly, the previous transformation rules cannot always generate the exact query corresponding to a requirement. However the goal of our approach is only to generate queries that are sufficiently close to the real ones in order to select a pertinent set of optimization structures.

4.3 Experimental Study

We have used the SSB Benchmark because it defines a set of 13 SQL queries with the corresponding requirements. If we denote by Q this set of queries and by Q_R the set of queries generated from the requirements. Our experiments consists in comparing the set of optimization structures SO and SO_R obtained with these two set of queries. We have conducted experiments with three optimization techniques: bitmap join indexes, materialized views and horizontal partitioning.

Table 1. Experiments with indexes and horizontal partitioning

Constr (GB)	Indexes with Q $(\mathcal{I})$	Indexes with Q_R $(\mathcal{I_R})$
1	d_year	d_year
3	d_year, d_yearmonth, s_region	d_year, d_yearmonth, s_region
5	d_year, d_yearmonth, c_region, s_region, s_nation, p_category	d_year, d_yearmonth, c_region, s_region, s_nation, p_category
	Partitioning Predicates $(\mathcal{HP})$	**Partitioning Predicates** $(\mathcal{HP_R})$
	d_year=1993, d_year=1994, s_nation='UNITED STATES', s_region='AMERICA', c_region='AMERICA', s_region='ASIA', s_region='EUROPE'	d_year=1993, d_year=1994, s_nation='UNITED STATES', s_region='AMERICA' c_region='AMERICA' s_region='ASIA' s_region='EUROPE'

Table 1 presents the results of our experiments[2]. These results show that the selected optimization structures are the same for the different constraints defined (storage space). Yet, the original and generated queries are not the same: the *group by* and *order by* clauses (10 queries) are not generated as well as complex predicates (5 queries). Despite these differences the selected optimization structures are the same because the *group by* and *order by* clauses do not play a preponderant role in the selection of optimization structures on these queries and because the complex predicates are not shared by other queries. Figure 3 shows the estimated query optimization obtained with the selected $\mathcal{MV}s$ and $\mathcal{HP}s$. This optimization is around 15% for materialized views and 40% for horizontal partitioning. Thus the selection of these optimization structures is not negligible which shows the interest of defining them as soon as possible.

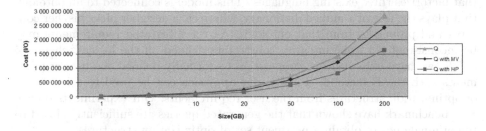

Fig. 3. Query optimization using Materialized Views and Horizontal Partitioning

5 Related Work

Several approaches have addressed the problem of integrating heterogeneous requirements. Laleau et al. [4] propose an approach to integrate the SysML language with the B method. This work extends the SysML language with a set of concepts belonging to different requirement engineering methods. This approach has two main steps: (1) an extension of the SysML language with basic concepts of the KAOS method (goal-oriented formalism) and (2) the definition of a set of rules to derive a formal specification from the extended SysML language. Lopez et al. [5] propose a metamodel to integrate a set of semi-formal formalisms in order to reuse the defined requirements. The defined metamodel supports six modeling techniques: scenarios, use cases, activity diagrams, data flow diagrams, tasks and workflows. Navarro et al. [6] propose a framework to integrate several modeling techniques found in the requirement engineering domain: traditional, Use Cases, Goal Oriented, Aspect Oriented and variability management. This integration is performed by using metamodeling techniques borrowed from the Model-driven engineering methodology (MDE). Brottier et al. [7] propose and

---

[2] The $\mathcal{MV}s$ are detailed at
http://www.lias-lab.fr/~sjean/dexa2014_LONG.pdf

MDE-based approach to integrate textual specifications of requirements. The main component of this approach is a requirement metamodel which is used to represent the integrated requirements. As we can see, these work focus on unifying specific requirement languages for a specific objective (e.g., formalize or analyze a set of requirements). Compared to these approaches, the originality of our work is twofold: (1) unifying both the vocabularies and a set of requirement languages while ensuring a high degree of autonomy to the designers and (2) studying the interest of requirements in the physical phase of the $\mathcal{DR}$ life cycle.

6   Conclusion

In this paper we have shown the interest of user requirements for different phases of the $\mathcal{DR}$ life cycle. We have first proposed a generic user requirement model that factorizes three existing languages. This model is connected to an ontology that plays the role of a unified dictionary, where each designer picks his/her concepts and properties from that ontology. As the integrated requirements may be inconsistent w.r.t the integrity constraints defined on the logical schema of a $\mathcal{DR}$, we have extended our proposition to formally defined the user requirements with the B method. Finally, our approach can be used to select a set of optimization structures from the user requirements. Our experiments on the SSB benchmark have shown that the generated queries are sufficiently closed to the original ones to obtain a pertinent set of optimization structures.

References

1. Golfarelli, M., Rizzi, S.: Data Warehouse Design: Modern Principles and Methodologies (2009)
2. Badia, A., Lemire, D.: A Call to Arms: Revisiting Database Design. SIGMOD Record 40(3), 61–69 (2011)
3. Rochfeld, A., Tardieu, H.: Merise: An information system design and development methodology. Information & Management 6, 143–159 (1983)
4. Laleau, R., Semmak, F., Matoussi, A., Petit, D., Hammad, A., Tatibouet, B.: A first attempt to combine SysML requirements diagrams and B. Innovations in Systems and Software Engineering (ISSE) 6(1-2), 47–54 (2010)
5. López, O., Laguna, M.A.: A Metamodel for Requirements Reuse. In: VII Jornadas de Ingeniería del Software y Bases de Datos, JISBD (2002)
6. Navarro, E., Mocholi, J., Letelier, P., Ramos, I.: A Metamodeling Approach for Requirements Specification. Journal of Computer Information Systems 46 (2006)
7. Brottier, E., Baudry, B., Traon, Y., Touzet, D., Nicolas, B.: Producing a Global Requirement Model from Multiple Requirement Specifications. In: EDOC (2007)

Inter-Data-Center Large-Scale Database Replication Optimization – A Workload Driven Partitioning Approach

Hong Min[1,*], Zhen Gao[3,*], Xiao Li[2], Jie Huang[3], Yi Jin[4,**], Serge Bourbonnais[2], Miao Zheng[5], and Gene Fuh[5]

[1] IBM T.J. Watson Research Center, Yorktown Heights, NY, USA
hongmin@us.ibm.com
[2] IBM Silicon Valley Lab, San Jose, CA, USA
{lixi,bourbon}@us.ibm.com
[3] School of Software Engineering, Tongji University, Shanghai, China
{gaozhen,huangjie}@tongji.edu.cn
[4] Pivotal Inc. Beijing, China
jinyi.smilodon@gmail.com
[5] IBM System and Technology Group
fuh@us.ibm.com, zhengm@cn.ibm.com

Abstract. Inter-data-center asynchronous middleware replication between active-active databases has become essential for achieving continuous business availability. Near real-time replication latency is expected despite intermittent peaks in transaction volumes. Database tables are divided for replication across multiple parallel replication consistency groups; each having a maximum throughput capacity, but doing so can break transaction integrity. It is often not known which tables can be updated by a common transaction. Independent replication also requires balancing resource utilization and latency objectives. Our work provides a method to optimize replication latencies, while minimizing transaction splits among a minimum of parallel replication consistency groups. We present a two-staged approach: a log-based workload discovery and analysis and a history-based database partitioning. The experimental results from a real banking batch workload and a benchmark OLTP workload demonstrate the effectiveness of our solution even for partitioning 1000s of database tables for very large workloads.

1 Introduction

Continuous availability (CA) is a critical aspect of business information technology resiliency. Enterprises normally maintain multiple active replicas for improving data availability and reducing data loss for the planned and unplanned outages. More enterprises and governments have realized that the data replicas should be geographically dispersed. Large-scale and long-distance data replication is a challenge especially under unprecedented application and data growths.

---

* Co-corresponding authors.
** Work done while employed by IBM.

H. Decker et al. (Eds.): DEXA 2014, Part II, LNCS 8645, pp. 417–432, 2014.
© Springer International Publishing Switzerland 2014

Various database replication technologies have been proposed for different purposes. High availability (HA) within a single data center employs data replication to maintain global transaction consistency [3] or to improve fault tolerance and system performance via transaction processing localization [1] [15] [17]. In an enterprise IT environment that consists of distant data centers across a wide area network (WAN), heterogeneous database architectures and active-active data serving configurations, data replication for continuous availability needs to address additional challenges. Our paper addresses an optimization problem in such an enterprise data replication setting.

Although some argue that integrating replication functionalities inside DBMS provides better replication performance, middleware-based replication solutions are preferable for supporting replications in cross-vendor heterogeneous database environments [11]. Industrial examples of such technology include IBM Infosphere Data Replication [21], Oracle GoldenGate [22], etc. One widely used approach is to capture committed data changes from DBMS recovery log and replicate to target DBMS. Replicating data after changes are committed at the source does not impact the response time of source-side applications. Thus, such an asynchronous solution (a.k.a. lazy replication) is widely applied in the context of WAN data replication.

However, lazy replication introduces data staleness and potential data loss in case of unplanned outages. Higher replication throughput implies shorter data stales and less data loss. Parallel replication is a desirable solution to increase the throughput by concurrently replicating changed data through multiple logical end-to-end replication channels. Such concurrent replication can potentially split a transaction's writeset among channels. Similar to DBMS snapshot consistency, point-in-time (PIT) snapshot consistency is provided via time-based coordination among replication channels [21]. PIT-consistency is a guarantee of replicated data having a consistent view with the source view at an instance of past time. Such a time delay in PIT-consistency is called PIT-consistency latency. PIT latency at the target DBMS is determined by both replication channel throughput and the duration between when the first element of a transaction's writeset is replicated and when the last element is replicated. Normally, the more replication channels a transaction's writeset is split into, the longer it takes to reach PIT-consistency. Over-provisioning with underutilized replication channels also introduces extra complexities and waste resources.

This paper proposes an automatic solution for addressing a partitioning problem in parallel middleware-based inter-data-center data replication. Partitioning database objects becomes a critical challenge in PIT latency reduction for parallel replication. First, databases and applications are often designed separately following DBMS data independence principles [2]. It is impractical to decouple database objects from application serving and replication prospective in most cases. Furthermore, new applications are continually deployed on existing databases and access patterns change as business requirements evolve. Giving the workload complexity, database object scales and resource constraints, it is challenging, error-prone and time-consuming for database administrators to understand the comprehensive picture of all the database activities and manually partition database for implementing parallel replication.

Our solution of partitioning database objects aims at employing minimum replication channels for achieving desired point-in-time consistency. We present a workload discovery and history-based partitioning approach based on the observation that similar workload characteristics and patterns reoccur in most business applications. The partition granularity is at DBMS object level such as tables and table partitions, which can reach up to thousands or tens of thousands in a large enterprise IT environment. Finer grained partitioning, such as at the row level, is less practical due to higher overhead in runtime replication coordination and DBMS contention resolution. Our approach discovers and analyzes the patterns from the DBMS recovery log, and makes partitioning recommendations using a proposed two-phased algorithm called Replication Partition Advisor (RPA)-algorithm. In the first phase, the algorithm finds a partitioning solution with the least replication channels such that the PIT latency is below a threshold tied to a service level agreement (SLA). The second phase refines the partitioning solution to minimize the number of transaction splits. Our approach is applicable to share-nothing, share-memory and share-disk databases [19]. The real-world workload evaluation and analysis demonstrate the effectiveness of our solution.

The rest of the paper is organized as follows. Section 2 introduces more background in inter-data-center parallel data replication. Section 3 describes briefly the workload profile collected for our RPA tool. Section 4 presents the RPA-algorithm. The experiment evaluations are presented in Section 5. We discuss related work in Section 6, ending with the conclusion in Section 7.

2 Background on Parallel Data Replication

Our work is applicable to an active-active configuration (where transactions can be executed at either site) presuming that proper transaction routing provides conflict prevention. For discussion simplicity, we present uni-directional replication in an active-query configuration (a.k.a. master-slave [8]) where update transactions are restricted to a designated master copy in one data center and read-only transactions are executed in other data center copies. Upon a failure on the active copy caused by disasters, one of the query copies assumes the master role and takes over the updates.

Fig. 1 illustrates a logical architecture of typical parallel lazy data replication between two database systems that potentially reside in two data centers. A parallel replication system can be modeled as a network $G(C \cup A, E)$ with a set of capture $C = \{c_1, c_2, ..., c_s\}$ and a set of apply $A = \{a_1, a_2, ..., a_r\}$. To replicate data changes, a capture agent, such as Capture1 in Fig. 1, captures the committed data changes from the database recovery logs at the source site, packs and sends it over a transport channel. The transport channels manage reliable data transfer between the two sites. An apply agent, such as Apply1 in Fig. 1, applies the changes to the target database. Each capture and apply agent can be attached to a different database node in a cluster. Within the capture and apply agent, whenever possible, multiple threads are used to handle the work with the protection of causal ordering.

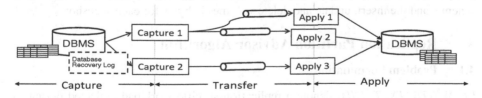

Fig. 1. Logical architecture of parallel lazy replication

The link $(a,c) \in E$ represents a *logical replication channel*, which is an end-to-end replication data path from a log change capture at the source site to a change apply at the target site. Three channels are shown in Fig. 1. A throughput capacity or bandwidth $BW(a,c)$ measures the maximum data throughput, in bytes/second, of a channel. The value is affected by all the involved components, e.g., source log reader, capture, network, apply, target database, etc. For simplicity, this paper assumes that the effective "bandwidth" is static. All changes within each database object (tables or partitions) are replicated by one channel and this is designated by preconfigured subscription policy. Each capture agent only captures the changes from its subscribed objects. Transported data changes at target site are subscribed by one or more apply agents on mutually disjoint sets of objects. The entire set of database objects within each replication channel is guaranteed to preserve serial transaction consistency. Hence, the set of database objects $TB=\{tb_1, tb_2,...,tb_k\}$ that are replicated within the same replication channel (a,c) is called a *consistency group* denoted as $cg(a,c,TB)$.

When a transaction's writeset is split into different consistency groups, the transaction is split into multiple sub-transactions that each is handled by a consistency group. The replication target eventually reaches point-in-time consistency. In the next two sections, we discuss our solution for optimizing the partitioning of database objects into minimum number of consistency groups to achieve a PIT latency goal.

3   Workload Profile and Replication Partition Advisor

Workload discovery is performed by our Workload Profiling Tool (WPT). This tool collects transaction information, called WPT workload profile, from DBMS recovery log for another tool, Replication Partition Advisor (RPA), for analysis. The log can either be a real time log or a history log. In the workload profile, all transactions are clustered based on their accessed tables. Each cluster is called a *transaction pattern*, which is characterized by a distinct set of tables or table partitions. Each transaction pattern contains all the transactions that access an identical set of tables or table partitions, but the access order and frequency can vary. For example, given a table set $S=\{A,B,C,D\}$, examples of possible transaction patterns are $P\{A,B,C\}$, $P\{A\}$, $P\{B,C,D\}$, etc. In our solution, a table partition is treated as a database object, just as any individual table. For simplicity, the rest of the paper only discusses tables.

Measuring the replication cost per operation depends on the operation type, the table definition and the actual logged column size and values. For example, replicating an update operation requires all the updated column values and the pre-updated key values. The total size of all these column values drives the costs of log capture at the source, network transmission, and operation replay at the target. WPT *workload profile* is defined as the workload patterns with the associated information, which consists of the snapshot time, transaction counts, the number of accessed tables, table schema, and the insert, update and delete volumes in bytes for each snapshot.

4   Replication Partition Advisor Algorithm

4.1   Problem Formulation

Let $WK(TB, TX, T, IUD)$ denote a replication-specific workload collected during a time window $T=\{t_0,t_0+dt,t_0+2\cdot dt,... t_0+v\cdot dt\}$, where dt is the sample collection interval; $TB=\{tb_1, tb_2,...,tb_n\}$ is a set of n replication objects (e.g., tables) whose changes are to be replicated and $TX=\{tx_1, tx_2,...,tx_k\}$ represents their transaction activities; and $IUD(TB,T)$ is the time series statistics of inserts, updates and deletes on the tables in a time window T. Given a parallel replication system $G(C \cup A,E)$, RPA-algorithm partitions all the replicated database objects TB to form a set of m mutually disjoint non-empty partitions $CG=\{cg_1,cg_2,...,cg_m\}$, where cg_i is a consistency group replicated by a particular channel $E(a,c)$. The objective is to find a solution such that m is minimal and the worst replication latency in CG is below a user-supplied threshold H.

For a particular replication channel, the PIT-consistency latency at a specific time point t_p is the difference between t_p and the source commit time for which all transactions to that point have been applied to the target at time t_p. The latency of each channel is directly related to the logical replication throughput capacity $BW(a,c)$ as well as the size of workload assigned to this channel. The workload size is defined as the number of replicated data bytes. For a specific channel, it can process at most $dt\cdot BW$ bytes within dt seconds. The residual workload will be delayed to the next intervals. Residual workload $RES_{cg,i}$ for a consistency group cg at time $t_0+i\cdot dt$ is the remaining work accumulated at $t_0+i\cdot dt$ that has not been consumed by cg_i. Thus, $RES_{cg,i}$ can be computed iteratively by:

$$RES_{cg,i} = max\{(RES_{cg,i-1} + \sum IUD(TB_{cg},t_0+(i-1)\cdot dt)-dt\cdot BW) ,0\} \tag{1}$$

Assuming data is consumed on a first-in-first-out basis, the PIT latency for cg at time $t_0+i\cdot dt$ is the time to process the accumulated residue and new activities at $t_0+i\cdot dt$:

$$PIT_{cg,i} = (RES_{cg,i} + \sum IUD(TB_{cg},t=t_0+i)) / (dt\cdot BW) \tag{2}$$

The maximum PIT latency PIT_{cg} of group cg during the time period is computed as:

$$PIT_{cg} = max\{ PIT_{cg,i} / i=0,1,2,...,v\} \tag{3}$$

The maximum PIT latency PIT_{CG-max} of a set of consistency groups CG is the highest value of PIT_{cg} among all consistency groups in CG.

The objectives of the partition optimization can be formulized as follows. Given a workload W, a parallel replication system G and its replication channel bandwidth $BW(a,c)$, and an SLA-driven PIT-consistency latency threshold H, the first objective function is defined as:

$$L = min\{|CG| | \forall CG : PIT_{CG-max} \le H\}, \tag{O1}$$

where $|CG|$ is the size of a consistency group set CG, i.e., the number of groups in the set. O1 is to find the partitioning solutions with the lowest number L of consistency groups such that the highest PIT-consistency latency of all the replication channels PIT_{CG-max} is less than or equal to H. Let P_L represent all the partition solutions of group size L and satisfy O1. The second objective is to find a partitioning solution with the minimized number of transaction splits.

$$T\_split = \arg\min_{CG\in P_L}\left\{ \sum_{tx\in TX} \sum_{i=1}^{L} tr^T (cg_i,tx) | tr^T (cg_i,tx) \in \{0,1\}\right\}, \tag{O2}$$

where $tr^T(cg_i,tx)$ is either 1 or 0 representing whether transaction tx has tables assigned to group cg_i or not. When all the tables in transaction tx is assigned to a single group, $tr^T(cg_i,tx)$ equals 0 for all groups except one. O2 seeks to find the partition solution in P_L such that the aggregated count is minimized. When no transaction split is required, T_split equals the total number of transaction instances in the workload.

4.2   RPA-algorithm Phase-1: Satisfying PIT Latency with the Least Groups

Our RPA algorithm consists of two phases: phase-1 is to find a solution that satisfies the first objective O1, and then phase-2 applies a transaction graph refinement approach to achieve the objective O2. The algorithm flow of phase-1 is listed below followed by a description.

RPA-algorithm Phase-1 Steps

1_1.Aggregate the total amount of work $W_{sum}= \sum IUD(TB,T)$ for all tables in
     $TB=\{tb_1, tb_2,...,tb_n\}$ and in time period $T=\{t_0,t_0+dt,t_0+2\cdot dt,... t_0+v\cdot dt\}$.
1_2.Compute the lower bound L_{lower} of the number of consistency groups $L_{lower}=$
     $W_{sum}/(BW\cdot v\cdot dt +BW\cdot H)$. Set initial CG number $L = L_{lower}$.
1_3.Sort all tables in TB in descending order by each table's peak activity $max(IUD)$
     and total activity $\sum (IUD)$, represented by TB_P and TB_T respectively.
1_4.Select a subset TB_{top} consisting of top tables from both list TB_P and list TB_T
1_5.Exhaust all the combinations of placing TB_{top} tables into L groups. Select the
     placement with the lowest maximum PIT latency and continue to next step.
1_6.Iterate through the rest tables in their descending order in TB_p. Test each table
     against each consistency group and compute potential maximum PIT latencies
     $PIT_{pmax\_i}, i = 1,2..L$, for each group. Place the table in the group with the low-
     est potential PIT_{pmax}. If $PIT_{pmax} > H$ for the selected group, stop and go to 1_8.
1_7.Compute the maximum $PIT_{max\_i}, i = 1,2...L$ for all consistency groups.
1_8.For all consistency groups. If $max(PIT_{max\_i}) > H$ for $i = 1,2...L$, increment
     $L=L+1$ and repeat steps 1_5 to 1_8 until $max(PIT_{max}) \leq H$. The last L is the min-
     imum number L_{min} of consistency groups.

Given bandwidth $BW(a,c)$ and a user specified PIT latency threshold H, the first two steps in Phase-1 obtain the lower bound L_{lower}, for the number of consistency groups. The lower bound describes the best case scenario: the workload volume distributes uniformly in both table and time dimensions, while the PIT latency reaches the highest at the end of the time window $t_0+v\cdot dt$ and the residual workload evenly spreads among all channels, i.e. $W_{sum} =L_{lower} \cdot BW\cdot(v\cdot dt+H)$. Starting with this lower bound L_{lower}, the process in steps *1_3* to *1_6* partitions the tables into L_{lower} groups. We then re-examine the actual maximum PIT-consistency latency of all groups in steps *1_7* and *1_8*. If the latency is higher than the threshold H, another round of partitioning using is performed with the number of groups incremented by 1.

For each fixed group number, the problem becomes to partition n tables into L consistency groups for PIT latency minimization, which is an NP-hard problem [6]. Given that the number of tables in a workload can reach thousands or even more, it is not realistic to exhaust all the partitioning combinations for finding the best among them. Instead, the greedy algorithm is introduced to resolve such a problem [7].

When applying the greedy algorithm, we use a two-step approach for improving the possibility of finding a *global optimal* solution instead of a local optimum. First, using the most active tables TB_{top} (selected in step 1_4), step 1_5 enumerates all the possibilities of partitioning them into L non-empty groups. The number of combinations for such a placement grows rapidly with the numbers of tables and groups. For avoiding an impractically high cost of step 1_5, the size of TB_{top} is determined based on a reasonable computation time on the system where RPA runs. The best choice from the exhaustive list of placements is the one with the lowest maximum PIT latency. Step 1_5 is then followed by a greedy procedure in step 1_6 that tests each of the rest tables against each consistency group and computes the group's potential new maximum PIT latency contributed by the table. The group with the lowest new maximum PIT latency is the target group for the table placement. The greedy iteration in step 1_6 uses a stronger heuristics for reaching the minimum number of consistency groups, even though it is possible that other partitioning schemes that satisfy objective $O1$ (Section 4.1) also exist. An added benefit is that this heuristics tends to generate consistency groups with less PIT latency skews among them.

Our approach is particularly effective when there are activity skews among the tables. In fact, such skews are common in real-world applications. Fig. 2 shows a customer workload analysis on how tables weight within the workload with respect to total and peak throughputs. A table with a higher *x-axis* value weights more in terms of total throughput than those with lower *x-axis* values. Such a table contributes more to the overall workload volume accumulation and channel saturation. A table with a higher *y-axis* value is more likely to contribute to higher PIT latency at its own peak time. As shown in Fig. 2, tables with higher peak or total throughputs constitute a small fraction in the entire workload. Based on this observation, *step 1_4* selects the top tables with higher total and peak throughputs for enumerative placement tests.

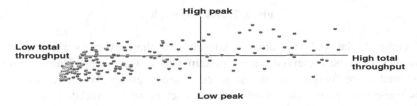

Fig. 2. Table activity distribution in a real-world banking application workload

Throughput Balancing: An Alternative to PIT-consistency Latency Minimization
Calculation of PIT-consistency latencies is impossible when quantified replication bandwidth is unavailable. In this case, the optimization goal of RPA-algorithm is adjusted to balance the peak volume and total volume given a targeted number of consistency groups. Instead of computing PIT latency, steps 1_5 and 1_6 choose the candidate group based on the accumulated peak volume and total volume after adding a new table. Both factors are positive correlated with the PIT latency. Total throughput based placement tries to balance utilizations of physical replication channels. Peak throughput based placement is for capping the highest workload volume among all channels. Understanding workload peaks also facilitates capacity planning and system configuration. This alternative is referred to as the throughput balancing algorithm (RPA-T-algorithm).

4.3   Transaction Split Reduction

RPA-algorithm phase-1 focuses on reducing PIT-consistency latency. This section describes phase-2, which attempts to reduce transaction splits for statistically increasing serial consistency in data replication.

Transaction Graphs. In RPA, we use an undirected weighted graph $TG(TB, TX, T, IUD)$ to model tables and their transaction relationships within a workload $WK(TB, TX, T, IUD)$. Each node in the graph represents a table in TB. For simplicity, the same notation $TB=\{tb_1, tb_2,....,tb_n\}$ is also used to represents the graph nodes. The weight of a node is the time series IUD statistics for the table in the workload profile. An edge $e(tb_i, tb_j)$ connecting two nodes tb_i and tb_j denotes that the there exists one or more transaction patterns that correlate both tables. The weight of the edge $|e(tb_i, tb_j)|$ is the total transaction instance counts from all the transaction patterns that involve both tables. Fig. 3 illustrates an example of a transaction graph with 19 tables. Table $T1$'s weight is associated with a time series statistics $\{234,21,654,2556,...\}$,which indicates data activities to be replicated at each time point for $T1$. The weight 731 of edge $e(T1,T2)$ means that there are 731 committed transactions involving both $T1$ and $T2$.

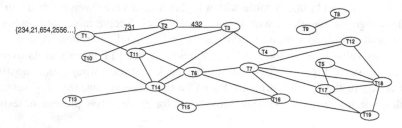

Fig. 3. Transaction graph

Because of relational constraints or other reasons, there are cases when transaction consistency must be preserved among certain tables. That means, these correlated tables need to be assigned to the same consistency group. When RPA-algorithm builds transaction graph for a workload, each set of such correlated tables is first merged into a single node with aggregated node statistics and edge weights

When partitioning a set of table nodes, RPA-algorithm groups the tables to form multiple clusters, which are possibly connected by edges. For a transaction instance that is split into q clusters, the number of edges (of weight 1) connecting these q clusters equals $q \cdot (q-1)/2$. This number monotonically increases with q when $q>1$. Hence, minimizing the number of split transactions, as formulated by O2 in Section 4.1, is equivalent to minimizing the number of edges, or aggregated edge weight. Equivalently, the problem of minimizing transaction splits is a graph partitioning problem, which is to divide a graph into two or more disconnected new graphs by removing a set of edges. As a classic partitioning problem, *minimum cut graph partitioning* is to remove a set of edges whose aggregated weight is minimal. A constraint for typical graph partitioning applications is to balance the total node weight of each partition. Differently, the target of our problem is to minimize the maximum PIT-consistency latency among all the groups, each corresponds to an individual consistency group.

General Graph Partitioning Algorithms. A graph partitioning problem, as an NP-complete problem in general, is typically solved by heuristics in practice. One widely used algorithm for two-way partitioning (bi-partitioning) is Kernighan-Lin algorithm (KL algorithm) [12]. It is an iterative improvement algorithm over two existing partitions. It seeks to reduce the total edge cut weight by iteratively swapping nodes in pairs between the two partitions. Fiduccia-Mattheyses algorithm [5] (FM algorithm) further enhances KL algorithm. By moving a node to a new group, it reduces its edge cut to the other partition while increasing its edge connection to its home partition. It also removes KL algorithm's restriction of moving nodes in pairs. The improved algorithm is referred to as KL-FM algorithm. For large graphs, multi-level bi-partitioning is often applied through graph coarsening and expansion [9]. The quality of their final solutions, which could be local optimum, is affected by the initial partitioning. Spectral solution [16] can find the global optimum by deriving partitions from the spectrum of the graph's adjacency matrix, but it does not fit our transaction graph model with time series statistics as node weights. Partitioning a graph into more than two partitions can be achieved via a sequence of recursive bi-partitioning. Refinement heuristics for k-way partitioned graph have also been developed [10].

Transaction Split Reduction by Consistency Group Refinement. Before introducing our RPA-algorithm phase-2, we first discuss how to reduce transaction splits between two already partitioned consistency groups by FM algorithms. This process is referred to as an algorithm for 2-CG refinement (CG-RF-2). The process refines the partition via node/table movement. Each move needs to ensure that the PIT-consistency latencies for both refined groups remain below PIT_{max} or within a specified margin around PIT_{max}.

Algorithm for 2-way CG refinement(CG-RF-2)

C_1 *Create graph representations for each input consistency groups cg_1 and cg_2*

C_2 *Compute PIT latencies PIT_{cg1} and PIT_{cg2} for cg_1 and cg_2, respectively. Define $PIT_{max}=max(PIT_{cg1}, PIT_{cg2})\cdot(1+\alpha)$ as the upper bound for margin α.*

C_3 *Compute the gain of each node. The gain for a node table tb_i, as defined in FM algorithm, is computed as the total edge weight between tb_i and all the nodes in the group that tb_i does not belong, subtracted by total edge weight between tb_i and all the nodes in the same group as tb_i, i.e. $g(tb_i)= \sum(|e(tb_i, tb_j)|)- \sum(|e(tb_i, tb_k)|)$ where tb_j belongs in the different group than tb_i, and tb_k belongs in the same group as tb_i. The intuition is that if $g(tb_i)$ is positive, moving tb_i from its current group to the other group reduces the edge cut between the two groups.*

C_4 *Find the node n_1 with the maximum gain g_1 and whose move from its current group to the other allows each group's PIT latency remain below the PIT_{max} value from C_2. Lock node n_1, mark its movement from its current group to the other as an element mv_1 and store in the moving list mv_list. In some cases, the gain of node n_1 is non-positive. However, it is still moved with the expectation that the move will allow the algorithm to "escape out of a local minimum".*

C_5 *Update the gains of all the nodes that are connected to n_1 due to its movement.*

C_6 *Repeat C_4 and C_5 for the rest of the nodes until all the nodes are locked. All movements are stored in $mv\_list\{ mv_1,...,mv_n\}$ in the order that they are found. The gains corresponding to these node moving steps is $\{g_1,g_2,...,g_n\}$.*

C_7   Find the best sequence of $mv_1, mv_2, \ldots, mv_k$ ($1 \leq k \leq n$) such that $\sum(\{g_1, g_2, \ldots, g_k\})$ is
  maximum and positive.
C_8   Mark the move of these k tables permanent. The refined groups are cg_1' & cg_2'.
C_9   Free all the locked nodes.
C_10 Repeat steps C_3 to C_9 until no move can be found in C_7.

The PIT latency upper bound in *C_2* is set to preserve the optimization objective and speed up the algorithm convergence. When the two input groups are produced by RPA-algorithm phase-1 and α is set to 0, CG-RF-2 algorithm preserves the same maximum PIT latency value from phase-1 while refining the groups for transaction split minimization. When $\alpha > 0$, the PIT latency constraint is relaxed and potentially more nodes are moved to reduce transaction split. Alternatively, a user-supplied PIT threshold *H* can be used as the constraint.

In some cases, the two-step procedure of bi-partitioning and refinement can be used recursively to create a higher number of partitions, given that the refinement constraint can be distributed along the recursion paths. Such an approach works for throughput balancing partitioning optimization, i.e. the alternative algorithm RPA-T. However, PIT-consistency latency is not a constraint measure that can be easily distributed while still guaranteeing convergence during recursive bi-partitioning. Therefore a non-recursive approach is needed.

RPA-algorithm Phase-2: K-way Consistency Group Refinement for Transaction Split Reduction. This section presents the phase-2 of our RPA algorithm for transaction split reduction. The algorithm (called CG-RF-k) is derived from the k-way refinement algorithm proposed by Karypis et al [10].

RPA-algorithm Phase-2 (CG-RF-k)

Ck_1 For the k consistency groups $cg_1, \ldots, cg_k$ created by RPA-algorithm phase-1,
  create the graph representation for the workload and these k partitions.
Ck_2 Iteration through all the nodes, find the set N_e of all the nodes that each has
  edge connections to other groups that it does not belong to. Compute the gain
  for each element in N_e, denote a gain as g(tb_i , cg_m) in which cg_m is a group
  that node (table) tb_i does not belong but has edge connections to one or more of
  its nodes. The gain is computed the same as algorithm CG-RF-2 step C_3.
Ck_3 Compose subset N_e' of N_e with nodes that only have positive gains.
Ck_4 For each node tb_i in N_e', test it with its connected groups for potential new PIT
  latencies. Among those groups whose potential new PIT latencies are below the
  user specified threshold H, select the group with the largest positive gain for tb_i
  to move into. If none of the group qualifies the PIT threshold requirement, do
  not move tb_i.
Ck_5 Update the gains of all the affected nodes due to the move of tb_i, including tb_i.
  Updates N_e' following the same criteria as in Ck_3.
Ck_6 Repeat steps Ck_4 and Ck_5 until there is no node in N_e'.

RPA-algorithm phase-2 starts its refinement process from the partitioning result of phase-1, which finds the minimum number of groups while satisfying maximum PIT-consistency latency threshold. Every node move seeks to reduce the positive gains, i.e. trading higher inter-group edge cut weight with lower intra-group edge cut weight. This process keeps reducing transaction split count until reaching the lowest.

5   Experiments and Analysis

We applied our work to a batch workload and an OLTP workload. The batch workload is from a banking business and we collected the WPT data from an offloaded production DBMS recovery log. For the OLTP workload, we expanded the schema of TPC-E benchmark [23] and simulated workload profile data for analysis. In both experiments, the analysis processes complete within minutes.

5.1   Transaction Split Avoidance Algorithm

For studying trade-offs between transaction split and replication latency or throughput balancing, we devised an algorithm Transaction Split Avoidance (TSA) that partitions database objects without allowing any transaction split. Using transaction graph, TSA algorithm is a modified tree traversal algorithm. Without getting into details, this algorithm works by repeatedly selecting an unassigned table node and grouping it with all the nodes that connect to it directly or indirectly.

5.2   Experiment with a Large Bank Batch Workload

This workload profile was collected from a database log representing a four-hour batch processing window with 1 minute sample interval. There are 824 tables with active statistics among a total of 2414 tables, and 5529 transaction patterns are discovered from 12.7 million transaction instances. The number of tables correlated by transaction patterns varies between 1 and 27 with histogram shown in Fig. 4.

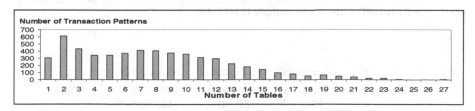

Fig. 4. Distribution of transaction patterns over the number of tables in a batch workload

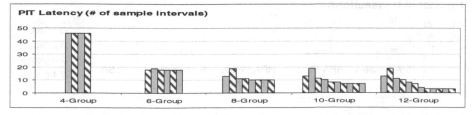

Fig. 5. Partitioning result of batch workload with RPA-algorithm phase-1

We apply RPA-algorithm with a replication bandwidth $BW= 5MB/second$. To put in prospective, this bandwidth is equivalent to insert 50K 100-byte records per second into a database. Starting from the lower bound of 3 consistency groups following step 1_2 of RPA-algorithm phase-1, Fig. 5 shows the maximum PIT-consistency latency of each group, in the unit of a sample interval, when the workload is partitioned into 4, 6,8,10 or 12 groups. As the number of consistency groups increases, the PIT latencies are reduced for each configuration. The reason that the three highest PIT latency values remain unchanged in 8-, 10- and 12-group cases is because these three groups are assigned with only one volume heavy table each. To further reduce point-in-time latency, single channel replication bandwidth has to be increased by improving the underline replication technologies in network, database, and replication software.

Next we apply both phase-1 and phase-2 of RPA-algorithm to reduce transaction splits for a given PIT latency threshold $H=60$ (1 hour). The lowest number of consistency group for this threshold is four from phase-1. Fig. 6 shows the result of phase-2. The first chart in Fig. 6 shows the maximum PIT latency of each consistency group using different variations of RPA-algorithm such as phase-1 only, phase-1 plus phase-2 with allowed increase in PIT latency within 0%, 10% and 20% margin, as labeled accordingly in the chart. The second chart in Fig. 6 shows the transaction split distribution in terms of number of groups, note that splitting into one group means no splitting. TSA algorithm, results are also provided for comparison.

The charts show that when phase-2 is used after phase-1, the percentage of non-splitting transactions increases from 70% with "RPA_Phase1" to 82%, 88% and 91% respectively for RPA_Phase1&2, RPA_Phase1&2-10% and RPA_Phase1&2-20%. With TSA algorithm, all the transactions are non-splitting; however the maximum PIT latency reaches unacceptably high of over 450 1-minute sample intervals. In addition to demonstrating that RPA-algorithm can effectively reduce transaction split, the result provides trade-offs study between transaction split and PIT latency.

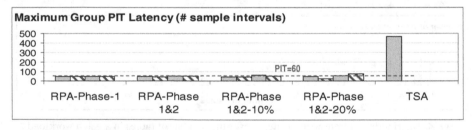

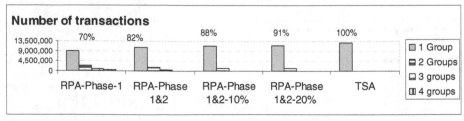

Fig. 6. Partition and transaction split results with RPA-algorithm phase-1 & phase-2 (4 CGs)

5.3   Experiment with an OLTP Workload

TPC-E is a newer OLTP data centric benchmark. Its processing is composed of both READ-ONLY and READ-WRITE transactions. Only the READ_WRITE transactions with data changes are used in our study. The TPC-E table schema consists of 33 tables, and 23 of which are actively updated during the transaction execution flows.

To simulate more complex real-world workloads, we expanded the schema by increasing the number of tables by 30x as well as increasing transaction correlations among the tables. Based on the augmented schema and workloads, as well as TPC-E specification on how the tables are updated, we generated a simulated workload profile data with 155 transaction patterns and over 6 million transactions

OLTP workloads usually update the smaller amount of data within the scope of a committed transaction. Since the volume is lower than the batch, we experiment with our alternative throughput balancing algorithm (RPA-T-algorithm) and to partition the tables and balance total throughput among 8 consistency groups.

The analyses of the partitioning results using RPA-T phase-1 and RPA-T phase-1&2 are shown in Table 1 and Fig. 7. To be more intuitive, relative standard deviation (RSTDEV=standard deviation/mean) is used to evaluate the effectiveness of throughput balancing among consistency groups, as listed in Table 1 for each algorithm. With no surprise, the RSTDEV value is near 0 (0.03%) for RPA-T phase-1 since it is optimized for balancing throughout; the RSTDEV value for TSA is very high (282%) since it does not address balancing. Fig. 7 offers a different view than Fig. 6 for analyzing how the transaction split is distributed. In Fig. 7, y-axis indicates the percentage of the total transactions that are contained within x number or less consistency groups, x being the label on x-axis. The percentage values on y-axis increase and reach 100% for eight consistency groups, i.e. all transactions are replicated within eight groups or less. An algorithm whose curve progresses to 100% slower than another means that a higher percentage of the transactions are split into more consistency groups when using this algorithm than using the other one. With TSA algorithm, none of the transactions are replicated with more than one consistency group. For RPA-T phase-1 algorithm, only a small number of transactions (0.0015%) are replicated in one group and 15% are replicated in one or two groups, etc.

Table 1. Throughput RSTDEV for different algorithm

	RPA-T	TSA	RPA-T phase-1&2 (throughput tradeoff		
	Phase-1		0%	1%	5%
RSTDEV	0.03%	282%	0.03%	1.15%	7.84%

Like RPA-algorithm, RPA-T phase-2 seeks to reduce transaction split count among consistency groups generated by RPA-T phase-1. Table 1 and Fig. 7 show that the RPA-T phase-1&2 (0%) curve progresses only marginally faster than RPA phase-1. Because the activities in this workload are uniformly distributed among different tables and along time dimension, by not allowing throughput trade-offs (0%), it limits the number of tables that can be moved during refinement. For further transaction split reduction, more trade-offs are needed on throughput balancing constraint. As observed from Fig. 7, with 1% and 5% allowed adjustment on throughputs constraint during each refinement step, there are significant increases in the number of transactions that are replicated using less consistency groups. For example, 49.2% and 84.0% of transactions are replicated with two consistency groups or less, respectively using RPA-T phase-1&2 (1%) and RPA-T phase-1&2 (5%). The trade-offs increase the throughput deviations among groups, e.g. to RSTDEV=1.15% for RPA-T phase-1&2 (1%) and RSTDEV=7.84% for RPA-T phase-1&2 (5%). Such deviation is less significant compared to the reduction in transaction splits.

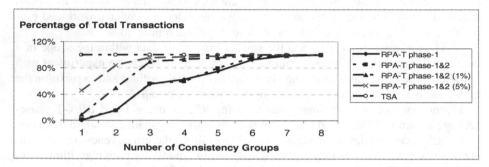

Fig. 7. Transaction split result for OLTP workload

6   Related Work

Database replication is a key technology and a challenging problem for achieving data serving high availability and disaster tolerance [8][11]. Prior works attempt to address various aspects of replication such as transaction consistency protocols, scalability, performance, etc. (e.g. [14] [13][18]) . As reported by Cecchet et al. [1], various challenges still exist when applying database replication in commercial business environments. Motivated by a real-world problem, this paper aims at optimizing middleware-based parallel data replication, especially in a long-distance multi-data-center setting. By filling a gap in understanding database objects affinities with transaction workloads, our work investigates how to group a large number of database objects to improve the performance with a constraint of user-specified PIT-consistency latency threshold. To the best of our knowledge, we are the first to propose an automatic design solution to this optimization problem.

We developed heuristics for using a greedy process [7] to achieve the first objective of minimizing the number of consistency groups with a PIT latency constraint. Based on practical analyses, an optimization technique is also proposed to improve the probability of finding a global optimal result. For reducing the transaction splits, which is the second optimization objective, we model the workload as a transaction graph and transform the problem to a graph partitioning problem. Finally, it is solved by our proposed heuristics based on the existing graph partitioning algorithms [5][12][10]. Both Schism work [4] and SWORD work [17] apply graph algorithms for finer-grain partitioning of tables horizontally across a distributed environment. They model tuples and transactions as graphs and use them, to determine the placements of data or works within a cluster. Instead, for resolving a partitioning problem in large-scale data replication across databases and data centers, our workload pattern driven approach focuses on modeling and analysis at database object levels. Common graph models and partitioning algorithms provided by existing software such as METIS [20] are not sufficient for our problem. The major reason is that the transaction graph model needs to support time series statistics and the computation of PIT-consistency latency is an iterative process.

7   Conclusion and Future Work

Large scale database replication is essential for achieving IT continuous availability. This paper presents a workload discovery and database replication partitioning approach to facilitate parallel inter-data-center data replication that is applicable to both share-nothing and share-disk databases. Our design and algorithms are demonstrated with a real customer batch workload and a simulated OLTP workload. In practice, the work has been applied to real-world business applications environment.

For future work, we plan to further fine-tune the optimization model for the replication stack. We are also interested in looking into how to further automate the cyclic flow of workload profile capturing and inter-database or inter-data-center data replication partitioning and re-adjustment.

Acknowledgements. We would like to thank Austin D'Costa and James Z. Teng for their insights.

References

1. Cecchet, E., Candea, G., Ailamaki, A.: Middleware-based database replication: the gaps between theory and practice. In: SIGMOD (2008)
2. Codd, E.F.: The relational model for database management: Version 2. Addison-Wesley (1990) ISBN 9780201141924
3. Corbett, J.C., et al.: Spanner: Google's globally-distributed database. In: OSDI (2012)
4. Curino, C., Jones, E., Zhang, Y., Madden, S.: Schism: a workload-driven approach to database replication and partitioning. VLDB (2010)
5. Fiduccia, C.M., Mattheyses, R.M.: A linear-time heuristic for improving network partitions. In: Proceedings of the 19th Design Automation Conference, pp. 175–181 (1982)
6. Garey, M.R., Johnson, D.S.: Computers and Intractability; A Guide to the Theory of NP-Completeness. W. H. Freeman & Co, New York (1990)
7. Graham, R.L.: Bounds on multiprocessing anomalies and related packing algorithms. In: AFIPS Spring Joint Computing Conference, pp. 205–217 (1972)
8. Gray, J., Helland, P., O'Neil, P.: The dangers of replication and a solution. In: SIGMOD (1996)
9. Karypis, G., Kumar, V.: A Fast and High Quality Multilevel Scheme for Partitioning Irregular Graphs. SIAM Journal on Scientific Computing 20(1), 359–392 (1998)
10. Karypis, G., Kumar, V.: Multilevel algorithms for multi-constraint graph partitioning. In: Proceedings of the 1998 ACM/IEEE conference on Supercomputing (1998)

11. Kemme, B., Jiménez-Peris, R., Patiño-Martínez, M.: Database Replication. Synthesis Lectures on Data Management. Morgan & Claypool Publishers (2010)
12. Kernighan, B.W., Lin, S.: An efficient heuristic procedure for partitioning graphs. Bell Systems Technical Journal 49, 291–307 (1970)
13. Lin, Y., Kemme, B., Patiño-Martínez, M., Jiménez-Peris, R.: Middleware based data replication providing snapshot isolation. In: SIGMOD (2005)
14. Patiño-Martinez, M., Jiménez-Peris, R., Kemme, B., Alonso, G.: MIDDLE-R: Consistent database replication at the middleware level. ACM TOCS 23(4) (2005)
15. Pavlo, A., Curino, C., Zdonik, S.B.: Skew-aware automatic database partitioning in shared-nothing, parallel OLTP systems. In: SIGMOD 2012 (2012)
16. Pothen, A., Simon, H.D., Liou, K.: Partitioning sparse matrices with eigenvectors of graphs. SIAM Journal on Matrix Analysis and Applications 11(3), 430–452 (1990)
17. Quamar, A., Kumar, K.A., Deshpande, A.: SWORD: scalable workload-aware data placement for transactional workloads. In: EDBT 2013 (2013)
18. Serrano, D., Patino-Martinez, M., Jimenez-Peris, R., Kemme, B.: Boosting Database Replication Scalability through Partial Replication and 1-Copy-Snapshot-Isolation. In: Proceedings of the 13th PRDC (2007)
19. Stonebraker, M.: The Case for Shared Nothing. IEEE Database Eng. Bull. 9(1), 4–9 (1986)
20. http://glaros.dtc.umn.edu/gkhome/views/metis
21. IBM Infosphere Data Replication, http://www-03.ibm.com/software/
22. Oracle GoldenGate, http://www.oracle.com/technetwork/middleware/goldengate/
23. http://www.tpc.org/tpce/

A First Attempt to Computing Generic Set Partitions: Delegation to an SQL Query Engine

Frédéric Dumonceaux, Guillaume Raschia, and Marc Gelgon

LINA (UMR CNRS 6241), Université de Nantes, France
firstname.lastname@univ-nantes.fr

Abstract. Partitions are a very common and useful way of organizing data, in data engineering and data mining. However, partitions currently lack efficient and generic data management functionalities. This paper proposes advances in the understanding of this problem, as well as elements for solving it. We formulate the task as efficient processing, evaluating and optimizing queries over set partitions, in the setting of relational databases. We first demonstrate that there is no trivial relational modeling for managing collections of partitions. We formally motivate a relational encoding and show that one cannot express all the operators of the partition lattice and set-theoretic operations as queries of the relational algebra. We provide multiple evidence of the inefficiency of FO queries. Our experimental results enforce this evidence. We claim that there is a strong requirement for the design of a dedicated system to manage set partitions, or at least to supplement an existing data management system, to which both data persistence and query processing could be delegated.

1 Introduction

Let us consider a set of items organized into disjoint groups. If these groups cover together the set of items, they form a partition of this set. Partitions are a very common and useful data structure manipulated in data engineering and mining. However, this structure is not yet a *first-class citizen*, in terms of data management functionalities. Our perspective is that further exploitation of clustering results would gain from the availability of (a) independence of the physical and logical representation of set partitions b.t.w. of an abstract layer (b) query facilities and (c) an optimized set partition query engine. In this paper, we address the issue of querying set partitions regardless of types, features and any other auxiliary information of the items. In other words, given a set Ω of items, our very basic assumption states that, given two distinct items x and y in Ω, we know whether they are in the same group ($x \in [y]$) or not. One does not make any other assumption on values of x and y, such like having identical or similar features. This *agnostic* position, being the most generic approach, makes possible combining partitions of the same set Ω of items with very different points of view (aka. feature spaces) w/o any trade-off or tricks to *a priori* merge both representations. As a motivating example, we shortly present hereunder a

H. Decker et al. (Eds.): DEXA 2014, Part II, LNCS 8645, pp. 433–440, 2014.

generic application scenario where storing and querying partitions can greatly support data exploration task.

In declarative query languages to multidimensional data $\Omega : \mathcal{A}$, coined tables, a partitioning task is usually performed explicitly by picking attributes from $\mathcal{A}$ and designing an *aggregation mapping* $\mathsf{agg} : 2^{\Omega} \times 2^{\mathcal{A}} \to 2^{2^{\Omega}} \times \mathbb{R}$, for which one wants to group tuples from Ω and ultimately get an aggregate value in $\mathbb{R}$ such that it underlies a partition of the ongoing table. As an example, we may think about the SQL `group-by` clause over a subset X of attributes in $\mathcal{A}$, with any usual aggregate additive function (min, max, sum, count). In this setting, every pair of tuples $(t, u) \in \Omega^2$ such like $t[X] = u[X]$ belongs to the same group and the aggregate function computes one single value for each group. By the way, the aggregation mapping yields to a disjoint assignment of tuples into groups and then, it denotes a partition $(P_{\mathsf{agg}} \in 2^{2^{\Omega}})$ of the whole table.

The above setting illustrates the need for management systems to support manipulations on set partitions. The baseline work direction consists in carrying out the mapping of partitions onto the well-established Codd's *relational model* [3]. This indeed provides mature systems with sophisticated query capabilities, query optimization policies and well-founded theoretical background. We then explore the relational encoding of set partitions as well as delegation of operators on partitions to the SQL engine.

2   Data Model(s)

This section motivates the relational encoding, formally defined in Section 2.3. The data model actually depends on combinatorial properties of set partitions. Set partitions admit two levels of nesting and constraints: a partition $P = \{a_1, a_2, \dots, a_n\}$ is indeed a set of sets (blocks a_i) of items where blocks satisfy both $\bigcup_{a_i \in P} a_i = \Omega$ and $a_i \cap a_j = \emptyset$ if $a_i \neq a_j$, with respect to the ground set Ω of items. For ease of reading, we shall denote a partition as $P = \diamondsuit \spadesuit \heartsuit \clubsuit \mid \infty \neg \mid \natural \sharp \flat$ where $\mid$ separates blocks of items. Thereafter, we use natural numbers $\mathbb{N}$ as the underlying domain for Ω without loss of generality.

2.1   Partition Algebra

The set of all partitions Π_{Ω} defined on the same ground set Ω is endowed with the *refinement relation* $\leq$ so that the *poset* $(\Pi_{\Omega}, \leq)$ is the well-known *partition lattice*. In this lattice, P refines Q, denoted $P \leq Q$, if and only if every block of P is a subset of a block in Q.

According to the algebraic definition of the partition lattice $(\Pi_{\Omega}, \wedge, \vee)$, there are semantically equivalent definitions for $\wedge$ (*meet*) and $\vee$ (*join*) operators by means of their *least upper bound* and *greatest lower bound*: $P \vee Q := \inf \{R \mid P \leq R \text{ and } Q \leq R\}$ and $P \wedge Q := \sup \{R \mid R \leq P \text{ and } R \leq Q\}$.

Example 1. Given $P = 123|456|78|9$ and $Q = 123|45|67|89$, then $P \wedge Q = 123|45|6|7|8|9$ et $P \vee Q = 123|456789$.

Besides, as partitions are also a family of sets, we would like to consider some set-theoretical operators which apply to blocks. Indeed, the very first idea behind partitions is that they are subsets of the *boolean lattice* 2^{Ω}. Further, we can apply some boolean operations on pairs of blocks: $P \cap Q := \{a \mid a \in P \wedge a \in Q\}$ and $P - Q := \{a \mid a \in P \wedge a \notin Q\}$.

The intersection operator is equivalent to computing either $P - (P - Q)$ or $Q - (Q - P)$. Both $-$ and $\cap$ are well-defined over partitions, given that they build partitions on support sets $\mathcal{S} \subseteq \Omega$. Indeed, only a subset of blocks from P composes the result set of $P \cap Q$ and $P - Q$ as well.

Finally, the set union operation $\cup$ is not expected to preserve mutual disjunction since blocks of distinct partitions may overlap, then it is not eligible in our study. To sum up, we are staying with four set partition operations: meet ($\wedge$), join ($\vee$), intersection ($\cap$), difference ($-$), that we would like to perform on generic set partitions.

2.2 Extensional Representation

For operational purpose, we are looking for an encoding schema that allows to perform boolean test over blocks (containment) and update block configurations. Then, *extension* refers to an equivalent representation where the same properties holds. Ultimately, we are concerned about the design of a mapping ε which conveys the same structure within each representation.

An extension of a set partition relies on the equivalence relation $\sim_{\Omega}$ on the ground set Ω (reflexive, symmetric and transitive). This extension simply emphasizes that, given two items $x, y \in \Omega$, x and y are in the same block of partition P iff $x \sim_P y$. Let us define the morphism $\phi : P \to \sim_P$ as $\phi(P) := \bigcup_{(a,a) \in P \times P} a \times a$. Such an extensional representation is the most expensive description as its space complexity is in $O(|\Omega|^2)$.

At this stage, we come up with a—very inefficient—extension b.t.w. of the equivalence relation counterpart of a partition. In this setting, performing a set-theoretic operation over partitions requires round tripping from the scope of the extensions to intentions and backward. This mechanism is described in the following equality:

$$\phi(P \; \mathsf{Op} \; Q) = \phi(\phi^{-1}(\sim_P) \; \mathsf{Op} \; \phi^{-1}(\sim_Q)), \quad \mathsf{Op} \in \{-, \cap\}.$$

Then, a key issue for an efficient encoding is to avoid such intensive enumeration of item-to-item associations while retrieving explicit item memberships when performing set operators $\{\cap, -\}$. Thankfully the *transitivity* property of the equivalence relation can be relaxed in our setting to a simple *reachability* property. Such a structure is optimal according to space complexity if it is a minimum spanning tree, *i.e.* $O(n)$-space, which entails every item with at least two edges (except for the leaves) and states that every item within a block is reachable by all the others. Furthermore, this tree-based representation can straightforwardly be traversed through a relational encoding since it is utmostly flattened.

2.3   Relational Encoding

As previously stated, the relational model natively supports partitions and provides a very straightforward encoding scheme for the tree-based representation of generic set partitions. We refer to the relational mapping as a *membership* encoding scheme. This encoding scheme represents the item-to-block membership within the relational model. Given a partition P and its related equivalence relation $\sim_P$; Assume a relation schema $M(\mathsf{elt} : \Omega, \mathsf{block} : \Omega)$ where columns elt and block are both items of Ω. The relational encoding scheme ε of set partitions is defined as:

$$\varepsilon : \Pi_\Omega \mapsto M(\Omega, \Omega)$$
$$P \ \to \mathbf{I}(M) := \{(x, y) \mid x \in [y]\}$$

In the above definition, we require each equivalent class $[y]$ of $\sim_P$ to have an highlighted item y that identifies the block. In conjunction with algebraic properties stated in Section 2.2, we arbitrarily decide to set the *minimum* item's value y as the anchor, assuming that $[y]$ has a deterministic and unique lower bound while the Path Independence condition [7] entails that ε must be defined as an extremal function, and thereafter, y could be assigned min or max value only.

3   Performing Operations

We focus in this section on the way to translate partition operators among $\{-, \cap, \wedge, \vee\}$ within relational queries over the membership encoding scheme ε.

We assume two partitions P and Q encoded *resp.* by relations M and N. For convenience, we do not distinguish relation schemes M and N from their respective instances. We also consider that P and Q are defined over the same ground set of objects Ω.

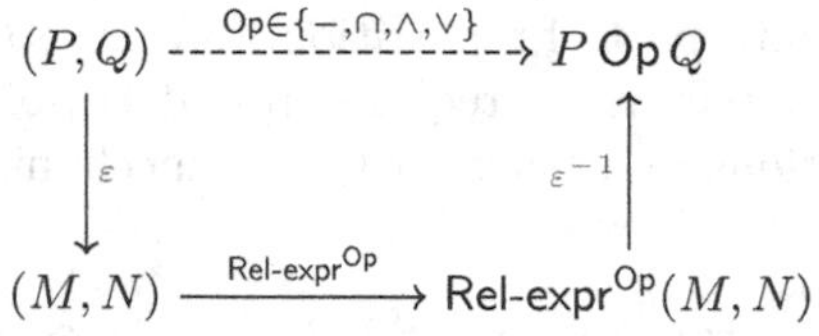

$$(P, Q) \xdashrightarrow{\mathsf{Op} \in \{-, \cap, \wedge, \vee\}} P \, \mathsf{Op} \, Q$$

$$\Big\downarrow{\varepsilon} \qquad\qquad \varepsilon^{-1}\Big\uparrow$$

$$(M, N) \xrightarrow{\mathsf{Rel\text{-}expr}^{\mathsf{Op}}} \mathsf{Rel\text{-}expr}^{\mathsf{Op}}(M, N)$$

Fig. 1. Relational Encoding Diagram

Figure 1 gives an overview of the challenge we are addressing in this Section. The main idea is to provide, for each partition operator Op, a query expression $\mathsf{Rel\text{-}expr}^{\mathsf{Op}}$ in any relational language, such that $P \, \mathsf{Op} \, Q$ is given by:

$$P \, \mathsf{Op} \, Q = \varepsilon^{-1}(\mathsf{Rel\text{-}expr}^{\mathsf{Op}}(\varepsilon(P), \varepsilon(Q)))$$

More precisely, three main issues have to be addressed to deal with set partition operators in the membership encoding : (i) block *comparison*, (ii) block *intersection* and (iii) block *identification*.

Firstly, set-theoretic operators $\{-, \cap\}$ on partitions require testing equality of M-blocks and N-blocks pairwise. We also have to decide whether or not, a given M-block overlaps an N-block, especially to perform the join $(\vee)$ operation that merges overlapping blocks. Those are block comparison operations, or predictates on blocks.

Secondly, performing the meet $(\wedge)$ operator involves calculating the set-theoretic intersection of pairs of M-blocks and N-blocks. The baseline mechanism requires computing the equivalence relation $\phi(P)$ resp. $\phi(Q)$ from M resp. N at query time. It is easily formulated as $\pi_{1,3}\sigma_{2=4}(M \times M)$ but it introduces severe performance pitfall (see Section 4).

Last but not least, every lattice operation (meet and join) requires the assignement of an anchor (block id) to the newly created blocks. Therefore, the process must iteratively fix block id's to minimum values.

Difference. The *Difference* operator in *Domain Relational Calculus* DRC is as follows:

$$\varepsilon(P - Q) := \{(x, y) : M(x, y) \wedge \forall z. \exists t. \neg (M(t, y) \leftrightarrow N(t, z))\}$$

where we build elt-block pairs (x, y) such that there is one such pair in M, and we can not find any block z in N that is equal to block y in M.

We then come up with the following algebraic expression:

$$\varepsilon(P - Q) := M - \pi_{1,2}(\sigma_{2=3}(M \times ((\mathbf{adom} \times \mathbf{adom}) - \\ \pi_{2,4}(\sigma_{1=3}(((\mathbf{adom} \times \mathbf{adom}) - N) \times M) \cup \\ \sigma_{1=3}(((\mathbf{adom} \times \mathbf{adom}) - M) \times N)))))$$

where $\mathbf{adom} = \mathrm{adom}(M) = \mathrm{adom}(N) = \pi_1(M) = \pi_1(N)$ since support sets of both partitions P and Q are equal. Hence, the naive evaluation of $\varepsilon(P - Q)$ query would require 3 full cross products, 3 equi-joins, 3 set differences and 1 union operation over relations of size in $O(|\mathbf{adom}|^2)$.

To follow on, since the equivalence $A \cap B \equiv A - (A - B)$ holds, then the set difference operator gives a proper definition for the *intersection* operator as well.

Meet. The *meet* operation $P \wedge Q$ translates in DRC as:

$$\varepsilon(P \wedge Q) := \{(x, y) : \exists z.(M(x, z) \wedge M(y, z) \wedge \exists t.(N(x, t) \wedge N(y, t) \wedge \\ \forall u.((M(u, z) \wedge N(u, t)) \rightarrow u \geq y)))\}$$

The above formula combines block intersection issue with block identification issue. Indeed, assigning a single y value as an anchor to each distinct (M-block$=z$, N-block$=t$) pair is operationally similar to assigning an anchor value to a set of equivalent objects. Actually, pairs (z, t) uniquely identify each block of the meet operator. It then requires recomputing the all equivalence relation from the membership encoding scheme.

Join. The *Join* operation $P \lor Q$ is not expressible in RA since a *transitive closure* needs to be performed [4]. Indeed, the union propagates to blocks each time there are pairwise overlapping blocks from P and Q, until we reach a *fixpoint*. Since it is not possible to *a priori* plan the number of iterations in the propagation, then there is not any RA expression that could compute $\varepsilon(P \lor Q)$.

Basically, we build the connexions between block id's within one partition, and thereafter we filter the result such that we stay with one single anchor for each set of equivalent block id's. To sum up, we first need to build an equivalence relation over blocks themselves, followed by the reducing step.

We then provide the sketch of the algorithm that performs the *Join* operation $\varepsilon(P \lor Q)$, where we mix for convenience procedural loop and DRC queries in Algorithm 1.

Algorithm 1. *Join* operation $\varepsilon(P \lor Q)$

$\theta^{(0)} \leftarrow \{(y,y) : \exists x. M(x,y)\}$ ▷ Init step

repeat

   $\theta^{(i+1)} \leftarrow \theta^{(i)} \cup \{(x,y) : \exists z.(\theta^{(i)}(x,z) \land \exists t.(M(t,z) \land \exists u.(N(t,u) \land \exists v.(N(v,u) \land M(v,y)))))\}$

   $i{+}{+}$

until $\theta^{(i+1)} = \theta^{(i)}$

$\theta^\omega \leftarrow \theta^{(i)}$

   return $\{(x,y) : \exists z.M(x,z) \land \theta^\omega(z,y) \land \forall t.(\theta^\omega(z,t) \to t \geq y)\}$

The above analysis serves the purpose of an implementation of $\varepsilon(P \lor Q)$ within regular R-DBMS. Knowing ANSI/ISO SQL3 introduces WITH RECURSIVE clause that extends RA features of SQL to mimic Datalog recursion, we are able to express a single SQL query as a *Common Table Expression* (CTE) statement to perform the $\lor$ operation over partitions.

4   Experiments

In this section, we report and discuss experimental results. The main conclusion confirms what we expected from the considerations exposed in previous Sections, *i.e.* that the SQL query engine essentially yields query processing times that are unbearably high for any non small-sized dataset. Experiments are conducted on randomly drawn partitions. Given two partitions P and Q, we assess the performance of $P \land Q$, $P - Q$ and $P \lor Q$ which are the legal baseline. With respect to the statement of the SQL queries for each operator, empirical considerations were taken into account. In that purpose, we undertook several attempts to tune up the query optimizer according to available join algorithms. Then, it seems that the default query plan computed by the optimizer is at last a good trade-off to achieve balanced performances. SQL query statements have been carefully designed to avoid usual drawbacks (useless subqueries, *etc.*) and ultimately, the execution plans have been reviewed for tracking sub-optimal evaluation strategies. We conducted experiments on a Windows XP (SP3) box powered by an Intel Q6600@2.4GHz CPU. We sent SQL queries into a PostgreSQL V9.1 R-DBMS.

Generating Partitions. In the following, let $\mathsf{sort}(P) = \tau$ be the distribution of size of blocks within a partition P and τ is a decreasing sorted list where $\tau[i]$ gives the size of block a_i in P.

Given n objects, we first draw partition P, then generate Q by applying random permutations on P. The distribution of size of blocks of raw partition P follows a power law. Indeed, a remarkable property of many natural or man-made phenomena is that, given a population, the frequency of its subpopulations may very often be well modeled by a *power law* [2]. Generation of such partitions is easily achieved with the Chinese Restaurant (stochastic) Process (CRP) [5]. The CRP draws a random partition over Ω. The expected number of blocks k grows as $O(\alpha \log n)$, where α is the scale parameter of the CRP. Next, some random permutations are performed on P to generate a list of partitions $(P^{(1)}, \dots, P^{(\ell)})$.

Results and Analysis. Results are reported for partitions P and Q defined over $n = 1\,000,\ 2\,000$ and $5\,000$ items. Although those numbers are quite low, we observed in experiments that query processing times prevent increasing n by a further order of magnitude. The number of blocks is set to range between 10 and 20. Overall, the largest blocks, generated under the CRP mechanism, are typically about half the size of the support set.

We report query processing time related to SQL implementations in Fig. 2. The boxplots describe the observed variability from a set of experiments, where the number of blocks ranges in $[10..20]$.

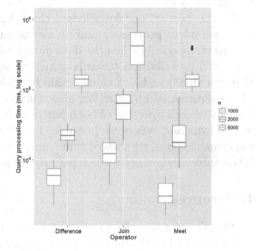

Fig. 2. Main operations $-$, $\vee$ and $\wedge$

Regarding genuine operators, the $P \wedge Q$ operation is slightly more expensive than $P - Q$ whereas $P \vee Q$ computation performs worst than all the others. Besides, operations on partitions of $5\,000$ objects are all much more costly (one order of magnitude) than those with $1\,000$ and $2\,000$ objects. Roughly, adding noise into partitions increases the root mean square of the execution time.

Ultimately, P op $P^{(\ell)}$, op $\in \{\wedge \vee -\}$ shows outlier runs w.r.t. the execution time, where $P^{(\ell)}$ is the farthest partition from P in the generation process. This observation is still emphasized by the growing size n of the support set. The very first conclusion is that query processing time rapidly becomes prohibitive, even for data sets with small to moderate size.

5   Related Work and Conclusion

There have been efforts for supporting an algebra of sets in SQL through set-comparison queries, involving several nested queries because each set-theoretic operator requires its semantical translation in predicate calculus restricted to existential quantifiers [6]. However, query evaluation has not been assessed and it does not cover partitions as sets of sets with mutual exclusion.

In another direction, there has been a 20 years line of research on *groupjoin* [9,1,8]. Though the merging of group-by and relational join allows for efficient computation of aggregate queries, groupjoin, whatever the implementation, has no extension to closure computation that is the critical issue of the partition join. Moreover, it can even not handle block identification by anchor since the aggregate value (min) must be associated to each and every object rather than one single representative. Roughly, the aggregate value is itself the parameter of the group-by clause.

To the best of our knowledge, there has been no proposal for *generic* set partition query processing and no assessment of how computation on partitions performs, especially when the data model follows a relational-based encoding. We provide a contribution towards achieving some relational modeling of partition through an object-block membership relation, so that it can handle set partitions of a collection of objects. We also studied its computing framework. We then translated each operator of both partition lattice and algebra of sets as relational algebra queries, wherever possible, Datalog query otherwise. Through several experimentations, we showed that computing operators over partitions is globally intractable when their underlying ground set is growing.

References

1. Akinde, M., Chatziantoniou, D., Johnson, T., Kim, S.: The MD-join: An operator for complex OLAP. In: Proc. ICDE, pp. 524–533 (2001)
2. Clauset, A., Shalizi, C.R., Newman, M.E.J.: Power-law distributions in empirical data. SIAM Review 51(4), 661–703 (2009)
3. Codd, E.F.: A relational model of data for large shared data banks. Commun. ACM 13, 377–387 (1970)
4. Dong, G., Libkin, L., Wong, L.: Local properties of query languages. Theor. Comput. Sci. 239, 277–308 (2000)
5. Goldwater, S., Griffiths, T.L., Johnson, M.: Producing power-law distributions and damping word frequencies with two-stage language models. J. Mach. Learn. Res. 12, 2335–2382 (2011)
6. Halpin, T., Morgan, T.: Information Modeling and Relational Databases, 2nd edn. Morgan Kaufmann Publishers Inc, San Francisco (2008)
7. Malishevski, A.V.: Path independence in serial–parallel data processing. Mathematical Social Sciences 27(3), 335–367 (1994)
8. Moerkotte, G., Neumann, T.: Accelerating queries with group-by and join by groupjoin. PVLDB 4(11), 843–851
9. von Bultzingsloewen, G.: Optimizing SQL queries for parallel execution. ACM SIGMOD Record 18(4), 17–22 (1989)

Algebra-Based Approach for Incremental Data Warehouse Partitioning

Rima Bouchakri[1,2], Ladjel Bellatreche[1], and Zoé Faget[1]

LIAS/ENSMA – Poitiers University, Futuroscope, France
LCSI/ESI – High School of Computer Science – Algiers, Algeria
{rima.bouchakri,bellatreche,zoe.faget}@ensma.fr

Abstract. Horizontal Data Partitioning is an optimization technique well suited to optimize star-join queries in Relational Data Warehouses. Most works focus on a static selection of a fragmentation schema. However, due to the evolution of data warehouses and the ad hoc nature of queries, the development of incremental algorithms for fragmentation schema selection has become a necessity. In this work, we present a Partitioning Algebra containing all operators needed to update a schema when a new query arrives. To identify queries which should trigger a schema update, we introduce the notion of query profiling.

1 Introduction

In the era of Big Data, there is a need to develop new models, data structures, algorithms, and tools for processing, analyzing, mining and understanding this huge amount of data using High Performance Computing ($\mathcal{HPC}$). The diversity of $\mathcal{HPC}$ platforms contributes in developing a new well established phase in the database life cycle which is the deployment phase. It consists in selecting the best platform to satisfy the requirements of end users in terms of query processing and data manageability. Horizontal Data Partitioning ($\mathcal{HDP}$) is a pre-condition for deploying a database/data warehouse on any platform: centralized [9], parallel [5], distributed [7], cloud [4], etc. The problem of $\mathcal{HDP}$ ($\mathcal{PHDP}$) has been largely studied in the literature in different database contexts: OLTP databases [7], data warehouses [1], scientific and statistical databases [8]. It consists in fragmenting a table, an index or a materialized view, into partitions (fragments), where each fragment contains a subset of tuples [9]. Many commercial (Oracle, DB2, SQLServer, Sybase) and academic DBMS (PostgreSQL and MySQL) implement it. Two main types of $\mathcal{HDP}$ are distinguished [3]: primary $\mathcal{HDP}$ and derived $\mathcal{HDP}$. In the first partitioning, a given table is partitioned based on its own attributes. The primary $\mathcal{HDP}$ optimizes selection operations and is used in rewriting queries in distributed and parallel databases [7]. When the result of a fragmentation of a given table is propagated to another table, this partitioning is called derived $\mathcal{HDP}$. This partitioning is feasible when a parent-child relationship exists between the two involved tables.

The formalization of the $\mathcal{PHDP}$ on relational data warehouses is as follows [1]: given a data warehouse schema, a set of a priori known queries, and a constraint

H. Decker et al. (Eds.): DEXA 2014, Part II, LNCS 8645, pp. 441–448, 2014.

limiting the total number of fragments, $\mathcal{PHDP}$ consists in fragmenting the fact table based on the partitioning schemes of the dimension tables such that the overall query processing cost is minimized and the final number of fragments does not exceed the constraint. The obtained fragments are disjoint and the union of all fragments belonging to a set of fragments is equal to the schema of its corresponding table.

By examining the literature, we get two main observations: (i) Most partitioning algorithms consider a well-known set of queries to perform a static selection. However, the ad-hoc nature of the OLAP and scientific queries calls for the development of incremental data partitioning algorithms. (ii) Partitioning algorithms use simple data structures mainly involving the attributes used to partition either table or a schema.

In the context of data warehouse, a few approaches dealing with the incremental $\mathcal{HDP}$ exist [6,10]. Authors in [10] propose a dynamic approach in distributed data warehouses design that is triggered for each change in the workload, which causes a high maintenance cost, and may be unnecessary where a change in the workload leads to the same schema. Authors in [6] define a refragmentation approach of relational centralized data warehouses where the fact table fragmentation[1] is performed using the 'By Range' Oracle mode on the same and only one foreign key of one dimension table. However, due to complex OLAP queries, a beneficial fragmentation should be performed according to many different dimensions tables.

Our vision is to propose an integrated solution for $\mathcal{PHDP}$ that satisfies these two objectives. We present a flexible encoding for any $\mathcal{HDP}$ schema and an algebra whose operators are applied on that coding to capture any change of the partitioning schema when the workload evolves. To manage workload evolution, we introduce the notion of *query profiling* which studies the impact of a new query on the encoding (expanding it, keeping it as it is, or reducing it). These operations will be managed by the proposed algebra.

This paper is organized as follows: Section 2 describes our algebra and its properties for managing any fragmentation schema in the dynamic context. Section 3 describes the management of the arrival of new queries by the use of our encoding and algebra. In section 4, we present in details our algorithm that uses the notion of query profiles. In Section 5, we conduct experiments to show the efficiency and effectiveness of our proposal. Section 6 concludes the paper.

2   Fragmentation Algebra

2.1   Flexible Encoding

Let us consider a $\mathcal{RDW}$ with a fact table F and d dimensions tables. A fragmentation schema is the result of the $\mathcal{RDW}$ partitioning process. It is defined on non-key dimension attributes $A = \{A_1, \cdots, A_n\}$. Each attribute A_i has a Domain, called $Dom(A_i)$. $Dom(A_i)$ can be partitioned into m_i sub-domains $Dom(A_i) = \{SD_1^i, SD_2^i, \cdots, SD_{m_i}^i\}$.

---

[1] In this paper, we use fragmentation and partitioning interchangeably.

Table 1. Maximal Fragmentation Schema MFS

A_1	SD_1^1	SD_2^1	$\cdots$	$SD_{m_1}^1$
A_2	SD_1^2	SD_2^2	$\cdots$	$SD_{m_2}^2$
$\cdots$	$\cdots$	$\cdots$	$\cdots$	.
A_n	SD_1^n	SD_2^n	$\cdots$	$SD_{m_n}^n$

Based on these notions, we define a *Data Structure* that represents the Maximal Fragmentation Schema MFS of dimension tables (table 1). The number of fact table fragments is then the product of the numbers of dimensions fragments $(\prod_1^n m_i)$. A similar encoding has been proposed in [2] and used with an incremental genetic algorithm. The main drawback of this work is that the genetic algorithm is executed every time changes on the workload occur. Note that a new query may cause an extension or a reduction of the fragmentation schema by adding/deleting attributes or sub-domains, or splitting/merging existing sub-domains. Also, new queries may have the same definition with the existing ones, hence should not trigger a new $\mathcal{HDP}$ selection. In order to determine the exact actions required after the arrival of a given query, we define a *Fragmentation Algebra* that contains all possible operations defined on a $\mathcal{HDP}$ schema.

2.2 Operators Description

Two types of transformations can be applied to FS. A *Reduction* (RFS) is conducted when attributes are removed or sub-domains are merged. An *Evolution* (EFS) is conducted when attributes are added or sub-domains are split. We define the set of operators needed for both transformations, which form our *Fragmentation Algebra*. We consider the attribute A_i, where n represents the number of different attributes, and m_i the number of sub-domains of A_i. Each operator takes a fragmentation schema FS as input and produces a fragmentation schema FS'

- $Add\_A(A_i, \{SD_{j_1}^i, \cdots, SD_{j_p}^i\})(FS)$: add the attribute A_i to the fragmentation schema FS including the set of sub-domains $\{SD_{j_1}^i, \cdots, SD_{j_p}^i\}$, which implies creating the set $Else_i = \{SD_1^i, \cdots, SD_{m_i}^i\} \setminus \{SD_{j_1}^i, \cdots, SD_{j_p}^i\}$.

- $Add\_SD(A_i, \{SD_{j_1}^i, \cdots, SD_{j_p}^i\})(FS)$: add to the attribute A_i a set of sub-domains $\{SD_{j_1}^i, \cdots, SD_{j_p}^i\}$ and delete it from the set $Else_i$.

- $Split\_Dom(A_i, \{SD_{j_1}^i, \cdots, SD_{j_p}^i\}, \{SD_{k_1}^i, \cdots, SD_{k_s}^i\})(FS)$: split the set of sub-domains $\{SD_{j_1}^i, \cdots, SD_{j_p}^i\}$ of the attribute A_i into two sets of sub-domains $\{SD_{k_1}^i, \cdots, SD_{k_s}^i\}$ and $\{SD_{j_1}^i, \cdots, SD_{j_p}^i\} \setminus \{SD_{k_1}^i, \cdots, SD_{k_s}^i\}$, where $\{k_1, \cdots, k_s\} \subset \{j_1, \cdots, j_p\}$.

- $Merge\_Dom(A_i, \{SD_{j_1}^i, \cdots, SD_{j_p}^i\}, \{SD_{k_1}^i, \cdots, SD_{k_s}^i\})(FS)$: merge the two sets $\{SD_{j_1}^i, \cdots, SD_{j_p}^i\}$ and $\{SD_{k_1}^i, \cdots, SD_{k_s}^i\}$ into one, where $\{j_1, \cdots, j_p\} \subset [1, m_i]$ and $\{k_1, \cdots, k_s\} \subset [1, m_i]$.

- $Del\_A(A_i)(FS)$: delete the attribute A_i from the $\mathcal{HDP}$ schema FS.

- $Del\_SD(A_i, \{SD_{j_1}^i, \cdots, SD_{j_p}^i\})(FS)$: delete from the attribute A_i the set containing the sub-domains $\{SD_{j_1}^i, \cdots, SD_{j_p}^i\}$ and include it in the set $Else_i$.

3 Queries Profiling

When new queries are executed on the $\mathcal{RDW}$, the fragmentation schema may be updated in order to take into account the workload changes. According to the executed queries, the fragmentation schema can be updated using a reduction RFS, an Evolution EFS or both. If the definition of the executed queries is similar to the current workload, no changes are required. As a consequence, we elaborate four query profiles:

Evolution queries: this profile describes queries that require the operations $Add\_A()$, $Add\_SD()$ and/or $Split\_Dom()$. When an Evolution query is executed on the $\mathcal{RDW}$, an EFS of the current schema is needed. As consequence, the number of facts table fragments will increase.

Reduction queries: this profile describes queries that require the operations $Del\_A()$, $Del\_SD()$ and $Merge\_Dom()$. When a Reduction query is executed on the $\mathcal{RDW}$, an RFS of the current schema is needed. Therefore, the number of facts table fragments will decrease.

Mixed queries: a Mixed query implies both Evolution and Reduction operations ($Add\_A()$, $Add\_SD()$, $Split\_Dom()$, $Del\_A()$, $Del\_SD()$, $Merge\_Dom()$). The number of facts table fragments can either increase or decrease.

Neutral queries: a Neutral query does not affect the current $\mathcal{RDW}$ fragmentation schema. Let us consider a selection predicate A_i op $\{SD^i_{j_1}, \cdots, SD^i_{j_p}\}$ in a query. This query is neutral if A_i appears in FS and all sub-domains, contained in $\{SD^i_{j_1}, \cdots, SD^i_{j_p}\}$, appear as a set in FS or the operations defined leaves the $\mathcal{HDP}$ schema unchanged.

4 Queries Profiling-Based Incremental Algorithm

We assume that a new query Q_i is executed on a partitioned $\mathcal{RDW}$ represented by a current $\mathcal{HDP}$ Schema called FS_t. In order to adapt the current fragmentation schema of the $\mathcal{RDW}$, we analyze Q_i using our Algebra in order to determine the query profile. According to the profile, we decide of the physical operations to perform in order to update the $\mathcal{RDW}$ schema. We present an algorithm that summarizes the Incremental selection of $\mathcal{RDW}$ schema using query profiling. This algorithm uses the classic $\mathcal{RDW}$ schema selection based on Genetic Algorithms [2]. Some functions are needed to implement our algorithm:

- *AnalyseQProfile(Algebra, Q_i)*: returns the profile of query Q_i.
- *ComputeNewFS(FS, Q_i)*: returns a new fragmentation Schema FS_{t+1} using the current Schema FS_t and the new query Q_i.
- *FragmentationSelectionGA (Q, FS, RDW, B)*: selects the best fragmentation Schema using Genetic Algorithm.
- *NBfragments(FS)*: returns the number of fact fragments of a given fragmentation Schema FS.

Incremental Selection of FS based on Queries Profiling
Input:

$Algebra$: a set of the Algebra operators
Q : the workload containing m queries
Q_i : the new executed query
FS : the current fragmentation schema of the RDW
RDW : data that compose the Cost Model used in the Genetic Algorithms
B : maximum number of fact fragments

Output: Fragmentation schema of the dimensions tables $NewFS$.
Begin

QueryProfil ← AnalyseQProfile(Algebra, Q_i);
if QueryProfil="Neutral" **then**
    Break; {*End Algorithm*}
end if
NewFS ← ComputeNewFS(FS, Q_i);
if QueryProfil="Evolution" or QueryProfil="Mixed" **then**
    if NBfragments($NewFS$)> B **then**
        $NewFS$ ← FragmentationSelectionGA($Q \cup \{Q_i\}$, FS, RDW, B);
    end if
end if
End

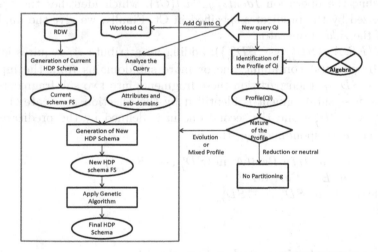

Fig. 1. Incremental Fragmentation Schema Selection based on Queries Profiling

We present in figure 1 an architecture that summarizes the steps to perform when the workload evolves. When a new query Q_i is executed on the RDW, its profile is determined based on the Algebra. If the query profile is "Neutral" or "Reduction", no selection and no implementation on the RDW are required, since cost gain will be marginal compared to the time needed to select and implement a new schema. However, if the query profile is "Evolution" or "Mixed", a new fragmentation schema $NewFS$ is computed. Then, if the $NewFS$ has a number of fact fragments that violates the constraint B, a new selection of fragmentation schema based on genetic algorithm is performed. Finally, the obtained fragmentation schema is implemented on the $\mathcal{RDW}$.

5   Portability under Oracle 11g

Our algebra operators are direclty deployed in Oracle11g. We choose Oracle11g, since it is the pionner that supports the referencial paritioning[2]. Due to lack of space, we can not detail here how operators are physically implement under Oracle 11g. The interested reader can refer to our technical document[3]. We give an example of the physical implementation of one Algebra's operator.

Since the physical operations depend on the Database Management System, we consider the $DBMS$ Oracle 11g[4]. Given a fragment P, let **Split(P, Ct)** the horizontal fragmentation of P into two disjoint fragments $P1$ and $P2$ according to the criteria Ct. The SQL syntax for this operation is:

```
ALTER TABLE table
  SPLIT PARTITION City VALUES ('Ct')
  INTO ( PARTITION table1, PARTITION table2 );
```

We define the operation $Identify\_Part(Ct)$, which identifies the fragment characterized by the predicates specified in Ct. Finally, we give the implementation of the Add_A operator.

$Add\_A(FS, A_i, \{SD_{j_1}^i, \cdots, SD_{j_p}^i\})$: adding an attribute A_i requires identifying the fragment(s) containing one or more sub-domains from the input set $\{SD_{j_1}^i, \cdots, SD_{j_p}^i\}$, then partition these fragments into two sub-fragments each, where the first sub-fragment is identified by the selection predicate A_i in the set $(SD_{j_1}^i, \cdots, SD_{j_p}^i)$ and the second one in is defined by the predicate A_i in $(Else_i)$. The corresponding algorithm is:

$Ens\_Part = Identify\_Part(A_i$ in $(SD_{j_1}^i, \cdots, SD_{j_p}^i))$
for each P in Ens_Part **do**
    Split(P, A_i in $(SD_{j_1}^i, \cdots, SD_{j_p}^i))$
end for

6   Experimentation under Oracle 11g

In order to evaluate our incremental selection based on Queries Profiling, we conduct experimental tests on a $\mathcal{RDW}$ schema of the APB1 benchmark[5] under the

---

[2] http://docs.oracle.com/cd/B10501_01/server.920/a96521/partiti.htm
[3] http://www.lias-lab.fr/~bellatreche/longversiondexa14.pdf
[4] Managing Partitioned Tables and Indexes
   http://docs.oracle.com/cd/B10501_01/server.920/a96521/partiti.htm
[5] http://www.olapcouncil.org/research/bmarkly.htm

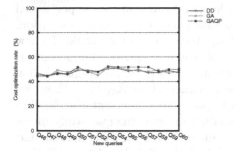

Fig. 2. Cost optimization rate

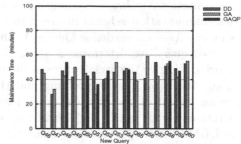

Fig. 3. Maintenance Time under Oracle11g

DBMS Oracle 11g. The $\mathcal{RDW}$ based on a star schema contains a facts table *Actvars* (24 786 000 tuples) and 4 dimension tables *Prodlevel* (9000 tuples), *Custlevel* (900 tuples), *Timelevel* (24 tuples) and *Chanlevel* (9 tuples). The genetic algorithm is implemented using the JAVA API *JGAP* (jgap.sourceforge.net). This study aim to evaluate the efficiency of the queries profiling performed using our fragmentation algebra. We conduct tests on a workload of 60 queries. The initial fragmentation schema of the $\mathcal{RDW}$ is obtained by a static selection using the first 45 queries of the workload with a constraint $B = 100$ (total number of final fragments). After that, we suppose that 15 new queries of different profiles (Neutral, Mixed, Evolution and Reduction) are successively executed on the $\mathcal{RDW}$. We perform two incremental selections: one based on the genetic algorithms that we presented in [2] GA and one based on the genetic algorithm with queries profiling that we name $GAQP$. We also implement an existing approach: the incremental fragmentation selection named DD based on the dynamic design of data warehouses proposed in [10] that we adapt in a centralized context. Figure 2 shows workload costs and optimization rates for each approach. Figure 3 shows maintenance times. According to Figure 2, the workload costs obtained by the three selections are similar, since all three are based on the horizontal fragmentation selection approach proposed in [1]. This shows queries profiling does not affect the quality of the selected schema. According to figure 3, we notice that the $GAQP$ selection gives a better maintenance time than the GA and DD selection (reduced by 52%). This is due to the fact that GA and DD select a new fragmentation schema after each new query. On the other hand, only 6 queries (mixed profiles) out of 15 trigger a new incremental selection for $GAPQ$. Therefore, according to the important parameter namely the maintenance time required to implement a new fragmentation schema on a partitioned $\mathcal{RDW}$, the $GAQP$ approach is better than the classic GA incremental selection and the existing approach DD.

7 Conclusion

This work deals with incremental selection of a horizontal data partitioning schema in the context of data warehouse modeled by a star schema. We propose a

Fragmentation Algebra containing all possible operations that can be performed on a fragmentation schema in order to take into account workload evolution. Using our Algebra, we define Queries Profiling. According to the profile of a new executed query, we determine if a selection of a new fragmentation schema is required. We give the architecture of the incremental selection of fragmentation schema based on queries profiling and the Fragmentation Algebra. Then, we give an insight of the physical operations required to implement the Algebra operations under Oracle 11g. Finally, we conduct an experimental study under the DBMS Oracle 11g to show efficiency of queries profiling. We showed that using queries profiling reduces by more than 50% the global maintenance time required to implement a new selection fragmentation schema on a partitioned *RDW*.

References

1. Bellatreche, L., Boukhalfa, K., Richard, P.: Referential horizontal partitioning selection problem in data warehouses: Hardness study and selection algorithms. International Journal of Data Warehousing and Mining 5(4), 1–23 (2009)
2. Bouchakri, R., Bellatreche, L., Faget, Z., Breß, S.: A coding template for handling static and incremental horizontal partitioning in data warehouses. Journal of Decision Systems (JDS) (to appear)
3. Ceri, S., Negri, M., Pelagatti, G.: Horizontal data partitioning in database design. In: ACM SIGMOD, pp. 128–136 (1982)
4. Curino, C., Jones, E.P.C., Popa, R.A., Malviya, N., Wu, E., Madden, S., Balakrishnan, H., Zeldovich, N.: Relational cloud: a database service for the cloud. In: CIDR, pp. 235–240 (2011)
5. Curino, C., Zhang, Y., Jones, E.P.C., Madden, S.: Schism: a workload-driven approach to database replication and partitioning. PVLDB 3(1), 48–57 (2010)
6. Derrar, H., Ahmed-Nacer, M., Boussaid, O.: Dynamic distributed data warehouse design. Journal of Intelligent Information and Database Systems 7 (2013)
7. Özsu, M.T., Valduriez, P.: Principles of Distributed Database Systems, 2nd edn. Prentice Hall International, Inc. (1999)
8. Papadomanolakis, S., Ailamaki, A.: AutoPart: Automating Schema Design for Large Scientific Databases Using Data Partitioning. In: SSDBM, pp. 383–392 (2004)
9. Sanjay, A., Narasayya, V.R., Yang, B.: Integrating vertical and horizontal partitioning into automated physical database design. In: ACM SIGMOD, pp. 359–370 (June 2004)
10. Tekaya, K.: Dynamic distributed data warehouse design. In: IRMA International Conference, pp. 1594–1598 (2007)

Multi-dimensional Sentiment Analysis for Large-Scale E-commerce Reviews

Lizhou Zheng[1,3], Peiquan Jin[1,3], Jie Zhao[2], and Lihua Yue[1,3]

[1] School of Computer Science and Technology,
University of Science and Technology of China, Hefei, China
[2] School of Business, Anhui University, 230601, Hefei, China
[3] Key Laboratory of Electromagnetic Space Information, Chinese Academy of Sciences, China
jpq@ustc.edu.cn

Abstract. E-commerce reviews reveal the customers' attitudes on the products, which are very helpful for customers to know other people's opinions on interested products. Meanwhile, producers are able to learn the public sentiment on their products being sold in E-commerce platforms. Generally, E-commerce reviews involve many aspects of products, e.g., appearance, quality, price, logistics, and so on. Therefore, sentiment analysis on E-commerce reviews has to cope with those different aspects. In this paper, we define each of those aspects as a dimension of product, and present a multi-dimensional sentiment analysis approach for E-commerce reviews. In particular, we employ a sentiment lexicon expanding mechanism to remove the word ambiguity among different dimensions, and propose an algorithm for sentiment analysis on E-commerce reviews based on rules and a dimensional sentiment lexicon. We conduct experiments on a large-scale dataset involving over 28 million reviews, and compare our approach with the traditional way that does not consider dimensions of reviews. The results show that the multi-dimensional approach reaches a precision of 95.5% on average, and outperforms the traditional way in terms of precision, recall, and F-measure.

Keywords: Sentiment analysis, E-commerce reviews, Dimension.

1 Introduction

Online E-commerce reviews are very useful for both customers and manufacturers. Almost all of the online trading websites, such as Amazon (http://www.amazon.com) and Taobao (http://www.taobao.com, the biggest E-commerce website in China), provide a reviewing system enabling customers to share their opinions towards one or more products. Those E-commerce reviews contain a lot of customers' opinions on certain products, and are helpful for other people to make decision before purchasing. Nevertheless, they are also helpful for the companies providing the products on the E-commerce platform, as the companies can know the public opinions on their products by conducting a sentiment analysis process on the online reviews.

Traditional sentiment analysis was usually performed in terms of a document-based granularity or a sentence-based granularity [1]. Their objective was to identify

H. Decker et al. (Eds.): DEXA 2014, Part II, LNCS 8645, pp. 449–463, 2014.

the positive/negative opinion expressed in the whole document or sentence. However, this approach is not suitable for the sentiment analysis on E-commerce reviews, as a piece of E-commerce review generally involves many aspects of product, such as appearance, quality, price, logistics, and so on. In this paper, we define each of these aspects as a dimension of a product. In general, a customer may be satisfied with some dimensions of a product but dislike some other features. Therefore, it is not suitable to conduct document-level or sentence-level sentiment analysis for E-commerce reviews. A better and more suitable way is to perform sentiment analysis on separated dimensions of products, i.e., to execute a multi-dimensional sentiment analysis on E-commerce reviews.

This paper is aimed at presenting a framework for multi-dimensional sentiment analysis on E-commerce reviews. Our study is based on a large scale of E-commerce reviews dataset consisting of 28 million of reviews. The challenges in multi-dimensional sentiment analysis lie in *dimensions mapping* and *sentiment word disambiguation*. The *dimension mapping* problem refers to mapping opinioned text blocks with right dimensions. For instance, the text block *"Nice look!"* will be mapped with the dimension *"appearance"*. The *sentiment word disambiguation* problem refers that a sentiment word may be connected with two or more dimensions. In this paper, we focus on those two problems and propose an effective framework to finally extract dimension-oriented opinions from E-commerce reviews. The major contributions of the paper are summarized as follows:

(1) We present a framework for multi-dimensional sentiment analysis on E-commerce reviews. In this framework, we propose effective ways to solve the dimension mapping and sentiment word disambiguation problems. (Section 3)

(2) We propose a new approach for removing the sentiment word ambiguity by introducing a sentiment lexicon expanding mechanism to construct the dimension-based sentiment lexicon, based on which a rule-based algorithm for multi-dimensional sentiment analysis is presented. (Section 4 and 5)

(3) We conduct experiments on a large-scale dataset including 28 million E-commerce reviews, and compare our algorithm with the traditional algorithm without dimensional consideration. The results in terms of precision, recall, and F-measure show that our algorithm is superior to the competitor algorithm. (Section 6).

2   Related Work

Opinion mining and sentiment analysis has been a hot research topic in recent years. Basically, sentiment analysis consists of five types of tasks [1], i.e., document-level sentiment analysis, sentence-level sentiment analysis, aspect-level sentiment analysis, comparative sentiment analysis, and sentiment lexicon acquisition. Different approaches have been proposed to realize the goals of those tasks. The first approach is called *semantic-based approach* [2, 8-12], which performs sentiment analysis on the basis of rules and sentiment lexicons, and the second approach is *machine-learning-based methods* [13-18], which regards sentiment analysis as a binary or multi-class classification task and uses common classification methods. Most of the previous works on sentiment analysis can be put in the above two types, but there are also some other works using different methods [3-7].

Aspect-based sentiment analysis is proposed to tackle with the cases when customers give opinions to several aspects of a product in one review. It aims to recognize all sentiment expressions within a given document or sentence and the aspects to which they refer. There are explicit and implicit features for each aspect in a review text [19]. Explicit features are words which are the aspect name or its synonyms, e.g., "外貌/appearance" and "外观/facade" can be regarded as explicit features for the aspect of "外观/appearance"). Implicit features are the words that do not refer to the aspect name or its synonyms. However, we can infer the exactly aspect through the implicit features. For example, given a review text "宝贝外观不错, 也很耐用, 就是非常贵. / The good looks nice and is durable as well but it's very expensive.", although the aspect name "appearance", "quality", and "price" do not appear in the review text, we can infer the aspects through analyzing the implicit feature words "looks", "durable", and "expensive". As to explicit aspects, one way to identify all explicit aspects in a corpus is to extract all noun phrases and keep those with a frequency over a given threshold [11, 19]. In order to find additional aspects, some researcher proposed to utilize the known sentiment expressions [20], and others used CRF [21, 22] or an association rules mining approach based on word co-occurrence to match implicit aspects with explicit ones.

Sentiment lexicon is almost the most important resource in opinions mining. How to construct sentiment lexicon has been a research focus in sentiment analysis. Generally, there are three types of approaches to constructing sentiment lexicon, namely the manual construction approach, the dictionary-based approach, and the corpus-based approach. The manual construction approach depends on some experts to define sentiment words and therefore is time-consuming and cannot suit various situations in different applications. In the dictionary-based approach, some widely-used word dictionary such as WordNet [23] is used to construct the sentiment lexicon. Those dictionaries contain many semantic relations among tokens, such as synonyms, antonyms, and hyponyms, which are helpful to determine the sentiment words. For example, in the literature [24], the synonym and antonym relations were used to extract sentiment words in which they assumed that synonyms have the same sentiment polarity. There are a lot of other works focusing on the dictionary-based approach for sentiment lexicon construction [25, 26]. For the third corpus-based approach, a large corpus is used in a training process to find a set of sentiment words [27-32]. Those words are then used as seeds to find more sentiment words and finally expect to make up a relatively complete sentiment lexicon. In 1997, the corpus-based idea for sentiment lexicon construction was first presented [27]. They used sentiment seeds and pre-defined conjunctions to identify more sentiment words. In [28], the semantic similarity between sentiment seeds and candidate sentiment word was introduced in sentiment lexicon construction.

3   General Framework

3.1   System Architecture

Figure 1 shows the framework of the multi-dimensional sentiment analysis on E-commerce reviews. Given a raw E-commerce review, we first conduct Part-of-Speech tagging as well as some other textual preprocessing on it. Next, we split the text into several text blocks by recognizing punctuations and employing syntactic analysis. The reason for employing this step is that a review sentence may be quite long, and each part of a long sentence (normally separate by punctuations like " , ", " 。 " or space) may contain attitudes of a customer towards different aspects of a product. By splitting a long sentence into several short ones, it is more efficiently and easier for us to map those short sentences into corresponding dimensions, and perform dimension-oriented sentiment analysis. After that, we map each of the short sentences into a proper dimension respectively. Here we use keyword matching as well as sentence similarity measurement to determine the dimension for each short sentence. Finally, we employ sentiment analysis on each short sentence under a specific dimension with the support of a dimensional sentiment lexicon. For the dimensional sentiment lexicon construction, we design a sentiment word expanding method to expand the sentiment word set and realize sentiment words disambiguation. It is an important part in our method since the quality of sentiment words affects a lot on the effectiveness of sentiment analysis.

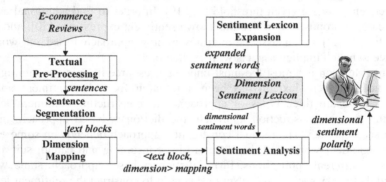

Fig. 1. The general framework of multi-dimensional sentiment analysis

3.2   Textual Pre-processing

In this step, we use NLPIR (http://nlpir.org/), a Chinese Part-of-Speech tagging toolkit, for both sentences partition and Part-of-Speech tagging task. Note that since many of the product reviews are short in length and contain new words, the precision of sentence partition and Part-of-Speech tagging is relatively low compared with its performance in normal texts. To alleviate the impact of this problem, we conduct some processing to improve the performance of sentence partition and Part-of-Speech tagging. Detail of our preprocessing is shown in Table 1.

Table 1. Textual pre-processing

Category	Description
User Dictionary	Manually collect online new words from Sogou Dictionary.
Word Combination	Combine consecutive entities ("nr"/ "person name", "ns"/ "location", "nt"/ "organization", etc.)
	Combine consecutive morphemes ("vg"/ "verb morphemes", "ng"/ "noun morphemes", etc.)
	Combine numeral and quantifier ("m", "q")
Stop Word	Remove stop words in the text.

3.3   Sentence Segmentation

The sentence segmentation procedure is used for splitting a long text into small text blocks, which is helpful for mining the opinions towards different dimensions. After a careful examination on the raw E-commerce reviews, we find that the raw E-commerce reviews can be generally divided into two types, as shown in Table 2. The first type includes reviews with a regular format, and each text block is separate by punctuations (mostly ", ", " 。 ", " ! ", and space). We simply split this type of reviews by recognizing punctuations. The other type contains no punctuations and spaces. We conduct a syntactic parser tree using Stanford Parser (http://nlp.stanford.edu) and extract each minimum subordinate clause as a text block.

Table 2. Examples of long sentences

Type	Example Sentence
With punctuations	蛮好看的，卖家发货很快很迅速。但是这次这个好像有点味道。(It looks nice, the shopkeeper makes a delivery very quickly, but this time it smells a little strange.)
Without punctuations	衣服很漂亮价钱也便宜我很喜欢 (The cloth is pretty good. The size is OK. The logistic is very fast as well.)

3.4   Dimension Mapping

Dimensions are typically domain-dependent. In this paper, the dimensions are predefined by the operating department of the E-commerce company. There are totally 570 predefined dimensions in our dataset, which covers almost all the aspects of different product categories since a dimension may appear in many categories (e.g., the dimensions "*质量/quality*" and "*商家服务/service of shopkeeper*" appear in almost all of the categories). For a given product review, we could obtain its category id directly and for a specific category, the candidate dimensions is provided in our dataset (the average count of candidate dimension is around 10). Thus the dimension mapping task could be regard as a multi-class classification problem. In our experiment, the dataset also provides some predefined key words for each dimension, which can be taken as explicit features in dimension mapping.

The simplest and most precise way for dimension mapping is to find key word in a short text and map the text into the dimension which the key word stands for. But since the initial key words are only a small part of feature for a dimension, for many short texts which do not contain a key word, this simple key word matching approach is no longer available. Here we use the co-occurrence information between words in short texts and initial key words to perform dimension mapping for short texts which contain none of the key words, since some implicit key words always appears together with one or more explicit key words. For example, "*expensive*" and "*cheap*" always appear with initial key word "*price*" in Chinese, thus even if those two words are not initial key words, we could infer that they are implicit key words of category "*price*" through their co-occurrence frequency with key word "*price*".

Given a short review sentence s, we can obtain its category c and the dimension set D_c of category c directly. For each word w_i in the short review sentence s and a dimension $d_j \in D_c$, along with d_j's initial key word set KW_j, we assign a probability score $wprob_{i,j}$ which describe the probability of word w_i belongs to category d_j (as shown in Equation 3.1).

$$\mathrm{wprob}_{i,j} = freq(w_i, d_j) * \sum_{kw_k \in KW_j} \frac{freq(w_i, kw_k, d_j)}{\sum_{d \in D} freq(w_i, kw_k, d)} \tag{3.1}$$

Here $freq(w_i, kw_k, d_j)$ means the frequency of word w in short sentence s appearing together with key word kw_k in dimension d_j. The former item is the frequency of word w_i appears in dimension d_j. And the latter item in the equation is a classical measurement when calculating co-occurrence between words. Here we omit the frequency of key word kw_k, i.e., $freq(kw_k)$, in the denominator, since for each word w_i in short sentence s and dimension d_j, $freq(kw_k)$ would have the same value.

And the probability of a short sentence s belonging to a dimension d_j, $sprob_{s,j}$, is calculated as the product of the probability of all words in sentence s belonging to dimension d_j (as shown in Equation 3.2).

$$sprob_{s,j} = \prod_{w_i \in s} wprob_{i,j} \tag{3.2}$$

Note that here we only retain nouns (tagged with "n.*"), verbs (tagged with "v.*"), and adjectives (tagged with "a.*") in a short sentence to calculate those probabilities, since other words contain few co-occurrence information and may bring in noise. Finally we map a short sentence s into a dimension d_j which has the maximum probability score with s.

$$d_j = \{d_j \mid \max_j(sprob_{s,j})\} \tag{3.3}$$

For some sentences of which the sentence probabilities under all dimensions are zero, we could not map them into a dimension individually. Therefore, we combine the sentence with its former one in the short sentence sequence obtained by sentence splitting process. The number of such sentences is quite small, so it affects little to our final sentiment analysis result.

4   Sentiment Lexicon Expansion

Sentiment lexicon is a crucial resource for both rule-based methods and machine learning-based methods on sentiment analysis. There are public sentiment word resources available in the Web, and the most widely used sentiment word lexicon in Chinese is HowNet Chinese polarity lexicon, which contains 4,500 positive sentiment words and 4,300 negative sentiment words. Another sentiment lexicon is NTUSD sentiment lexicon, which contains 2,800 positive and 8,200 negative sentiment words. Only employing those sentiment lexicons may face two typical problems when performing sentiment analysis on product reviews. The first problem is that different dimensions of a product may contain different sentiment words. The two public lexicons mentioned above could only cover a small part of the sentiment words appearing in a specific dimension. Besides, one sentiment word may represent different polarities under various dimensions. For example "硬/ hard, solid" is a positive word when modifying building materials but could be a negative word when describe food. Both HowNet and NTUSD lexicons are not able to distinguish different polarities under different dimensions. This brings difficulty for sentiment analysis even if the text is short and contains only one aspect of a product. Therefore, we employ a semi-auto mechanism to enlarge sentiment words under each category. By employing this step, we can significantly improve the precision of sentiment analysis.

Algorithm. *Expand_DimSentiLexicon()*

Input: (1) set of dimension $D = \{D_1, D_2, ..., D_d\}$
      (2) set of seed words $SWS = \{SW_1, SW_2, ..., SW_d\}$, where SW_d represents the seed set of D_d.
      (3) the corpus under each dimension $DCorpus = \{DCorpus_1, DCorpus_2, ..., Dcorpus_d\}$.
      (4) seed word co-occurrence score a, word frequency word score β.
Output: the dimensional sentiment lexicon $DSenti_d$ for each of the dimension d.
Preliminary: n is the count of dimension.

1: **for** each dimension D_d in D **do**
      /* *The seed word-based procedure* */
2:    **while true**
2:      **for** each word w_i in corpus $DCorpus_d$ **do**
3:        Calculate the co-occurrence score sc_i of w_i;
4:        Create word set $TWS = \{w_1, w_2, w_3\}$ where w_j is word with top-3 sc and $sc > a$;
5:      **if** $|TWS| == 0$ or $|D_d| > 20$ **then** break;
6:      **else** $SW_d = SW_d \cup TWS$;
7:    **end while**
      /* *The frequency filtering-based procedure* */
8:    **for** each word w_i in corpus $DCorpus_d$ **do**
9:      Calculate the $DFICF_{i,d}$ of w_i;
10:   Create word set $FWS = \{w_1, w_2, ..., w_{20}\}$ where w_j is word with top-20 $DFICF$ and $DFICF_{j,d} > \beta$;
      /* *Manually examine* */
11:   Create candidate sentiment word set $CSet_d = SW_d \cup FWS$;
12:   Manually examine each word in $CSet_d$ and form lexicon $senti_d$;
13: $DSenti = \{dsenti_1, dsenti_2, ..., dsenti_d\}$;
14: **return** $DSenti$;

End *Expand_DimSentiLexicon*

Fig. 2. Algorithm for sentiment lexicon expansion

Our algorithm of constructing dimensional sentiment lexicon contains two parts, as shown in Fig. 2. For each dimension, there are two procedures to extract the final sentiment lexicon of the dimension. One is based on seed sentiment words and another is based on filtering high frequency words. Note that we only retain nouns, verbs, adjectives for candidate sentiment words in our method.

The seed-word-based procedure starts with some manually collected seed sentiment words which could be of the same polarity under different dimensions (e.g., positive words like "好/good" and "不错/nice", negative words like "糟糕/terrible" and "惨不忍睹/too horrible to look at", etc.). Based on the co-occurrence information with seed sentiment words, we could enlarge the original seed words iteratively, in each round of iteration we put top-3 words with the highest co-occurrence score into the seed word set. The procedure ends when the size of seed word set is larger than 20 or the co-occurrence score of each *Top-K* words are below a threshold. The co-occurrence score in this procedure is defined as sc_i for a word w_i (as shown in Equation 4.1).

$$sc_i = \frac{freq(w_i, sw)}{freq(w_i)} \quad (4.1)$$

At the end of the iteration, all the words in the seed word set are put into the candidate sentiment word set. And the polarities of the words are the same with the seed word.

Another procedure of our method is based on filtering high frequency word. Here we calculate the *DF*ICF* score (similar to the classical *TF*IDF* measurement [32], here *DF* means dimension frequency, and *ICF* means inverse corpus frequency) under a specific dimension d for each word in d. The *DF*ICF* score of a word w_i under d is defined as Equation 4.2.

$$DF * ICF_{i,d} = \frac{\mid sentence_{i,d} \mid}{\mid sentence_d \mid} * \log \frac{\mid sentence_{corpus} \mid}{\mid sentence_{i,corpus} \mid} \quad (4.2)$$

Here $\mid sentence_{i,d} \mid$ is the count of sentence contains word w_i in dimension d, and $\mid sentence_d \mid$ is the count of sentence under dimension d. $\mid sentence_{corpus} \mid$ is the count of sentence under the whole corpus, and $\mid sentence_{i,corpus} \mid$ is the count of sentence contains word w_i in the whole corpus. For each dimension, we put the *Top*-20 words with highest *DF*ICF* scores into the candidate sentiment word set. To maximize the precision of sentiment word extraction since the quality of sentiment word is crucial for short sentence sentiment analysis, we import some manual work at the last of the sentiment word extraction work. Here we manually examine the correctness of polarity of each candidate word obtained from seed-word-based procedure and label the polarity of each candidate word obtained from filtering-based procedure. It a relative tough task but we do not think it is a trivial work since an incorrectness of a sentiment word may lead to tens or hundreds of errors in the sentiment analysis work.

5   Sentiment Analysis

The basic idea in this paper to perform sentiment analysis on product reviews is to conduct dimension-oriented analysis. Since the text blocks after sentence splitting process are very short in length and only contain customer opinions towards one aspect (dimension) of a product, we employ a directly method which is based on rules and sentiment lexicon. The difference of our work compared to previous works is that we combine both dimension sentiment lexicon and public sentiment lexicon together. Besides, we use multiple clause-rules to improve the performance of sentiment analysis. The detailed algorithm for sentiment analysis is shown in Fig. 3. There are three steps in our method, namely *Strong Rule Recognition*, *Sentiment Word Polarity Recognition*, and *Adversative Clause Recognition*, which are described as follow.

Algorithm. *Sentiment_Analysis (s)*

Input: (1) Short sentence *s* with its dimension *d*.

    (2) Dimensional sentiment lexicon *DSenti* = {*dsenti₁, dsenti₂, ..., dsentid*} and public sentiment lexicon *PSenti*.

    (3) Strong rule set *RSet* = {<*rule₁, pol₁*>, <*rule₂, pol₂*>, ..., <*ruleₙ, polₙ*>} where <*rulei, poli*> indicates the *i*-th rule and its polarity.

Output: the sentiment polarity of *s*.

/* *Strong rule recognition* */

1:   **for** each <*rulei, poli*> in *RSet* **do if** *s* matches *rulei* **then return** *poli*;

/* *Sentiment word polarity recognition* */

2:   **for** each word *wi* in sentence s **do**

3:      **if** *wi* in *dsentid* **then** //match dimensional sentiment lexicon

4:         get *pol(wi)* by dimensional sentiment lexicon;

5:      **else if** *wi* in *PSenti* **then** //match public sentiment lexicon

6:         get *pol(wi)* by public sentiment lexicon;

7:      **if** negative word exists in the previous three words of *wi* **then** *pol(wi)* = - *pol(wi)*;

8:      **if** degree word exists in the previous one word of *wi* **then** *pol(wi)* = *degree_weight* * *pol(wi)*;

9:   **end for**

10: *pol(s)* = Σ*pol(wi)*;

/* *Adversative clause recognition* */

11: **if** sentence *s* matches adversative clauses **then** *pol(s)* = - *pol(s)*;

12: **return** *pol(s)*

End *Sentiment_Analysis*

Fig. 3. Algorithm for short sentence sentiment analysis

5.1   Strong Rule Recognition

Table 3. Examples of strong rules

Polarity	Example Rules
positive	希望...保持/继续/一直...。 （hope that ... keep/always ...） 与... 想象的一样。(it's the same ... as imagine)
negative	除了...有点/有些/有一点...。(it's ... except for a little ... in ...) 有些/有点过于...了。(it's excessively too ...)
neutral	不知道/不知 ... 是否/会不会 ... (I don't know whether ...)

There are strong rules through which we could imply the polarity of a short sentence in high precision. The strong rules show the same polarity under every dimension. So at first, we directly obtain the polarity of some short sentences by matching manually collected rules, examples of the rules are listed in Table 3.

5.2   Sentiment Word Polarity Recognition

In this step we recognize sentiment word polarity using sentiment lexicon as well as negation words and degree words in a short sentence. In our work, there are two lexicons available, one is the public sentiment lexicon HowNet Chinese Sentiment

Lexicon and NTUSD Chinese Sentiment Lexicon, and another is the dimensional sentiment lexicon we constructed. In our method, the dimensional sentiment lexicon has a high priority towards public sentiment lexicon (HowNet and NTUSD). Given a short text and its dimension name, we first examine the dimension sentiment lexicon and then public lexicon to see if any word in the sentence appears in the lexicons. For a word which appears in both of the lexicons, we take the polarity in dimension sentiment lexicon as the polarity of the word.

Some sentiment words may be prefixed by negation words, and thus the polarity of the words should be inversed in that case. Here we examine the previous words of a sentiment word with a window of size=3 in which no other sentiment word exists to see if there exists negation words.

The degree word in front of a sentiment word describes the intensity of polarity. We employ the HowNet Chinese Degree Phrase Set to measure this intensity. The HowNet Chinese Degree Phrase Set contains 219 degree phrases and those phrases are divided into six levels based on intensity, the six levels are: "极其/ extreme, 最 /most"(with 69 words), "很/ very" (with 42 words), "较/ more" (with 37 words), "稍/ a little" (with 29 words), "欠/ insufficiently" (with 12 words) and "超/ over" (with 30 words)。 We assign weights for those degree phrases based on their intensity level.

After all, for a short sentence s, we calculate the linear weighted sum of the sentiment words appearing in s, and the polarity of sentence s is based on the sum value (with positive polarity when above zero, negative polarity when below zero and neutral polarity when equal zero).

5.3   Adversative Clauses Recognition

In the case of adversative clause, where transition words like "but", "however", "nevertheless" in English and "但是", "然而", "除了" in Chinese appears, the polarity of a sentence should be inversed since customer is expressing an opposite opinion using those transition words. In our work we collected transition words as well as rules to recognize those adversative clauses. For a short sentence containing such words or rules, the polarity we obtained from previous step will be reversed.

6   Performance Evaluation

6.1   Dataset

To illustrate our method more clearly, here we make a description of our dataset. The real dataset we use for evaluation is obtained from online E-Commerce website. The dataset contains 28 million user reviews. Each of the review belongs to a category (e.g., the same with the category of the product) and the category of each review is known in our dataset. There are totally more than 8000 categories. For each category, there are multiple dimensions which describe different aspects of the products in that category. There are 570 dimensions in our dataset, each dimension may exist in more than one categories. The dimension list of each category is also available in our dataset. For example, a product "2134231" (suppose it is a T-shirt) belongs to only one category "5432", the category "5432" contains five dimensions "price", "appearance", "material", "style", and "quality". The dimension "price" could be appear in hundreds of categories including category "5432", the other four categories is of similar format. For each of the dimension, there are several key words which describe the dimension (aspect of products). Those key words could be regard as seed words in the dimension mapping task. The data structure of our dataset is shown in Fig. 4 and Table 4.

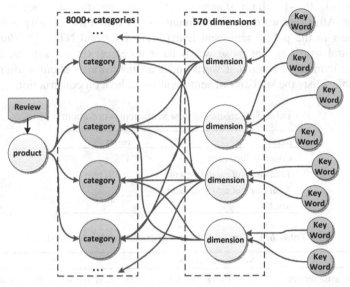

Fig. 4. Relation of *Product*, *Category* and *Dimension* in the dataset

Table 4. Attributes describing the dataset

Name	Type
review_id	*Long Integer*
product_id	*Long Integer*
category_id	*Long Integer*
dimension_name	*String*
content	*String*

So given a customer review along with its product id, we first get the category of the review, and then we get the dimension set of the category. In the dimension step we map each short sentence obtained from sentence splitting step into one of the dimensions in the dimension set and finally adopt sentiment analysis on each short sentence under a dimension.

6.2 Dimensional Sentiment Lexicon Construction

By performing dimensional sentiment lexicon construction, we tackle the problem of sentiment word ambiguity. There are two procedures in the process of lexicon construction. One is seed word-based approach and another approach is based on filtering high frequency words. For each of the method, we obtained candidate sentiment word set which has a maximum size of 25. Since the two candidate word set may have duplicated words, the number of words in the combination candidate sentiment word set is below 40. As shown in Table 5, the average size of the candidate sentiment word set is around 13. In order to maximize the precision of the sentiment words extracted from the corpus, we perform manual work on the final candidate sentiment word set. Here we examine or label the polarity of candidate words into +1, 0, -1 which stands for positive, neutral and negative polarity respectively. After that we discard sentiment words which the polarity is neutral and do not appear in the public sentiment lexicon HowNet and NTUSD. The reason to discard neutral words is that those words have no effect on our sentiment analysis. The average number of sentiment word under a dimension is about 4 after manually labeled. Table 5 lists the statistics of sentiment word lexicon construction.

Table 5. Statistics of new sentiment word count

items	count
Count of dimension	423
Total count of candidate word	5605
Total count of new sentiment word	2218

Table 6. Examples of new sentiment words expanded

Dimension	New Sentiment Words
物流服务/ logistic service	快速 (fast), 特快 (very fast), 火箭 (rocket), 蜗牛 (snail), etc.
商家服务 / manufacturer service	没得说 (no problem), 令人感动 (impressive), 尽职尽责 (responsible), etc.
质量/ quality	有保证 (guaranteed), 旧旧的 (old), 料足(full weight), 没得说 (no problem), etc.
款式/ style	卡哇伊 (kawaii), 百搭(matching everything), 好搭配 (easy to match), 潮 (fashion), etc.
尺码/ size	显肥 (look fat), 显胖 (look fat), 偏窄 (a bit narrow), 偏长 (a bit long), etc.

Note that the count of dimension in Table 5, i.e., 423, is less than the total count of dimension in our dataset (570). That is because in some dimensions we cannot get even one candidate word since the frequency of words is below the threshold as we describe in Section 5.1. In Table 6, we list some of the sentiment words we extract under those three dimensions. Note that some new sentiment words appear in more than one dimension, which is reasonable since some sentiment words may modify several aspect of a product.

6.3 Sentiment Analysis

In order to measure the performance of sentiment analysis, we randomly sample 9,000 results of sentiment analysis on short texts to show the performance of our method. We compare our method with the approach which does not contain the dimensional sentiment word lexicon (only utilize the HowNet and NTUSD lexicon), and the results show that the method of employing dimensional sentiment lexicon can raise the *precision*, *recall* and *F-measure* for each type of polarities (*positive*, *negative*, and *neutral*). Figure 5 to 7 shows the precision, recall, and F-measure on the sampling 9,000 short reviews.

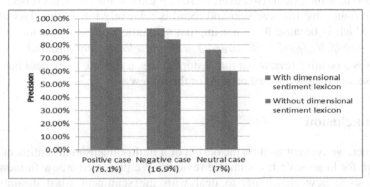

Fig. 5. Precision of our method and the algorithm without dimensional sentiment lexicon

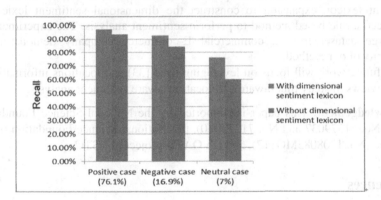

Fig. 6. Recall of our method and the algorithm without dimensional sentiment lexicon

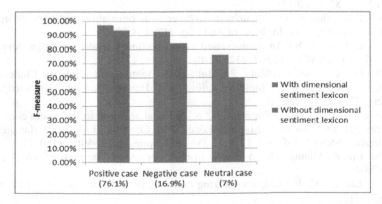

Fig. 7. F-measure of our method and the algorithm without dimensional sentiment lexicon

The result of the sentiment analysis shows that our approach which employs the dimensional sentiment lexicon performs better than that of do not contains a dimensional sentiment lexicon. That proves our method is able to solve the problem of sentiment word ambiguity when describing different aspects of a product (dimension) to some extent. However, there still exist some problems reflected by the result. We found that the precision and recall is relatively low in the class of neutral reviews, which is because there are implicit semantic express by customer. For the review "老公穿着跟模特一样 *Dressed with the clothes, my husband looks like a model*", it is a positive review but the customer uses a rhetoric to express her opinion. In this case, our method labeled neutral for the review.

7   Conclusion

In this paper, we concentrated on the design of a multi-dimensional sentiment analysis framework for large-scale E-commerce reviews. We presented a new framework and proposed some new algorithms to deal with the sentiment word disambiguation problem as well as the sentiment polarity annotation issue. In particular, we used sentiment lexicon expansion to construct the dimensional sentiment lexicon, and introduced a rule-based method to perform sentiment analysis. The experiments on a real large dataset from a commercial E-commerce company demonstrated the superiority of our method.

Our future work will focus on leveraging time [33] and locations information [34] in the reviews to study time-aware and location-aware sentiment analysis.

Acknowledgement. This paper is supported by the National Science Foundation of China (No. 61379037 and No. 71273010), the National Science Foundation of Anhui Province (No. 1208085MG117), and the OATF project in USTC.

References

1. Feldman, R.: Techniques and applications for sentiment analysis. Communications of the ACM 56(4), 82–89 (2013)
2. Turney, R.D.: Thumbs up or thumbs down?: semantic orientation applied to unsupervised classification of reviews. In: Proc. of ACL, pp. 417–424 (2002)
3. Brody, S., Elhadad, N.: An unsupervised aspect-sentiment model for online reviews. In: Proc. of HLT-NAACL, pp. 804–812 (2010)
4. Fu, X., Liu, G., Guo, Y., Wang, Z.: Multi-aspect sentiment analysis for Chinese online social reviews based on topic modeling and HowNet lexicon. Knowledge-Based Systems 37, 186–195 (2013)
5. Hogenboom, A., Boon, F., Frasincar, F.: A statistical approach to star rating classification of sentiment. In: Casillas, J., Martínez-López, F.J., Corchado, J.M. (eds.) Management of Intelligent Systems. AISC, vol. 171, pp. 251–260. Springer, Heidelberg (2012)
6. Hu, M., Liu, B.: Mining opinion features in customer reviews. In: Proc. of AAAI, pp. 755–760 (2004)
7. Hu, M., Liu, B.: Mining and summarizing customer reviews. In: Proc. of KDD, pp. 168–177 (2004)
8. Xu, J., Ding, Y., Wang, X., Wu, G.Y.: Identification of Chinese finance text using machine learning method. Proc. of SMC, 455–459 (2008)

9. Hu, M., Liu, B.: Opinion extraction and summarization on the web. In: Proc. of AAAI, pp. 1621–1624 (2006)
10. Ding, X., Liu, B., Yu, P.S.: A holistic lexicon-based approach to opinion mining. In: Proc. of WSDM, pp. 231–240 (2008)
11. Popescu, A., Etzioni, O.: Extracting product features and opinions from reviews. In: Natural Language Processing and Text Mining, pp. 9–28. Springer, London (2007)
12. Yu, H., Hatzivassiloglou, V.: Towards answering opinion questions: Separating facts from opinions and identifying the polarity of opinion sentences. In: Proc. of EMNLP, pp. 129–136 (2003)
13. Mullen, T., Collier, N.: Sentiment analysis using support vector machines with diverse Information sources. In: Proc. of EMNLP, pp. 412–418 (2004)
14. Pang, B., Lee, L., Vaithyanathan, S.: Thumbs up?: sentiment classification using machine learning techniques. In: Proc. of ACL, pp. 79–86 (2002)
15. Rushdi-Saleh, M., Martín-Valdivia, M.T., Ráez, A.M., López, L.A.U.: Experiments with SVM to classify opinions in different domains. Expert Systems with Applications 38(12), 14799–14804 (2011)
16. Zhang, Z., Ye, Q., Zhang, Z., Li, Y.: Sentiment classification of Internet restaurant reviews written in Cantonese. Expert Systems with Applications 38(6), 7674–7682 (2011)
17. Pang, B., Lee, L.: Opinion mining and sentiment analysis. Foundations and Trends in Information Retrieval 2(1-2), 1–135 (2008)
18. Whitelaw, C., Garg, N., Argamon, S.: Using appraisal groups for sentiment analysis. In: Proc. of CIKM, pp. 625–631 (2005)
19. Liu, B.: Sentiment analysis and subjectivity. In: Handbook of natural language processing (2010)
20. Wu, Y., Zhang, Q., Huang, X., Wu, L.: Phrase dependency parsing for opinion mining. In: Proc. of EMNLP, pp. 1533–1541 (2009)
21. Lafferty, J., McCallum, A., Pereira, F.: Conditional random fields: Probabilistic models for segmenting and labeling sequence data. In: Proc. of ICML, pp. 282–289 (2001)
22. Jakob, N., Gurevych, I.: Extracting opinion targets in a single-and cross-domain setting with conditional random fields. In: Proc. of EMNLP, pp. 1035–1045 (2010)
23. Fellbaum, C.: WordNet: An electronic lexical database, http://www.cogsci.princ
24. Kim, S.M., Hovy, E.: Determining the sentiment of opinions. Proc. of ACL 1367 (2004)
25. Esuli, A., Sebastiani, F.: Determining the semantic orientation of terms through gloss classification. In: Proc. of CIKM, pp. 617–624 (2005)
26. Rao, D., Ravichandran, D.: Semi-supervised polarity lexicon induction. In: Proc. of ACL, pp. 675–682 (2009)
27. Hatzivassiloglou, V., McKeown, K.R.: Predicting the semantic orientation of adjectives. In: Proc. of ACL, pp. 174–181 (1997)
28. Turney, P., Littman, M.: Measuring praise and criticism: Inference of semantic orientation from association. ACM Transactions on Information Systems 21(4), 315–346 (2003)
29. Ding, X., Liu, B.: Resolving object and attribute coreference in opinion mining. In: Proc. of ACL, pp. 268–276 (2010)
30. Qiu, G., Liu, B., Bu, J., Chen, C.: Opinion word expansion and target extraction through double propagation. Computational Linguistics 37(1), 9–27 (2011)
31. Li, F., Pan, S.J., Jin, O., Yang, Q., Zhu, X.: Cross-domain co-extraction of sentiment and topic lexicons. In: Proc. of ACL, pp. 410–419 (2012)
32. Jin, P., Li, X., Chen, H., Yue, L.: CT-Rank: a time-aware ranking algorithm for web search. Journal of Convergence Information Technology 5(6), 99–111 (2010)
33. Zhao, X., Jin, P., Yue, L.: Automatic Temporal Expression Normalization with Reference Time Dynamic-Choosing. In: Proc. of COLING, pp. 1498–1506 (2010)
34. Zhang, Q., Jin, P., Lin, S., Yue, L.: Extracting Focused Locations for Web Pages. In: Wang, L., Jiang, J., Lu, J., Hong, L., Liu, B. (eds.) WAIM 2011. LNCS, vol. 7142, pp. 76–89. Springer, Heidelberg (2012)

The Entity-Flow Perspective in Business Process Models

Giorgio Bruno

Politecnico di Torino, Torino, Italy
giorgio.bruno.@polito.it

Abstract. The research reported in this paper is grounded on the principles of the artifact-centric approach and stresses the importance of the entity-flow perspective in business process models with the help of a notation named ACTA. ACTA process models are based on a number of entity-flow patterns, which help readers understand the operation of the process before looking at the task descriptions. In particular, these patterns emphasize choice situations such as those related to the grouping or matching of input entities.

Keywords: business processes, information models, entity flow, life cycles, tasks, choices.

1   Introduction

The workflow perspective has permeated the definition of most notations for business process models as well as the implementation of the software support systems. The emphasis is put on the ordering of the units of work (called tasks), which is achieved through a detailed control flow; when a task has been completed, the control flow determines the next one to be performed. This viewpoint fits repetitive, standardized processes, whose models look like flow charts: the process is an orchestrator of tasks whose implementation is outside the scope of the workflow system. For example, in BPMN [4], the process elements representing tasks have the purpose of exchanging input and output parameters with external activities. The input and output parameters are mapped to process variables.

In the domain of information systems, whose purpose is to enable human participants and/or machines to perform work using information [2], workflow models appear as technological tools and lack strong links with the ontological items, i.e., the tasks and the business entities.

A recent line of research has led to the development of the artifact-centric approach: it emphasizes the business entities, called artifacts, which represent concrete and self-describing chunks of information used to run a business [14]. Processes are structured around the artifact life cycles, which show how the artifacts evolve over time by means of transitions from one state to another. The business activities, which are responsible for the state transitions, may be introduced in a subsequent step of analysis along with the rules governing their execution.

The research reported in this paper is grounded on the principles of the artifact-centric approach and stresses the importance of the entity-flow perspective in business

H. Decker et al. (Eds.): DEXA 2014, Part II, LNCS 8645, pp. 464–471, 2014.

process models with the help of a notation named ACTA. The entity flow encompasses the entities flowing through the process and acted on by its tasks.

In ACTA, the entity flow plays a major role in the activation of tasks: in fact, the availability of suitable input entities is an essential condition for their execution. On the contrary, in the workflow perspective, the control flow, which is based on the completion events of tasks, is the primary means for the activation of tasks.

This shift of focus is needed in situations that are not adequately supported by the functionalities of typical workflow systems, although they are very common in industrial applications [15]. An example taken from build-to-order processes is the many-to-many mapping between customer orders and supplier orders. Customer orders may be distributed over a number of supplier orders and a supplier order may be associated with a number of customer orders. The placement of a supplier order is not directly caused by the placement of a customer order: in fact, an officer of the purchasing department may wait until a number of customer orders have been placed so as to collect them in one supplier order.

A further advantage of the entity-flow perspective is the possibility of providing high-level descriptions of tasks, which indicate their effects and the constraints that affect their execution in a declarative way based on post-conditions and pre-conditions, respectively. To this end, an information model describing the structure of entities along with their relationships and attributes is added to the process model.

ACTA process models come in two flavors: compact models and compound ones. In a compact model the life cycles of all the entities involved are integrated. This monolithic representation makes the coordination of life cycles easier to represent. On the contrary, a compound model is made up of separate life cycles with synchronization points. This paper focuses on compact models while compound ones have been illustrated in a previous paper [5].

ACTA process models are based on a number of entity-flow patterns, which help readers understand the operation of the process before looking at the task descriptions. In particular, these patterns emphasize choice situations such as those related to the grouping or matching of input entities.

The organization of this paper is as follows. Section 2 presents an overview of ACTA along with the process model of a build-to-order process; section 3 introduces a number of entity-flow patterns and section 4 explains the role of companion information models. Then, sections 5 and 6 present the related work and the conclusion, respectively.

2   Overview of ACTA

This section presents an overview of the ACTA notation with the help of a simplified build-to-order process named HandleOrders whose requirements are as follows.

The process is run by a selling organization and enables customers to place customer orders and suppliers to fulfill supplier orders. Each customer is served by a staff member called account manager. Customer orders are made up of lines and each line refers to a product type. In the information system of the organization, product types are associated with the suppliers able to provide them.

An account manager may reject a customer order or may handle it by associating its lines with compatible open supplier orders. An open supplier order is generated by an account manager, who may close it after collecting a number of customer order lines. Account managers may also cancel an open supplier order as long as it is empty. Closed supplier orders are fulfilled by the intended suppliers. An account manager can fulfill a customer order when all the supplier orders including some of its lines have been fulfilled. Customers are notified of the rejection or fulfillment of their orders.

The complexity of the process is due to the many-to-many mapping between customer orders and supplier ones.

An ACTA model is made up of two components, i.e., a process model and a companion information model. The process model of the example is shown in Fig.1; it integrates the life cycles of three entity types: COrder (Customer Order), SOrder (Supplier Order) and Line.

Process models consist of tasks, state nodes and connections. Tasks are depicted as rectangles and state nodes as circles; connections are oriented links drawn from tasks to state nodes or from state nodes to tasks.

The annotations in Fig.1 point out the entity-flow patterns that characterize ACTA models; they are discussed in the next section.

State nodes and connections define the entity flow of the process. State nodes represent entities of a specific type in a given state: the first label is the type name and the second is the state name. The initial state is named i. A state node (or simply node) is referred to with the type name followed by the state name between parentheses, e.g. COrder (i).

In general, state nodes (tasks) may have several input and output links, which connect them to their input and output tasks (state nodes). The input (output) types of a task are the types of its input (output) nodes.

The input state nodes of a task indicate which entities may cause its execution. Tasks may take one or more entities from each input state node and may deliver one or more entities to each output state node; the number of entities is shown next to the corresponding link and is called weight of the link. The default weight is 1 and is omitted. Weight n indicates that the number of entities is not predefined but is determined during the execution of the task.

Tasks are divided into human tasks and automatic ones; the former are told apart because they have a role label. The role label indicates the role of the process participant who is entitled to perform the task; the actual participant is called task performer and may be specific or generic. In process HandleOrders, participants are divided into three roles: customers, suppliers and account managers. The process has no automatic tasks.

In most cases, task performers are specific participants who bear relationships with the entities to be acted on. For example, task rejectCOrder is meant to be carried out by the account manager serving the customer who placed the order that is going to be rejected. If, instead, the qualifier "any" is written after the role name, any member of the role is entitled to perform the task; for this reason, task placeCOrder can be carried out by any customer.

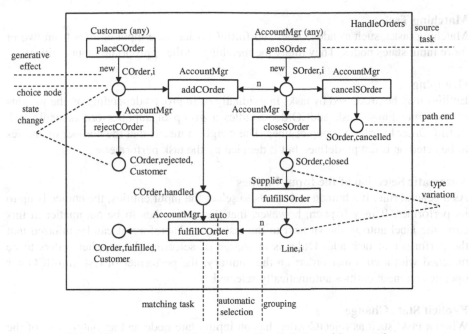

Fig. 1. The compact model of process HandleOrders with annotations

3   Entity-Flow Patterns

ACTA process models are based on a number of entity-flow patterns, which are exemplified in Fig.1. These patterns help readers understand the operation of the process before looking at the task descriptions.

Source Task
Source tasks have only output links: they inject entities into the flow. Human source tasks, like placeCOrder and genSOrder, may be carried out by the intended performers when they want to.

Generative Effect
If a task outputs a new entity, i.e., an entity generated during its execution, the label "new" is shown next to the corresponding link. The state of the output node must be the initial one. Therefore, it can be inferred that the effects of tasks placeCOrder and genSOrder are the generation of a customer order and of a supplier order, respectively.

Choice (confluence) Node
If a state node has two or more output links, the entities exiting the node may follow different paths or, in other terms, they may be handled with different tasks. In process HandleOrders, there are two choice nodes, i.e., COrder (i) and SOrder (i); the latter is also a confluence node in that it has more than one input link.

Matching Task

Matching tasks, such as addCOrder and fulfillCOrder, take input entities from two or more input states nodes. They imply the matching of the appropriate input entities.

Grouping

Entities may be processed by tasks individually or in groups depending on the weights of the input links. Task addCOrder handles a group of supplier orders while task fulfillCOrder handles a group of lines. The weight n means that the number of entities to be acted on is not predefined but is decided by the task performer.

Automatic Selection of the Input Entities

As a general rule, if a human task needs to select the input entities, the choice is up to its performer. It may happen, however, that the choice has to be automatic; in this case, the label auto is shown next to the task icon. Therefore, it can be inferred that the performer of task addCOrder is in charge of selecting the supplier orders to be matched with a customer order; on the contrary, the performer of task fulfillCOrder operates on input entities automatically selected.

Explicit State Change

When a task, such as rejectCOrder, has an input state node and an output one of the same type, it determines an explicit state change in the entities exiting the input node.

Path End

State nodes having no output links, such as COrder (rejected), mark the end of a path and are called final states. If a final state node has an additional label (the third one), which is a role name, then the node is a notifier: when an entity enters the node, a notification is sent to the intended participant. Therefore, if a customer order is rejected, the customer who issued it is notified.

Type Variation in the Entity Flow

It takes place when there is a mismatch between the input and output types of a task. For example, task fulfillSOrder takes a supplier order and delivers a number of lines. The focus of the process is shifted to the lines and the input supplier order can be thought of as transferred to an implicit (and hidden) state node. Even if the lines are not generated by the task, they enter the initial state of their life cycle; it is only at this point that they become important to the process.

4   Information Models

Information models complement process models by showing the types of the entities involved along with their attributes and relationships. The information model of process HandleOrders is shown in Fig.2. Attributes are not considered in this paper.

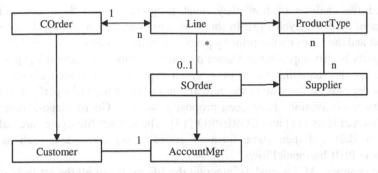

Fig. 2. The information model of process HandleOrders

ACTA information models are similar to Unified Modeling Language (UML) class models [17], but, in addition, they may specify mandatory properties, i.e., the attributes and associations which must be set when a new entity is generated.

In general, a process is in charge of handling several entities, which are divided into managed entities and contextual ones, on the basis of their impact on the process. The managed entities are those forming the entity flow, while contextual entities provide contextual information. In process HandleOrders, customer orders, lines and supplier orders are managed entities while participants (customers, account managers and suppliers) and product types are contextual entities. The corresponding types are called managed types and contextual ones, respectively. The types of the participants in the process are called role types.

The relationships between contextual types are called contextual relationships and the others are called managed relationships.

Mandatory relationships are important in that they determine which associations must be set when a new entity is generated. For example, a new customer order needs to be connected to the customer entity representing the customer who placed it. The relationship COrder-Customer is shown as an oriented link: the connection constraint holds for the source of the link. The default multiplicity is one on the destination end and is * (which means zero or many) on the source end. Default multiplicities may be omitted. Symbol n means 1 or many.

Relationship COrder-Line is mandatory on both sides as a customer order is made up of lines generated during its construction; the link is shown with arrows on both ends.

The multiplicity 0..1 on the SOrder end of relationship Line-SOrder is needed in that the lines of a customer order that has been rejected are not connected to any supplier order.

Contextual types cannot be the sources of mandatory relationships since they are not instantiated by the process.

5   Related Work

The artifact-centric approach [14] emphasizes the identification of the key business entities (called artifacts) used to run a business. Artifacts are associated with business goals and the analysis of how they must progress toward them determines the definition of their life cycles [3]. In this way, communication among stakeholders is facilitated and the focus on the primary purposes of the business is stressed [6].

Flexibility is also improved as shown by the case-handling approach [1]: a process is the recipe for handling cases of a given type, e.g. insurance claims, and its evolution depends on the state of the case and not only on the tasks performed [10].

A number of notations have been proposed, such as Guard-Stage-Milestone [7], PHILharmonicFlows [11] and COREPRO [13]. The artifact life cycles are defined in separate models and then some form of coordination is needed, such as macro processes in PHILharmonicFlows.

On the contrary, ACTA models integrate the life cycles of all the artifacts involved in the process. This monolithic representation makes the coordination of life cycles easier to represent; however, because of its size, the resulting model may be more difficult to grasp.

Integrated models have also been presented in other papers [8,9]. The purpose of ACTA, however, is to leverage the entity flow so as to represent situations that are very common in industrial applications but that are not adequately supported by the functionalities of typical workflow systems: an example is the many-to-many mapping between customer orders and supplier ones in build-to-order processes. That issue has also been addressed with an extension [12] to BPMN; the solution, however, presents some limitations due to the rigidity of the control flow in BPMN.

While the data flow is a kind of "afterthought" in business process models [16], such as BPMN, in ACTA the entity flow supersedes the control flow in order to improve flexibility: the life cycles may evolve independently of each other until activities calling for synchronization are needed. In addition, human choices, such as the selection of the task with which an input entity can be handled, the choice of a group of input entities, or the matching between entities of different types, are emphasized.

6   Conclusion

This paper has stressed the importance of the entity-flow perspective in business process models with the help of the ACTA notation. The emphasis on the entity flow is essential to improve flexibility: in particular, a number of choice situations - such as those concerning the selection of the task with which to handle an input entity (when two or more tasks are allowed), the grouping of similar entities, and the matching between entities of different types - can be represented in a form easy to understand by non-specialists.

A further advantage is the possibility of providing high-level descriptions of tasks in terms of their intended effects as well as the constraints that affect their execution, in a declarative way based on post-conditions and pre-conditions, respectively. To this end, an information model describing the structure of the entities along with their relationships and attributes is added to the process model.

This paper has presented a number of entity-flow patterns, which help readers understand the operation of the process before looking at the task descriptions.

References

1. van der Aalst, W.M.P., Weske, M., Grünbauer, D.: Case handling: a new paradigm for business process support. Data & Knowledge Engineering 53, 129–162 (2005)
2. Alter, S.: Defining information systems as work systems: implications for the IS field. European Journal of Information Systems 17, 448–469 (2008)
3. Bhattacharya, K., et al.: A model-driven approach to industrializing discovery processes in pharmaceutical research. IBM Systems Journal 44(1), 145–162 (2005)
4. BPMN, Business Process Model and Notation, V.2.0.2 (2011), http://www.omg.org/spec/BPMN/2.0.2/ (retrieved February 6, 2014)
5. Bruno, G.: Coordination issues in artifact-centric business process models. In: Decker, H., Lhotská, L., Link, S., Basl, J., Tjoa, A.M. (eds.) DEXA 2013, Part I. LNCS, vol. 8055, pp. 209–223. Springer, Heidelberg (2013)
6. Chao, T., et al.: Artifact-Based Transformation of IBM Global Financing. In: Dayal, U., Eder, J., Koehler, J., Reijers, H.A. (eds.) BPM 2009. LNCS, vol. 5701, pp. 261–277. Springer, Heidelberg (2009)
7. Hull, R., et al.: Introducing the Guard-Stage-Milestone Approach for Specifying Business Entity Lifecycles (Invited Talk). In: Bravetti, M. (ed.) WS-FM 2010. LNCS, vol. 6551, pp. 1–24. Springer, Heidelberg (2011)
8. Kucukoguz, E., Su, J.: On lifecycle constraints of artifact-centric workflows. In: Bravetti, M. (ed.) WS-FM 2010. LNCS, vol. 6551, pp. 71–85. Springer, Heidelberg (2011)
9. Kumaran, S., Liu, R., Wu, F.Y.: On the duality of information-centric and activity-centric models of business processes. In: Bellahsène, Z., Léonard, M. (eds.) CAiSE 2008. LNCS, vol. 5074, pp. 32–47. Springer, Heidelberg (2008)
10. Künzle, V., Reichert, M.: Towards object-aware process management systems: issues, challenges, benefits. In: Halpin, T., Krogstie, J., Nurcan, S., Proper, E., Schmidt, R., Soffer, P., Ukor, R. (eds.) BPMDS/EMMSAD 2009. LNBIP, vol. 29, pp. 197–210. Springer, Heidelberg (2009)
11. Künzle, V., Reichert, M.: PHILharmonicFlows: towards a framework for object-aware process management. Journal of Software Maintenance and Evolution: Research and Practice 23(4), 205–244 (2011)
12. Meyer, A., Pufahl, L., Fahland, D., Weske, M.: Modeling and enacting complex data dependencies in business processes. In: Daniel, F., Wang, J., Weber, B. (eds.) BPM 2013. LNCS, vol. 8094, pp. 171–186. Springer, Heidelberg (2013)
13. Müller, D., Reichert, M., Herbst, J.: Data-driven modeling and coordination of large process structures. In: Meersman, R., Tari, Z. (eds.) OTM 2007, Part I. LNCS, vol. 4803, pp. 131–149. Springer, Heidelberg (2007)
14. Nigam, A., Caswell, N.S.: Business artifacts: an approach to operational specification. IBM Systems Journal 42(3), 428–445 (2003)
15. Sadiq, S., Orlowska, M., Sadiq, W., Schulz, K.: When workflows will not deliver: The case of contradicting work practice. In: 8th Int. Conference on Business Information Systems (2005)
16. Sanz, J.L.C.: Entity-centric operations modeling for business process management - A multidisciplinary review of the state-of-the-art. In: 6th IEEE Int. Symposium on Service Oriented System Engineering, pp. 152–163. IEEE Press, New York (2011)
17. UML. Unified Modeling Language, V.2.4.1 (2011), http://www.omg.org/spec/UML/2.4.1/ (retrieved February 6, 2014)

Projections of Abstract Interorganizational Business Processes

Julius Köpke, Johann Eder, and Markus Künstner

Department of Informatics-Systems, Alpen-Adria Universität Klagenfurt, Austria
{julius.koepke,johann.eder,markus.kuenstner}@aau.at
http://isys.uni-klu.ac.at

Abstract. Distributed interorganizational business processes are constituted by cooperating private processes of different partners. These private processes must be aligned to fulfill the overall goal of the (virtual) interorganizational process. Process views are considered as an excellent concept for modeling such interorganizational processes by providing a good balance between the information needs and privacy requirements of the partners. We follow a top-down development approach, where first an abstract global process is designed. Then each task of the global process is assigned to one of the partners. Finally, a partitioning procedure is applied to generate views of the global process for each partner. The partners can then develop or adopt their private processes based on their views. We present a novel algorithm for the automatic partitioning of any block-structured process with any arbitrary assignment of partners and have proven its correctness.

1 Introduction

Interorganizational cooperation between business partners is nowadays considered to be crucial for increasing productivity. In recent years many approaches were developed to support process cooperation with web technologies [6,16,2] in general and in particular with SOA-based protocols between business partners [1]. There are two main approaches: Orchestrations where a local (private) process is executed and controlled by a single orchestration engine and communicates with external services, that can be entry or exit points of partner processes; or choreographies which define a communication protocol (message exchanges) between collaborating business partners, which can be interpreted as a decentralized global process [13].

Process views such as the ones discussed in [14] and [5,2,11] are an appropriate approach to represent the externally observable behavior of business processes and to balance the request for privacy and loose coupling between processes with the communication demands for collaboration. Most approaches for process views follow a bottom up or *global as view* approach [2,11,14]. A view is derived from a private process definition which is actually instantiated and executed in a process engine at runtime. The integration of such cooperating views can constitute an interorganizational business process. Views can also be used the

H. Decker et al. (Eds.): DEXA 2014, Part II, LNCS 8645, pp. 472–479, 2014.

other way round - to distribute the steps of a global process definition. In this top-down or *local as view* approach first an abstract process is defined as a global interorganizational process. The activities contained in this process definition are then distributed onto the involved partners. Since the process definition is abstract, it means that none of the steps are executed *globally* but that each step which is defined in the global process is executed by one of the participating processes. The projection of the global process onto a particular participant defines a view, i.e. a workflow derived from the abstract global process. The views of the global process specify the obligations of the partners (execution of steps defined in the global process) and the externally observable behavior of the private processes (choreography). The p2p approach presented in [20,17] is an example for such a top-down modeling approach. As the construction of views on a global process by projection requires the introduction of communication steps and the distribution of data, in particular data needed for control flow decisions it is quite different from the construction of views on local processes which only abstract from the process by deletion and aggregation of steps [14,2,11].

In this paper, we present an automatic process partitioning algorithm that can be used in a top-down development approach. Starting with a global process each step of the process is assigned to one of the partners. In a next step, the global process with partner assignments is partitioned fully automatically into views for each partner. We could prove that the union of the generated views correctly realizes the fully distributed execution of the global process. Due to space limitations, we cannot present the proof and all algorithms here and refer the reader to [9].

2   Process Model

We specify the process definitions we consider in this approach with a meta-model shown in Figure 1. It supports the capabilities of block-structured workflow nets (called full blocked workflows by the WfMC supporting sequence, PAR split / join, XOR split/join, and LOOP-split/join as control flow patterns [19,15]. We focus on block-structured workflow-nets as they prevent typical flaws of unstructured business processes dealing with data [3] and are also in line with the WS-BPEL [12] standard. Block-structured workflow-nets can be represented either as graphs or as hierarchies of (abstract) activities, where the transformation between both representations is trivial. We only discuss the essential parts of the meta-model here and refer to [9,4] for details. A process is defined by steps, where steps can be control-steps, activity-steps and communication steps. Control-steps can be of type XOR, PAR, LOOP or SEQ and have two references to arbitrary steps, namely *stepA* and *stepB* (except loops that only have *stepA*). Activity-steps refer to some predefined activity. Partner information can be assigned to each step in a process in order to define which partner is responsible for the execution. Control-steps may have two different partners assigned, where the first one is responsible for the split and the last one is responsible for the join. Activity- and communication steps also have two slots for partners for generality reasons but both must always refer to the same partner. Communication

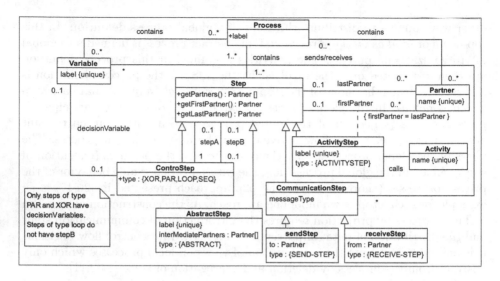

Fig. 1. Workflow Meta-Model

steps can be of type *sendStep* or *receiveStep* and are used to pass control- and data-flow between different partners in the distributed execution.

XOR- and LOOP- steps in local processes have some Boolean expression that formalizes the condition. In a distributed process, this gets problematic because it requires that all partners have exactly the same data values in order to take the same decisions. This results on the one hand in additional synchronization overhead and it may even be impossible, if some or all of the variables in the Boolean expression are only locally available for some partner. Therefore, we use a different approach: Each condition may only refer to a single Boolean variable. This variable can easily be distributed to all relevant partners. Any process can be transformed to an equivalent process that adheres to this requirement by simply adding an additional step that writes the result of each complex Boolean expression to a variable.

3   Projecting a Global Process to Views for Each Partner

As in [17,20] we propose the following procedure for defining interorganizational workflows: (1) define a global (abstract) workflow, (2) assign each step in the workflow to a partner (3) partition the process, (4) create or adopt private processes based on the generated views. The global process is executed by the distributed execution of the private processes. While the work in [17,20,18] concentrates on the question whether a private process correctly implements a (public) view (is in accordance with) our aim is to provide a partitioning procedure that does not impose any restrictions on the global process and on the partner assignments. Additional design rationales are: (1) the resulting views should be as simple as possible, and (b) unnecessary message exchange should be avoided.

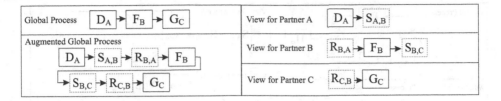

Fig. 2. Partitioning of Sequences

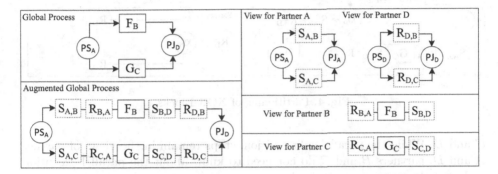

Fig. 3. Partitioning of Parallel Blocks

E.g. a partner should only know about control structures and receive only messages that are absolutely required. In the following we discuss our partitioning method by examples for each control structure. Our approach operates in two phases: First, the global process is augmented with communication steps that realize the interorganizational control-flow and variable passing. In a next step, views for the partners are created by projection.

Sequence Blocks: In Figure 2 an example for the partitioning of a sequence is shown. The global process contains the steps D_A, F_B and G_C in a sequence. D is defined to be executed by A, F is executed by B and G is executed by C. In order to support a distributed execution the global process is first augmented with pairs of send- and receive- steps whenever a step is followed by another step, which is assigned to a different partner. In particular in the example send- and receive- steps are inserted between D_A and F_B and between F_B and G_C in the augmented global process. The corresponding views for each partner are shown on the right. They are created by simply projecting only steps that are assigned to the specific target partner.

Parallel Blocks: An example for the augmentation and partitioning of a parallel block is shown in Figure 3. Partner A is responsible for the parallel split. The partners B and C execute their tasks F and G in parallel. Partner D is responsible for the synchronization. The process is augmented by send- and receive- steps between partner A and B, and A and C and finally between B and D and

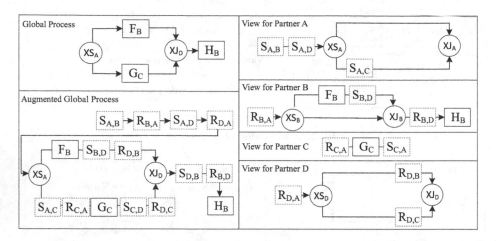

Fig. 4. Partitioning of XOR-Blocks

C and D. While the parallel split and join steps remain in the views of partner A and D, partners B and C do not need to know about the parallel execution and get only sequences in their views.

XOR-Blocks: XOR-blocks are based on some decision variable. The owner of the XOR-split takes the decision. All partners who need to execute the XOR split themselves are informed about the decision by a message transporting the decision variable. In the example in Figure 4 partner A is responsible for the XOR split and distributes the decision variable to B and D by send/receive-steps before the split. Partner C does not need to know about the decision because he does not need to have an XOR-block in his view. After the XOR-split the control-flow must be forwarded. This is realized by send- and receive-steps between A and C. There is no such communication required between A and B because B can proceed as soon as the value of the decision variable was received. The XOR join is executed by partner D. Therefore, like for parallel join nodes the last partner in each branch needs to pass the control-flow to the join partner. However, the join partner requires the decision variable from partner A to know whether the incoming message will arrive from B or from C.

In the example partner C has a sequence of steps in his view rather than an XOR-Block. This simplification is possible because C takes exclusively part in one branch of the XOR-block (and in no other step before, after, or parallel to this XOR-block). Therefore, C does not need to know about the XOR-block at all. Such a simplification is not possible for partner B because B must know whether step F needs to be executed before step H or not.

LOOP-Blocks: Loops are partitioned into views in analogy to XOR-blocks. The first partner in the loop (assigned to the LOOP-split node) is responsible for the decision. Therefore, the first partner distributes the decision variable to the partners that take part in the loop. An example is shown in Figure 5. As

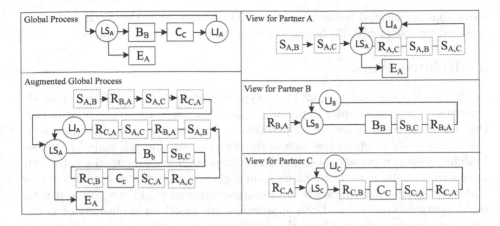

Fig. 5. Partitioning of Loop-Blocks

for $XOR-$blocks, the first partner in the loop body (B) does not need to be explicitly called by A because it receives the decision variable from A and can directly execute the LOOP-split and the activity step. After each iteration of the loop, the current version of the decision variable is distributed to all relevant partners by a sequence of send- and receive- steps that are added directly before the LOOP-join.

4   Automatic Generation of Views

We have implemented an algorithm that automatically derives views for any partner assignments for any block-structured input process and have proven its correctness. Due to space limitations, we cannot provide all details here and refer the reader to [9]. The algorithm operates in three phases:

Augmentation: The automatic augmentation procedure operates on the hierarchic process representation as discussed in Section 2. It traverses the input process recursively and adds send- and receive pairs to support the distributed execution by applying the augmentation patterns discussed in Section 3.

Projection: Views for each partner are created by projecting the steps of the augmented global process to the specific partners. To create a view for a partner p only those steps are added that are assigned to p or that are control-steps. Abstract steps are projected themselves by removing all references to partners others than p.

Cleanup: Finally, irrelevant constructs are removed from the views: (1) If a $PAR - Block$ has one empty branch in the view, then the $PAR - Block$ is replaced by a sequence. (2) If an $XOR - Block$ has one empty branch and the first partner of the corresponding XOR-Block in the global process is not responsible for the split, and the partner of the view is not assigned to any other block that is executed before or after the XOR-block, then the $XOR - block$ is replaced by a sequence. (3) Sequences with empty slots are eliminated.

5   Related Work

Workflow view mechanisms [5,2,11,14] typically allow to hide and aggregate elements of a process in order to provide a good balance between the information that needs to be shared for the cooperation and privacy concerns of the partners. Most process view approaches such as the ones discussed in [14] and [2,11] allow to define views on private processes. In contrast, we have presented a top-down approach that allows to derive views from a global process. This scenario is also addressed by the p2p approach to interorganizational workflows [20] and with multi party contracts [18]. Both approaches are based on extensions of petrinets and therefore abstract from the data perspective using indeterminism. Additionally, they impose requirements on the global process that do not allow to automatically derive correct views for any possible partners assignment. Therefore, it is up to the designer to create a global process that can be partitioned correctly based on the requirements for partner assignments. We do not need to impose such restrictions and can automatically correctly partition any valid process with any arbitrary partner assignments and we operate on deterministic models.

Another field of related research is the top-down interaction modeling of choreographies including data (message contents) which is addressed in approaches such as [8,10]. However, they have a different scope: In interaction modeling, the entities of concern for projection are limited to messages while ignoring tasks. In our case not messages but processes are partitioned and the exchanged messages are generated automatically.

6   Conclusions

We presented a novel process partitioning method for the top down development of distributed interorganizational processes. The general idea is that the cooperating partners first agree on a global business process. In a next step, each step of the global process is assigned to one of the partners. Finally, our partitioning algorithm is used to automatically derive process views of each partner. The views contain the (abstract) tasks that should be realized by the partners and the required control structures and communication steps that specify the distributed execution (choreography) of the global process. In contrast to existing approaches, we can guarantee that every full-blocked process can correctly be partitioned for any arbitrary partner assignment and we generate deterministic processes by explicitly addressing the distribution of decision variables. The partitioning algorithm is an essential module in designing p2p processes. A view derived for a particular partner can be used as skeleton for the local process. Such a top-down view can also be used for checking the compatibility of a local process (or of the public interface of it) with the interorganizational process.

References

1. Alonso, G., et al.: Web Services: Concepts, Architectures and Applications. Springer, Berlin (October 2003)
2. Chebbi, I., Dustdar, S., Tata, S.: The view-based approach to dynamic inter-organizational workflow cooperation. Data Knowl. Eng. 56(2), 139–173 (2006)
3. Combi, C., Gambini, M.: Flaws in the flow: The weakness of unstructured business process modeling languages dealing with data. In: Meersman, R., Dillon, T., Herrero, P. (eds.) OTM 2009, Part I. LNCS, vol. 5870, pp. 42–59. Springer, Heidelberg (2009)
4. Eder, J., Gruber, W.: A meta model for structured workflows supporting workflow transformations. In: Manolopoulos, Y., Návrat, P. (eds.) ADBIS 2002. LNCS, vol. 2435, pp. 326–339. Springer, Heidelberg (2002)
5. Eder, J., Kerschbaumer, N., Köpke, J., Pichler, H., Tahamtan, A.: View-based interorganizational workflows. In: Proc. 12th Int. Conf. Computer Syst. and Tech. CompSysTech 2011, pp. 1–10. ACM (2011)
6. Groiss, H., Eder, J.: Workflow systems for inter-organizational business processes. ACM SIGGroup Bulletin 18, 23–26 (1997)
7. Hollingsworth, D.: The workflow reference model (1995)
8. Knuplesch, D., Pryss, R., Reichert, M.: Data-aware interaction in distributed and collaborative workflows: Modeling, semantics, correctness. In: CollaborateCom, pp. 223–232. IEEE (2012)
9. Köpke, J., Eder, J., Künstner, M.: Implementing projections of abstract interorganizational business processes. Technical report, Universität Klagenfurt - ISYS (2014),
http://isys.uni-klu.ac.at/PDF/2014-Impl-Process-Partitioning.pdf
10. Nguyen, H., Poizat, P., Zaidi, F.: Automatic skeleton generation for data-aware service choreographies. In: 2013 IEEE 24th International Symposium on Software Reliability Engineering (ISSRE), pp. 320–329 (2013)
11. Norta, A., Eshuis, R.: Specification and verification of harmonized business-process collaborations. Information Systems Frontiers 12(4), 457–479 (2010)
12. OASIS. OASIS Web Services Business Process Execution Language (WSBPEL) TC. Technical report, "OASIS" (April 2007)
13. Peltz, C.: Web services orchestration and choreography. IEEE Computer 36(10), 46–52 (2003)
14. Smirnov, S., et al.: Business process model abstraction: a definition, catalog, and survey. Distrib Parallel Dat 30(1), 63–99 (2012)
15. van der Aalst, W.M.P.: Verification of workflow nets. In: Azéma, P., Balbo, G. (eds.) ICATPN 1997. LNCS, vol. 1248, pp. 407–426. Springer, Heidelberg (1997)
16. van der Aalst, W.M.P.: Process-oriented architectures for electronic commerce and interorganizational workflow. Information Systems 24(8), 115–126 (1999)
17. van der Aalst, W.M.P.: Inheritance of interorganizational workflows: How to agree to disagree without loosing control? IT and Management 4(4), 345–389 (2003)
18. van der Aalst, W.M.P., et al.: Multiparty contracts: Agreeing and implementing interorganizational processes. Comput. J. 53(1), 90–106 (2010)
19. van der Aalst, W.M.P., et al.: Workflow patterns. Distrib. Parallel Databases 14(1), 5–51 (2003)
20. van der Aalst, W.M.P., Weske, M.: The P2P approach to interorganizational workflows. In: Dittrich, K.R., Geppert, A., Norrie, M. (eds.) CAiSE 2001. LNCS, vol. 2068, p. 140. Springer, Heidelberg (2001)

Author Index